Lecture Notes in Mathematics 2046

For further volumes:
http://www.springer.com/series/304

Catherine Donati-Martin • Antoine Lejay
Alain Rouault
Editors

Séminaire de Probabilités XLIV

Springer

Editors
Catherine Donati-Martin
Université de Versailles-St-Quentin
Versailles
France

Alain Rouault
Université de Versailles-St-Quentin
Versailles
France

Antoine Lejay
Nancy-Université, INRIA
Vandoeuvre-lès-Nancy
France

ISBN 978-3-642-27460-2 ISBN 978-3-642-27461-9 (eBook)
DOI 10.1007/978-3-642-27461-9
Springer Heidelberg New York Dordrecht London

Lecture Notes in Mathematics ISSN print edition: 0075-8434
ISSN electronic edition: 1617-9692

Library of Congress Control Number: 2012933110

Mathematics Subject Classification (2010): 60-XX, 60JXX, 60J60, 60J10, 60J65, 60J55, 46L54

Printed on acid-free paper

Springer is part of Springer Science+Business Media (www.springer.com)

Preface

As usual, some of the contributions to this 44th Séminaire de Probabilités were exposed during the Journées de Probabilités held in Dijon in June 2010. The other ones come from spontaneous submissions or were solicited by the editors. The traditional and historical themes of the Séminaire are present, such as stochastic calculus, local times and excursions and martingales. Some subjects already largely present in the previous volumes are stil here: free probability, rough paths, limit theorems for general processes (here fractional Brownian motion and polymers) and large deviations. Finally, this volume explores new topics, including variable length Markov chains and peacoks. We hope that the whole volume is a good sample of the main streams of current research on probability and stochastic processes, in particular those active in France.

We remind that the web site of the Séminaire is

http://portail.mathdoc.fr/SemProba/

and that all the articles of the Séminaire from Volume I in 1967 to Volume XXXVI in 2002 are freely accessible from the web site

http://www.numdam.org/numdam-bin/feuilleter?j=SPS

We thank the Cellule Math Doc for hosting all these articles within the NUMDAM project.

Catherine Donati-Martin
Antoine Lejay
Alain Rouault

Contents

Context Trees, Variable Length Markov Chains and Dynamical Sources

Peggy Cénac, Brigitte Chauvin, Frédéric Paccaut, and Nicolas Pouyanne

Abstract Infinite random sequences of letters can be viewed as stochastic chains or as strings produced by a source, in the sense of information theory. The relationship between Variable Length Markov Chains (VLMC) and probabilistic dynamical sources is studied. We establish a probabilistic frame for context trees and VLMC and we prove that any VLMC is a dynamical source for which we explicitly build the mapping. On two examples, the "comb" and the "bamboo blossom", we find a necessary and sufficient condition for the existence and the uniqueness of a stationary probability measure for the VLMC. These two examples are detailed in order to provide the associated Dirichlet series as well as the generating functions of word occurrences.

P. Cénac (✉)
Université de Bourgogne, Institut de Mathématiques de Bourgogne IMB UMR 5584 CNRS, 9 avenue Alain Savary - BP 47870, 21078 DIJON CEDEX, Bourgogne, France
e-mail: peggy.cenac@u-bourgogne.fr

B. Chauvin
INRIA Rocquencourt, project Algorithms, Domaine de Voluceau B.P.105, 78153 Le Chesnay CEDEX, France

Laboratoire de Mathématiques de Versailles, CNRS, UMR 8100, Université de Versailles - St-Quentin, 45 avenue des Etats-Unis, 78035 Versailles CEDEX, France
e-mail: chauvin@math.uvsq.fr

F. Paccaut
LAMFA, CNRS, UMR 6140, Université de Picardie Jules Verne, 33 rue Saint-Leu, 80039 Amiens, France
e-mail: frederic.paccaut@u-picardie.fr

N. Pouyanne
Laboratoire de Mathématiques de Versailles, CNRS, UMR 8100, Université de Versailles - St-Quentin, 45 avenue des Etats-Unis, 78035 Versailles CEDEX, France
e-mail: pouyanne@math.uvsq.fr

C. Donati-Martin et al. (eds.), *Séminaire de Probabilités XLIV*, Lecture Notes in Mathematics 2046, DOI 10.1007/978-3-642-27461-9_1,

Keywords Variable length Markov chains • Dynamical systems of the interval • Dirichlet series • Occurrences of words • Probabilistic dynamical sources

AMS Classification: 60J05, 37E05

1 Introduction

Our objects of interest are infinite random sequences of letters. One can imagine DNA sequences (the letters are A, C, G, T), bits sequences (the letters are 0, 1) or any random sequence on a finite alphabet. Such a sequence can be viewed as a stochastic chain or as a string produced by a source, in the sense of information theory. We study this relation for the so-called Variable Length Markov Chains (VLMC).

From now on, we are given a finite alphabet $\mathcal{A}$. An infinite random sequence of letters is often considered as a chain $(X_n)_{n\in\mathbb{Z}}$, i.e. an $\mathcal{A}^{\mathbb{Z}}$-valued random variable. The X_n are the *letters* of the chain. Equivalently such a chain can be viewed as a random process $(U_n)_{n\in\mathbb{N}}$ that takes values in the set $\mathcal{L} := \mathcal{A}^{-\mathbb{N}}$ of left-infinite words[1] and that grows by addition of a letter on the right at each step of discrete time. The $\mathcal{L}$-valued processes we consider are Markovian ones. The evolution from $U_n = \ldots X_{-1}X_0X_1\ldots X_n$ to $U_{n+1} = U_nX_{n+1}$ is described by the transition probabilities $\mathbf{P}(U_{n+1} = U_n\alpha|U_n)$, $\alpha \in \mathcal{A}$.

In the context of chains, the point of view has mainly been a statistical one until now, going back to Harris [14] who speaks of *chains of infinite order* to express the fact that the production of a new letter depends on a finite but unbounded number of previous letters. Comets et al. [7] and Gallo and Garcia [11] deal with chains of infinite memory. Rissanen [23] introduces a class of models where the transition from the word U_n to the word $U_{n+1} = U_nX_{n+1}$ depends on U_n through a finite suffix of U_n and he calls this relevant part of the past a *context*. Contexts can be stored as the leaves of a so-called *context tree* so that the model is entirely defined by a family of probability distributions indexed by the leaves of a context tree. In this paper, Rissanen develops a near optimal universal data compression algorithm for long strings generated by non independent information sources. The name VLMC is due to Bühlmann and Wyner [5]. It emphasizes the fact that the length of memory needed to predict the next letter is a not necessarily bounded function of the sequence U_n. An overview on VLMC can be found in Galves and Löcherbach [12].

We give in Sect. 2 a complete probabilistic definition of VLMC. Let us present here a foretaste, relying on the particular form of the transition probabilities $\mathbf{P}(U_{n+1} = U_n\alpha|U_n)$. Let $\mathcal{T}$ be a *saturated tree* on $\mathcal{A}$, which means that every internal node of the tree—i.e. a word on $\mathcal{A}$—has exactly $|\mathcal{A}|$ children. With each leaf c of the tree, also called a *context*, is associated a probability distribution q_c on $\mathcal{A}$. The basic fact is that any left-infinite sequence can thus be "plugged in" a

[1] In the whole text, $\mathbb{N}$ denotes the set of nonnegative integers.

unique context of the tree $\mathscr{T}$: any U_n can be uniquely written $U_n = \dots\overline{c}$, where, for any word $c = \alpha_1 \cdots \alpha_N$, $\overline{c}$ denotes the reversed word $\overline{c} = \alpha_N \cdots \alpha_1$. In other terms, for any n, there is a unique context c in the tree $\mathscr{T}$ such that $\overline{c}$ is a suffix of U_n; this word is denoted by $c = \overleftarrow{\text{pref}}\,(U_n)$. We define the VLMC associated with these data as the $\mathscr{L}$-valued homogeneous Markov process whose transition probabilities are, for any letter $\alpha \in \mathscr{A}$,

$$\mathbf{P}(U_{n+1} = U_n\alpha | U_n) = q_{\overleftarrow{\text{pref}}\,(U_n)}(\alpha).$$

When the tree is finite, the final letter process $(X_n)_{n\geq 0}$ is an ordinary Markov chain whose order is the height of the tree. The case of infinite trees is more interesting, providing concrete examples of non Markov chains.

In the example of Fig. 1, the context tree is finite of height 4 and, for instance, $\mathbf{P}(U_{n+1} = U_n 0 | U_n = \cdots 0101110) = q_{011}(0)$ because $\overleftarrow{\text{pref}}\,(\cdots 0101\mathbf{110}) = \mathbf{011}$ (read the word $\cdots 0101110$ right-to-left and stop when finding a context).

In information theory, one considers that words are produced by a *probabilistic source* as developed in Vallée and her group papers (see Clément et al. [6] for an overview). In particular, a *probabilistic dynamical source* is defined by a coding function $\rho : [0,1] \to \mathscr{A}$, a mapping $T : [0,1] \to [0,1]$ having suitable properties and a probability measure μ on $[0,1]$. These data being given, the dynamical source produces the $\mathscr{A}$-valued random process $(Y_n)_{n\in\mathbb{N}} := (\rho(T^n\xi))_{n\in\mathbb{N}}$, where ξ is a μ-distributed random variable on $[0,1]$. On the right side of Fig. 1, one can see the graph of some T, a subdivision of $[0,1]$ in two subintervals $I_0 = \rho^{-1}(0)$ and $I_1 = \rho^{-1}(1)$ and the first three real numbers x, Tx and T^2x, where x is a realization

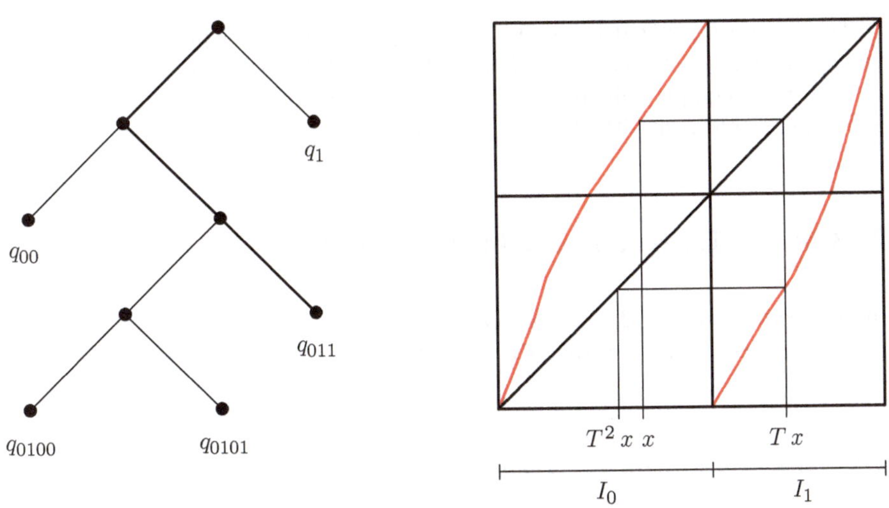

Fig. 1 Example of probabilized context tree (*on the left*) and its corresponding dynamical system (*on the right*)

of the random variable ξ. The right-infinite word corresponding to this example has 010 as a prefix.

We prove in Theorem 1 that every stationary VLMC is a dynamical source. More precisely, given a stationary VLMC, $(U_n)_{n\in\mathbb{N}}$ say, we construct explicitly a dynamical source $(Y_n)_{n\in\mathbb{N}}$ such that the letter processes $(X_n)_{n\in\mathbb{N}}$ and $(Y_n)_{n\in\mathbb{N}}$ are symmetrically distributed, which means that for any finite word w of length $N+1$, $\mathbf{P}(X_0 \dots X_N = w) = \mathbf{P}(Y_0 \dots Y_N = \overline{w})$. In Fig. 1, the dynamical system together with Lebesgue measure on $[0,1]$ define a probabilistic source that corresponds to the stationary VLMC defined by the drawn probabilized context tree.

The previous result is possible only when the VLMC is stationary. The question of existence and uniqueness of a stationary distribution arises naturally. We give a complete answer in two particular cases (Propositions 1 and 4 in Sect. 4) and we propose some tracks for the general case. Our two examples are called the "infinite comb" and the "bamboo blossom"; they can be visualized in Figs. 6 and 7, respectively pages 17 and 28. Both have an infinite branch so that the letter process of the VLMC is non Markovian. They provide quite concrete cases of infinite order chains where the study can be completely handled. We first exhibit a necessary and sufficient condition for existence and uniqueness of a stationary measure. Then the dynamical system is explicitly built and drawn. In particular, for some suitable data values, one gets in this way examples of intermittent sources.

Quantifying and visualizing repetitions of patterns is another natural question arising in combinatorics on words. *Tries*, *suffix tries* and *digital search trees* are usual convenient tools. The analysis of such structures relies on the generating functions of the word occurrences and on the Dirichlet series attached to the sources. In both examples, these computations are performed.

The paper is organized as follows. Section 2 is devoted to the precise definition of variable length Markov chains. In Sect. 3 the main result Theorem 1 is established. In Sect. 4, we complete the paper with our two detailed examples: "infinite comb" and "bamboo blossom". The last section gathers some prospects and open problems.

2 Context Trees and Variable Length Markov Chains

In this section, we first define *probabilized context trees*; then we associate with a probabilized context tree a so-called *variable length Markov chain (VLMC)*.

2.1 Words and Context Trees

Let $\mathscr{A}$ be a finite alphabet, i.e. a finite ordered set. Its cardinality is denoted by $|\mathscr{A}|$. For the sake of shortness, our results in the paper are given for the alphabet $\mathscr{A} = \{0, 1\}$ but they remain true for any finite alphabet. Let

$$\mathscr{W} = \bigcup_{n \geq 0} \mathscr{A}^n$$

be the set of all finite words over $\mathscr{A}$. The concatenation of two words $v = v_1 \dots v_M$ and $w = w_1 \dots w_N$ is $vw = v_1 \dots v_M w_1 \dots w_N$. The empty word is denoted by $\emptyset$. Let

$$\mathscr{L} = \mathscr{A}^{-\mathbb{N}}$$

be the set of left-infinite sequences over $\mathscr{A}$ and

$$\mathscr{R} = \mathscr{A}^{\mathbb{N}}$$

be the set of right-infinite sequences over $\mathscr{A}$. If k is a nonnegative integer and if $w = \alpha_{-k} \cdots \alpha_0$ is any finite word on $\mathscr{A}$, the reversed word is denoted by

$$\overline{w} = \alpha_0 \cdots \alpha_{-k}.$$

The *cylinder based on* w is defined as the set of all left-infinite sequences having w as a suffix:

$$\mathscr{L}w = \{s \in \mathscr{L},\ \forall j \in \{-k, \cdots, 0\},\ s_j = \alpha_j\}.$$

By extension, the reversed sequence of $s = \cdots \alpha_{-1}\alpha_0 \in \mathscr{L}$ is $\overline{s} = \alpha_0\alpha_{-1} \cdots \in \mathscr{R}$. The set $\mathscr{L}$ is equipped with the σ-algebra generated by all cylinders based on finite words. The set $\mathscr{R}$ is equipped with the σ-algebra generated by all cylinders $w\mathscr{R} = \{r \in \mathscr{R}, w \text{ is a prefix of } r\}$.

Let $\mathscr{T}$ be a tree, *i.e.* a subset of $\mathscr{W}$ satisfying two conditions:

- $\emptyset \in \mathscr{T}$
- $\forall u, v \in \mathscr{W}, uv \in \mathscr{T} \Longrightarrow u \in \mathscr{T}$

This corresponds to the definition of rooted planar trees in algorithmics. Let $\mathscr{C}^F(\mathscr{T})$ be the set of *finite leaves* of $\mathscr{T}$, i.e. the nodes of $\mathscr{T}$ without any descendant:

$$\mathscr{C}^F(\mathscr{T}) = \{u \in \mathscr{T}, \forall j \in \mathscr{A}, uj \notin \mathscr{T}\}.$$

An infinite word $u \in \mathscr{R}$ such that any finite prefix of u belongs to $\mathscr{T}$ is called an *infinite leaf* of $\mathscr{T}$. Let us denote the set of infinite leaves of $\mathscr{T}$ by

$$\mathscr{C}^I(\mathscr{T}) = \{u \in \mathscr{R}, \forall v \text{ prefix of } u, v \in \mathscr{T}\}.$$

Let $\mathscr{C}(\mathscr{T}) = \mathscr{C}^F(\mathscr{T}) \cup \mathscr{C}^I(\mathscr{T})$ be the set of all *leaves* of $\mathscr{T}$. The set $\mathscr{T} \setminus \mathscr{C}^F(\mathscr{T})$ is constituted by the *internal nodes* of $\mathscr{T}$. When there is no ambiguity, $\mathscr{T}$ is omitted and we simply write $\mathscr{C}, \mathscr{C}^F$ and $\mathscr{C}^I$.

Definition 1. A tree is *saturated* when each internal node w has exactly $|\mathscr{A}|$ children, namely the set $\{w\alpha, \alpha \subset \mathscr{A}\} \subset \mathscr{T}$.

Definition 2. (Context tree) A *context tree* is a saturated tree having a finite or countable set of leaves. The leaves are called *contexts*.

Definition 3. (Probabilized context tree) A *probabilized context tree* is a pair

$$\left(\mathcal{T}, (q_c)_{c\in\mathcal{C}(\mathcal{T})}\right)$$

where $\mathcal{T}$ is a context tree over $\mathcal{A}$ and $(q_c)_{c\in\mathcal{C}(\mathcal{T})}$ is a family of probability measures on $\mathcal{A}$, indexed by the countable set $\mathcal{C}(\mathcal{T})$ of all leaves of $\mathcal{T}$.

Examples. See Fig. 1 for an example of finite probabilized context tree with five contexts. See Fig. 6 for an example of infinite probabilized context tree, called the infinite comb.

Definition 4. A subset $\mathcal{K}$ of $\mathcal{W}\cup\mathcal{R}$ is a *cutset* of the complete $|\mathcal{A}|$-ary tree when both following conditions hold

(i) no word of $\mathcal{K}$ is a prefix of another word of $\mathcal{K}$
(ii) $\forall r\in\mathcal{R}, \exists u\in\mathcal{K}$, u prefix of r.

Condition (i) entails uniqueness in (ii). Obviously a tree $\mathcal{T}$ is saturated if and only if the set of its leaves $\mathcal{C}$ is a cutset. Take a saturated tree, then

$$\forall r\in\mathcal{R},\ \text{either } r\in\mathcal{C}^I \text{ or } \exists! u\in\mathcal{W},\ u\in\mathcal{C}^F,\ u \text{ prefix of } r. \tag{1}$$

This can also be said on left-infinite sequences:

$$\forall s\in\mathcal{L},\ \text{either } \overline{s}\in\mathcal{C}^I \text{ or } \exists! w\in\mathcal{W},\ \overline{w}\in\mathcal{C}^F,\ w \text{ suffix of } s. \tag{2}$$

In other words:

$$\mathcal{L}=\bigcup_{\overline{s}\in\mathcal{C}^I}\{s\}\cup\bigcup_{\overline{w}\in\mathcal{C}^F}\mathcal{L}w. \tag{3}$$

This partition of $\mathcal{L}$ will be extensively used in the sequel. Both cutset properties (1) and (2) will be used in the paper, on $\mathcal{R}$ for trees, on $\mathcal{L}$ for chains. Both orders of reading will be needed.

Definition 5. (Prefix function) Let $\mathcal{T}$ be a saturated tree and $\mathcal{C}$ its set of contexts. For any $s\in\mathcal{L}$, $\overleftarrow{\text{pref}}\,(s)$ denotes the unique context $\alpha_1\ldots\alpha_N$ such that $s=\ldots\alpha_N\ldots\alpha_1$. The map

$$\overleftarrow{\text{pref}}\,:\mathcal{L}\to\mathcal{C}$$

is called the *prefix function*. For technical reasons, this function is extended to

$$\overleftarrow{\text{pref}}\,:\mathcal{L}\cup\mathcal{W}\to\mathcal{T}$$

in the following way:

- if $\overline{w}\in\mathcal{T}$ then $\overleftarrow{\text{pref}}\,(w)=\overline{w}$;

- if $\overline{w} \in \mathcal{W} \setminus \mathcal{T}$ then $\overleftarrow{\text{pref}}\,(w)$ is the unique context $\alpha_1 \dots \alpha_N$ such that w has $\alpha_N \dots \alpha_1$ as a suffix.

Note that the second item of the definition is also valid when $\overline{w} \in \mathcal{C}$. Moreover $\overleftarrow{\text{pref}}\,(w)$ is always a context except when $\overline{w}$ is an internal node.

2.2 *VLMC Associated with a Context Tree*

Definition 6. (VLMC) Let $(\mathcal{T}, (q_c)_{c\in\mathcal{C}})$ be a probabilized context tree. The associated *Variable Length Markov Chain* (VLMC) is the order 1 Markov chain $(U_n)_{n\geq 0}$ with state space $\mathcal{L}$, defined by the transition probabilities

$$\forall n \geq 0,\ \forall \alpha \in \mathcal{A},\ \mathbf{P}\left(U_{n+1} = U_n\alpha | U_n\right) = q_{\overleftarrow{\text{pref}}\,(U_n)}(\alpha)\,. \tag{4}$$

Remark 1. As usually, we speak of *the* Markov chain defined by the transition probabilities (4), because these data together with the distribution of U_0 define a unique $\mathcal{L}$-valued Markov random process $(U_n)_{n\geq 0}$ (see for example Revuz [22]).

The rightmost letter of the sequence $U_n \in \mathcal{L}$ will be denoted by X_n so that

$$\forall n \geq 0,\ U_{n+1} = U_n X_{n+1}.$$

The final letter process $(X_n)_{n\geq 0}$ is *not* Markov of any finite order as soon as the context tree has at least one infinite context. As already mentioned in the introduction, when the tree is finite, $(X_n)_{n\geq 0}$ is a Markov chain whose order is the height of the tree, i.e. the length of its longest branch. The vocable VLMC is somehow confusing but commonly used.

Definition 7. (SVLMC) Let $(U_n)_{n\geq 0}$ be a VLMC. When a stationary probability measure on $\mathcal{L}$ exists and when it is the initial distribution, we say that $(U_n)_{n\geq 0}$ is a *Stationary Variable Length Markov Chain* (SVLMC).

Remark 2. In the literature, the name VLMC is usually applied to the chain $(X_n)_{n\in\mathbb{Z}}$. There exists a natural bijective correspondence between $\mathcal{A}$-valued chains $(X_n)_{n\in\mathbb{Z}}$ and $\mathcal{L}$-valued processes $(U_n = U_0X_1 \dots X_n, n \geq 0)$. Consequently, finding a stationary probability for the chain $(X_n)_{n\in\mathbb{Z}}$ is equivalent to finding a stationary probability for the process $(U_n)_{n\geq 0}$.

3 Stationary Variable Length Markov Chains

The existence and the uniqueness of a stationary measure for two examples of VLMC will be established in Sect. 4. In the present section, we assume that a stationary measure π on $\mathcal{L}$ exists and we consider a π-distributed VLMC. In the preliminary Sect. 3.1, we show how the stationary probability of finite words can be expressed as a function of the data and the values of π at the tree nodes. In Sect. 3.2, the main theorem is proved.

3.1 General Facts on Stationary Probability Measures

For the sake of shortness, when π is a stationary probability for a VLMC, we write $\pi(w)$ instead of $\pi(\mathscr{L}w)$, for any $w \in \mathscr{W}$:

$$\pi(w) = \mathbf{P}(U_0 \in \mathscr{L}w) = \mathbf{P}(X_{-(|w|-1)} \dots X_0 = w). \tag{5}$$

Extension of notation q_u for internal nodes. The VLMC is defined by its context tree $\mathscr{T}$ together with a family $(q_c)_{c\in\mathscr{C}}$ of probability measures on $\mathscr{A}$ indexed by the contexts of the tree. When u is an internal node of the context tree, we extend the notation q_u by

$$q_u(\alpha) = \begin{cases} \dfrac{\pi(\overline{u}\alpha)}{\pi(\overline{u})} \text{ if } \pi(\overline{u}) \neq 0 \\ \\ 0 \text{ if } \pi(\overline{u}) = 0 \end{cases} \tag{6}$$

for any $\alpha \in \mathscr{A}$. Thus, in any case, π being stationary, $\pi(\overline{u}\alpha) = \pi(\overline{u})q_u(\alpha)$ as soon as $\overline{u}$ is an internal node of the context tree. With this notation, the stationary probability of any cylinder can be expressed by the following simple formula (8).

Lemma 1. *Consider a SVLMC defined by a probabilized context tree and let π denote any stationary probability measure on $\mathscr{L}$. Then,*

(i) *for any finite word $w \in \mathscr{W}$ and for any letter $\alpha \in \mathscr{A}$,*

$$\pi(w\alpha) = \pi(w) q_{\overleftarrow{\text{pref}}(w)}(\alpha). \tag{7}$$

(ii) *For any finite word $w = \alpha_1 \dots \alpha_N \in \mathscr{W}$,*

$$\pi(w) = \prod_{k=0}^{N-1} q_{\overleftarrow{\text{pref}}(\alpha_1\dots\alpha_k)}(\alpha_{k+1}) \tag{8}$$

(if $k = 0$, $\alpha_1 \dots \alpha_k$ denotes the empty word $\emptyset$, $\overleftarrow{\text{pref}}(\emptyset) = \emptyset$, $q_\emptyset(\alpha) = \pi(\alpha)$ and $\pi(\emptyset) = \pi(\mathscr{L}) = 1$).

Proof. *(i)* If $\overline{w}$ is an internal node of the context tree, then $\overleftarrow{\text{pref}}(w) = \overline{w}$ and the formula comes directly from the definition of $q_{\overline{w}}$. If not, $\pi(w\alpha) = \pi(U_1 \in \mathscr{L}w\alpha)$ by stationarity; because of Markov property,

$$\pi(w\alpha) = \mathbf{P}(U_0 \in \mathscr{L}w)\mathbf{P}(U_1 \in \mathscr{L}w\alpha | U_0 \in \mathscr{L}w) = \pi(w) q_{\overleftarrow{\text{pref}}(w)}(\alpha).$$

Finally, *(ii)* follows from *(i)* by a straightforward induction. □

Remark 3. When $\mathscr{A} = \{0, 1\}$ and π is any stationary probability of a SVLMC, then, for any natural number n, $\pi(10^n) = \pi(0^n 1)$. Indeed, on one hand, by disjoint union, $\pi(0^n) = \pi(0^{n+1}) + \pi(10^n)$. On the other hand, by stationarity,

$$\begin{aligned}\pi(0^n) &= \mathbf{P}(X_1 \dots X_n = 0^n) = \mathbf{P}(X_0 \dots X_{n-1} = 0^n)\\ &= \mathbf{P}(X_0 \dots X_n = 0^{n+1}) + \mathbf{P}(X_0 \dots X_n = 0^n 1) = \pi(0^{n+1}) + \pi(0^n 1).\end{aligned}$$

These equalities lead to the result. Of course, symmetrically, $\pi(01^n) = \pi(1^n 0)$ under the same assumptions.

3.2 Dynamical System Associated with a VLMC

We begin with a general presentation of a probabilistic dynamical source in Sect. 3.2.1. Then we build step by step partitions of the interval [0, 1] (Sect. 3.2.2) and a mapping (Sect. 3.2.3) based on the stationary measure of a given SVLMC. It appears in Sect. 3.2.4 that this particular mapping keeps Lebesgue measure invariant. All these arguments combine to provide in the last Sect. 3.2.5 the proof of Theorem 1 which allows us to see a VLMC as a dynamical source.

In the whole section, I stands for the real interval [0, 1] and the Lebesgue measure of a Borelian J will be denoted by $|J|$.

3.2.1 General Probabilistic Dynamical Sources

Let us present here the classical formalism of probabilistic dynamical sources (see Clément et al. [6]). It is defined by four elements:

- A topological partition of I by intervals $(I_\alpha)_{\alpha \in \mathscr{A}}$
- A coding function $\rho : I \to \mathscr{A}$, such that, for each letter α, the restriction of ρ to I_α is equal to α
- A mapping $T : I \to I$
- A probability measure μ on I

Such a dynamical source defines an $\mathscr{A}$-valued random process $(Y_n)_{n \in \mathbb{N}}$ as follows. Pick a random real number x according to the measure μ. The mapping T yields the orbit $(x, T(x), T^2(x), \dots)$ of x. Thanks to the coding function, this defines the right-infinite sequence $\rho(x)\rho(T(x))\rho(T^2(x))\cdots$ whose letters are $Y_n := \rho(T^n(x))$ (see Fig. 2).

For any finite word $w = \alpha_0 \dots \alpha_N \in \mathscr{W}$, let

$$B_w = \bigcap_{k=0}^{N} T^{-k} I_{\alpha_k}$$

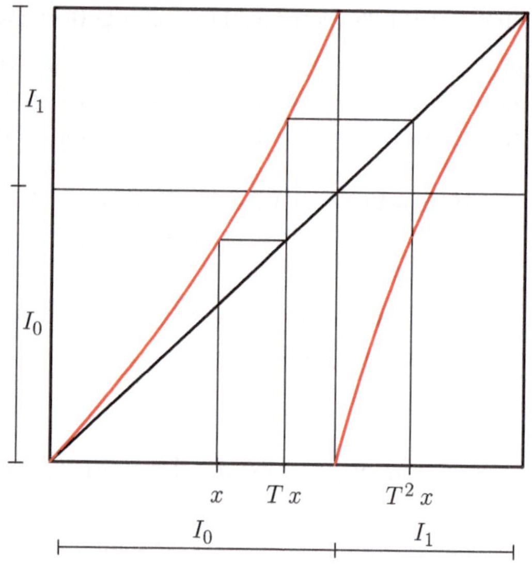

Fig. 2 The graph of a mapping T, the intervals I_0 and I_1 that code the interval I by the alphabet $\mathscr{A} = \{0, 1\}$ and the first three points of the orbit of an $x \in I$ by the corresponding dynamical system

be the Borelian set of real numbers x such that the sequence $(Y_n)_{n\in\mathbb{N}}$ has w as a prefix. Consequently, the probability that the source emits a sequence of symbols starting with the pattern w is equal to $\mu(B_w)$. When the initial probability measure μ on I is T-invariant, the dynamical source generates a stationary $\mathscr{A}$-valued random process which means that for any $n \in \mathbb{N}$, the random variable Y_n is $\rho \circ \mu$-distributed.

The following classical examples often appear in the literature: let $p \in]0, 1[$, $I_0 = [0, 1-p]$ and $I_1 =]1-p, 1]$. Let $T : I \to I$ be the only function which maps linearly and increasingly I_0 and I_1 onto I (see Fig. 3 when $p = 0.65$, left side). Then, starting from Lebesgue measure, the corresponding probabilistic dynamical source is Bernoulli: the Y_n are i.i.d. and $\mathbf{P}(Y_0 = 1) = p$. In the same vein, if T is the mapping drawn on the right side of Fig. 3, starting from Lebesgue measure, the $\{0, 1\}$-valued process $(Y_n)_{n\in\mathbb{N}}$ is Markov and stationary, with transition matrix

$$\begin{pmatrix} 0.4 & 0.6 \\ 0.7 & 0.3 \end{pmatrix}.$$

The assertions on both examples are consequences of Thales theorem. These two basic examples are particular cases of Theorem 1.

3.2.2 Ordered Subdivisions and Ordered Partitions of the Interval

Definition 8. A family $(I_w)_{w\in\mathscr{W}}$ of subintervals of I indexed by all finite words is said to be an *$\mathscr{A}$-adic subdivision* of I whenever

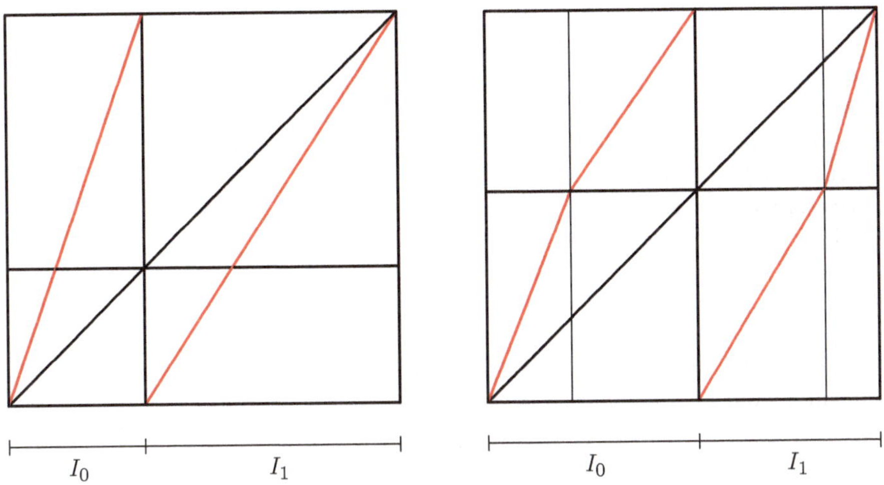

Fig. 3 Mappings generating a Bernoulli source and a stationary Markov chain of order 1. In both cases, Lebesgue measure is the initial one

(i) for any $w \in \mathscr{W}$, I_w is the disjoint union of $I_{w\alpha}$, $\alpha \in \mathscr{A}$;
(ii) for any $(\alpha, \beta) \in \mathscr{A}^2$, for any $w \in \mathscr{W}$,

$$\alpha < \beta \Longrightarrow \forall x \in I_{w\alpha},\ \forall y \in I_{w\beta},\ x < y.$$

Remark 4. For any integer $p \geq 2$, the usual p-adic subdivision of I is a particular case of $\mathscr{A}$-adic subdivision for which $|\mathscr{A}| = p$ and $|I_w| = p^{-|w|}$ for any finite word $w \in \mathscr{W}$. For a general $\mathscr{A}$-adic subdivision, the intervals associated with two k-length words need not have the same length.

The inclusion relations between the subintervals I_w of an $\mathscr{A}$-adic subdivision are thus coded by the prefix order in the complete $|\mathscr{A}|$-ary planar tree. In particular, for any $w \in \mathscr{W}$ and for any cutset $\mathscr{K}$ of the complete $|\mathscr{A}|$-ary tree,

$$I_w = \bigcup_{v \in \mathscr{K}} I_{wv}$$

(this union is a disjoint one; see Sect. 2.1 for a definition of a cutset).

We will use the following convention for $\mathscr{A}$-adic subdivisions: we require the intervals I_v to be open on the left side and closed on the right side, except the ones of the form I_{0^n} that are compact. Obviously, if μ is any probability measure on $\mathscr{R} = \mathscr{A}^{\mathbb{N}}$, there exists a unique $\mathscr{A}$-adic subdivision of I such that $|I_w| = \mu(w\mathscr{R})$ for any $w \in \mathscr{W}$.

Given an $\mathscr{A}$-adic subdivision of I, we extend the notation I_w to right-infinite words by

$$\forall r \in \mathscr{R},\ I_r = \bigcap_{\substack{w \in \mathscr{W} \\ w \text{ prefix of } r}} I_w.$$

Definition 9. A family $(I_v)_{v \in V}$ of subintervals of I indexed by a totally ordered set V is said to define an *ordered topological partition* of I when

(i) $I = \bigcup_{v \in V} \mathrm{cl}(I_v)$,
(ii) for any $v, v' \in V$, $v \neq v' \Longrightarrow \mathrm{int}(I_v) \cap \mathrm{int}(I_{v'}) = \emptyset$,
(iii) for any $v, v' \in V$,

$$v \leq v' \Longrightarrow \forall x \in I_v,\ \forall x' \in I_{v'},\ x \leq x'$$

where $\mathrm{cl}(I_v)$ and $\mathrm{int}(I_v)$ stand respectively for the closure and the interior of I_v. We will denote

$$I = \sum_{v \in V} \uparrow I_v.$$

We will use the following fact: if $I = \sum_{v \in V} \uparrow I_v = \sum_{v \in V} \uparrow J_v$ are two ordered topological partitions of I indexed by the same countable ordered set V, then $I_v = J_v$ for any $v \in V$ as soon as $|I_v| = |J_v|$ for any $v \in V$.

3.2.3 Definition of the Mapping T

Let $(U_n)_{n \geq 0}$ be a SVLMC, defined by its probabilized context tree $(\mathscr{T}, (q_c)_{c \in \mathscr{C}})$ and a stationary[2] probability measure π on $\mathscr{L}$. We first associate with π the unique $\mathscr{A}$-adic subdivision $(I_w)_{w \in \mathscr{W}}$ of I, defined by:

$$\forall w \in \mathscr{W},\ |I_w| = \pi(\overline{w}),$$

(recall that if $w = \alpha_1 \ldots \alpha_N$, $\overline{w}$ is the reversed word $\alpha_N \ldots \alpha_1$ and that $\pi(\overline{w})$ denotes $\pi(\mathscr{L}\overline{w})$).

We consider now three ordered topological partitions of I.

- The *coding partition*:
  It consists in the family $(I_\alpha)_{\alpha \in \mathscr{A}}$:

$$I = \sum_{\alpha \in \mathscr{A}} \uparrow I_\alpha = I_0 + I_1.$$

- The *vertical partition*:
  The countable set of finite and infinite contexts $\mathscr{C}$ is a cutset of the $\mathscr{A}$-ary tree. The family $(I_c)_{c \in \mathscr{C}}$ thus defines the so-called vertical ordered topological

[2] Note that this construction can be made replacing π by any probability measure on $\mathscr{L}$.

partition

$$I = \sum_{c \in \mathscr{C}} \uparrow I_c.$$

- The *horizontal partition*:
  $\mathscr{A}\mathscr{C}$ is the set of leaves of the context tree $\mathscr{A}\mathscr{T} = \{\alpha w,\ \alpha \in \mathscr{A},\ w \in T\}$. As before, the family $(I_{\alpha c})_{\alpha c \in \mathscr{A}\mathscr{C}}$ defines the so-called horizontal ordered topological partition

$$I = \sum_{\alpha c \in \mathscr{A}\mathscr{C}} \uparrow I_{\alpha c}.$$

Definition 10. The mapping $T : I \to I$ is the unique left continuous function such that:
(i) the restriction of T to any $I_{\alpha c}$ is affine and increasing;
(ii) for any $\alpha c \in \mathscr{A}\mathscr{C}$, $T(I_{\alpha c}) = I_c$.

The function T is always increasing on I_0 and on I_1. When $q_c(\alpha) \neq 0$, the slope of T on an interval $I_{\alpha c}$ is $1/q_c(\alpha)$. Indeed, with formula (7), one has

$$|I_{\alpha c}| = \pi(\overline{c}\alpha) = q_c(\alpha)\pi(\overline{c}) = |I_c| q_c(\alpha).$$

When $q_c(\alpha) = 0$ and $|I_c| \neq 0$, the interval $I_{\alpha c}$ is empty so that T is discontinuous at $x_c = \pi(\{s \in \mathscr{L},\ \overline{s} \leq c\})$ ($\leq$ denotes here the alphabetical order on $\mathscr{R}$). Note that $|I_c| = 0$ implies $|I_{\alpha c}| = 0$. In particular, if one assumes that all the probability measures q_c, $c \in \mathscr{C}$, are nontrivial (i.e. as soon as they satisfy $q_c(0)q_c(1) \neq 0$), then T is continuous on I_0 and I_1. Furthermore, $T(I_0) = \mathrm{cl}(T(I_1)) = I$ and for any $c \in \mathscr{C}$, $T^{-1}I_c = I_{0c} \cup I_{1c}$ (see Fig. 4).

Example: the four flower bamboo. The *four flower bamboo* is the VLMC defined by the finite probabilized context tree of Fig. 5. There exists a unique stationary measure π under conditions which are detailed later, in Example 3. We represent on Fig. 5 the mapping T built with this π, together with the respective subdivisions

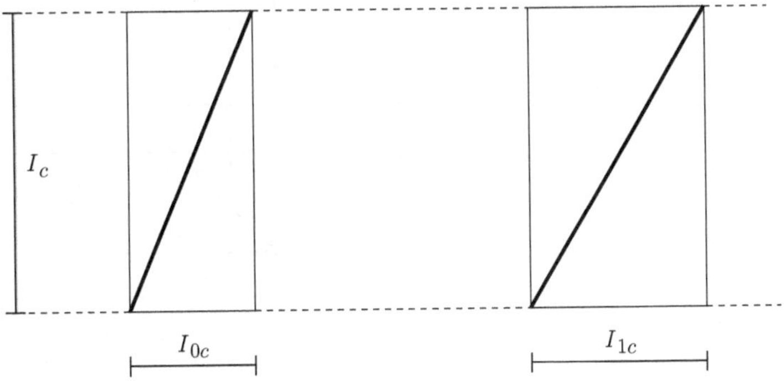

Fig. 4 Action of T on horizontal and vertical partitions. On this figure, c is any context and the alphabet is $\{0, 1\}$

of x-axis and y-axis by the four I_c and the eight $I_{\alpha c}$. The x-axis is divided by both coding and horizontal partitions; the y-axis is divided by both coding and vertical partitions. This figure has been drawn with the following data on the four flower bamboo: $q_{00}(0) = 0.4$, $q_{010}(0) = 0.6$, $q_{011}(0) = 0.8$ and $q_1(0) = 0.3$.

3.2.4 Properties of the Mapping T

The following key lemma explains the action of the mapping T on the intervals of the $\mathscr{A}$-adic subdivision $(I_w)_{w \in \mathscr{W}}$. More precisely, it extends the relation $T(I_{\alpha c}) = I_c$, for any $\alpha c \in \mathscr{A}\mathscr{C}$, to any finite word.

Lemma 2. *For any finite word $w \in \mathscr{W}$ and any letter $\alpha \in \mathscr{A}$, $T(I_{\alpha w}) = I_w$.*

Proof. Assume first that $w \notin \mathscr{T}$. Let then $c \in \mathscr{C}$ be the unique context such that c is a prefix of w. Because of the prefix order structure of the $\mathscr{A}$-adic subdivision $(I_v)_v$, one has the first ordered topological partition

$$I_c = \sum_{\substack{v \in \mathscr{W},\ |v|=|w| \\ c \text{ prefix of } v}} \uparrow I_v \tag{9}$$

(the set of indices is a cutset in the tree of c descendants). On the other hand, the same topological partition applied to the finite word αw leads to

$$I_{\alpha c} = \sum_{\substack{v \in \mathscr{W},\ |v|=|w| \\ c \text{ prefix of } v}} \uparrow I_{\alpha v}.$$

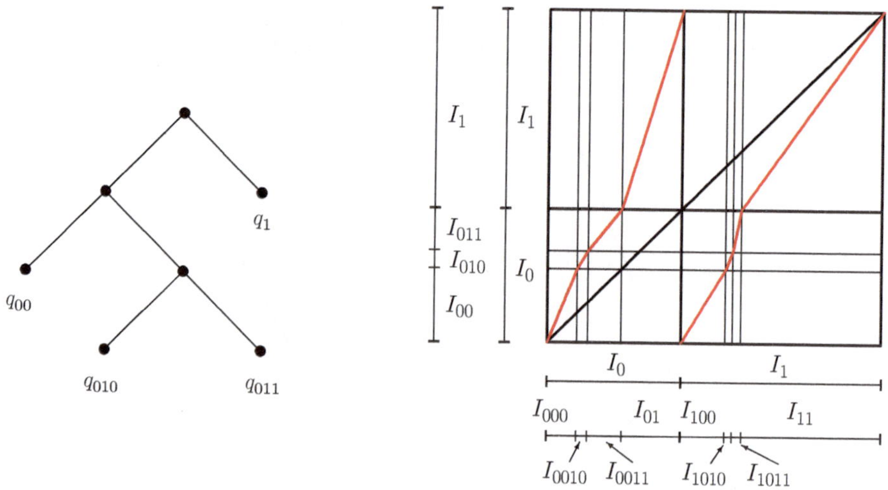

Fig. 5 On the *left*, the four flower bamboo context tree. On the *right*, its mapping together with the coding, the vertical and the horizontal partitions of [0, 1]

Taking the image by T, one gets the second ordered topological partition

$$I_c = \sum_{\substack{v \in \mathscr{W},\ |v|=|w| \\ c \text{ prefix of } v}} \uparrow T(I_{\alpha v}). \tag{10}$$

Now, if c is a prefix of a finite word v, $I_{\alpha v} \subseteq I_{\alpha c}$ and the restriction of T to $I_{\alpha c}$ is affine. By Thales theorem, it comes

$$|T(I_{\alpha v})| = |I_{\alpha v}|.\frac{|I_c|}{|I_{\alpha c}|}.$$

Since π is a stationary measure for the VLMC,

$$|I_{\alpha c}| = \pi(\overline{c}\alpha) = q_c(\alpha)\pi(\overline{c}) = |I_c|q_c(\alpha).$$

Furthermore, one has $\pi(\overline{v}\alpha) = q_c(\alpha)\pi(\overline{v})$. Finally, $|T(I_{\alpha v})| = |I_v|$. Relations (9) and (10) are two ordered countable topological partitions, the components with the same indices being of the same length: the partitions are necessarily the same. In particular, because w belongs to the set of indices, this implies that $T(I_{\alpha w}) = I_w$.

Assume now that $w \in \mathscr{T}$. Since the set of contexts having w as a prefix is a cutset of the tree of the descendants of w, one has the disjoint union

$$I_{\alpha w} = \bigcup_{\substack{c \in \mathscr{C} \\ w \text{ prefix of } c}} I_{\alpha c}.$$

Taking the image by T leads to

$$T(I_{\alpha w}) = \bigcup_{c \in \mathscr{C}, w \text{ prefix of } c} I_c = I_w$$

and the proof is complete. □

Remark 5. The same proof shows in fact that if w is any finite word, $T^{-1}I_w = I_{0w} \cup I_{1w}$ (disjoint union).

Lemma 3. *For any $\alpha \in \mathscr{A}$, for any context $c \in \mathscr{C}$, for any Borelian set $B \subseteq I_c$,*

$$|I_\alpha \cap T^{-1}B| = |B|q_c(\alpha).$$

Proof. It is sufficient to show the lemma when B is an interval. The restriction of T to $I_{\alpha c}$ is affine and $T^{-1}I_c = I_{0c} \cup I_{1c}$. The result is thus due to Thales Theorem. □

Corollary 1. *If T is the mapping associated with a SVLMC, Lebesgue measure is invariant by T, i.e. $|T^{-1}B| = |B|$ for any Borelian subset of I.*

Proof. Since $B = \bigcup_{c\in\mathscr{C}} B \cap I_c$ (disjoint union), it suffices to prove that $|T^{-1}B| = |B|$ for any Borelian subset of I_c where c is any context. If $B \subseteq I_c$, because of Lemma 3,

$$|T^{-1}B| = |I_0 \cap T^{-1}B| + |I_1 \cap T^{-1}B| = |B|(q_c(0) + q_c(1)) = |B|.$$

□

3.2.5 SVLMC as Dynamical Source

We now consider the stationary probabilistic dynamical source $((I_\alpha)_{\alpha\in\mathscr{A}}, \rho, T, |.|)$ built from the SVLMC. It provides the $\mathscr{A}$-valued random process $(Y_n)_{n\in\mathbb{N}}$ defined by

$$Y_n = \rho(T^n\xi)$$

where ξ is a uniformly distributed I-valued random variable and ρ the coding function. Since Lebesgue measure is T-invariant, all random variables Y_n have the same law, namely $\mathbf{P}(Y_n = 0) = |I_0| = \pi(0)$.

Definition 11. Two $\mathscr{A}$-valued random processes $(V_n)_{n\in\mathbb{N}}$ and $(W_n)_{n\in\mathbb{N}}$ are called *symmetrically distributed* whenever for any $N \in \mathbb{N}$ and for any finite word $w \in \mathscr{A}^{N+1}$, $\mathbf{P}(W_0W_1\dots W_N = w) = \mathbf{P}(V_0V_1\dots V_N = \overline{w})$.

In other words, $(V_n)_{n\in\mathbb{N}}$ and $(W_n)_{n\in\mathbb{N}}$ are symmetrically distributed if and only if for any $N \in \mathbb{N}$, the random words $W_0W_1\dots W_N$ and $V_NV_{N-1}\dots V_0$ have the same distribution.

Theorem 1. *Let $(U_n)_{n\in\mathbb{N}}$ be a SVLMC and π a stationary probability measure on $\mathscr{L}$. Let $(X_n)_{n\in\mathbb{N}}$ be the process of final letters of $(U_n)_{n\in\mathbb{N}}$. Let $T : I \to I$ be the mapping defined in Sect. 3.2.3. Then,*

(i) Lebesgue measure is T-invariant.

(ii) If ξ is any uniformly distributed random variable on I, the processes $(X_n)_{n\in\mathbb{N}}$ and $(\rho(T^n\xi))_{n\in\mathbb{N}}$ are symmetrically distributed.

Proof. *(i)* has been already stated and proven in Corollary 1.

(ii) As before, for any finite word $w = \alpha_0\dots\alpha_N \in \mathscr{W}$, let $B_w = \bigcap_{k=0}^{N} T^{-k}I_{\alpha_k}$ be the Borelian set of real numbers x such that the right-infinite sequence $(\rho(T^nx))_{n\in\mathbb{N}}$ has w as a prefix. By definition, $B_\alpha = I_\alpha$ if $\alpha \in \mathscr{A}$. More generally, we prove the following claim: *for any $w \in \mathscr{W}$, $B_w = I_w$*. Indeed, if $\alpha \in \mathscr{A}$ and $w \in \mathscr{W}$, $B_{\alpha w} = I_\alpha \cap T^{-1}B_w$; thus, by induction on the length of w, $B_{\alpha w} = I_\alpha \cap T^{-1}I_w = I_{\alpha w}$, the last equality being due to Lemma 2. There is now no difficulty in finishing the proof: if $w \in \mathscr{W}$ is any finite word of length $N + 1$, then $\mathbf{P}(X_0\dots X_N = \overline{w}) = \pi(\overline{w}) = |I_w|$. Thus, because of the claim, $\mathbf{P}(X_0\dots X_N = \overline{w}) = |B_w| = \mathbf{P}(Y_0\dots Y_N = w)$. This proves the result. □

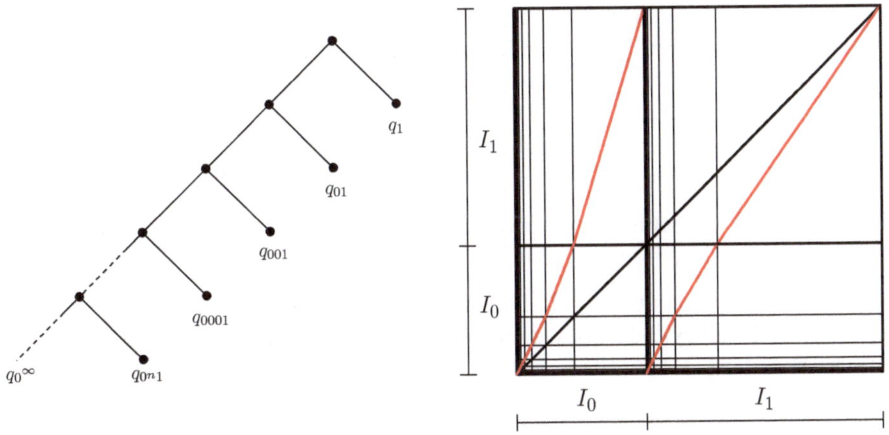

Fig. 6 Infinite comb probabilized context tree (*on the left*) and the associated dynamical system (*on the right*)

4 Examples

4.1 The Infinite Comb

4.1.1 Stationary Probability Measures

Consider the probabilized context tree given on the left side of Fig. 6. In this case, there is one infinite leaf 0^∞ and countably many finite leaves 0^n1, $n \in \mathbb{N}$. The data of a corresponding VLMC consists thus in probability measures on $\mathscr{A} = \{0, 1\}$:

$$q_{0^\infty} \text{ and } q_{0^n1},\ n \in \mathbb{N}.$$

Suppose that π is a stationary measure on $\mathscr{L}$. We first compute $\pi(w)$ (notation (5)) as a function of $\pi(1)$ when $\overline{w}$ is any context or any internal node. Because of formula (7), $\pi(10) = \pi(1)q_1(0)$ and an immediate induction shows that, for any $n \geq 0$,

$$\pi(10^n) = \pi(1)c_n, \tag{11}$$

where $c_0 = 1$ and, for any $n \geq 1$,

$$c_n = \prod_{k=0}^{n-1} q_{0^k1}(0). \tag{12}$$

The stationary probability of a reversed context is thus necessarily given by formula (11). Now, if 0^n is any internal node of the context tree, we need going down along the branch in $\mathcal{T}$ to reach the contexts; using then the disjoint union $\pi(0^{n+1}) = \pi(0^n) - \pi(10^n)$, by induction, it comes for any $n \geq 0$,

$$\pi(0^n) = 1 - \pi(1) \sum_{k=0}^{n-1} c_k. \tag{13}$$

The stationary probability of a reversed internal node of the context tree is thus necessarily given by formula (13).

It remains to compute $\pi(1)$. The countable partition of the whole probability space given by all cylinders based on leaves in the context tree (Formula (3)) implies $1 - \pi(0^\infty) = \pi(1) + \pi(10) + \pi(100) + \ldots$, i.e.

$$1 - \pi(0^\infty) = \sum_{n \geq 0} \pi(1) c_n. \tag{14}$$

This leads to the following statement that covers all cases of existence, uniqueness and nontriviality for a stationary probability measure for the infinite comb. In the generic case (named *irreducible* case hereunder), we give a necessary and sufficient condition on the data for the existence of a stationary probability measure; moreover, when a stationary probability exists, it is unique. The *reducible* case is much more singular and gives rise to nonuniqueness.

Proposition 1. (Stationary probability measures for an infinite comb)
Let $(U_n)_{n \geq 0}$ be a VLMC defined by a probabilized infinite comb.
(i) Irreducible case. *Assume that $q_{0^\infty}(0) \neq 1$.*

(i.a) Existence. *The Markov process $(U_n)_{n \geq 0}$ admits a stationary probability measure on $\mathcal{L}$ if and only if the numerical series $\sum c_n$ defined by (12) converges.*

(i.b) Uniqueness. *Assume that the series $\sum c_n$ converges and denote*

$$S(1) = \sum_{n \geq 0} c_n. \tag{15}$$

Then, the stationary probability measure π on $\mathcal{L}$ is unique; it is characterized by

$$\pi(1) = \frac{1}{S(1)} \tag{16}$$

and formula (11), (13) and (8).

Furthermore, π is trivial if and only if $q_1(0) = 0$, in which case it is defined by $\pi(1^\infty) = 1$.

(ii) Reducible case. *Assume that $q_{0^\infty}(0) = 1$.*

(ii.a) If the series $\sum c_n$ diverges, then the trivial probability measure π on $\mathcal{L}$ defined by $\pi(0^\infty) = 1$ is the unique stationary probability.

(ii.b) If the series $\sum c_n$ *converges, then there is a one parameter family of stationary probability measures on* $\mathcal{L}$*. More precisely, for any* $a \in [0,1]$*, there exists a unique stationary probability measure* π_a *on* $\mathcal{L}$ *such that* $\pi_a(0^\infty) = a$*. The probability* π_a *is characterized by* $\pi_a(1) = \frac{1-a}{S(1)}$ *and formula (11), (13) and (8).*

Furthermore, π_a *is non trivial except in the two following cases:*

- $a = 1$*, in which case* π_1 *is defined by* $\pi_1(0^\infty) = 1$*;*
- $a = 0$ *and* $q_1(0) = 0$*, in which case* π_0 *is defined by* $\pi_0(1^\infty) = 1$*.*

Proof. *(i)* Assume that $q_{0^\infty}(0) \neq 1$ and that π is a stationary probability measure. By definition of probability transitions, $\pi(0^\infty) = \pi(0^\infty)q_{0^\infty}(0)$ so that $\pi(0^\infty)$ necessarily vanishes. Thus, thanks to (14), $\pi(1) \neq 0$, the series $\sum c_n$ converges and formula (16) is valid. Moreover, when $\overline{w}$ is any context or any internal node of the context tree, $\pi(w)$ is necessarily given by formula (16), (11) and (13). This shows that for any finite word w, $\pi(w)$ is determined by formula (8). Since the cylinders $\mathcal{L}w$, $w \in \mathcal{W}$ span the σ-algebra on $\mathcal{L}$, there is at most one stationary probability measure. This proves the *only if* part of *(i.a)*, the uniqueness and the characterization claimed in *(i.b)*.

Conversely, when the series converges, formula (16), (11), (13) and (8) define a probability measure on the semiring spanned by cylinders, which extends to a stationary probability measure on the whole σ-algebra on $\mathcal{L}$ (see Billingsley [3] for a general treatment on semirings, σ-algebra, definition and characterization of probability measures). This proves the *if* part of *(i.a)*. Finally, the definition of c_n directly implies that $S(1) = 1$ if and only if $q_1(0) = 0$. This proves the assertion of *(i.b)* on the triviality of π.

(ii) Assume that $q_{0^\infty}(0) = 1$. Formula (14) is always valid so that the divergence of the series $\sum c_n$ forces $\pi(1)$ to vanish and, consequently, any stationary measure π to be the trivial one defined by $\pi(0^\infty) = 1$.

Besides, with the assumption $q_{0^\infty}(0) = 1$, one immediately sees that this trivial probability is stationary, proving *(ii.a)*.

To prove *(ii.b)*, assume furthermore that the series $\sum c_n$ converges and let $a \in [0,1]$. As before, any stationary probability measure π is completely determined by $\pi(1)$. Moreover, the probability measure defined by $\pi_a(1) = \frac{1-a}{S(1)}$, formula (11), (13) and (8) and in a way extended to the whole σ-algebra on $\mathcal{L}$ is clearly stationary. Because of formula (14), it satisfies

$$\pi_a(0^\infty) = 1 - \pi_a(1)S(1) = a.$$

This proves the assertion on the one parameter family. Finally, π_a is trivial only if $\pi_a(1) \in \{0,1\}$. If $a = 1$ then $\pi_a(1) = 0$ thus π_1 is the trivial probability that only charges 0^∞. If $a = 0$ then $\pi_a(1) = 1/S(1)$ is nonzero and it equals 1 if and only if $S(1) = 1$, i.e. if and only if $q_1(0) = 0$, in which case π_0 is the trivial probability that only charges 1^∞. □

Remark 6. This proposition completes previous results which give sufficient conditions for the existence of a stationary measure for an infinite comb. For instance, in

Galves and Löcherbach [12], the intervening condition is

$$\sum_{k\geq 0} q_{0^k 1}(1) = +\infty,$$

which is equivalent with our notations to $c_n \rightarrow 0$. Note that if $\sum c_n$ is divergent, then the only possible stationary distribution is the trivial Dirac measure δ_{0^∞}.

4.1.2 The Associated Dynamical System

The vertical partition is made of the intervals $I_{0^n 1}$ for $n \geq 0$. The horizontal partition consists in the intervals $I_{00^n 1}$ and $I_{10^n 1}$, for $n \geq 0$, together with two intervals coming from the infinite context, namely I_{0^∞} and I_{10^∞}. In the irreducible case, $\pi(0^\infty) = 0$ and these two last intervals become two accumulation points of the partition: 0 and $\pi(0)$. The following lemma is classical and helps understand the behaviour of the mapping T at these accumulation points.

Lemma 4. *Let $f : [a,b] \rightarrow \mathbb{R}$ be continuous on $[a,b]$, differentiable on $]a,b[\setminus D$ where D is a countable set. The function f admits a right derivative at a and*

$$f_r'(a) = \lim_{\substack{x\rightarrow a, x>a \\ x\notin D}} f'(x)$$

as soon as this limit exists.

Corollary 2. *If $(q_{0^n 1}(0))_{n\in\mathbb{N}}$ converges, then T is differentiable at 0 and $\pi(0)$ (with a possibly infinite derivative) and*

$$T_r'(0) = \lim_{n\rightarrow+\infty} \frac{1}{q_{0^n 1}(0)}, \quad T_r'(\pi(0)) = \lim_{n\rightarrow+\infty} \frac{1}{q_{0^n 1}(1)}.$$

When $(q_{0^n 1}(0))_{n\in\mathbb{N}}$ converges to 1, $T_r'(0) = 1$. In this case, 0 is an indifferent fixed point and $T_r'(\pi(0)) = +\infty$. The mapping T is a slight modification of the so-called Wang map (Wang [27]). The statistical properties of the Wang map are quite well understood (Lambert et al. [17]). The corresponding dynamical source is said intermittent.

4.1.3 Dirichlet Series

For a stationary infinite comb, the Dirichlet series is defined on a suitable vertical open strip of $\mathbb{C}$ as

$$\Lambda(s) = \sum_{w\in\mathscr{W}} \pi(w)^s.$$

In the whole section we suppose that $\sum c_n$ is convergent. Indeed, if it is divergent then the only stationary measure is the Dirac measure δ_{0^∞} and $\Lambda(s)$ is never defined.

The computation of the Dirichlet series is tractable because of the following formula: for any finite words $w, w' \in \mathscr{W}$,

$$\pi(w1w')\pi(1) = \pi(w1)\pi(1w'). \tag{17}$$

This formula, which comes directly from formula (8), is true because of the very particular form of the contexts in the infinite comb. It is the expression of its renewal property. The computation of the Dirichlet series is made in two steps.

Step 1. A finite word either does not contain any 1 or is of the form $w10^n$, $w \in \mathscr{W}$, $n \geq 0$. Thus,

$$\Lambda(s) = \sum_{n\geq 0} \pi(0^n)^s + \sum_{n\geq 0} \sum_{w\in\mathscr{W}} \pi(w10^n)^s.$$

Because of formula (17) and (16), $\pi(w10^n) = S(1)\pi(w1)\pi(10^n)$. Let us denote

$$\Lambda_1(s) = \sum_{w\in\mathscr{W}} \pi(w1)^s.$$

With this notation and formula (11) and (13),

$$\Lambda(s) = \frac{1}{S(1)^s} \sum_{n\geq 0} R_n^s + \Lambda_1(s) \sum_{n\geq 0} c_n^s$$

where R_n stands for the rest

$$R_n = \sum_{k\geq n} c_k. \tag{18}$$

Step 2. It consists in the computation of Λ_1. A finite word having 1 as last letter either can be written 0^n1, $n \geq 0$ or is of the form $w10^n1$, $w \in \mathscr{W}$, $n \geq 0$. Thus it comes,

$$\Lambda_1(s) = \sum_{n\geq 0} \pi(0^n1)^s + \sum_{n\geq 0} \sum_{w\in\mathscr{W}} \pi(w10^n1)^s.$$

By formula (17) and (11), $\pi(w10^n1) = \pi(w1)c_n q_{0^n1}(1) = \pi(w1)(c_n - c_{n+1})$, so that

$$\Lambda_1(s) = \frac{1}{S(1)^s} \sum_{n\geq 0} c_n^s + \Lambda_1(s) \sum_{n\geq 0} (c_n - c_{n+1})^s$$

and

$$\Lambda_1(s) = \frac{1}{S(1)^s} \cdot \frac{\sum_{n\geq 0} c_n^s}{1 - \sum_{n\geq 0}(c_n - c_{n+1})^s}.$$

Putting results of both steps together, we obtain the following proposition.

Proposition 2. *With notations (12), (15) and (18), the Dirichlet series of a source obtained from a stationary infinite comb is*

$$\Lambda(s) = \frac{1}{S(1)^s}\left[\sum_{n\geq 0} R_n^s + \frac{\left(\sum_{n\geq 0} c_n^s\right)^2}{1-\sum_{n\geq 0}(c_n - c_{n+1})^s}\right].$$

Remark 7. The analytic function $\frac{\left(\sum_{n\geq 0} c_n^s\right)^2}{1-\sum_{n\geq 0}(c_n - c_{n+1})^s}$ is always singular for $s = 1$ because its denominator vanishes while its numerator is a convergent series.

Examples. (1) Suppose that $0 < a < 1$ and that $q_{0^n1}(0) = a$ for any $n \geq 0$. Then $c_n = a^n$, $R_n = \frac{a^n}{1-a}$ and $S(1) = \frac{1}{1-a}$. For such a source, the Dirichlet series is

$$\Lambda(s) = \frac{1}{1-[a^s + (1-a)^s]}.$$

In this case, the source is memoryless: all letters are drawn independently with the same distribution. The Dirichlet series of such sources have been extensively studied in Flajolet et al. [8] in the realm of asymptotics of average parameters of a trie.

(2) Extension of Example 1: take $a, b \in]0,1[$ and consider the probabilized infinite comb defined by

$$q_{0^n1}(0) = \begin{cases} a \text{ if } n \text{ is even,} \\ b \text{ if } n \text{ is odd.} \end{cases}$$

After computation, the Dirichlet series of the corresponding source under the stationary distribution turns out to have the explicit form

$$\Lambda(s) = \frac{1}{1-(ab)^s}\left[1 + \left(\frac{a+ab}{1+a}\right)^s + \left(\frac{1-ab}{1+a}\right)^s \right.$$
$$\left. \times \frac{(1+a^s)^2}{1-(ab)^s-(1-a)^s-a^s(1-b)^s}\right].$$

The configuration of poles of Λ depends on arithmetic properties (approximation by rationals) of the logarithms of ab, $1-a$ and $a(1-b)$. The poles of such a series are the same as in the case of a memoryless source with an alphabet of three letters, see Flajolet et al. [8]. This could be extended to a family of examples.

(3) Let $\alpha > 2$. We take data $q_{0^n1}(0)$, $n \geq 0$ in such a way that $c_0 = 1$ and, for any $n \geq 1$,

$$c_n = \zeta(n,\alpha) := \frac{1}{\zeta(\alpha)}\sum_{k\geq n}\frac{1}{k^\alpha},$$

where ζ is the Riemann function. Since $c_n \in \mathcal{O}(n^{1-\alpha})$ when n tends to infinity, there exists a unique stationary probability measure π on $\mathcal{L}$. One obtains

$$S(1) = 1 + \frac{\zeta(\alpha-1)}{\zeta(\alpha)}$$

and, for any $n \geq 1$,

$$R_n = \frac{\zeta(\alpha-1)}{\zeta(\alpha)}\zeta(n,\alpha-1) - (n-1)\zeta(n,\alpha).$$

In particular, $R_n \in \mathcal{O}(n^{2-\alpha})$ when n tends to infinity. The final formula for the Dirichlet series of this source is

$$\Lambda(s) = \frac{1}{S(1)^s}\left[\sum_{n\geq 0} R_n^s + \frac{\left(\sum_{n\geq 0} c_n^s\right)^2}{1-\frac{\zeta(\alpha s)}{\zeta(\alpha)^s}}\right].$$

(4) One case of interest is when the associated dynamical system has an indifferent fixed point (see Sect. 4.1.2), for example when

$$q_{0^n 1}(0) = \left(1 - \frac{1}{n+2}\right)^{\alpha},$$

with $1 < \alpha < 2$. In this situation, $c_n = (1+n)^{-\alpha}$ and

$$\Lambda(s) = \sum_{n\geq 1}\zeta(n,\alpha)^s + \frac{\zeta(\alpha s)^2}{\zeta(\alpha)^s}\cdot\frac{1}{1-\sum_{n\geq 1}\frac{1}{n^{\alpha s}}\left[1-\left(1-\frac{1}{n+1}\right)^{\alpha}\right]}.$$

4.1.4 Generating Function for the Exact Distribution of Word Occurrences in a Sequence Generated by a Comb

The behaviour of the entrance time into cylinders is a natural question arising in dynamical systems. There exists a large literature on the asymptotic properties of entrance times into cylinders for various kind of systems, symbolic or geometric; see Abadi and Galves [1] for an extensive review on the subject. Most of the results deal with an exponential approximation of the distribution of the first entrance time into a small cylinder, sometimes with error terms. The most up-to-date result on this framework is Abadi and Saussol [2] in which the hypothesis are made only in terms of the mixing type of the source (so-called α-mixing). We are here interested in exact distribution results instead of asymptotic behaviours.

Several studies in probabilities on words are based on generating functions. For example one may cite Régnier [20], Reinert et al. [21], Stefanov and Pakes [25]. For i.i.d. sequences, Blom and Thorburn [4] give the generating function of the first occurrence of a word, based on a recurrence relation on the probabilities. This result is extended to Markovian sequences by Robin and Daudin [24]. Nonetheless, other approaches are considered: one of the more general techniques is the so-called Markov chain embedding method introduced by Fu [9] and further developed by Fu and Koutras [10], Koutras [16]. A martingale approach (see Gerber and Li [13], Li [18], Williams [28]) is an alternative to the Markov chain embedding method. These two approaches are compared in Pozdnyakov et al. [19].

We establish results on the exact distribution of word occurrences in a random sequence generated by a comb (or a bamboo in Sect. 4.2.4). More precisely, we make explicit the generating function of the random variable giving the rth occurrence of a k-length word, for any word w such that $\overline{w}$ is not an internal node of $\mathcal{T}$.

Let us consider the process $X = (X_n)_{n\geq 0}$ of final letters of $(U_n)_{n\geq 0}$, in the particular case of a SVLMC defined by an infinite comb. Let $w = w_1 \dots w_k$ be a word of length $k \geq 1$. We say that w occurs at position $n \geq k$ in the sequence X if the word w ends at position n:

$$\{w \text{ at } n\} = \{X_{n-k+1} \dots X_n = w\} = \{U_n \in \mathcal{L}w\}.$$

Let us denote by $T_w^{(r)}$ the position of the rth occurrence of w in X and $\Phi_w^{(r)}$ its generating function:

$$\Phi_w^{(r)}(x) := \sum_{n\geq 0} \mathbf{P}\left(T_w^{(r)} = n\right) x^n.$$

The following notation is used in the sequel: for any finite word $u \in \mathcal{W}$, for any finite context $c \in \mathcal{C}$ and for any $n \geq 0$,

$$q_c^{(n)}(u) = \mathbf{P}\left(X_{n-|u|+1} \dots X_n = u | X_{-(|c|-1)} \dots X_0 = \overline{c}\right).$$

These quantities may be computed in terms of the data q_c. Proposition 3 generalizes results of Robin and Daudin [24].

Proposition 3. *For a SVLMC defined by an infinite comb, with the above notations, for a word w such that $\overline{w}$ is non internal node, the generating function of its first occurrence is given, for $|x| < 1$, by*

$$\Phi_w^{(1)}(x) = \frac{x^k \pi(w)}{(1-x)S_w(x)}$$

and the generating function of its rth occurrence is given, for $|x| < 1$, by

$$\Phi_w^{(r)}(x) = \Phi_w^{(1)}(x)\left(1 - \frac{1}{S_w(x)}\right)^{r-1},$$

where

$$S_w(x) = C_w(x) + \sum_{j=k}^{\infty} q^{(j)}_{\overleftarrow{\text{pref}}\,(w)}(w)x^j,$$

$$C_w(x) = 1 + \sum_{j=1}^{k-1} \mathbf{1}_{\{w_{j+1}\dots w_k = w_1\dots w_{k-j}\}} q^{(j)}_{\overleftarrow{\text{pref}}\ (w)} \left(w_{k-j+1}\dots w_k\right) x^j.$$

Remark 8. The term $C_w(x)$ is a generalization of the probabilized autocorrelation polynomial defined in Jacquet and Szpankowski [15] in the particular case when the $(X_n)_{n\geq 0}$ are independent and identically distributed. For a word $w = w_1 \dots w_k$ this polynomial is equal to

$$c_w(x) = \sum_{j=0}^{k-1} c_{j,w} \frac{1}{\pi(w_1 \dots w_{k-j})} x^j,$$

where $c_{j,w} = 1$ if the $k-j$-length suffix of w is equal to its $k-j$-length prefix, and is equal to zero otherwise. When the $(X_n)_{n\geq 0}$ are independent and identically distributed, we have

$$\sum_{j=1}^{k-1} \mathbf{1}_{\{w_{j+1}\dots w_k = w_1\dots w_{k-j}\}} q^{(j)}_{\overleftarrow{\text{pref}}\ (w)} \left(w_{k-j+1}\dots w_k\right) x^j = \sum_{j=1}^{k-1} c_{j,w} \frac{\pi(w)}{\pi\left(w_1 \dots w_{k-j}\right)} x^j$$

that is

$$C_w(x) = \pi(w) c_w(x).$$

Proof. We first deal with $w = 10^{k-1}$, that is the only word w of length k such that $\overline{w} \in \mathscr{C}$. For the sake of shortness, we will denote by p_n the probability that $T_w^{(1)} = n$. From the obvious decomposition

$$\{w \text{ at } n\} = \{T_w^{(1)} = n\} \cup \{T_w^{(1)} < n \text{ and } w \text{ at } n\}, \quad \text{(disjoint union)}$$

it comes by stationarity of π

$$\pi(w) = p_n + \sum_{z=k}^{n-1} p_z \mathbf{P}\left(X_{n-k+1} \dots X_n = w | T_w^{(1)} = z\right).$$

Due to the renewal property of the comb, the conditional probability can be rewritten

$$\begin{cases} \mathbf{P}\left(X_{n-k+1} \dots X_n = w | X_{z-k+1} \dots X_z = w\right) & \text{if } z \leq n-k \\ 0 & \text{if } z > n-k \end{cases}$$

the second equality being due to the lack of possible auto-recovering in w. Consequently, we have

$$\pi(w) = p_n + \sum_{z=k}^{n-k} p_z q_{\overline{w}}^{(n-z)}(w).$$

Hence, for $x < 1$, it comes

$$\frac{x^k \pi(w)}{1-x} = \sum_{n=k}^{+\infty} p_n x^n + \sum_{n=k}^{+\infty} x^n \sum_{z=k}^{n-k} p_z q_{\overline{w}}^{(n-z)}(w),$$

so that

$$\frac{x^k \pi(w)}{1-x} = \Phi_w^{(1)}(x) \left(1 + \sum_{j=k}^{\infty} x^j q_{\overline{w}}^{(j)}(w) \right),$$

which leads to

$$\Phi_w^{(1)}(x) = \frac{x^k \pi(w)}{(1-x) S_w(x)}.$$

Note that when $w = 10^{k-1}$, $C_w(x) = 1$.

Proceeding in the same way for the rth occurrence, from the decomposition

$$\{w \text{ at } n\} = \{T_w^{(1)} = n\} \cup \{T_w^{(2)} = n\} \cup \ldots \cup \{T_w^{(r)} = n\} \cup \{T_w^{(r)} < n \text{ and } w \text{ at } n\},$$

and denoting $p(n,\ell) = \mathbf{P}(T_w^{(\ell)} = n)$, the following recursive equation holds:

$$\pi(w) = p_n + p(n,2) + \ldots + p(n,r) + \sum_{z=k}^{n-1} \mathbf{P}\left(T_w^{(r)} = z \text{ and } w \text{ at } n\right).$$

Again, by splitting the last term into two terms and using the non-overlapping structure of w, one gets

$$\pi(w) = p_n + p(n,2) + \ldots + p(n,r) + \sum_{z=k}^{n-k} p_z q_{\overline{w}}^{(n-z)}(w).$$

From this recursive equation, proceeding exactly in the same way, one gets for the generating function, for $x < 1$,

$$\Phi_w^{(r)}(x) = \Phi_w^{(1)}(x) \left(1 - \frac{1}{S_w(x)} \right)^{r-1}.$$

Let us now consider the case of words w such that $\overline{w} \notin \mathscr{T}$, that is the words w such that $w_j = 1$ for at least one integer $j \in \{2, \ldots, k\}$. We denote by i the last position of a 1 in w, that is $\overleftarrow{\text{pref}}\,(w) = 0^{k-i}1$. Once again we have

$$\pi(w) = p_n + \sum_{z=k}^{n-1} p_z \mathbf{P}\left(X_{n-k+1} \ldots X_n = w \mid T_w^{(1)} = z\right).$$

When $z \le n-k$, due to the renewal property, the conditional probability can be rewritten as

$$\mathbf{P}\left(X_{n-k+1}\dots X_n = w | T_w^{(1)} = z\right) = q^{(n-z)}_{\overleftarrow{\text{pref}}\,(w)}(w).$$

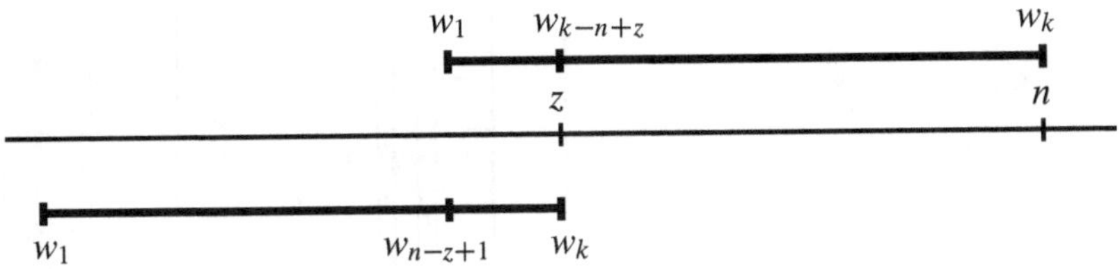

When $z > n-k$ (see figure above),

$$\mathbf{P}\left(w \text{ at } n | T_w^1 = z\right) = \mathbf{1}_{\{w_{n-z+1}\dots w_k = w_1\dots w_{k-n+z}\}} q^{(n-z)}_{\overleftarrow{\text{pref}}\,(w)}(w_{k-n+z+1}\dots w_k),$$

this equality holding if $n-k+i \ne z$. But when $z = n-k+i$, because the first occurrence of w is at z, necessarily $w_k = 1$ and hence $i = k$, and $z = n$ which contradicts $z < n$. Consequently for $z = n-k+i$ we have

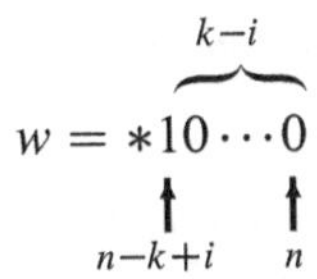

$$\mathbf{P}\left(X_{n-k+1}\dots X_n = w | T_w^1 = z\right) = 0 = \mathbf{1}_{\{w_{n-z+1}\dots w_k = w_1\dots w_{k-n+z}\}}.$$

Finally one gets

$$\pi(w) = p_n + \sum_{z=1}^{n-k} p_z q^{(n-z)}_{\overleftarrow{\text{pref}}\,(w)}(w)$$
$$+ \sum_{z=n-k+1}^{n-1} p_z \mathbf{1}_{\{w_{n-z+1}\dots w_k = w_1\dots w_{k-n+z}\}} q^{(n-z)}_{\overleftarrow{\text{pref}}\,(w)}(w_{k-n+z+1}\dots w_k),$$

and hence

$$\Phi_w^{(1)}(x) = \frac{x^k \pi(w)/(1-x)}{1 + \sum_{j=k}^{\infty} x^j q^{(j)}_{\overleftarrow{\text{pref}}\,(w)}(w) + \sum_{j=1}^{k-1} x^j \mathbf{1}_{\{w_{j+1}\dots w_k = w_1\dots w_{k-j}\}} q^{(j)}_{\overleftarrow{\text{pref}}\,(w)}(w_{k-j+1}\dots w_k)}.$$

Proceeding exactly in the same way by induction on r, we get the expression of Theorem 3 for the r-th occurrence. □

Remark 9. The case of internal nodes $w = 0^k$ is more intricate, due to the absence of any symbol 1 allowing a renewal argument. Nevertheless, for the forthcoming

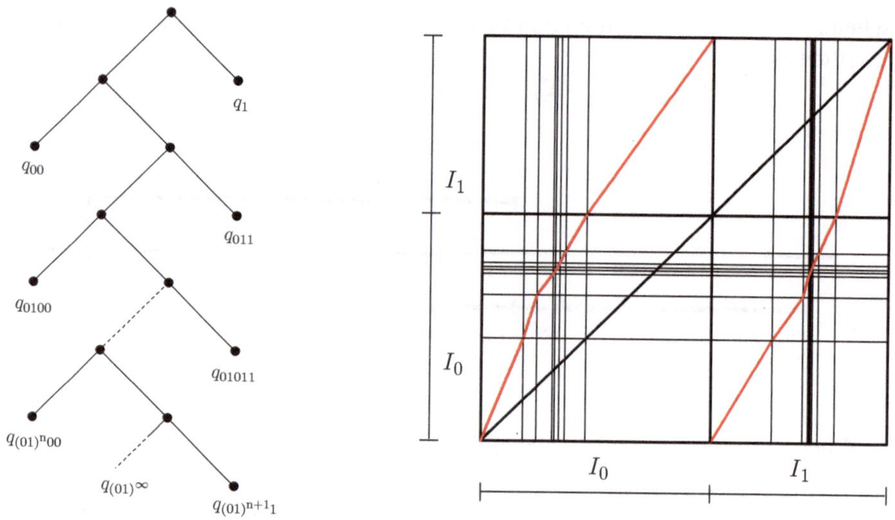

Fig. 7 Bamboo blossom probabilized context tree (*on the left*) and the associated dynamical system (*on the right*)

applications, we will not need the explicit expression of the generating function of such words occurrences.

4.2 *The Bamboo Blossom*

4.2.1 Stationary Probability Measures

Consider the probabilized context tree given by the left side of Fig. 7. The data of a corresponding VLMC consist in probability measures on $\mathscr{A}$ indexed by the two families of finite contexts

$$(q_{(01)^n1})_{n\geq 0} \text{ and } (q_{(01)^n00})_{n\geq 0}$$

together with a probability measure on the infinite context $q_{(01)^\infty}$.

As before, assuming that π is a stationary probability measure on $\mathscr{L}$, we compute the probabilities of any $\pi(w)$, $\overline{w}$ being an internal node or $\overline{w}$ being a context, as functions of the data and of both $\pi(1)$ and $\pi(00)$. Determination of stationary probabilities of cylinders based on both contexts 1 and 00 then leads to assumptions that guarantee existence and uniqueness of such a stationary probability measure.

Computation of $\pi(w), \overline{w}$ Context
Two families of cylinders, namely $\mathscr{L}1(10)^n$ and $\mathscr{L}00(10)^n$, correspond to contexts. For any $n \geq 0$, $\pi(1(10)^{n+1}) = \pi(1(10)^n)q_{(01)^n1}(1)q_1(0)$ and $\pi(00(10)^{n+1}) =$

$\pi(00(10)^n)q_{(01)^n00}(1)q_1(0)$. A straightforward induction implies thus that for any $n \geq 0$,

$$\begin{cases} \pi(1(10)^n) = \pi(1)c_n(1) \\ \\ \pi(00(10)^n) = \pi(00)c_n(00) \end{cases} \tag{19}$$

where $c_0(1) = c_0(00) = 1$ and

$$\begin{cases} c_n(1) = q_1(0)^n \prod_{k=0}^{n-1} q_{(01)^k 1}(1) \\ \\ c_n(00) = q_1(0)^n \prod_{k=0}^{n-1} q_{(01)^k 00}(1) \end{cases}$$

for any $n \geq 1$.

Computation of $\pi(w)$, $\overline{w}$ Internal Node

Two families of cylinders, $\mathscr{L}0(10)^n$ and $\mathscr{L}(10)^n$, correspond to internal nodes. By disjoint union of events, they are related by

$$\begin{cases} \pi(0(10)^n) = \pi((10)^n) - \pi(1(10)^n) \\ \pi((10)^{n+1}) = \pi(0(10)^n) - \pi(00(10)^n) \end{cases}$$

for any $n \geq 0$. By induction, this leads to: $\forall n \geq 0$,

$$\begin{cases} \pi(0(10)^n) = 1 - \pi(1)S_n(1) - \pi(00)S_{n-1}(00) \\ \\ \pi((10)^n) = 1 - \pi(1)S_{n-1}(1) - \pi(00)S_{n-1}(00) \end{cases} \tag{20}$$

where $S_{-1}(1) = S_{-1}(00) = 0$ and, for any $n \geq 0$,

$$\begin{cases} S_n(1) = \sum_{k=0}^n c_k(1) \\ \\ S_n(00) = \sum_{k=0}^n c_k(00). \end{cases}$$

These formula give, by quotients, the conditional probabilities on internal nodes defined by (6) and appearing in formula (8).

Computation of $\pi(1)$ and of $\pi(00)$

The context tree defines a partition of the set $\mathscr{L}$ of left-infinite sequences (see (3)). In the case of bamboo blossom, this partition implies

$$1 - \pi((10)^\infty) = \sum_{n\geq 0} \pi(1(10)^n) + \sum_{n\geq 0} \pi(00(10)^n) \tag{21}$$

$$= \sum_{n\geq 0} \pi(1)c_n(1) + \sum_{n\geq 0} \pi(00)c_n(00). \tag{22}$$

We denote

$$\begin{cases} S(1) = \sum_{n\geq 0} c_n(1) \\ S(00) = \sum_{n\geq 0} c_n(00) \in [1, +\infty]. \end{cases}$$

Note that the series $S(1)$ always converges. Indeed, the convergence is obvious if $q_1(0) \neq 1$; otherwise, $q_1(0) = 1$ and $q_1(1) = 0$, so that any $c_n(1)$, $n \geq 1$ vanishes and $S(1) = 1$. In the same way, the series $S(00)$ is finite as soon as $q_1(0) \neq 1$.

Proposition 4. (Stationary measure on a bamboo blossom)
Let $(U_n)_{n\geq 0}$ be a VLMC defined by a probabilized bamboo blossom context tree.

(i) Assume that $q_1(0) \neq 1$, then the Markov process $(U_n)_{n\geq 0}$ admits a stationary probability measure on $\mathcal{L}$ which is unique if and only if $S(1) - S(00)(1+q_1(0)) \neq 0$.

(ii) Assume that $q_1(0) = 1$.

(ii.a) If $S(00) = \infty$, then $(U_n)_{n\geq 0}$ admits $\pi = \frac{1}{2}\delta_{(10)^\infty} + \frac{1}{2}\delta_{(10)^\infty 1}$ as unique stationary probability measure on $\mathcal{L}$.

(ii.b) If $S(00) < \infty$, then $(U_n)_{n\geq 0}$ admits a one parameter family of stationary probability measures on $\mathcal{L}$.

Proof. *(i)* Assume that $q_1(0) \neq 1$ and that π is a stationary probability measure. Applying (7) gives

$$\pi((10)^\infty) = q_1(0)q_{(01)^\infty}(1)\pi((10)^\infty) \tag{23}$$

and consequently $\pi((10)^\infty) = 0$. Therefore, (21) becomes $S(1)\pi(1) + S(00)\pi(00) = 1$. We get another linear equation on $\pi(1)$ and $\pi(00)$ by disjoint union of events: $\pi(0) = 1 - \pi(1) = \pi(10) + \pi(00) = \pi(1)q_1(0) + \pi(00)$. Thus $\pi(1)$ and $\pi(00)$ are solutions of the linear system

$$\begin{cases} S(1)\pi(1) + S(00)\pi(00) = 1 \\ [1 + q_1(0)]\,\pi(1) + \pi(00) = 1. \end{cases} \tag{24}$$

This system has a unique solution if and only if the determinantal assumption

$$S(1) - S(00)\,[1 + q_1(0)] \neq 0$$

is fulfilled, which is a very light assumption (if this determinant happens to be zero, it suffices to modify one value of some q_u, u context for the assumption to be satisfied). Otherwise, when the determinant vanishes, System (24) is reduced to its second equation, so that it admits a one parameter family of solutions. Indeed,

$$1 \le S(1) \le 1 + q_1(0)(1 - q_1(0)) + \sum_{n \ge 2} q_1(0)^n (1 - q_1(0)) = 1 + q_1(0)$$

and $S(00) \ge 1$, so that $S(1) - S(00)(1 + q_1(0)) = 0$ implies that $S(1) = 1 + q_1(0)$ and $S(00) = 1$. In any case, System (24) has at least one solution, which ensures the existence of a stationary probability measure with formula (20), (19) and (8) by a standard argumentation. Assertions on uniqueness are straightforward.

(ii) Assume that $q_1(0) = 1$. This implies $q_1(1) = 0$ and consequently $S(1) = 1$. Thus, $\pi(1)$ and $\pi(00)$ are solutions of

$$\begin{cases} \pi(1) + S(00)\pi(00) = 1 - \pi((10)^\infty) \\ \\ 2\pi(1) + \pi(00) = 1. \end{cases} \tag{25}$$

so that, since $S(00) \ge 1$, the determinantal condition $S(1) - S(00)(1 + q_1(0)) \ne 0$ is always fulfilled.

(ii.a) When $S(00) = \infty$, $\pi(00) = 0$, $\pi(1) = \frac{1}{2}$ and $\pi((10)^\infty) = \frac{1}{2}$. This defines uniquely a stationary probability measure π. Because of (23), $q_{(01)^\infty}(1) = 1$ so that $\pi((10)^\infty 1) = \pi((10)^\infty)) = \frac{1}{2}$. This shows that $\pi = \frac{1}{2}\delta_{(10)^\infty} + \frac{1}{2}\delta_{(10)^\infty 1}$.

(ii.b) When $S(00) < \infty$, if we fix the value $a = \pi((10)^\infty)$, System (25) has a unique solution that determines in a unique way the stationary probability measure π_a. □

4.2.2 The Associated Dynamical System

The vertical partition is made of the intervals $I_{(01)^n 00}$ and $I_{(01)^n 1}$ for $n \ge 0$. The horizontal partition consists in the intervals $I_{0(01)^n 00}$, $I_{1(01)^n 00}$, $I_{0(01)^n 1}$ and $I_{1(01)^n 1}$ for $n \ge 0$, together with the two intervals coming from the infinite context, namely $I_{0(01)^\infty}$ and $I_{1(01)^\infty}$. If we make an hypothesis to ensure $\pi((10)^\infty) = 0$, then these two last intervals become two accumulation points of the horizontal partition, a_0 and a_1. The respective positions of the intervals and the two accumulation points are given by the alphabetical order

$$0(01)^{n-1}00 < 0(01)^n 00 < 0(01)^\infty < 0(01)^n 1 < 0(01)^{n-1}1$$
$$1(01)^{n-1}00 < 1(01)^n 00 < 1(01)^\infty < 1(01)^n 1 < 1(01)^{n-1}1$$

Lemma 5. *If $(q_{(01)^n 00}(0))_{n\in\mathbb{N}}$ and $(q_{(01)^n 1}(0))_{n\in\mathbb{N}}$ converge, then T is right and left differentiable in a_0 and a_1 – with possibly infinite derivatives – and*

$$T'_\ell(a_0) = \lim_{n\to\infty} \frac{1}{q_{(01)^n 00(0)}}, \quad T'_r(a_0) = \lim_{n\to\infty} \frac{1}{q_{(01)^n 1(0)}}$$

$$T'_\ell(a_1) = \lim_{n\to\infty} \frac{1}{q_{(01)^n 00(1)}}, \quad T'_r(a_1) = \lim_{n\to\infty} \frac{1}{q_{(01)^n 1(1)}}.$$

Proof. We use Lemma 4. □

4.2.3 Dirichlet Series

As for the infinite comb, the Dirichlet series of a source generated by a stationary bamboo blossom can be explicitly computed as a function of the SVLMC data. For simplicity, we assume that the generic Condition *(i)* of Proposition 4 is satisfied. An internal node is of the form $(01)^n$ or $(01)^n0$ while a context writes $(01)^n00$ or $(01)^n1$. Therefore, by disjoint union,

$$\Lambda(s) = A(s) + \sum_{n\geq 0, w\in\mathcal{W}} \pi(w00(10)^n)^s + \sum_{n\geq 0, w\in\mathcal{W}} \pi(w1(10)^n)^s$$

where

$$A(s) = \sum_{n\geq 0} \pi((10)^n)^s + \sum_{n\geq 0} \pi(0(10)^n)^s$$

is explicitly given by formula (20) and (24). Because of the renewal property of the bamboo blossom, formula (7) leads by two straightforward inductions to $\pi(w00(10)^n) = \pi(w00)c_n(00)$ and $\pi(w1(10)^n) = \pi(w1)c_n(1)$ for any $n \geq 0$. This implies that

$$\Lambda(s) = A(s) + \Lambda_{00}(s)\sum_{n\geq 0} c_n^s(00) + \Lambda_1(s)\sum_{n\geq 0} c_n^s(1)$$

where

$$\Lambda_{00}(s) = \sum_{w\in\mathcal{W}} \pi(w00)^s \text{ and } \Lambda_1(s) = \sum_{w\in\mathcal{W}} \pi(w1)^s.$$

It remains to compute both Dirichlet series Λ_{00} and Λ_1, which can be done by a similar procedure.

By disjoint union of finite words,

$$\Lambda_{00}(s) = A_{00}(s) + \sum_{n\geq 0, w\in\mathcal{W}} \pi(w00(10)^n00)^s + \sum_{n\geq 0, w\in\mathcal{W}} \pi(w1(10)^n00)^s \quad (26)$$

where

$$A_{00}(s) = \sum_{n\geq 0} \pi((10)^n00)^s + \sum_{n\geq 0} \pi(0(10)^n00)^s$$

and

$$\Lambda_1(s) = A_1(s) + \sum_{n\geq 0, w\in\mathcal{W}} \pi(w00(10)^n1)^s + \sum_{n\geq 0, w\in\mathcal{W}} \pi(w1(10)^n1)^s \quad (27)$$

with

$$A_1(s) = \sum_{n\geq 0} \pi((10)^n 1)^s + \sum_{n\geq 0} \pi(0(10)^n 1)^s.$$

Computation of A_1 and A_{00}
By disjoint union and formula (7),

$$\pi((10)^{n+1}00) = \pi(0(10)^n 00) - \pi(00(10)^n) q_{(01)^n 00}(0) q_{00}(0), \; n \geq 0$$

and

$$\pi(0(10)^n 00) = \pi((10)^n 00) - \pi(1(10)^n) q_{(01)^n 1}(0) q_{00}(0), \; n \geq 1$$

where $\pi(00(10)^n)$ and $\pi(1(10)^n)$ are already computed probabilities of contexts (19). Since $\pi(000) = \pi(00)q_{00}(0)$, one gets recursively $\pi((10)^n 00)$ and $\pi(0(10)^n 00)$ from these two relations as functions of the data. This computes A_{00}. A very similar argument leads to an explicit form of A_1.

Ultimate Computation of Λ_1 and Λ_{00}
Start with (26) and (27). As above, for any $n \geq 0$, by induction and with formula (7),

$$\pi(w00(10)^n 00) = \pi(w00) c_n(00) q_{(01)^n 00}(0) q_{00}(0).$$

In the same way, but only when $n \geq 1$,

$$\pi(w1(10)^n 00) = \pi(w1) c_n(1) q_{(01)^n 1}(0) q_{00}(0).$$

Similar computations lead to similar formula for $\pi(w00(10)^n 1)$ and $\pi(w1(10)^n 1)$, for any $n \geq 0$. So, (26) and (27) lead to

$$\Lambda_{00}(s) = A_{00}(s) + \Lambda_{100}(s) + \Lambda_{00}(s) B_{00}(s) + \Lambda_1(s) B_1(s) \tag{28}$$

where $B_{00}(s)$ and $B_1(s)$ are explicit functions of the data and where

$$\Lambda_{100}(s) = \sum_{w\in\mathcal{W}} \pi(w100).$$

As above, after disjoint union of words, splitting by formula (7) and double induction, one gets

$$\Lambda_{100}(s) = A_{100}(s) + \Lambda_{00}(s) C_{00}(s) + \Lambda_1(s) C_1(s)$$

where $A_{100}(s)$, $C_{00}(s)$ and $C_1(s)$ are explicit series, functions of the data. Replacing Λ_{100} by this value in formula (28) leads to a first linear equation between $\Lambda_{00}(s)$ and $\Lambda_1(s)$. A second linear equation between them is obtained from (27) by similar arguments. Solving the system one gets with both linear equations gives an explicit form of $\Lambda_{00}(s)$ and $\Lambda_1(s)$ as functions of the data, completing the expected computation.

4.2.4 Generating Function for the Exact Distribution of Word Occurrences in a Sequence Generated by a Bamboo Blossom

Let us consider the process $X = (X_n)_{n\geq 0}$ of final letters of $(U_n)_{n\geq 0}$ in the particular case of a SVLMC defined by a bamboo blossom. We only deal with finite words w such that $\overline{w}$ is not an internal node, i.e. $\overline{w}$ is a finite context or $\overline{w} \notin \mathscr{T}$. One can see that such a word of length $k > 1$ can be written in the form $*11(10)^\ell 1^p$ or $*00(10)^\ell 1^p$, with $p \in \{0, 1\}$ and $\ell \in \mathbb{N}$, where $*$ stands for any finite word.

Proposition 5. *For a SVLMC defined by a bamboo blossom, with notations of Sect. 4.1.4, the generating function of the first occurrence of a finite word $w = w_1 \dots w_k$ is given for $|x| < 1$ by*

$$\Phi_w^{(1)}(x) = \frac{x^k \pi(w)}{(1-x)S_w(x)}$$

and the generating function of the rth occurrence of w is given by

$$\Phi_w^{(r)}(x) = \Phi_w^{(1)}(x)\left(1 - \frac{1}{S_w(x)}\right)^{r-1},$$

where

(i) *if w is of the form $*00(10)^\ell$ or $*11(01)^\ell 0$, with $\ell \in \mathbb{N}$, $S_w(x)$ is defined in Proposition 3 and*

(ii) *if w is of the form $*00(10)^\ell 1$, $\ell \in \mathbb{N}$,*

$$S_w(x) = C_w(x) + \sum_{j=k}^{\infty} q^{(j)}_{1(01)^\ell 00}(w)x^j,$$

$$C_w(x) = 1 + \sum_{j=1}^{k-1} \mathbf{1}_{\{w_{j+1}\dots w_k = w_1 \dots w_{k-j}\}} q^{(j)}_{1(01)^\ell 00}\left(w_{j+1}\dots w_k\right) x^j.$$

and if w is of the form $*11(01)^\ell$, $\ell \in \mathbb{N}$,

$$S_w(x) = C_w(x) + \sum_{j=k}^{\infty} q^{(j)}_{(10)^\ell 11}(w)x^j,$$

$$C_w(x) = 1 + \sum_{j=1}^{k-1} \mathbf{1}_{\{w_{j+1}\dots w_k = w_1 \dots w_{k-j}\}} q^{(j)}_{(10)^\ell 11}\left(w_{j+1}\dots w_k\right) x^j.$$

Proof. **(i)** We first deal with the words w such that

$$\overleftarrow{\mathrm{pref}}\,(w) = (01)^\ell 00 \quad \text{or} \quad \overleftarrow{\mathrm{pref}}\,(w) = (01)^\ell 1.$$

Let us denote by p_n the probability that $T_w^{(1)} = n$. Proceeding exactly in the same way as for Proposition 3, from the decomposition

$$\pi(w) = p_n + \sum_{z=k}^{n-1} p_z \mathbf{P}\left(X_{n-k+1} \dots X_n = w | T_w^{(1)} = z\right),$$

and due to the renewal property of the bamboo, one has

$$\pi(w) = p_n + \sum_{z=k}^{n-k} p_z \mathbf{P}\left(U_n \in \mathscr{L}w \middle| U_z \in \mathscr{L}\,\mathrm{suff}(w)\right)$$
$$+ \sum_{z=n-k+1}^{n-1} p_z \mathbf{1}_{\{w_{n-z+1}\dots w_k = w_1 \dots w_{k-n+z}\}} \mathbf{P}\left(U_n \in \mathscr{L}w \middle| U_z \in \mathscr{L}\,\mathrm{suff}(w)\right)$$

where suff(w) is the suffix of w equal to the reversed word of $\overleftarrow{\mathrm{pref}}\,(w)$. Hence, for $x < 1$, it comes

$$\frac{x^k \pi(w)}{1-x} = \sum_{n=k}^{+\infty} p_n x^n + \sum_{n=k}^{+\infty} x^n \sum_{z=k}^{n-k} p_z q^{(n-z)}_{\overleftarrow{\mathrm{pref}}\,(w)}(w)$$
$$+ \sum_{n=k}^{+\infty} x^n \sum_{z=n-k+1}^{n-1} p_z \mathbf{1}_{\{w_{n-z+1}\dots w_k = w_1 \dots w_{k-n+z}\}} q^{(n-z)}_{\overleftarrow{\mathrm{pref}}\,(w)}(w)$$

which leads to the expression of $\Phi_w^{(1)}(x)$ given in Proposition 3. The rth occurrence can be derived exactly in the same way from the decomposition

$$\{w \text{ at } n\} = \{T_w^{(1)} = n\} \cup \{T_w^{(2)} = n\} \cup \dots \cup \{T_w^{(r)} = n\} \cup \{T_w^{(r)} < n \text{ and } w \text{ at } n\}.$$

(ii) In the particular case of words $w = *00(10)^\ell 1$, the main difference is that the context 1 is not sufficient for the renewal property. The computation relies on the equality

$$\mathbf{P}\left(X_{n-k+1} \dots X_n = w | T_w^{(1)} = z\right)$$
$$= \mathbf{P}\left(X_{n-k+1} \dots X_n = w | X_{z-2\ell-2} \dots X_z = 00(10)^\ell 1\right).$$

The sketch of the proof remains the same replacing $q_{\overleftarrow{\mathrm{pref}}\,(w)}(w)$ by $q_{1(01)^\ell 00}(w)$. The case $w = *11(01)^\ell$ is analogous. □

5 Some Remarks, Extensions and Open Problems

5.1 Stationary Measure for a General VLMC

Infinite comb and bamboo blossom are two instructive but very particular examples, close to renewal processes. Nevertheless, we think that an analogous of Propositions 1 or 4 can be written for a VLMC defined by a general context tree with a finite or countable number of infinite branches.

In order to generalize the proofs, it is clear that formula (8) in Lemma 1 is crucial. In this formula, for a given finite word $w = \alpha_1 \dots \alpha_N \in \mathscr{W}$ it is important to check whether the subwords $\overleftarrow{\text{pref}}\ (\alpha_1 \dots \alpha_k), k < N$, are internal nodes of the tree or not. Consequently, the following concept of *minimal context* is natural.

Definition 12. (Minimal context) Define the following binary relation on the set of the finite contexts as follows:

$$\forall u, v \in \mathscr{C}^F, \quad u \prec v \Longleftrightarrow \exists w, w' \in \mathscr{W}, \ v = wuw'$$

(in other words u is a sub-word of v). This relation is a partial order. In a context tree, a finite context is called *minimal* when it is minimal for this partial order on contexts.

Remark 10. **(Alternative definition of a minimal context)** Let $\mathscr{T}$ be a context tree. Let $c = \alpha_N \dots \alpha_1$ be a finite context of $\mathscr{T}$. Then c is minimal if and only if $\forall k \in \{1, \dots, N-1\}$, $\overleftarrow{\text{pref}}\ (\alpha_1 \dots \alpha_k) \notin \mathscr{C}^F(\mathscr{T})$.

Example 1. In the infinite comb, the only minimal context is 1. In the bamboo blossom, the minimal contexts are 1 and 00.

Remark 11. There exist some trees with infinitely many infinite leaves and a finite number of minimal contexts. Take the infinite comb and at each 0^k branch another infinite comb. In such a tree, the finite leaf 10 is the only minimal context.

Nonetheless, a tree with a finite number of infinite contexts has necessarily a finite number of minimal contexts.

As one can see for the infinite comb or for the bamboo blossom (see Sects. 4.1.1 and 4.2.1), minimal contexts play a special role in the computation of stationary probability measures. First of all, when π is a stationary probability measure and w a finite word such that $\overline{w} \notin \mathscr{T}$, formula (8) implies that $\pi(w)$ is a rational monomial of the data $q_c(\alpha)$ and of the $\pi(u)$ where $\overline{u}$ belongs to $\mathscr{T}$. This shows that any stationary probability is determined by its values on the nodes of the context tree. In both examples, we compute these values as functions of the data and of the $\pi(m)$, where $\overline{m}$ are minimal contexts, and we finally write a rectangular linear system satisfied by these $\pi(m)$. Assuming that this system has maximal rank can be viewed as making an irreducibility condition for the Markov chain on $\mathscr{L}$. We conjecture that this situation happens in any case of VLMC.

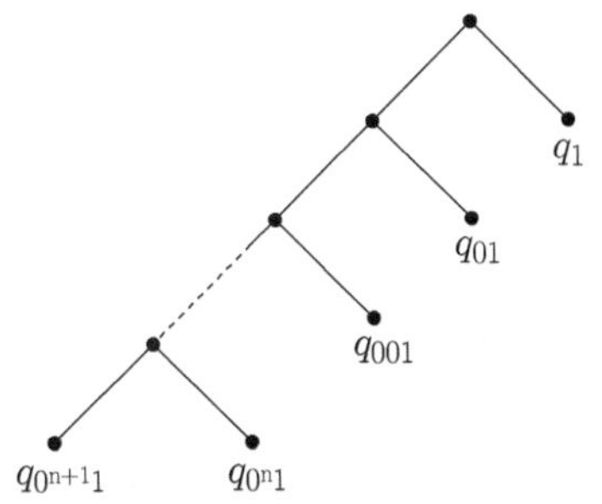

Fig. 8 $(n+1)$-teeth comb probabilized context tree

In the following example, we detail the above procedure, in order to understand how the two main principles (the partition (3) and the disjoint union) give the linear system leading to the irreducibility condition.

Example 2. Let $\mathcal{T}$ be a probabilized context tree corresponding to Fig. 8 (finite comb with $n+1$ teeth). There are two minimal contexts: 1 and 0^{n+1}. Assume that π is a stationary probability measure on $\mathcal{L}$. Like in the case of the infinite comb, the probability of a word that corresponds to a teeth is $\pi(10^k) = \pi(1)c_k$, $0 \le k \le n$ where c_k is the product defined by (12). Also, the probabilities of the internal nodes and of the handle are

$$\pi(0^k) = 1 - \pi(1)S_{k-1}, \quad 0 \le k \le n+1,$$

where $S_p := \sum_{j=0}^{p} c_j$. By means of these formula, π is determined by $\pi(1)$.

In order to compute $\pi(1)$, one can proceed as follows. First, by the partition principle (3), we have $1 = \pi(0^{n+1}) + \pi(1)\sum_{k=0}^{n} c_k$. Secondly, by disjoint union,

$$\pi(0^{n+1}) = \pi(0^{n+2}) + \pi(10^{n+1}) = \pi(0^{n+1})q_{0^{n+1}}(0) + \pi(10^n)q_{0^n1}(0).$$

This implies the linear relation between both minimal contexts probabilities:

$$\begin{cases} \pi(0^{n+1}) + S_n\pi(1) = 1 \\ q_{0^{n+1}}(1)\pi(0^{n+1}) - q_{0^n1}(0)c_n\pi(1) = 0. \end{cases}$$

In particular, this leads to the irreducibility condition $q_{0^{n+1}}(1)S_n + q_{0^n1}(0)c_n \neq 0$ for the VLCM to admit a stationary probability measure. One can check that this irreducibility condition is the classical one for the corresponding $\mathcal{A}$-valued Markov chain of order $n+1$.

Example 3. Let $\mathcal{T}$ be a probabilized context tree corresponding to Fig. 5 (four flower bamboo). This tree provides another example of computation procedure using formula (7) and (8), the partition principle (3) and the disjoint union. This VLMC admits a unique stationary probability measure if the determinantal condition

$$q_{00}(1)[1 + q_1(0)] + q_1(0)^2 q_{010}(0) + q_1(0)q_1(1)q_{011}(0) \neq 0$$

is satisfied; it is fulfilled if none of the q_c is trivial.

5.2 Tries

In a first kind of problems, n words independently produced by a source are inserted in a trie. There are results on the classical parameters of the trie (size, height, path length) for a dynamical source Clément et al. ([6]), which rely on the existence of a spectral gap for the underlying dynamical system. We would like to extend these results to cases when there is no spectral gap, as may be guessed in the infinite comb example.

Another interesting application consists in producing a suffix trie from *one* sequence coming from a VLMC dynamical source, and analyzing its parameters. For his analysis, Szpankowski [26] puts some mixing assumptions (called strong α-mixing) on the source. A first direction consists in trying to find the mixing type of a VLMC dynamical source. In a second direction, we plan to use the generating function for the occurrence of words to improve these results.

Acknowledgements We are very grateful to Antonio Galves, who introduced us to the challenging VLMC topics. We warmly thank Brigitte Vallée for valuable and stormy discussions.

References

1. M. Abadi, A. Galves, Inequalities for the occurrence times of rare events in mixing processes. The state of the art. Markov Proc. Relat. Field. **7**(1), 97–112 (2001)
2. M. Abadi, B. Saussol, Stochastic processes and their applications. **121**(2), 314–323
3. P. Billingsley, *Probability and Measure*, 3rd edn. Wiley Series in Probability and Mathematical Statistics (Wiley, New York, 1995)
4. G. Blom, D. Thorburn, How many random digits are required until given sequences are obtained? J. Appl. Probab. **19**, 518–531 (1982)
5. P. Bühlmann, A.J. Wyner, Variable length markov chains. Ann. Statist. **27**(2), 480–513 (1999)
6. J. Clément, P. Flajolet, B. Vallee, Dynamical sources in information theory: Analysis of general tries. Algorithmica **29**, 307–369 (2001)
7. F. Comets, R. Fernandez, P. Ferrari, Processes with long memory: Regenerative construction and perfect simulation. Ann. Appl. Prob. **12**(3), 921–943 (2002)
8. P. Flajolet, M. Roux, B. Vallée, Digital trees and memoryless sources: From arithmetics to analysis. DMTCS Proc. **AM**, 233–260 (2010)
9. J.C. Fu, Bounds for reliability of large consecutive-k-out-of-n : f system. IEEE Trans. Reliab. **35**, 316–319 (1986)
10. J.C. Fu, M.V. Koutras, Distribution theory of runs: A markov chain approach. J. Amer. Statist. Soc. **89**, 1050–1058 (1994)
11. S. Gallo, N. Garcia, Perfect simulation for stochastic chains of infinite memory: Relaxing the continuity assumptions. pp. 1–20 (2010) [arXiv:1105.5459v1]
12. A. Galves, E. Löcherbach, Stochastic chains with memory of variable length. TICSP Series **38**, 117–133 (2008)
13. H. Gerber, S. Li, The occurrence of sequence patterns in repeated experiments and hitting times in a markov chain. Stoch. Process. Their Appl. **11**, 101–108 (1981)
14. T.E. Harris, On chains of infinite order. Pac. J. Math. **5**, 707–724 (1955)
15. P. Jacquet, W. Szpankowski, Autocorrelation on words and its applications. Analysis of suffix trees by string-ruler approach. J. Combin. Theor. **A.66**, 237–269 (1994)

16. M.V. Koutras, Waiting Times and Number of Appearances of Events in a Sequence of Discrete Random Variables. *Advances in Combinatorial Methods and Applications to Probability and Statistics*, Stat. Ind. Technol., (Birkhäuser Boston, Boston, 1997), pp. 363–384
17. A. Lambert, S. Siboni, S. Vaienti, Statistical properties of a nonuniformly hyperbolic map of the interval. J. Stat. Phys. **72**(5/6), 1305–1330 (1993)
18. S.-Y.R. Li, A martingale approach to the study of occurrence of sequence patterns in repeated experiments. Ann. Probab. **8**(6): 1171–1176 (1980)
19. V. Pozdnyakov, J. Glaz, M. Kulldorff, J.M. Steele, A martingale approach to scan statistics. Ann. Inst. Statist. Math. **57**(1), 21–37 (2005)
20. M. Régnier, A unified approach to word occurrence probabilities. Discrete Appl. Math. **104**, 259–280 (2000)
21. G. Reinert, S. Schbath, M.S. Waterman, Probabilistic and statistical properties of words: An overview. J. Comput. Biol. **7**(1/2), 1–46 (2000)
22. D. Revuz, *Markov Chains*. (North-Holland Mathematical Library, Amsterdam, 1984)
23. J. Rissanen, A universal data compression system. IEEE Trans. Inform. Theor. **29**(5), 656–664 (1983)
24. S. Robin, J.J. Daudin, Exact distribution of word occurrences in a random sequence of letters. J. Appl. Prob. **36**, 179–193 (1999)
25. V. Stefanov, A.G. Pakes, Explicit distributional results in pattern formation. Ann. Appl. Probab. **7**, 666–678 (1997)
26. W. Szpankowski, A generalized suffix tree and its (un)expected asymptotic behaviors. SIAM J. Comput. **22**(6), 1176–1198 (1993)
27. X-J. Wang, Statistical physics of temporal intermittency. Phys. Rev. A **40**(11), 6647–6661 (1989)
28. D. Williams, *Probability with Martingales*. Cambridge Mathematical Textbooks (Cambridge University Press, Cambridge, 1991)

Martingale Property of Generalized Stochastic Exponentials

Aleksandar Mijatović, Nika Novak, and Mikhail Urusov

Abstract For a real Borel measurable function b, which satisfies certain integrability conditions, it is possible to define a stochastic integral of the process $b(Y)$ with respect to a Brownian motion W, where Y is a diffusion driven by W. It is well-known that the stochastic exponential of this stochastic integral is a local martingale. In this paper we consider the case of an arbitrary Borel measurable function b where it may not be possible to define the stochastic integral of $b(Y)$ directly. However the notion of the stochastic exponential can be generalized. We define a non-negative process Z, called *generalized stochastic exponential*, which is not necessarily a local martingale. Our main result gives deterministic necessary and sufficient conditions for Z to be a local, true or uniformly integrable martingale.

Keywords Generalized stochastic exponentials • Local martingales vs. true martingales • One-dimensional diffusions

AMS Classification: 60G44, 60G48, 60H10, 60J60

A. Mijatović
Department of Statistics, University of Warwick, Coventry, England, CV4 8JY, UK
e-mail: a.mijatovic@warwick.ac.uk

N. Novak (✉)
University of Ljubljana, Ljubljana, Slovenia

Department of Mathematics, Imperial College London, London, UK
e-mail: nika.novak@fmf.uni-lj.si

M. Urusov
Institute of Mathematical Finance, Ulm University, Ulm, Germany
e-mail: mikhail.urusov@uni-ulm.de

C. Donati-Martin et al. (eds.), *Séminaire de Probabilités XLIV*, Lecture Notes in Mathematics 2046, DOI 10.1007/978-3-642-27461-9_2,

1 Introduction

A *stochastic exponential* of X is a process $\mathscr{E}(X)$ defined by

$$\mathscr{E}(X)_t = \exp\left\{X_t - X_0 - \frac{1}{2}\langle X\rangle_t\right\}$$

for some continuous local martingale X, where $\langle X\rangle$ denotes a quadratic variation of X. It is well-known that the process $\mathscr{E}(X)$ is also a continuous local martingale. Sufficient conditions for the martingale property of $\mathscr{E}(X)$ have been studied extensively in the literature because this question appears naturally in many situations. Novikov's and Kazamaki's sufficient conditions (see [6, 9]) for $\mathscr{E}(X)$ to be a martingale are particularly well-known. Novikov's condition consists in $\mathbb{E}\exp\{(1/2)\langle X\rangle_t\} < \infty$, $t \in [0,\infty)$. Kazamaki's condition consists in that $\exp\{(1/2)X\}$ should be a submartingale. Novikov's criterion is of narrower scope than Kazamaki's one but often easier to apply. In this respect let us note that none of conditions $\mathbb{E}\exp\{(1/2-\varepsilon)\langle X\rangle_t\} < \infty$ (with $\varepsilon > 0$) is sufficient for $\mathscr{E}(X)$ to be a martingale (see Liptser and Shiryaev [7, Chap. 6]). For a further literature review see the bibliographical notes in the monographs Karatzas and Shreve [5, Chap. 3], Liptser and Shiryaev [7, Chap. 6], Protter [10, Chap. III], and Revuz and Yor [11, Chap. VIII].

In the case of one-dimensional processes, necessary and sufficient conditions for the process $\mathscr{E}(X)$ to be a martingale were recently studied by Engelbert and Senf in [3], Blei and Engelbert in [1] and Mijatović and Urusov in [8]. In [3] X is a general continuous local martingale and the characterisation is given in terms of the Dambis–Dubins–Schwartz time-change that turns X into a Brownian motion. In [1] X is a strong Markov continuous local martingale and the condition is deterministic, expressed in terms of the speed measure of X.

In [8] the local martingale X is of the form $X_t = \int_0^t b(Y_u)\,\mathrm{d}W_u$ for some measurable function b and a one-dimensional diffusion Y with drift μ and volatility σ driven by a Brownian motion W. In order to define the stochastic integral X, an assumption that the function $\frac{b^2}{\sigma^2}$ is locally integrable on the entire state space of the process Y is required. Under this restriction the characterization of the martingale property of $\mathscr{E}(X)$ is studied in [8], where the necessary and sufficient conditions are deterministic and are expressed in terms of functions μ, σ and b only.

In the present paper we consider an arbitrary Borel measurable function b. In this case the stochastic integral X can only be defined on some subset of the probability space. However, it is possible to define a non-negative possibly discontinuous process Z, known as a generalized stochastic exponential, on the entire probability space. It is a consequence of the definition that, if the function b satisfies the required local integrability condition, the process Z coincides with $\mathscr{E}(X)$. We show that the process Z is not necessarily a local martingale. In fact Z is a local martingale if and only if it is continuous. We find a deterministic necessary and sufficient condition for Z to be a local martingale, which is expressed in terms of local integrability of the quotient $\frac{b^2}{\sigma^2}$ multiplied by a linear function. We also characterize the processes Z that are true martingales and/or uniformly integrable martingales. All the necessary and sufficient conditions are deterministic and are given in terms of functions μ, σ and b.

The paper is structured as follows. In Sect. 2 we define the notion of generalized stochastic exponential and study its basic properties. The main results are stated in Sect. 3, where we give a necessary and sufficient condition for the process Z defined by (8) and (12) to be a local martingale, a true martingale or a uniformly integrable martingale. Finally, in Sect. 4 we prove Theorem 3 that is central in obtaining the deterministic characterisation of the martingale property of the process Z. Appendix A contains an auxiliary fact that is used in Sect. 2.

2 Definition of Generalized Stochastic Exponential

Let $J = (l, r)$ be our state space, where $-\infty \leq l < r \leq \infty$. Let us define a J-valued diffusion Y on a probability space $(\Omega, \mathscr{F}, (\mathscr{F}_t)_{t\in[0,\infty)}, \mathbb{P})$ driven by a stochastic differential equation

$$dY_t = \mu(Y_t)\,dt + \sigma(Y_t)\,dW_t, \quad Y_0 = x_0 \in J,$$

where W is a $(\mathscr{F}_t)$-Brownian motion, and μ, σ are real, Borel measurable functions defined on J that satisfy the Engelbert–Schmidt conditions

$$\sigma(x) \neq 0 \quad \forall x \in J, \tag{1}$$

$$\frac{1}{\sigma^2}, \frac{\mu}{\sigma^2} \in L^1_{\text{loc}}(J). \tag{2}$$

With $L^1_{\text{loc}}(J)$ we denote the class of locally integrable functions, i.e. real Borel measurable functions defined on J that are integrable on every compact subset of J. Engelbert–Schmidt conditions guarantee existence of a weak solution that might exit the interval J and is unique in law (see [5, Chap. 5]). Denote by ζ the exit time of Y. In addition, we assume that the boundary points are absorbing, i.e. the solution Y stays at the boundary point at which it exits on the set $\{\zeta < \infty\}$. Let us note that we assume that $(\mathscr{F}_t)$ is generated neither by Y nor by W.

We would like to define a process X as a stochastic integral of a process $b(Y)$ with respect to Brownian motion W, where $b : J \to \mathbb{R}$ is an arbitrary Borel measurable function. Before further discussion, we should establish if the stochastic integral can be defined.

Define a set

$$A = \{x \in J;\ \tfrac{b^2}{\sigma^2} \notin L^1_{\text{loc}}(x)\},$$

where $L^1_{\text{loc}}(x)$ denotes a space of real Borel measurable functions f such that $\int_{x-\varepsilon}^{x+\varepsilon} |f(y)|\mathrm{d}y < \infty$ for some $\varepsilon > 0$. Then A is closed and its complement is a union of open intervals. Furthermore we can define maps α and β on $J \setminus A$ so that

$$\alpha(x), \beta(x) \in A \cup \{l, r\} \quad \text{and} \quad x \in (\alpha(x), \beta(x)) \subseteq J \setminus A. \tag{3}$$

In other words $\alpha(x)$ is the point in $A \cup \{l, r\}$ that is closest to x from the left side and $\beta(x)$ is the closest point in $A \cup \{l, r\}$ from the right side. Let us also note that the equality

$$L^1_{\text{loc}}(I) = \bigcap_{x \in I} L^1_{\text{loc}}(x) \tag{4}$$

holds for any interval I by a simple compactness argument.

For any $x, y \in J$ we define the stopping times

$$\tau_x = \inf\{t \geq 0;\ Y_t = x\}, \tag{5}$$

$$\tau_{x,y} = \tau_x \wedge \tau_y, \tag{6}$$

with the convention $\inf \emptyset = \infty$, where $c \wedge d := \min\{c, d\}$. Define the stopping time $\zeta^A = \zeta \wedge \tau_A$, where

$$\tau_A = \inf\{t \geq 0;\ Y_t \in A\}. \tag{7}$$

Then for all $t \geq 0$ we have

$$\int_0^t b^2(Y_u)\,\mathrm{d}u < \infty \ \ \mathbb{P}\text{-a.s. on } \{t < \zeta^A\}.$$

This follows from Proposition 1 given in Appendix A and the fact that a continuous process Y on $\{t < \zeta^A\}$ reaches only values in an open interval that is a component of the complement of A, where $\frac{b^2}{\sigma^2}$ is locally integrable (note that (4) is applied here).

Let us define $A_n = \{x \in J;\ \rho(x, A \cup \{l, r\}) \leq \frac{1}{n}\}$, where $\rho(x, y) = |\arctan x - \arctan y|$, $x, y \in \bar{J}$, and set $\zeta_n^A = \inf\{t \geq 0;\ Y_t \in A_n\}$. Since $\zeta_n^A < \zeta^A$ on the set $\{\zeta^A < \infty\}$, we have $\int_0^{t \wedge \zeta_n^A} b^2(Y_u)\mathrm{d}u < \infty$ $\mathbb{P}$-a.s. Thus, we can define the stochastic integral $\int_0^{t \wedge \zeta_n^A} b(Y_u)\mathrm{d}W_u$ for every n. Since $\int_0^{t \wedge \zeta_n^A} b(Y_u)\mathrm{d}W_u$ and $\int_0^{t \wedge \zeta_{n+1}^A} b(Y_u)\mathrm{d}W_u$ coincide on $\{t < \zeta_n^A\}$ and $\zeta_n^A \nearrow \zeta^A$, we can define $\int_0^{t \wedge \zeta^A} b(Y_u)\mathrm{d}W_u$ as a $\mathbb{P}$-a.s limit of integrals

$$\int_0^{t \wedge \zeta^A} b(Y_u)\mathrm{d}W_u = \lim_{n \to \infty} \int_0^{t \wedge \zeta_n^A} b(Y_u)\mathrm{d}W_u \text{ on } \{t < \zeta^A\} \cup \left\{ \int_0^{\zeta^A} b^2(Y_u)\mathrm{d}u < \infty \right\}.$$

In the case where A is not empty or Y exits the interval J, the stochastic exponential cannot be defined. However, we can define a *generalized stochastic exponential* Z in the following way for every $t \in [0, \infty)$:

$$Z_t = \begin{cases} \exp\{\int_0^t b(Y_u)\mathrm{d}W_u - \frac{1}{2}\int_0^t b^2(Y_u)\mathrm{d}u\} \text{ if } t < \zeta^A \\ \exp\{\int_0^\zeta b(Y_u)\mathrm{d}W_u - \frac{1}{2}\int_0^\zeta b^2(Y_u)\mathrm{d}u\} \text{ if } t \geq \zeta^A = \zeta,\ \int_0^\zeta b^2(Y_u)\,\mathrm{d}u < \infty \\ 0 \text{ if } t \geq \zeta^A = \tau_A \text{ or } t \geq \zeta^A = \zeta, \int_0^\zeta b^2(Y_u)\,\mathrm{d}u = \infty \end{cases} \tag{8}$$

The different behaviour of Z on $\{t \geq \zeta^A = \zeta\}$ and $\{t \geq \zeta^A = \tau_A\}$ follows from the fact that after the exit time ζ the process Y is stopped, while this does not happen after τ_A. The definition of the set A and Proposition 1 imply that the integral $\int_0^t b^2(Y_u)\,\mathrm{d}u$ is infinite on the event $\{t > \tau_A\}$ $\mathbb{P}$-a.s. Therefore, we set $Z = 0$ on the set $\{t \geq \zeta^A = \tau_A\}$.

Let us define the processes

$$\bar{Z}_t = \exp\left\{\int_0^{t\wedge\zeta^A} b(Y_u)\,\mathrm{d}W_u - \frac{1}{2}\int_0^{t\wedge\zeta^A} b^2(Y_u)\,\mathrm{d}u\right\}, \tag{9}$$

where we set $\bar{Z}_t = 0$ for $t \geq \zeta^A$ on $\{\zeta^A < \infty, \int_0^{\zeta^A} b^2(Y_u)\mathrm{d}u = \infty\}$. The process $\bar{Z}$ has continuous trajectories: continuity at time ζ^A on the set $\{\zeta^A < \infty, \int_0^{\zeta^A} b^2(Y_u)\,du = \infty\}$ follows from the Dambis–Dubins–Schwarz theorem on stochastic intervals; see Theorem 1.6 and Exercise 1.18 [11, Chap. V]. Note further that $\bar{Z}_t$ is strictly positive on the event $\{t < \zeta^A\} \cup \left\{\int_0^{\zeta^A} b^2(Y_u)\mathrm{d}u < \infty\right\}$ $\mathbb{P}$-a.s. and is equal to zero on its complement.

Lemma 1. *The process $\bar{Z} = (\bar{Z}_t)_{t\geq 0}$ defined in (9) is a continuous local martingale.*

Remark 1. Fatou's lemma and Lemma 1 imply that the process $\bar{Z}$ is a continuous non-negative supermartingale.

Proof. If the starting point x_0 of the diffusion Y is in A, then $\tau_A = 0$ $\mathbb{P}$-a.s and $\bar{Z}$ is constant and hence a local martingale. If $x_0 \in J \setminus A$, then pick a decreasing (resp. increasing) sequence $(\alpha_n)_{n\in\mathbb{N}}$ (resp. $(\beta_n)_{n\in\mathbb{N}}$) in the open interval $(\alpha(x_0), \beta(x_0))$ such that $\alpha_n \searrow \alpha(x_0)$ (resp. $\beta_n \nearrow \beta(x_0)$), where $\alpha(x_0), \beta(x_0)$ are defined in (3). Assume also that $\alpha_n < x_0 < \beta_n$ for all $n \in \mathbb{N}$. Note that $\tau_{\alpha_n,\beta_n} \nearrow \zeta^A$ $\mathbb{P}$-a.s.

The process $M^n = (M_t^n)_{t\in[0,\infty)}$, defined by

$$M_t^n = \int_0^{t\wedge\tau_{\alpha_n,\beta_n}} b(Y_u)\mathrm{d}W_u \qquad \text{for all} \quad t \in [0,\infty),$$

is a local martingale. Therefore its stochastic exponential $\mathscr{E}(M^n)$ is also a local martingale and, since $\tau_{\alpha_n,\beta_n} < \zeta^A$ $\mathbb{P}$-a.s., the equality $\bar{Z}_{t\wedge\tau_{\alpha_n,\beta_n}} = \mathscr{E}(M^n)_t$ holds $\mathbb{P}$-a.s. for all $t \geq 0$. For any $m \in \mathbb{N}$ define the stopping time $\eta_m = \inf\{t \geq 0; \bar{Z}_t \geq m\}$. The stopped process $\bar{Z}^{\eta_m\wedge\tau_{\alpha_n,\beta_n}}$ is a bounded local martingale and hence a martingale. Furthermore note that $\eta_m \nearrow \infty$ $\mathbb{P}$-a.s. as m tends to infinity. To prove that $\bar{Z}^{\eta_m}$ is a martingale for each $m \in \mathbb{N}$, note that the process $\bar{Z}$ is stopped at ζ^A, which implies the almost sure limit $\lim_{n\to\infty} \bar{Z}_t^{\eta_m \wedge \tau_{\alpha_n,\beta_n}} = \bar{Z}_t^{\eta_m}$ for every $t \in [0,\infty)$. Since $\bar{Z}_t^{\eta_m \wedge \tau_{\alpha_n,\beta_n}} \le m$ $\mathbb{P}$-a.s., the conditional dominated convergence theorem implies the martingale property of $\bar{Z}^{\eta_m}$ for every $m \in \mathbb{N}$. This concludes the proof. □

Define the process $S = (S_t)_{t\in[0,\infty)}$ by

$$S_t = \exp\left\{\int_0^{\tau_A} b\left(Y_u\right) \mathrm{d}W_u - \frac{1}{2}\int_0^{\tau_A} b^2\left(Y_u\right) \mathrm{d}u\right\} \mathbf{1}_{\{t\ge\tau_A, \int_0^{\tau_A} b^2(Y_u)\mathrm{d}u<\infty\}}. \tag{10}$$

Note that for any $t \in [0,\infty)$, $\mathbb{P}$-a.s. on the set $\{\tau_A \le t\}$ we have $\tau_A < \zeta$ and hence the integrals in (10) are well-defined. We can express Z as

$$Z = \bar{Z} - S. \tag{11}$$

It is clear from this representation that Z is not necessarily a continuous process. Moreover, since that paths of S are non-decreasing, we have

$$\mathbb{E}[Z_t|\mathscr{F}_s] \le \bar{Z}_s - \mathbb{E}[S_t|\mathscr{F}_s] \le \bar{Z}_s - S_s = Z_s.$$

It follows that Z is a non-negative supermartingale and we can define

$$Z_\infty = \lim_{t\to\infty} Z_t. \tag{12}$$

Note further that if $x_0 \in A$, we have $Z \equiv 0$.

A path of the process Z defined by (8) and (12) is equal to a path of a stochastic exponential if $\zeta^A = \infty$. Otherwise, if $\zeta^A < \infty$, it has one of the following forms:

1. $\tau_A < \zeta$ and $\int_0^{\tau_A} b^2(Y_t)\mathrm{d}t < \infty$ (see Fig. 1).
2. $\zeta^A < \infty$ and $\int_0^{\zeta^A} b^2(Y_t)\mathrm{d}t = \infty$ (see Fig. 2).
3. $\zeta < \tau_A$ and $\int_0^{\zeta} b^2(Y_t)\mathrm{d}t < \infty$ (see Fig. 3).

3 Main Results

The case $A = \emptyset$ was studied by Mijatović and Urusov in [8]. We generalize their result for the case where $A \ne \emptyset$.

3.1 The Case $A = \emptyset$

In this case we have

$$\frac{b^2}{\sigma^2} \in L^1_{\mathrm{loc}}(J). \tag{13}$$

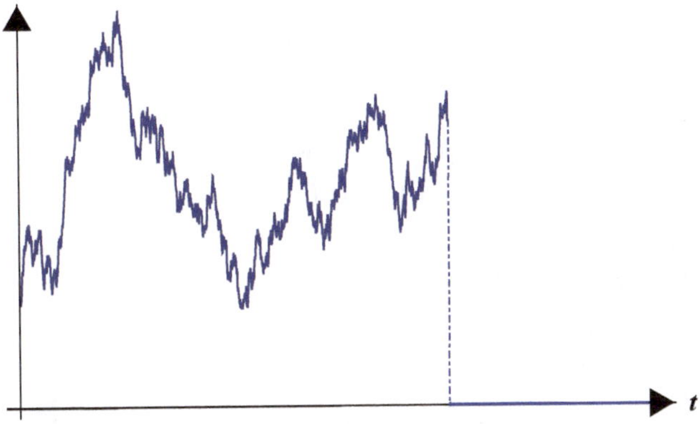

Fig. 1 If $\tau_A < \zeta$, then the process Z is positive up to time τ_A and is equal to zero afterwards. If the integral $\int_0^{\tau_A} b^2(Y_t)\mathrm{d}t$ is finite, then Z_t approaches a positive value as t approaches τ_A. Therefore, there is a jump at $t = \tau_A$

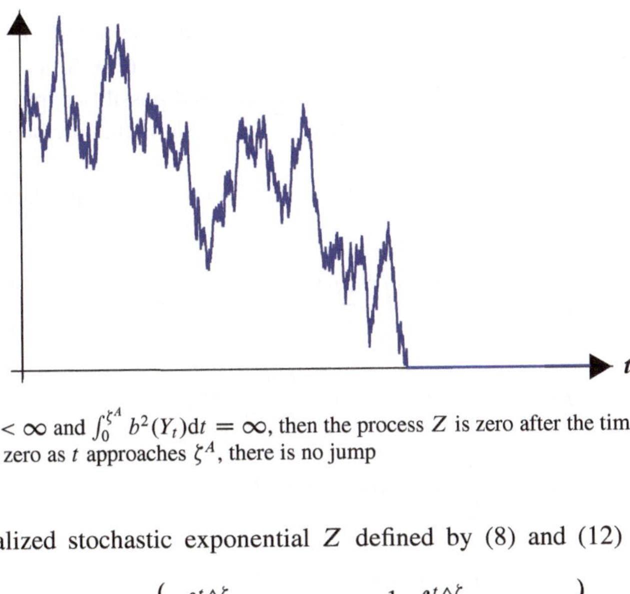

Fig. 2 If $\zeta^A < \infty$ and $\int_0^{\zeta^A} b^2(Y_t)\mathrm{d}t = \infty$, then the process Z is zero after the time ζ^A. Since the limit of Z_t is zero as t approaches ζ^A, there is no jump

The generalized stochastic exponential Z defined by (8) and (12) can now be written as

$$Z_t = \exp\left\{\int_0^{t\wedge\zeta} b\,(Y_u)\,\mathrm{d}W_u - \frac{1}{2}\int_0^{t\wedge\zeta} b^2\,(Y_u)\,\mathrm{d}u\right\},$$

where we set $Z_t = 0$ for $t \geq \zeta$ on $\{\zeta < \infty, \int_0^{\zeta} b^2\,(Y_u)\,\mathrm{d}u = \infty\}$. Note that in this case Z is a local martingale by Lemma 1, since $\zeta^A = \zeta$ and hence $Z = \bar{Z}$.

Let us now define an auxiliary J-valued diffusion $\widetilde{Y}$ governed by the SDE

$$\mathrm{d}\widetilde{Y}_t = (\mu + b\sigma)\left(\widetilde{Y}_t\right)\mathrm{d}t + \sigma\left(\widetilde{Y}_t\right)\mathrm{d}\widetilde{W}_t, \quad \widetilde{Y}_0 = x_0,$$

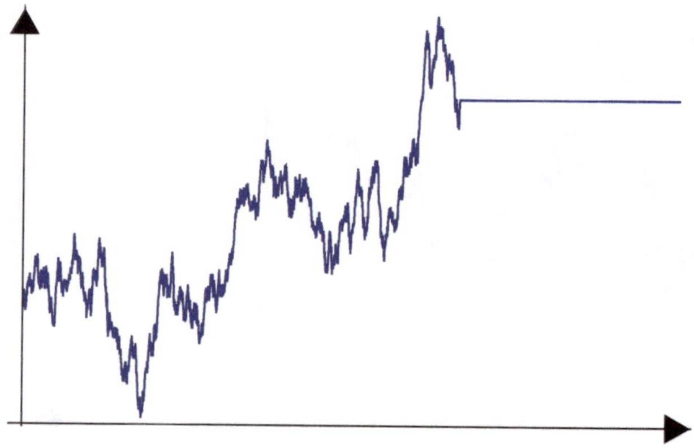

Fig. 3 If $\zeta < \tau_A$, the process Z is stopped after the exit time. Since $\int_0^\zeta b^2(Y_t)\mathrm{d}t$ is finite, Z_t is equal to a positive constant for $t \geq \zeta$

on some probability space $(\widetilde{\Omega}, \widetilde{\mathscr{F}}, (\widetilde{\mathscr{F}}_t)_{t\in[0,\infty)}, \widetilde{\mathbb{P}})$. The coefficients $\mu + b\sigma$ and σ satisfy Engelbert–Schmidt conditions since $\frac{b}{\sigma} \in L^1_{\mathrm{loc}}(J)$ (this follows from (13)). Hence the SDE has a weak solution, unique in law and possibly explosive. As with diffusion Y, we denote by $\widetilde{\zeta}$ the exit time of $\widetilde{Y}$ and assume that the boundary points are absorbing.

For an arbitrary $c \in J$ we define the scale functions $s, \widetilde{s}$ and their derivatives $\rho, \widetilde{\rho}$:

$$\begin{aligned} \rho(x) &= \exp\left\{-\int_c^x \frac{2\mu(y)}{\sigma^2(y)}\,\mathrm{d}y\right\}, \quad x \in J, \\ \widetilde{\rho}(x) &= \rho(x)\exp\left\{-\int_c^x \frac{2b(y)}{\sigma(y)}\,\mathrm{d}y\right\}, \quad x \in J, \\ s(x) &= \int_c^x \rho(y)\,\mathrm{d}y, \quad x \in \bar{J}, \\ \widetilde{s}(x) &= \int_c^x \widetilde{\rho}(y)\,\mathrm{d}y, \quad x \in \bar{J}. \end{aligned} \tag{14}$$

For any $a \in (l, r]$, $L^1_{\mathrm{loc}}(a-)$ denotes the set of all Borel measurable functions $f : J \to \mathbb{R}$ such that there exists $b < a, b \in J$, with $\int_b^a |f(x)|\,\mathrm{d}x < \infty$. Similarly, we define the set $L^1_{\mathrm{loc}}(a+)$ for any $a \in [l, r)$.

We say that the endpoint r is *good* if

$$s(r) < \infty \quad \text{and} \quad \frac{(s(r) - s)b^2}{\rho\sigma^2} \in L^1_{\mathrm{loc}}(r-).$$

It is equivalent to show that

$$\widetilde{s}(r) < \infty \quad \text{and} \quad \frac{(\widetilde{s}(r) - \widetilde{s})b^2}{\widetilde{\rho}\sigma^2} \in L^1_{\text{loc}}(r-).$$

The endpoint l is *good* if

$$s(l) > -\infty \quad \text{and} \quad \frac{(s - s(l))b^2}{\rho\sigma^2} \in L^1_{\text{loc}}(l+),$$

or equivalently

$$\widetilde{s}(l) > -\infty \quad \text{and} \quad \frac{(\widetilde{s} - \widetilde{s}(l))b^2}{\widetilde{\rho}\sigma^2} \in L^1_{\text{loc}}(l+).$$

If an endpoint is not good, we say it is *bad*. The good and bad endpoints were introduced in [8], where one can also find the proof of equivalences above.

We will use the following terminology:

$\widetilde{Y}$ *exits at* r means $\widetilde{\mathbb{P}}(\widetilde{\zeta} < \infty, \lim_{t \nearrow \widetilde{\zeta}} \widetilde{Y}_t = r) > 0$;
$\widetilde{Y}$ *exits at* l means $\widetilde{\mathbb{P}}(\widetilde{\zeta} < \infty, \lim_{t \nearrow \widetilde{\zeta}} \widetilde{Y}_t = l) > 0$.

Define

$$\widetilde{v}(x) = \int_c^x \frac{\widetilde{s}(x) - \widetilde{s}(y)}{\widetilde{\rho}(y)\sigma^2(y)} \, \mathrm{d}y, \quad x \in J, \tag{15}$$

and

$$\widetilde{v}(r) = \lim_{x \nearrow r} \widetilde{v}(x), \quad \widetilde{v}(l) = \lim_{x \searrow l} \widetilde{v}(x) \tag{16}$$

(note that $\widetilde{v}$ is decreasing on $(l, c]$ and increasing on $[c, r)$).

Feller's test for explosions (see [5, Chap. 5, Theorem 5.29]) tells us that:

1. $\widetilde{Y}$ exits at the boundary point r if and only if

$$\widetilde{v}(r) < \infty.$$

   It is equivalent to check (see [2, Chap. 4.1])

$$\widetilde{s}(r) < \infty \quad \text{and} \quad \frac{\widetilde{s}(r) - \widetilde{s}}{\widetilde{\rho}\sigma^2} \in L^1_{\text{loc}}(r-).$$

2. $\widetilde{Y}$ exits at the boundary point l if and only if

$$\widetilde{v}(l) < \infty,$$

which is equivalent to

$$\widetilde{s}(l) > -\infty \quad \text{and} \quad \frac{\widetilde{s}-\widetilde{s}(l)}{\widetilde{\rho}\sigma^2} \in L^1_{\text{loc}}(l+).$$

Remark 2. The endpoint r (resp. l) is bad whenever one of the processes Y and $\widetilde{Y}$ exits at r (resp. l) and the other does not.

The following theorems are reformulations of Theorems 2.1 and 2.3 in [8].

Theorem 1. *Let the functions μ, σ and b satisfy conditions* (1), (2) *and* (13). *Then the process Z is a martingale if and only if $\widetilde{Y}$ does not exit at the bad endpoints.*

Theorem 2. *Let the functions μ, σ and b satisfy conditions* (1), (2) *and* (13). *Then Z is a uniformly integrable martingale if and only if one of the conditions $(a)-(d)$ below is satisfied:*

(a) *$b = 0$ a.e. on J with respect to the Lebesgue measure;*
(b) *r is good and $\widetilde{s}(l) = -\infty$;*
(c) *l is good and $\widetilde{s}(r) = \infty$;*
(d) *l and r are good.*

3.2 The Case $A \neq \emptyset$

In the rest of the paper we assume that

$$x_0 \notin A,$$

since $x_0 \in A$ implies that $Z \equiv 0$. The following example shows that even when A is not empty we can get a martingale or a uniformly integrable martingale defined by (8) and (12).

Example 1. (i) Let us consider the state space $J = \mathbb{R}$, the coefficients of the SDE $\mu = 0, \sigma = 1$, the starting point of the diffusion $x_0 > 0$ and the function b such that $b(x) = \frac{1}{x}$ for $x \in \mathbb{R} \setminus \{0\}$ and $b(0) = 0$. Then $A = \{0\}$ and $Y_t = W_t$, $W_0 = x_0$. Using Itô's formula and the fact that Brownian motion does not exit at infinity, we get for $t < \tau_0$

$$\begin{aligned} Z_t &= \exp\left\{\int_0^t \frac{1}{W_u}\,\mathrm{d}W_u - \frac{1}{2}\int_0^t \frac{1}{W_u^2}\,\mathrm{d}u\right\} \\ &= \frac{1}{x_0} W_t \end{aligned}$$

and $Z_t = 0$ for $t \geq \tau_0$. Hence, $Z_t = \frac{1}{x_0} W_{t\wedge\tau_0}$ that is a martingale.

(ii) Using the same functions μ, σ and b as above on a state space $J = (-\infty, x_0 + 1)$ we get

$$Z_t = \frac{1}{x_0} W_{t \wedge \tau_{0,x_0+1}},$$

which is a uniformly integrable martingale.

Let for any $x \in J \setminus A$ the points $\alpha(x), \beta(x) \in A \cup \{l, r\}$ be as in (3). Then $\frac{b^2}{\sigma^2} \in L^1_{\text{loc}}(\alpha(x), \beta(x))$. Therefore, on $(\alpha(x), \beta(x))$ functions μ, σ and b satisfy the same conditions as in the previous subsection.

For any starting point $x_0 \in J \setminus A$ we can define an auxiliary diffusion $\widetilde{Y}$ with state space $(\alpha(x_0), \beta(x_0))$ driven by the SDE

$$\mathrm{d}\widetilde{Y}_t = (\mu + b\sigma)\left(\widetilde{Y}_t\right)\mathrm{d}t + \sigma\left(\widetilde{Y}_t\right)\mathrm{d}\widetilde{W}_t, \quad \widetilde{Y}_0 = x_0,$$

on some probability space $(\widetilde{\Omega}, \widetilde{\mathscr{F}}, (\widetilde{\mathscr{F}}_t)_{t\in[0,\infty)}, \widetilde{\mathbb{P}})$. There exists a unique weak solution of this equation since coefficients satisfy the Engelbert–Schmidt conditions.

As in the previous subsection we can define good and bad endpoints. Let functions $\rho, \widetilde{\rho}, s, \widetilde{s}$ and $\widetilde{v}$ be defined by (14), (15) and (16) with $c \in (\alpha(x_0), \beta(x_0))$. We say that the endpoint $\beta(x_0)$ is *good* if

$$s(\beta(x_0)) < \infty \quad \text{and} \quad \frac{(s(\beta(x_0)) - s)b^2}{\rho\sigma^2} \in L^1_{\text{loc}}(\beta(x_0)-).$$

It is equivalent to show and sometimes easier to check that

$$\widetilde{s}(\beta(x_0)) < \infty \quad \text{and} \quad \frac{(\widetilde{s}(\beta(x_0)) - \widetilde{s})b^2}{\widetilde{\rho}\sigma^2} \in L^1_{\text{loc}}(\beta(x_0)-).$$

The endpoint $\alpha(x_0)$ is *good* if

$$s(\alpha(x_0)) > -\infty \quad \text{and} \quad \frac{(s - s(\alpha(x_0)))b^2}{\rho\sigma^2} \in L^1_{\text{loc}}(\alpha(x_0)+),$$

or equivalently

$$\widetilde{s}(\alpha(x_0)) > -\infty \quad \text{and} \quad \frac{(\widetilde{s} - \widetilde{s}(\alpha(x_0)))b^2}{\widetilde{\rho}\sigma^2} \in L^1_{\text{loc}}(\alpha(x_0)+).$$

If an endpoint is not good, we say it is *bad*.

The following theorem plays a key role in all that follows.

Theorem 3. *Let $\alpha : J \setminus A \to A \cup \{l, r\}$ be the function defined by* (3) *and assume that the starting point $x_0 \in J \setminus A$ of diffusion Y satisfies $\alpha(x_0) > l$. Denote $\alpha_0 = \alpha(x_0) \in A$. Then:*

(*a*) $(x-\alpha_0)\frac{b^2}{\sigma^2}(x) \in L^1_{\mathrm{loc}}(\alpha_0+) \iff \int_0^{\tau_{\alpha_0}} b^2(Y_t)\,\mathrm{d}t < \infty$ $\mathbb{P}$-*a.s. on* $\{\tau_{\alpha_0} = \tau_A < \infty\}$;
(*b*) $(x-\alpha_0)\frac{b^2}{\sigma^2}(x) \notin L^1_{\mathrm{loc}}(\alpha_0+) \iff \int_0^{\tau_{\alpha_0}} b^2(Y_t)\,\mathrm{d}t = \infty$ $\mathbb{P}$-*a.s. on* $\{\tau_{\alpha_0} = \tau_A < \infty\}$.

Remark 3. (i) The assumption $\alpha(x_0) > l$ means that Theorem 3 deals with the situation where the set A contains points to the left of the starting point x_0.
(ii) Clearly, Theorem 3 has its analogue for $\beta_0 = \beta(x_0) < r$, i.e. the case when A contains points to the right of x_0. The deterministic criterion in this case takes the form

$$(\beta_0 - x)\frac{b^2}{\sigma^2}(x) \in L^1_{\mathrm{loc}}(\beta_0-)$$
$$\iff \int_0^{\tau_{\beta_0}} b^2(Y_t)\,\mathrm{d}t < \infty \ \mathbb{P}\text{-a.s. on } \{\tau_{\beta_0} = \tau_A < \infty\}, \tag{17}$$
$$(\beta_0 - x)\frac{b^2}{\sigma^2}(x) \notin L^1_{\mathrm{loc}}(\beta_0-)$$
$$\iff \int_0^{\tau_{\beta_0}} b^2(Y_t)\,\mathrm{d}t = \infty \ \mathbb{P}\text{-a.s. on } \{\tau_{\beta_0} = \tau_A < \infty\}. \tag{18}$$

(iii) Note that $\mathbb{P}(\tau_{\alpha_0} = \tau_A < \infty) > 0$. Indeed, if $\beta_0 = r$, then $\{\tau_{\alpha_0} = \tau_A < \infty\} = \{\tau_{\alpha_0} < \infty\}$; if $\beta_0 < r$, then $\{\tau_{\alpha_0} = \tau_A < \infty\} = \{\tau_{\alpha_0} < \tau_{\beta_0}\}$. In both cases $\mathbb{P}(\tau_{\alpha_0} = \tau_A < \infty) > 0$ by [2, Theorem 2.11].
(iv) Since the diffusion Y starts at x_0, $\mathbb{P}$-a.s. we have the following implications

$$\alpha_0, \beta_0 \in A \Longrightarrow \tau_A = \tau_{\alpha_0,\beta_0}, \tag{19}$$
$$\alpha_0 \in A, \beta_0 = r \Longrightarrow \tau_A = \tau_{\alpha_0}, \tag{20}$$
$$\alpha_0 = l, \beta_0 \in A \Longrightarrow \tau_A = \tau_{\beta_0}. \tag{21}$$

The case $\alpha_0 = l, \beta_0 = r$ cannot occur since $A \neq \emptyset$.
(v) Note that right-hand sides of the equivalences in (*a*) and (*b*) in Theorem 3 are not negations of each other. If it does not hold that the integral $\int_0^{\tau_{\alpha_0}} b^2(Y_t)\,\mathrm{d}t$ is finite $\mathbb{P}$-a.s. on the set $\{\tau_{\alpha_0} = \tau_A < \infty\}$, then it must be infinite on some subset of $\{\tau_{\alpha_0} = \tau_A < \infty\}$ of positive probability, which may be strictly smaller than $\mathbb{P}(\tau_{\alpha_0} = \tau_A < \infty)$.

For any starting point $x_0 \in J \setminus A$ of diffusion Y, define the set

$$B(x_0) = \left\{ x \in J \cap \{\alpha_0, \beta_0\};\ \int_0^{\tau_x} b^2(Y_t)\,\mathrm{d}t = \infty \ \mathbb{P}\text{-a.s. on } \{\tau_x = \tau_A < \infty\} \right\}.$$

Note that $B(x_0)$ is contained in A and that it contains at most two points. Theorem 3 implies that $\alpha_0 \in B(x_0)$ if and only if the deterministic condition in (b) is satisfied.

Similarly $\beta_0 \in B(x_0)$ is equivalent to the deterministic condition in (18). Therefore Theorem 3 yields a deterministic description of the set $B(x_0)$.

We can now give a deterministic characterisation for a generalized stochastic exponential Z to be a local martingale and a true martingale.

Theorem 4. *(i) The generalized stochastic exponential Z is a local martingale if and only if $\alpha(x_0), \beta(x_0) \in B(x_0) \cup \{l, r\}$.*

(ii) The generalized stochastic exponential Z is a martingale if and only if Z is a local martingale and at least one of the conditions (a)-(b) below is satisfied and at least one of the conditions (c)-(d) below is satisfied:

(a) $\widetilde{Y}$ does not exit at $\beta(x_0)$, i.e. $\widetilde{v}(\beta(x_0)) = \infty$ or, equivalently, we have

$$\widetilde{s}(\beta(x_0)) = \infty \quad or \quad \left(\widetilde{s}(\beta(x_0)) < \infty \; and \; \frac{\widetilde{s}(\beta(x_0)) - \widetilde{s}}{\widetilde{\rho}\sigma^2} \notin L^1_{\text{loc}}(\beta(x_0)-)\right);$$

(b) $\beta(x_0)$ is good,

(c) $\widetilde{Y}$ does not exit at $\alpha(x_0)$, i.e. $\widetilde{v}(\alpha(x_0)) = -\infty$ or, equivalently, we have

$$\widetilde{s}(\alpha(x_0)) = -\infty \quad or \quad \left(\widetilde{s}(\alpha(x_0)) > -\infty \; and \; \frac{\widetilde{s} - \widetilde{s}(\alpha(x_0))}{\widetilde{\rho}\sigma^2} \notin L^1_{\text{loc}}(\alpha(x_0)+)\right);$$

(d) $\alpha(x_0)$ is good.

Remark 4. Part *(ii)* of Theorem 4 says that Z is a martingale if and only if the $(\alpha(x_0), \beta(x_0))$-valued process $\widetilde{Y}$ can exit only at the good endpoints.

Proof. (i) We can write $Z = \bar{Z} - S$ as in (11). The process $\bar{Z}$ is a continuous local martingale by Lemma 1. Suppose that Z is a local martingale. Then S can be written as a sum of two local martingales and therefore, it is also a local martingale. It follows that S is a supermartingale (since it is non-negative). Since $\zeta^A > 0$ (we are assuming that $x_0 \in J \setminus A$) and $S_0 = 0$, S should be almost surely equal to 0. By definition (10) of S this happens if and only if $\mathbb{P}\left(\tau_A < \infty, \int_0^{\tau_A} b^2(Y_u) du < \infty\right) = 0$, which is by the definition of the set $B(x_0)$ and (19)–(21) equivalent to $\alpha(x_0), \beta(x_0) \in B(x_0) \cup \{l, r\}$.

(ii) To get at least a local martingale S needs to be zero $\mathbb{P}$-a.s. Then $Z = \bar{Z}$. Since the values of Y on $[0, \zeta^A)$ do not exit the interval $(\alpha(x_0), \beta(x_0))$, the conditions of Theorem 1 are satisfied and the result follows. □

Similarly, we can characterize uniformly integrable martingales. We can use characterization in Theorem 2 for the process $\bar{Z}$ defined by (9). As above, for $\alpha(x_0), \beta(x_0) \in B(x_0) \cup \{l, r\}$ the process Z defined by (8) and (12) coincides with $\bar{Z}$. Otherwise, Z is not even a local martingale.

Theorem 5. *The process Z is a uniformly integrable martingale if and only if Z is a local martingale and at least one of the conditions $(a)-(d)$ below is satisfied:*

(a) $b = 0$ a.e. on $(\alpha(x_0), \beta(x_0))$ with respect to the Lebesgue measure;

(*b*) $\alpha(x_0)$ *is good and* $\tilde{s}(\beta(x_0)) = \infty$*;*
(*c*) $\beta(x_0)$ *is good and* $\tilde{s}(\alpha(x_0)) = -\infty$*;*
(*d*) $\alpha(x_0)$ *and* $\beta(x_0)$ *are good.*

The following remark simplifies the application of Theorems 4 and 5 in specific situations.

Remark 5. If $\alpha(x_0) \in B(x_0)$, then $\alpha(x_0)$ is not a good endpoint. Indeed, if $s(\alpha(x_0)) > -\infty$, then we can write

$$\frac{(s(x) - s(\alpha(x_0)))b^2(x)}{\rho(x)\sigma^2(x)} = \frac{(s(x) - s(\alpha(x_0)))}{(x - \alpha(x_0))\rho(x)}(x - \alpha(x_0))\frac{b^2}{\sigma^2}(x).$$

The first fraction is bounded away from zero, since it is continuous for $x > \alpha(x_0)$ and has a limit equal to 1 as x approaches $\alpha(x_0)$. Since $\alpha(x_0) \in B(x_0)$, (b) of Theorem 3 implies $(x - \alpha(x_0))\frac{b^2}{\sigma^2}(x) \notin L^1_{\text{loc}}(\alpha(x_0)+)$. Therefore $\frac{(s-s(\alpha(x_0)))b^2}{\rho\sigma^2} \notin L^1_{\text{loc}}(\alpha(x_0)+)$ and the conclusion follows. Similarly, $\beta(x_0) \in B(x_0)$ implies that $\beta(x_0)$ is not a good endpoint.

4 Proof of Theorem 3

For the proof of Theorem 3 we first consider the case of a Brownian motion. Let W be a Brownian motion with $W_0 = x_0$. Denote by $L^y_t(W)$ the local time of W at time t and level y. Let us consider a Borel function $b : \mathbb{R} \to \mathbb{R}$ and set

$$A = \{x \in \mathbb{R} \ : \ b^2 \notin L^1_{\text{loc}}(x)\}.$$

We assume that $x_0 \notin A$ and define α_0, β_0 $(\alpha_0 < \beta_0)$ so that

$$\alpha_0, \beta_0 \in A \cup \{-\infty, \infty\} \quad \text{and} \quad x_0 \in (\alpha_0, \beta_0) \subseteq \mathbb{R} \setminus A.$$

We additionally assume that $\alpha_0 > -\infty$. Below we use the notations τ^W_x, $\tau^W_{x,y}$, and τ^W_A for the stopping times defined by (5), (6) and (7) respectively with Y replaced by W.

Lemma 2. *If* $(x - \alpha_0)b^2(x) \in L^1_{\text{loc}}(\alpha_0+)$*, then*

$$\int_0^{\tau^W_{\alpha_0}} b^2(W_t)\,dt < \infty \ \mathbb{P}\text{-a.s. on } \{\tau^W_{\alpha_0} = \tau^W_A\}.$$

Remark 6. In the setting of Lemma 2 we have $\mathbb{P}(\tau^W_{\alpha_0} < \infty) = 1$ since Brownian motion reaches every level in finite time almost surely. Therefore the events $\{\tau^W_{\alpha_0} = \tau^W_A\}$ and $\{\tau^W_{\alpha_0} = \tau^W_A < \infty\}$ are equal. Cf. with the formulation of Theorem 3.

Proof. Let $(\beta_n)_{n\in\mathbb{N}}$ be an increasing sequence such that $x_0 < \beta_n < \beta_0$ and $\beta_n \nearrow \beta_0$. By [11, Chap. VII, Corollary 3.8] we get

$$\mathbb{E}\left[\int_0^{\tau_{\alpha_0}^W \wedge \tau_{\beta_n}^W} b^2(W_t)\,\mathrm{d}t\right] = 2\tfrac{\beta_n - x_0}{\beta_n - \alpha_0}\int_{\alpha_0}^{x_0}(y-\alpha_0)b^2(y)\,\mathrm{d}y + 2\tfrac{x_0-\alpha_0}{\beta_n-\alpha_0}\int_{x_0}^{\beta_n}(\beta_n - y)b^2(y)\,\mathrm{d}y$$

for every β_n. Both integrals are finite since $b^2 \in L^1_{\text{loc}}(\alpha_0, \beta_0)$ and $(x-\alpha_0)b^2(x) \in L^1_{\text{loc}}(\alpha_0+)$. Thus, $\mathbb{E}[\int_0^{\tau_{\alpha_0}^W \wedge \tau_{\beta_n}^W} b^2(W_t)\,\mathrm{d}t] < \infty$ and therefore $\int_0^{\tau_{\alpha_0}^W \wedge \tau_{\beta_n}^W} b^2(W_t)\,\mathrm{d}t < \infty$ almost surely for every n. It remains to note that $\mathbb{P}$-a.s. on $\{\tau_{\alpha_0}^W = \tau_A^W\}$ we have $\tau_{\alpha_0}^W < \tau_{\beta_n}^W$ for sufficiently large n. This concludes the proof. □

Lemma 3. *If* $\int_0^{\tau_{\alpha_0}^W} b^2(W_t)\,\mathrm{d}t < \infty$ *on a set* U *with* $\mathbb{P}(U) > 0$, *then* $(x-\alpha_0)b^2(x) \in L^1_{\text{loc}}(\alpha_0+)$.

Proof. The idea of the proof comes from [4]. Using the occupation times formula we can write

$$\int_0^{\tau_{\alpha_0}^W} b^2(W_t)\,\mathrm{d}t = \int_{\alpha_0}^{\infty} b^2(y)L^y_{\tau_{\alpha_0}^W}(W)\,\mathrm{d}y \geq \int_{\alpha_0}^{x_0} b^2(y)L^y_{\tau_{\alpha_0}^W}(W)\,\mathrm{d}y.$$

Let us define a process $R_y = \frac{1}{y-\alpha_0}L^y_{\tau_{\alpha_0}^W}(W)$. Then R is positive and we have

$$\int_0^{\tau_{\alpha_0}^W} b^2(W_t)\,\mathrm{d}t \geq \int_{\alpha_0}^{x_0} R_y(y-\alpha_0)b^2(y)\,\mathrm{d}y. \tag{22}$$

By [11, Chap. VI, Proposition 4.6], Laplace transform of R_y is

$$\mathbb{E}[\exp\{-\lambda R_y\}] = \frac{1}{1+2\lambda} \quad \text{for every } y.$$

Hence, every random variable R_y has exponential distribution with $\mathbb{E}[R_y] = 2$.

Denote by L an indicator function of a measurable set. We can write

$$\mathbb{E}[LR_y] = \mathbb{E}\left[L\int_0^\infty \mathbf{1}_{\{R_y>u\}}\mathrm{d}u\right] = \int_0^\infty \mathbb{E}[L\mathbf{1}_{\{R_y>u\}}]\mathrm{d}u.$$

By Jensen's inequality we get a lower bound for the integrand

$$\begin{aligned}\mathbb{E}[L\mathbf{1}_{\{R_y>u\}}] &= \mathbb{E}[(L-\mathbf{1}_{\{R_y\le u\}})^+]\\ &\ge (\mathbb{E}[L]-\mathbb{P}[R_y\le u])^+\\ &= (\mathbb{E}[L]+e^{-\frac{u}{2}}-1)^+.\end{aligned}$$

Hence,

$$\mathbb{E}[LR_y] \ge \int_0^\infty (\mathbb{E}[L]+e^{-\frac{u}{2}}-1)^+ \mathrm{d}u = C, \tag{23}$$

where C is a strictly positive constant if $\mathbb{E}[L]$ is strictly positive.

Let us suppose that we can choose L, so that $\mathbb{E}[L\int_0^{\tau_{\alpha_0}^W} b^2(W_t)\,\mathrm{d}t]$ is finite. Using Fubini's Theorem and inequalities (22) and (23), we get

$$\mathbb{E}\left[L\int_0^{\tau_{\alpha_0}^W} b^2(W_t)\,\mathrm{d}t\right] \ge \int_{\alpha_0}^{x_0} \mathbb{E}[LR_y](y-\alpha_0)b^2(y)\,\mathrm{d}y \ge C\int_{\alpha_0}^{x_0}(y-\alpha_0)b^2(y)\,\mathrm{d}y.$$

Therefore, $(y-\alpha_0)b^2(y)\in L^1_{\mathrm{loc}}(\alpha_0+)$ if we can find an indicator function L such that $\mathbb{E}[L]$ is strictly positive and $\mathbb{E}[L\int_0^{\tau_{\alpha_0}^W} b^2(W_t)\,\mathrm{d}t]$ is finite.

Since $\int_0^{\tau_{\alpha_0}^W} b^2(W_t)\,\mathrm{d}t<\infty$ on a set with positive measure, such L exists. Indeed, denote by L_n an indicator function of the set $U_n=\{\int_0^{\tau_{\alpha_0}^W} b^2(W_t)\,\mathrm{d}t\le n\}$. Then, for every integer n, we have $\mathbb{E}[L_n\int_0^{\tau_{\alpha_0}^W} b^2(W_t)\,\mathrm{d}t]<\infty$. Since the sequence $(U_n)_{n\in\mathbb{N}}$ is increasing, $U\subseteq\bigcup_{n\in\mathbb{N}}U_n$ and $\mathbb{P}(U)>0$, there exists an integer N such that $\mathbb{P}(U_N)>0$ and therefore $\mathbb{E}[L_N]>0$. □

Now we return to the setting of Sect. 2.

Proof of Theorem 3. Suppose that $\mu\equiv 0$ and σ is arbitrary. Since Y_t is a continuous local martingale, by the Dambis–Dubins–Schwarz theorem we have $Y_t=B_{\langle Y\rangle_t}$ for a Brownian motion B with $B_0=x_0$, defined possibly on an enlargement of the initial probability space. Using the substitution $u=\langle Y\rangle_t$, we get

$$\int_0^{\tau_{\alpha_0}^Y} b^2(Y_t)\,\mathrm{d}t = \int_0^{\tau_{\alpha_0}^Y}\frac{b^2}{\sigma^2}(Y_t)\,\mathrm{d}\langle Y\rangle_t = \int_0^{\langle Y\rangle_{\tau_{\alpha_0}^Y}}\frac{b^2}{\sigma^2}(B_u)\,\mathrm{d}u \quad \mathbb{P}\text{-a.s.} \tag{24}$$

Furthermore, it is easy to see from $Y_t=B_{\langle Y\rangle_t}$ that

$$\langle Y\rangle_{\tau_{\alpha_0}^Y}=\tau_{\alpha_0}^B \quad \mathbb{P}\text{-a.s. on} \quad \{\tau_{\alpha_0}^Y<\infty\} \tag{25}$$

and

$$\{\tau_{\alpha_0}^Y=\tau_A^Y<\infty\}\subseteq\{\tau_{\alpha_0}^B=\tau_A^B\} \quad \mathbb{P}\text{-a.s.} \tag{26}$$

Now (24)–(26) imply the theorem in the case $\mu\equiv 0$ and σ arbitrary.

It only remains to prove the general case when both μ and σ are arbitrary. Let $\widetilde{Y}_t=s(Y_t)$, where s is the scale function of Y. Then $\widetilde{Y}$ satisfies SDE

$$d\widetilde{Y}_t = \widetilde{\sigma}\left(\widetilde{Y}_t\right) dW_t,$$

where $\widetilde{\sigma}(x) = s'(q(x))\sigma(q(x))$ and q is the inverse of s.

Define $\widetilde{b} = b \circ q$. Since s is strictly increasing, $s(\alpha_0) > -\infty$ by the assumption $\alpha_0 > l$, and $\widetilde{Y}_{\tau_{\alpha_0}} = s(Y_{\tau_{\alpha_0}}) = s(\alpha_0)$, it follows that the equality $\tau^Y_{\alpha_0} = \tau^{\widetilde{Y}}_{s(\alpha_0)}$ holds $\mathbb{P}$-a.s. Then we have

$$\int_0^{\tau^Y_{\alpha_0}} b^2(Y_t)\, dt = \int_0^{\tau^{\widetilde{Y}}_{s(\alpha_0)}} \widetilde{b}^2\left(\widetilde{Y}_t\right) dt.$$

Besides, for some small positive ε we have

$$\int_{s(\alpha_0)}^{s(\alpha_0+\varepsilon)} \frac{\widetilde{b}^2(x)}{\widetilde{\sigma}^2(x)} (x - s(\alpha_0))\, dx = \int_{\alpha_0}^{\alpha_0+\varepsilon} \frac{b^2(y)}{\sigma^2(y)} \frac{s(y) - s(\alpha_0)}{s'(y)}\, dy.$$

The fraction $\frac{s(y)-s(\alpha_0)}{s'(y)(y-\alpha_0)}$ is continuous for $y > \alpha_0$ and tends to 1 as $y \searrow \alpha_0$. Hence it is bounded and bounded away from zero on $(\alpha_0, \alpha_0 + \varepsilon]$. It follows that $(x - \alpha_0)\frac{b^2}{\sigma^2}(x) \in L^1_{\text{loc}}(\alpha_0+)$ if and only if $(x - s(\alpha_0))\frac{\widetilde{b}^2}{\widetilde{\sigma}^2}(x) \in L^1_{\text{loc}}(s(\alpha_0)+)$. Then the result follows from the first part of the proof. □

Remark 7. It is interesting to note that, in fact, both sets in (26) are $\mathbb{P}$-a.s. equal. One can prove the reverse inclusion using the Engelbert–Schmidt construction of solutions of SDEs and the fact that $\alpha_0 > l$.

A Appendix

Let Y be a J-valued diffusion starting from x_0 with a drift μ and volatility σ that satisfy the Engelbert–Schmidt conditions. Let $b : J \to \mathbb{R}$ be a Borel measurable function and let $(c, d) \subseteq J$, $c < x_0 < d$. Recall that, for any $x, y \in J$, the stopping times τ_x and $\tau_{x,y}$ are defined in (5) and (6). We now extend the definition in (6) by setting $\tau_{c,d} := \zeta$ if $c = l, d = r$; $\tau_{c,d} := \tau_c \wedge \zeta$ if $c > l, d = r$; $\tau_{c,d} := \zeta \wedge \tau_d$ if $c = l, d < r$.

Proposition 1. *(i) The condition*

$$\frac{b^2}{\sigma^2} \in L^1_{\text{loc}}(c, d)$$

implies that for all $t \in [0, \infty)$,

$$\int_0^t b^2(Y_u)\, du < \infty \ \mathbb{P}\text{-a.s. on } \{t < \tau_{c,d}\}.$$

(ii) For any $\alpha \in J$ such that $\frac{b^2}{\sigma^2} \notin L^1_{\text{loc}}(\alpha)$ we have

$$\int_0^t b^2(Y_u)\,du = \infty \ \mathbb{P}\text{-a.s. on } \{\tau_\alpha < t < \zeta\}.$$

Remark 8. Let us note that $\mathbb{P}(\tau_\alpha < \infty) > 0$ by [2, Theorem 2.11]. Clearly, $\{\tau_\alpha < \infty\} = \{\tau_\alpha < \zeta\}$. Hence there exists $t \in [0, \infty)$ such that $\mathbb{P}(\tau_\alpha < t < \zeta) > 0$.

Proof. (i) Using the occupation times formula, $\mathbb{P}$-a.s. we get

$$\int_0^t b^2(Y_u)\mathrm{d}u = \int_0^t \frac{b^2}{\sigma^2}(Y_u)\mathrm{d}\langle Y\rangle_u = \int_J \frac{b^2}{\sigma^2}(y)L_t^y(Y)\mathrm{d}y, \quad t \in [0, \zeta). \tag{27}$$

$\mathbb{P}$-a.s. on the set $\{t < \tau_{c.d}\}$ the function $y \mapsto L_t^y(Y)$ is cádlág (see [11, Chap. VI, Theorem 1.7] with a compact support in the interval (c, d). Now the first statement follows from (27).

(ii) We have

$$\int_{\alpha-\varepsilon}^{\alpha+\varepsilon} \frac{b^2}{\sigma^2}(y)\mathrm{d}y = \infty \quad \text{for all } \varepsilon > 0.$$

By [2, Theorem 2.7], we have for any $t \geq 0$

$$L_t^\alpha(Y) > 0 \text{ and } L_t^{\alpha-}(Y) > 0 \ \mathbb{P}\text{-a.s. on} \quad \{\tau_\alpha < t < \zeta\}.$$

Then, $\mathbb{P}$-a.s. on $\{\tau_\alpha < t < \zeta\}$, there exists $\varepsilon > 0$ such that the function $y \mapsto L_t^y(Y)$ is bounded away from zero on the interval $(\alpha - \varepsilon, \alpha + \varepsilon)$. It follows from (27) that $\int_0^t b^2(Y_u)\,du = \infty$ $\mathbb{P}$-a.s on $\{\tau_\alpha < t < \zeta\}$. This concludes the proof. □

Acknowledgements We would like to thank the anonymous referee for a thorough reading of the paper and many useful suggestions, which significantly improved the paper.

References

1. S. Blei, H.-J. Engelbert, On exponential local martingales associated with strong Markov continuous local martingales. Stoch. Process. Their Appl. **119**(9), 2859–2880 (2009)
2. A.S. Cherny, H.-J. Engelbert, *Singular Stochastic Differential Equations*. Lecture Notes in Mathematics, vol. 1858 (Springer, Berlin, 2005)
3. H.-J. Engelbert, T. Senf, On functionals of a Wiener process with drift and exponential local martingales, in *Stochastic Processes and Related Topics (Georgenthal, 1990)*. Mathematical Research, vol. 61 (Akademie, Berlin, 1991), pp. 45–58
4. T. Jeulin, *Semi-martingales et grossissement d'une Filtration*. Lecture Notes in Mathematics, vol. 833 (Springer, Berlin, 1980)
5. I. Karatzas, S.E. Shreve, *Brownian Motion and Stochastic Calculus*. Graduate Texts in Mathematics, vol. 113, 2nd edn. (Springer, Berlin, 1991)

6. N. Kazamaki, On a problem of Girsanov. Tôhoku Math. J. **29**(4), 597–600 (1977)
7. R.S. Liptser, A.N. Shiryaev, *Statistics of Random Processes, I.* Applications of Mathematics (New York), vol. 5, 2nd edn. (Springer, Berlin, 2001)
8. A. Mijatović, M. Urusov, On the martingale property of certain local martingales. *Probability Theory and Related Fields*, **152**(1):1–30 (2012)
9. A.A. Novikov, A certain identity for stochastic integrals. Theor. Probab. Appl. **17**, 761–765 (1972)
10. P.E. Protter, *Stochastic Integration and Differential Equations*. Stochastic Modelling and Applied Probability, vol. 21, 2nd edn. (Springer, Berlin, 2005) Version 2.1 (Corrected third printing)
11. D. Revuz, M. Yor, in *Continuous Martingales and Brownian Motion. Grundlehren der Mathematischen Wissenshaften*, vol. 293, 3rd edn. (Springer, Berlin, 1999)

Some Classes of Proper Integrals and Generalized Ornstein–Uhlenbeck Processes

Andreas Basse-O'Connor, Svend-Erik Graversen, and Jan Pedersen

Abstract We give necessary and sufficient conditions for existence of proper integrals from 0 to infinity or from minus infinity to 0 of one exponentiated Lévy process with respect to another Lévy process. The results are related to the existence of stationary generalized Ornstein–Uhlenbeck processes. Finally, in the square integrable case the Wold-Karhunen representation is given.

Keywords Stochastic integration • Lévy processes • Generalized Ornstein–Uhlenbeck processes

1 Introduction

Let $(\xi, \eta) = (\xi_t, \eta_t)_{t \in \mathbb{R}}$ denote a bivariate Lévy process indexed by $\mathbb{R}$ satisfying $\xi_0 = \eta_0 = 0$, that is, (ξ, η) is defined on a probability space $(\Omega, \mathscr{F}, P)$, has càdlàg paths and stationary independent increments. We are interested in the two integrals

$$(a): \int_0^\infty e^{-\xi_{s-}} \, d\eta_s \qquad \text{and} \qquad (b): \int_{-\infty}^0 e^{\xi_{s-}} \, d\eta_s. \tag{1}$$

The first of these has been thoroughly studied, see e.g. [5, 6, 9, 13, 15, 18], where it is treated as an improper integral, i.e. as the a.s. limit as $t \to \infty$ of $\int_0^t e^{-\xi_{s-}} \, d\eta_s$. Recall that Erickson and Maller, [9] Theorem 2, give necessary and sufficient conditions

A.B.-O'Connor
Department of Mathematics, The University of Tennessee, 227 Ayres Hall, 1403 Circle Drive, Knoxville, TN 37996-1320, USA
e-mail: aboc@math.utk.edu

S.-E. Graversen · J. Pedersen (✉)
Department of Mathematical Sciences, University of Aarhus, Ny Munkegade, 8000 Aarhus C, Denmark
e-mail: matseg@imf.au.dk; jan@imf.au.dk

C. Donati-Martin et al. (eds.), *Séminaire de Probabilités XLIV*, Lecture Notes in Mathematics 2046, DOI 10.1007/978-3-642-27461-9_3,

in terms of the Lévy-Khintchine triplet of (ξ, η) for the existence of (1)(a) in the improper sense. In the following this integral is considered as a semimartingale integral up to infinity in the sense of e.g. [7] or [2], which we can think of as a proper integral. Theorem 1 shows that the conditions given by Erickson and Maller are also necessary and sufficient for the existence of (1)(a) in the proper sense.

To the best of our knowledge the second integral has previously only been studied in special cases, in particular when ξ is deterministic. As we shall see in the next section, η is a so-called increment semimartingale in the natural filtration of (ξ, η), that is, the least right-continuous and complete filtration to which (ξ, η) is adapted. Integration with respect to increment semimartingales has been studied by the authors in [2], and we use the results obtained there to give necessary and sufficient conditions for the existence of (1)(b); see Theorem 1.

As an application, generalized Ornstein–Uhlenbeck processes (and some generalizations hereof) are considered. Recall that a càdlàg process $V = (V_t)_{t\in\mathbb{R}}$ is a generalized Ornstein–Uhlenbeck process if it satisfies

$$V_t = e^{-(\xi_t - \xi_s)} V_s + e^{-\xi_t} \int_s^t e^{\xi_{u-}} \, d\eta_u \quad \text{for } s < t. \tag{2}$$

See [3, 4, 14] for a thorough study of these processes and their generalizations and references to theory and applications. Assuming $\xi_t \to \infty$ as $t \to \infty$ a.s., Theorem 2 shows that a necessary and sufficient condition for the existence of a stationary V is that (1)(b) exists, and in this case V is represented as

$$V_t = e^{-\xi_t} \int_{-\infty}^t e^{\xi_{u-}} \, d\eta_u \quad \text{for } t \in \mathbb{R}. \tag{3}$$

This result complements Theorem 2.1 in [14] where the stationary distribution is expressed in terms of an integral from 0 up to infinity. Finally, assuming second moments, Theorem 3 gives the Wold-Karhunen representation of V.

2 Integration with Respect to Increment Semimartingales

In this section we first recall a few general results related to integration with respect to increment semimartingales. Afterwards we specialize to integration with respect to η.

Let $(\mathscr{F}_t)_{t\in\mathbb{R}}$ denote a filtration satisfying the usual conditions of right-continuity and completeness. Recall from [2] that a càdlàg $\mathbb{R}$-valued process $Z = (Z_t)_{t\in\mathbb{R}}$ is called an increment semimartingale with respect to $(\mathscr{F}_t)_{t\in\mathbb{R}}$ if for all $s \in \mathbb{R}$ the process $(Z_{t+s} - Z_s)_{t\geq 0}$ is an $(\mathscr{F}_{s+t})_{t\geq 0}$-semimartingale in the usual sense. Equivalently, by Example 4.1 in [2], Z is an increment semimartingale if and only if it induces an $L^0(P)$-valued Radon measure on the predictable σ-field $\mathscr{P}$. Note that in general an increment semimartingale is not adapted. Let $\mu^Z = \mu^Z(\omega; dt \times dx)$ denote the jump measure of Z defined as

$$\mu^Z(A) = \#\{s \in \mathbb{R} : (s, \Delta Z_s) \in A\} \quad \text{for } A \in \mathscr{B}(\mathbb{R} \times \mathbb{R}_0),$$

where $\mathbb{R}_0 = \mathbb{R} \setminus \{0\}$, and let (B, C, ν) denote the triplet of Z; see [2]. That is, $\nu = \nu(\omega; \mathrm{d}t \times \mathrm{d}x)$ is the predictable compensator of μ^Z in the sense of [12], Theorem II.1.8. Moreover, $B = B(\omega; \mathrm{d}t)$ is a random signed measure on $\mathbb{R}$ of finite total variation on compacts satisfying that $t \mapsto B((s, s+t])$ is an $(\mathscr{F}_{s+t})_{t\geq 0}$-predictable process for all $s \in \mathbb{R}$ and $C = C(\omega; \mathrm{d}t)$ is a random positive measure on $\mathbb{R}$ which is finite on compacts. Finally, for all $s < t$ we have

$$Z_t - Z_s = Z_t^c - Z_s^c + \int_{(s,t]\times\{|x|\leq 1\}} x\,[\mu^Z(\mathrm{d}u \times \mathrm{d}x) - \nu(\mathrm{d}u \times \mathrm{d}x)]$$
$$+ \int_{(s,t]\times\{|x|>1\}} x\,\mu^Z(\mathrm{d}u \times \mathrm{d}x) + B((s,t])$$

where $Z^c = (Z_t^c)_{t\in\mathbb{R}}$ is a continuous increment local martingale and for all $s \in \mathbb{R}$ the quadratic variation of $(Z_{s+t}^c - Z_s^c)_{t\geq 0}$ is $C((s, s+t])$, $t \geq 0$. Choose a predictable random positive measure $\lambda = \lambda(\omega; \mathrm{d}t)$ on $\mathbb{R}$ which is finite on compacts, a real-valued predictable process $b = b_t(\omega)$, a positive predictable process $c = c_t(\omega)$, and a transition kernel $K = K(t, \omega; \mathrm{d}x)$ from $(\mathbb{R} \times \Omega, \mathscr{P})$ into $(\mathbb{R}, \mathscr{B}(\mathbb{R}))$ satisfying $\int_{\mathbb{R}} (1 \wedge x^2)\, K(t; \mathrm{d}x) < \infty$ and $K(t; \{0\}) = 0$ for all $t \in \mathbb{R}$ such that

$$B(\mathrm{d}t) = b_t \lambda(\mathrm{d}t), \quad C(\mathrm{d}t) = c_t \lambda(\mathrm{d}t), \quad \nu(\mathrm{d}t \times \mathrm{d}x) = K(t; \mathrm{d}x)\lambda(\mathrm{d}t).$$

As shown in [2] a necessary and sufficient condition for the existence of $\int_{\mathbb{R}} \phi_s \, \mathrm{d}Z_s$ is that we have the following:

$$\int_{\mathbb{R}} \left| \phi_s b_s + \int_{\mathbb{R}} \left[\tau(\phi_s x) - \phi_s \tau(x)\right] K(s; \mathrm{d}x) \right| \lambda(\mathrm{d}s) < \infty, \tag{4}$$

$$\int_{\mathbb{R}} \phi_s^2 c_s \, \lambda(\mathrm{d}s) < \infty, \qquad \int_{\mathbb{R}} \int_{\mathbb{R}} (1 \wedge (\phi_s x)^2)\, K(s; \mathrm{d}x)\, \lambda(\mathrm{d}s) < \infty, \tag{5}$$

where $\tau : \mathbb{R} \to \mathbb{R}$ is a truncation function, i.e. it is bounded, measurable and satisfies $\tau(x) = x$ in a neighborhood of 0. Moreover, when these conditions are satisfied the process $\int_{-\infty}^t \phi_s \, \mathrm{d}Z_s$, $t \in \mathbb{R}$, is a semimartingale up to infinity. Here we use the usual convention that for a measurable subset A of $\mathbb{R}$, $\int_A \phi_s \, \mathrm{d}Z_s := \int_{\mathbb{R}} \phi_s 1_A(s) \, \mathrm{d}Z_s$, and $\int_s^t := \int_{(s,t]}$ for $s < t$.

Let us turn to integration with respect to η where (ξ, η) is a bivariate Lévy process indexed by $\mathbb{R}$ with $\xi_0 = \eta_0 = 0$, that is, η plays the role of Z from now on. Denote the Lévy-Khintchine triplet of ξ_1 by $(\gamma_\xi, \sigma_\xi^2, m_\xi)$, and let $\sigma_{\xi,\eta}$ denote the covariance of the Gaussian components of ξ and η at time $t = 1$. A similar notation will be used for all other Lévy processes. Let $(\mathscr{F}_t^{(\xi,\eta)})_{t\in\mathbb{R}}$ denote the natural filtration of (ξ, η). Note that $(\eta_t)_{t\geq 0}$ is a Lévy process in $(\mathscr{F}_t^{(\xi,\eta)})_{t\geq 0}$, i.e. for all $0 \leq s < t$, $\eta_t - \eta_s$ is independent of $\mathscr{F}_s^{(\xi,\eta)}$. Using this it is easily seen that the increment semimartingale triplet of η in $(\mathscr{F}_t^{(\xi,\eta)})_{t\in\mathbb{R}}$ is, for $t > 0$, given by $\lambda(\mathrm{d}t) = \mathrm{d}t$ and $(b_t, c_t, K(t; \mathrm{d}x)) = (\gamma_\eta, \sigma_\eta^2, m_\eta(\mathrm{d}x))$. Thus, (4)–(5) provide necessary and sufficient conditions that $\int_0^\infty \phi_s \, d\eta_s$ exists for an arbitrary $(\mathscr{F}_t^{(\xi,\eta)})_{t\in\mathbb{R}}$-predictable process ϕ. This in particular includes (1)(a). When it comes to integrals involving the negative half axis, such as (1)(b), the situation is more complicated since η is not a Lévy process in $(\mathscr{F}_t^{(\xi,\eta)})_{t\in\mathbb{R}}$ (see [2], Sect. 5). In fact, a priori it is not even clear that η is an increment semimartingale in the natural filtration of (ξ,η) (or in any other filtration). However, using an enlargement of filtration result due to Protter and Jacod, see [11], Theorem 5.3 in [2] shows that this is indeed the case, and the triplet of η is calculated in an enlarged filtration. Theorem 5.5 in [2] provides sufficient conditions for integrals of the form $\int_{\mathbb{R}} \phi_s \, d\eta_s$ to exist and as this result will be used throughout the paper we rephrase it as a remark.

Remark 1. Assume $E[\eta_1^2] < \infty$. Then for any $(\mathscr{F}_t^{(\xi,\eta)})_{t\in\mathbb{R}}$-predictable process having a.a. paths locally bounded the integral $\int_{\mathbb{R}} \phi_s \, d\eta_s$ exists if $\int_{\mathbb{R}} (|\phi_s| + \phi_s^2) \, ds < \infty$ a.s.

3 Main Results

As above let (ξ,η) denote a bivariate Lévy process indexed by $\mathbb{R}$ with $\xi_0 = \eta_0 = 0$. To avoid trivialities assume that none of them are identically equal to 0. Often we will assume that $\xi_t \to \infty$ a.s. as $t \to \infty$ because if this fails, Theorem 2 in [9] shows that $\int_0^t e^{-\xi_{s-}} \, d\eta_s$ does not converge as $t \to \infty$ a.s., implying that (1)(a) does not exist. We need the function A_ξ defined, cf. [9] and [14], as

$$A_\xi(x) = \max\{1, m_\xi((1,\infty))\} + \int_1^x m_\xi((y,\infty)) \, dy, \quad x \geq 1. \tag{6}$$

To study (1)(b) we follow Lindner and Maller [14] and introduce $(L_t)_{t\geq 0}$ given by

$$L_t = \eta_t + \sum_{0<s\leq t} (e^{-\Delta\xi_s} - 1)\Delta\eta_s - t\sigma_{\xi,\eta} \quad \text{for } t \geq 0. \tag{7}$$

The process $(L_t, \xi_t)_{t\geq 0}$ is then a bivariate Lévy process in the $(\mathscr{F}_t^{(\xi,\eta)})_{t\geq 0}$-filtration (see [14], Proposition 2.3) and for all $t \geq 0$ we have

$$\int_{-t}^0 e^{\xi_{s-}} \, d\eta_s \stackrel{\mathscr{D}}{=} e^{-\xi_t} \int_0^t e^{\xi_{s-}} \, d\eta_s \stackrel{\mathscr{D}}{=} \int_0^t e^{-\xi_{s-}} \, dL_s. \tag{8}$$

(Here $\stackrel{\mathscr{D}}{=}$ denotes equality in distribution.) Indeed, the second equality follows from [14], Proposition 2.3. To prove the first equality note that

$$\int_{-t}^0 e^{\xi_{s-}} \, d\eta_s = e^{-(\xi_0 - \xi_{-t})} \int_{-t}^0 e^{\xi_{s-} - \xi_{-t}} \, d\eta_s,$$

which by the stationary increments has the same law as

$$e^{-(\xi_t-\xi_0)}\int_0^t e^{\xi_{s-}-\xi_0}\,\mathrm{d}\eta_s = e^{-\xi_t}\int_0^t e^{\xi_{s-}}\,\mathrm{d}\eta_s.$$

The existence of (1)(a) and (1)(b) is characterized in the following. All integrals over infinite intervals are defined in the proper sense of [2], implying in particular that they exist as improper integrals.

Theorem 1. *Assume* $\xi_t \to \infty$ *as* $t \to \infty$ *a.s.*

(1) The following statements are equivalent:

(a) The integral (1)(a) *exists.*
(b) $\int_0^t e^{-\xi_{s-}}\,\mathrm{d}\eta_s$ *converges in distribution as* $t \to \infty$.
(c) We have

$$\int_{[-e,e]^c} \frac{\log|y|}{A_\xi(\log|y|)}\, m_\eta(\mathrm{d}y) < \infty. \tag{9}$$

(2) The following statements are equivalent:

(a) The integral (1)(b) *exists.*
(b) $\int_{-t}^0 e^{\xi_{s-}}\,\mathrm{d}\eta_s$ *converges in distribution as* $t \to \infty$.
(c) We have

$$\int_{[-e,e]^c} \frac{\log|y|}{A_\xi(\log|y|)}\, m_L(\mathrm{d}y) < \infty. \tag{10}$$

If (1)(b) *exists then* $\int_{-\infty}^0 e^{\xi_{s-}}\,\mathrm{d}\eta_s \overset{\mathscr{D}}{=} \int_0^\infty e^{-\xi_{s-}}\mathrm{d}L_s$.

It should be noted that (1c) coincides with the condition in [9], Theorem 2, for (1)(a) to exist as an improper integral. In the case when $\xi_t \to \infty$ as $t \to \infty$ a.s, (1c) implies (2c); this is shown in [14], where also further interesting relations between these conditions can be found.

Proof. Throughout the proof assume $\xi_t \to \infty$ as $t \to \infty$ a.s. The equivalence between (1b) and (1c) is given in [14], Proposition 2.4. Using (8) it thus follows that (2b) and (2c) are equivalent. Since, as noted above, existence of integrals in the proper sense implies existence as improper integrals it is obvious that (1a) implies (1b) and (2a) implies (2b). We prove the remaining assertions in a few steps.

Step 1. Assume there is an $\epsilon > 0$ such that $m_\eta(\{|x| > \epsilon\}) = 0$. (That is, η has no big jumps, implying in particular square integrability). We show that in this case (1)(a) and (1)(b) both exist.

Note that if $E[|\xi_1|] < \infty$ then by the law of large numbers $\xi_t/t \to E[\xi_1]$ as $t \to \infty$ a.s. and in this case $E[\xi_1] > 0$. Recall that we assume $\xi_t \to \infty$ as $t \to \infty$ a.s. It follows from Kesten's trichotomy theorem (see e.g. [8], Theorem 4.4) that if

$E[|\xi_1|] = \infty$ then $\lim_{t\to\infty} \xi_t/t = \infty$ a.s. Thus, there is a $\mu \in (0,\infty]$ such that $\xi_s/s \to \mu$ as $s \to -\infty$ and $\xi_t/t \to \mu$ as $t \to \infty$ a.s. In particular

$$\int_{-\infty}^{0} (e^{\xi_s} + e^{2\xi_s})\,\mathrm{d}s + \int_0^{\infty} (e^{-\xi_t} + e^{-2\xi_t})\,\mathrm{d}t$$
$$= \int_{-\infty}^{0} (e^{s(\xi_s/s)} + e^{2s(\xi_s/s)})\,\mathrm{d}s + \int_0^{\infty} (e^{-t(\xi_t/t)} + e^{-2t(\xi_t/t)})\,\mathrm{d}t < \infty$$

which by Remark 1 implies the existence of (1)(a) and (1)(b).

Step 2. Assume η is a compound Poisson process, that is,

$$m_\eta(\mathbb{R}) < \infty, \quad \sigma_\eta^2 = 0, \quad \gamma_\eta = \int_{|x|\le 1} x\, m_\eta(\mathrm{d}x)$$

and η_t is given by $\eta_t = \sum_{0<s\le t} \Delta\eta_s$ for all $t > 0$ a.s. Assume in addition that (1c) holds. We show that in this case $\sum_{s>0} e^{-\xi_{s-}}|\Delta\eta_s| < \infty$ a.s. (This clearly implies that (1)(a) exists and equals $\sum_{s>0} e^{-\xi_{s-}}\Delta\eta_s$).

For this purpose we first prove

$$\int_0^{\infty}\int_{\mathbb{R}} 1 \wedge (e^{-\xi_s}|x|)\, m_\eta(\mathrm{d}x)\,\mathrm{d}s < \infty \quad \text{a.s.} \tag{11}$$

The integral in (11) can be written as $\int_0^\infty g(\xi_s)\,\mathrm{d}s$, where, for $y \in \mathbb{R}$, $g(y) = \int_{\mathbb{R}}(1 \wedge (e^{-y}|x|))\, m_\eta(\mathrm{d}x)$. Therefore, since g is non-increasing it follows from [9], Theorem 1, that (11) is satisfied if and only if

$$\int_{(1,\infty)} \frac{y}{A_\xi(y)}\, |\mathrm{d}g(y)| < \infty. \tag{12}$$

Simple manipulations show that

$$g(y) = m_\eta(\mathbb{R}) - \int_{-\infty}^{y}\int_{|x|\le e^z} |x|\, m_\eta(\mathrm{d}x) e^{-z}\,\mathrm{d}z \tag{13}$$

and hence the integral in (12) equals

$$\int_{(1,\infty)} \frac{y}{A_\xi(y)} \int_{|x|\le e^y} |x|\, m_\eta(\mathrm{d}x) e^{-y}\,\mathrm{d}y. \tag{14}$$

Since A_ξ is non-decreasing we can use Fubini to rewrite and dominate this integral as

$$\int_{[-e,e]^c} |x| \int_{\log|x|}^{\infty} \frac{ye^{-y}}{A_\xi(y)}\,\mathrm{d}y\, m_\eta(\mathrm{d}x) \leq \int_{[-e,e]^c} \frac{|x|}{A_\xi(\log|x|)} \int_{\log|x|}^{\infty} ye^{-y}\,\mathrm{d}y\, m_\eta(\mathrm{d}x)$$
$$= \int_{[-e,e]^c} \frac{|x|}{A_\xi(\log|x|)} e^{-\log|x|}(1+\log|x|)\, m_\eta(\mathrm{d}x),$$

which is finite by assumption. Now, by [7], Lemma 3.4, (11) is equivalent to $\sum_{s>0}\left(1 \wedge (e^{-\xi_{s-}}|\Delta\eta_s|)\right) < \infty$ a.s. Hence $\sum_{s>0} e^{-\xi_{s-}}|\Delta\eta_s| < \infty$ a.s.

Step 3. Proof that (1c) implies (1a). Decompose $(\eta_t)_{t\geq 0}$ as $\eta_t = \eta_t^1 + \eta_t^2$ where $\eta_t^2 = \sum_{0<s\leq t} \Delta\eta_s 1_{\{|\Delta\eta_s|>1\}}$, that is, η^2 contains all jumps of magnitude larger than 1. By Step 1, $\int_0^\infty e^{-\xi_{s-}}\,\mathrm{d}\eta_s^1$ exists and by Step 2, $\int_0^\infty e^{-\xi_{s-}}\,\mathrm{d}\eta_s^2$ exists if (1c) is fulfilled.

Step 4. We first prove that (2c) implies (2a). As in the proof of Step 2 we may and will assume that η is a compound Poisson process. In this case $(L_t)_{t\geq 0}$ in (7) is a compound Poisson process as well. By definition of $(L_t)_{t\geq 0}$ and Step 2 we have

$$\sum_{0<s<\infty} e^{-\xi_s}|\Delta\eta_s| = \sum_{0<s<\infty} e^{-\xi_{s-}}\,|\Delta L_s| < \infty \quad \text{a.s.}$$

Since $(\xi_t,\eta_t)_{t\geq 0} \stackrel{\mathscr{D}}{=} (-\xi_{(-t)-}, -\eta_{(-t)-})_{t\geq 0}$ it follows that

$$\sum_{0<s<\infty} e^{-\xi_s}|\Delta\eta_s| \stackrel{\mathscr{D}}{=} \sum_{-\infty<s<0} e^{\xi_{s-}}|\Delta\eta_s|.$$

Thus, the right-hand side is finite a.s., implying that the integral $\int_{-\infty}^0 e^{\xi_{s-}}\,\mathrm{d}\eta_s$ exists and equals $\sum_{-\infty<s<0} e^{\xi_{s-}}\Delta\eta_s$.

Finally, if (1)(b) exists then the condition in (10) is satisfied implying by (1) that $\int_0^\infty e^{-\xi_{s-}}\,\mathrm{d}L_s$ exists. From (8) it follows that

$$\int_{-\infty}^0 e^{\xi_{s-}}\,d\eta_s = \lim_{t\to\infty}\int_{-t}^0 e^{\xi_{s-}}\,d\eta_s \stackrel{\mathscr{D}}{=} \lim_{t\to\infty}\int_0^t e^{-\xi_{s-}}\,\mathrm{d}L_s = \int_0^\infty e^{-\xi_{s-}}\,\mathrm{d}L_s$$

where the first and third equality signs hold a.s. □

Next we use the above theorem to study generalized Ornstein–Uhlenbeck processes. For this, consider a bivariate Lévy process $(U,\eta) = (U_t,\eta_t)_{t\in\mathbb{R}}$ with $U_0 = \eta_0 = 0$ and assume that none of them are identically equal to 0. Assume in addition that $m_U(\{-1\}) = 0$, meaning that U has no jumps of size -1. Following [2], Sect. 5.2, we introduce an extended filtration $(\mathscr{F}_t^{(U,\eta),\mathrm{ex}})_{t\in\mathbb{R}}$ which is defined as $\mathscr{F}_t^{(U,\eta),\mathrm{ex}} = \mathscr{F}_t^{(U,\eta)}$ for $t \geq 0$ and

$$\mathscr{F}_t^{(U,\eta),\mathrm{ex}} = \mathscr{F}_t^{(U,\eta)} \vee \sigma(\mu^{(U,\eta)}((t,0]\times A) : A \in \mathscr{B}(\mathbb{R}^2)) \quad \text{for } t < 0,$$

where $\mu^{(U,\eta)}$ is the jump measure of (U,η). By [2], Theorem 5.3, η and U are increment semimartingales in the extended filtration, ensuring a well-defined integration theory with respect to these processes.

Since the index set is $\mathbb{R}$ rather than $\mathbb{R}_+$, we define the stochastic exponential of U, $\mathscr{E}(U) = (\mathscr{E}(U)_t)_{t\in\mathbb{R}}$, as the càdlàg process satisfying $\mathscr{E}(U)_0 = 1$ and

$$\frac{\mathscr{E}(U)_t}{\mathscr{E}(U)_s} = e^{(U_t-U_s)-\frac{(t-s)}{2}\sigma_U^2} \prod_{s<u\leq t} (1+\Delta U_u)e^{-\Delta U_u} \quad \text{for } s<t. \tag{15}$$

Put differently, $\mathscr{E}(U)$ is given by

$$\mathscr{E}(U)_t = \begin{cases} e^{U_t-\frac{t}{2}\sigma_U^2} \prod_{0<u\leq t}(1+\Delta U_u)e^{-\Delta U_u} & \text{for } t\geq 0, \\ e^{-U_t+\frac{t}{2}\sigma_U^2} \prod_{t<u\leq 0}(1+\Delta U_u)e^{-\Delta U_u} & \text{for } t\leq 0. \end{cases} \tag{16}$$

Equation (15) shows that for $s\in\mathbb{R}$, $(\mathscr{E}(U)_{t+s}/\mathscr{E}(U)_s)_{t\geq 0}$, is the usual stochastic exponential (with index set $\mathbb{R}_+$) of $(U_{t+s}-U_s)_{t\geq 0}$, cf. [12], II.8.

To describe $\mathscr{E}(U)$ in a more convenient way we follow [4] and introduce two important auxiliary processes $N = (N_t)_{t\in\mathbb{R}}$ and $\xi = (\xi_t)_{t\in\mathbb{R}}$ as follows. Let $N_0 = \xi_0 = 0$ and

$$N_t - N_s = \mu^U((s,t]\times(-\infty,-1]) \quad \text{for } s<t \tag{17}$$

$$\xi_t - \xi_s = -(U_t-U_s) + \frac{(t-s)}{2}\sigma_U^2 + \sum_{s<u\leq t}[\Delta U_u - \log|1+\Delta U_u|] \quad \text{for } s<t. \tag{18}$$

These are essentially the definitions given in [4] except that the index set is $\mathbb{R}$ rather than $\mathbb{R}_+$, and our ξ corresponds to the process called $\widehat{U}$ there. As noted in [4], $N_t - N_s$ is the number of jumps in U of size less than -1 on the interval $(s,t]$. Moreover, N and ξ are both Lévy processes.

Lemma 1. *The process* $(\xi_t, N_t)_{t\in\mathbb{R}}$ *is* $(\mathscr{F}_t^{(U,\eta),\mathrm{ex}})_{t\in\mathbb{R}}$*-adapted.*

Proof. For $t\geq 0$ we let $s=0$ in (17) and (18), which trivially shows that (N_t,ξ_t) is $\mathscr{F}_t^U$-measurable and hence also $\mathscr{F}_t^{(U,\eta),\mathrm{ex}}$-measurable. For $t<0$ note that (use $t=0$ and $s=t$ in (17)) $-N_t = \mu^U((t,0]\times(-\infty,-1])$ implying that N_t is $\mathscr{F}_t^{(U,\eta),\mathrm{ex}}$-measurable by definition of this σ-field. Moreover, by a standard argument,

$$\int_{(t,0]\times\mathbb{R}} \phi(x)\,\mu^U(\mathrm{d}u\times\mathrm{d}x) \tag{19}$$

is $\mathscr{F}_t^{(U,\eta),\mathrm{ex}}$-measurable for all measurable $\phi:\mathbb{R}\to\mathbb{R}$ for which (19) exists. In particular $\sum_{t<u\leq 0}[\Delta U_u - \log|1+\Delta U_u|]$ is measurable, implying by (18) that ξ_t is measurable with respect to $\mathscr{F}_t^{(U,\eta),\mathrm{ex}}$. □

The importance of N and ξ is due to the fact that $\mathscr{E}(U)$ is given as

$$\frac{\mathscr{E}(U)_t}{\mathscr{E}(U)_s} = (-1)^{N_t - N_s} e^{-(\xi_t - \xi_s)} \quad \text{for } s < t. \tag{20}$$

By Lemma 1 this shows that $\mathscr{E}(U)$ is $(\mathscr{F}_t^{(U,\eta),\mathrm{ex}})_{t\in\mathbb{R}}$-adapted. When U does not have jumps of size less than -1, we have

$$\frac{\mathscr{E}(U)_t}{\mathscr{E}(U)_s} = e^{-(\xi_t - \xi_s)} \quad \text{for } s < t. \tag{21}$$

In this case [12], II.8, shows that for all $s \in \mathbb{R}$ and $t \geq 0$, $U_{t+s} - U_s = \mathscr{L}\mathrm{og}(e^{-(\xi_{\cdot+s} - \xi_s)})_t$, where $\mathscr{L}\mathrm{og}$ denotes the stochastic logarithm.

Finally, we need the process $L^* = (L_t^*)_{t\in\mathbb{R}}$ defined as $L_0^* = 0$ and

$$\begin{aligned} L_t^* - L_s^* &= \eta_t - \eta_s + ([\eta, U]_t - [\eta, U]_s) & (22)\\ &= \eta_t - \eta_s + \sum_{s<u\leq t} \Delta U_u \Delta\eta_u + (t-s)\sigma_{U,\eta} \quad \text{for } s < t. & (23) \end{aligned}$$

It follows from (18) that when $\mathscr{E}(U_t) = e^{-\xi_t}$ for $t \in \mathbb{R}$ (that is, U has no jumps of size less than -1), then $L_t^* = L_t$, $t \geq 0$, where the latter is defined in (7). Clearly, L^* is determined by (U, η). Conversely, η is determined by (U, L^*) since

$$\eta_t - \eta_s = L_t^* - L_s^* - \sum_{s<u\leq t} \frac{\Delta U_u \Delta L_u^*}{1 + \Delta U_u} - (t-s)\sigma_{U,L^*} \quad \text{for } s < t.$$

Note that since L^* differs from η only by a term which is of bounded variation on compacts and η is an increment semimartingale in the extended filtration, so is L^*. Similarly it follows that $(L_t)_{t\geq 0}$ is a semimartingale in $(\mathscr{F}_t^{(U,\eta)})_{t\geq 0}$.

In the following we consider càdlàg processes $V = (V_t)_{t\in\mathbb{R}}$ satisfying

$$\begin{aligned} V_t &= \frac{\mathscr{E}(U)_t}{\mathscr{E}(U)_s}\left(V_s + \int_s^t \frac{\mathscr{E}(U)_s}{\mathscr{E}(U)_{u-}}\,\mathrm{d}\eta_u\right) & (24)\\ &= \frac{\mathscr{E}(U)_t}{\mathscr{E}(U)_s} V_s + \mathscr{E}(U)_t \int_s^t [\mathscr{E}(U)_{u-}]^{-1}\,\mathrm{d}\eta_u \quad \text{for } s < t. & (25) \end{aligned}$$

Remark 2. Assume $V = (V_t)_{t\in\mathbb{R}}$ is càdlàg and $(\mathscr{F}_t^{(U,\eta),\mathrm{ex}})_{t\in\mathbb{R}}$-adapted. Then V is given by (24) if and only if it satisfies the linear stochastic differential equation

$$V_t = V_s + (L_t^* - L_s^*) + \int_s^t V_{u-}\,\mathrm{d}U_u \quad \text{for } s < t. \tag{26}$$

For a proof, see [4], Proposition 3.2, or [10], Theorem VI(6.8). A detailed study of stationary solutions to (26), including the nasty case $m_U(\{-1\}) > 0$, can be found in [4].

In the case when $\mathscr{E}(U_t) = e^{-\xi_t}$ for $t \in \mathbb{R}$, (24) and (25) reduce to so-called *generalized Ornstein–Uhlenbeck processes*:

$$V_t = e^{-(\xi_t - \xi_s)} \left(V_s + \int_s^t e^{\xi_{u-} - \xi_s} \, \mathrm{d}\eta_u \right) \tag{27}$$

$$= e^{-(\xi_t - \xi_s)} V_s + e^{-\xi_t} \int_s^t e^{\xi_{u-}} \, \mathrm{d}\eta_u \quad \text{for } s < t. \tag{28}$$

Theorem 2. *(1) Assume $\xi_t \to \infty$ as $t \to \infty$ a.s. The integral $\int_{-\infty}^0 \mathscr{E}(U)_{u-}^{-1} \, \mathrm{d}\eta_u$ exists if and only if $\int_{-\infty}^0 e^{\xi_{u-}} \, \mathrm{d}\eta_u$ exists. We have $\int_{-\infty}^0 \mathscr{E}(U)_{u-}^{-1} \, \mathrm{d}\eta_u \overset{\mathscr{D}}{=} \int_0^\infty \mathscr{E}(U)_{u-} \, \mathrm{d}L_u^*$ in case of existence. Moreover, there is a stationary càdlàg process $V = (V_t)_{t \in \mathbb{R}}$ satisfying (24) if and only if $\int_{-\infty}^0 \mathscr{E}(U)_{u-}^{-1} \, \mathrm{d}\eta_u$ exists. In this case $V = (V_t)_{t \in \mathbb{R}}$ is uniquely determined as*

$$V_t = \mathscr{E}(U)_t \int_{-\infty}^t \mathscr{E}(U)_{u-}^{-1} \, \mathrm{d}\eta_u, \quad t \in \mathbb{R}. \tag{29}$$

(2) Assume $\xi_t \to -\infty$ as $t \to \infty$. The integrals $\int_0^\infty \mathscr{E}(U)_{u-}^{-1} \, \mathrm{d}\eta_u$ and $\int_0^\infty e^{\xi_{u-}} \, \mathrm{d}\eta_u$ exist at the same time. There is a stationary càdlàg process $V = (V_t)_{t \in \mathbb{R}}$ satisfying (24) if and only if $\int_0^\infty \mathscr{E}(U)_{u-}^{-1} \, \mathrm{d}\eta_u$ exists. In this case $V = (V_t)_{t \in \mathbb{R}}$ is uniquely determined as

$$V_t = -\mathscr{E}(U)_t \int_t^\infty \mathscr{E}(U)_{u-}^{-1} \, \mathrm{d}\eta_u, \quad t \in \mathbb{R}. \tag{30}$$

Recall that necessary and sufficient conditions that the integrals $\int_0^\infty e^{\xi_{u-}} \, \mathrm{d}\eta_u$ and $\int_{-\infty}^0 e^{\xi_{u-}} \, \mathrm{d}\eta_u$ exist are given in Theorem 1. In particular, $\int_{-\infty}^0 e^{\xi_{u-}} \, \mathrm{d}\eta_u$ and $\int_0^\infty e^{-\xi_{u-}} \, \mathrm{d}L_u$ exist at the same time. This is also equivalent to the existence of $\int_0^\infty e^{-\xi_{u-}} \, \mathrm{d}L_u^*$; indeed, it is easily verified that $|\Delta L_t| = |\Delta L_t^*|$ for all $t > 0$, and thus the condition in Theorem 1(1c) with $\eta = L$ is equivalent to the one with $\eta = L^*$. Moreover, as in the proof of the first part of (1) below it follows that $\int_0^\infty e^{-\xi_{u-}} \, \mathrm{d}L_u^*$ and $\int_0^\infty \mathscr{E}(U)_{u-} \, \mathrm{d}L_u^*$ exist at the same time.

The above conditions for existence of V are also given in [4], Theorem 2.1, and so is the representation (30); thus, (2) is completely contained in [4] (except that we use proper integrals) but it is restated here for completeness. When $\xi_t = \lambda t$ for some non-zero constant λ, (27) and (28) simplify to a usual Ornstein–Uhlenbeck process; see e.g. [1] and [16]. In this case, (29) and (30) are well known representations of stationary Ornstein–Uhlenbeck processes cf. e.g. [16], Theorem 55.

The case when ξ_t does not converge to $\pm\infty$ as $t \to \infty$ is treated in Theorem 2.1 of [4].

Proof. (1) Since $\mathscr{E}(U)_{u-}^{-1}$ and $e^{\xi_{u-}}$ only differ by a factor of absolute value 1 the two proper integrals exist at the same time. The identity in distribution follows as in the proof of the last assertion of Theorem 1(2), where instead of (8) we use that for $t \geq 0$,

$$\int_{-t}^{0} \mathscr{E}(U)_{u-}^{-1}\, \mathrm{d}\eta_u \overset{\mathscr{D}}{=} \mathscr{E}(U)_t \int_0^t \mathscr{E}(U)_{u-}^{-1}\, \mathrm{d}\eta_u \overset{\mathscr{D}}{=} \int_0^t \mathscr{E}(U)_{u-}\, \mathrm{d}L_u^*, \qquad (31)$$

where the first equality follows as in the proof of (8) and the second comes from Lemma 3.1 in [4].

Assume V is stationary and satisfies (24). Letting $s \to -\infty$ and using that $\xi_s \to -\infty$ a.s. it follows from (25) and (20) that $\int_s^0 \mathscr{E}(U)_{u-}^{-1}\, \mathrm{d}\eta_u$ converges in distribution. From (31) it follows that $\int_0^t \mathscr{E}(U)_{u-}\, \mathrm{d}L_u^*$ converges in distribution as $t \to \infty$. Theorem 3.6 in [4] shows that the condition in Theorem 1(2c) (with L replaced by L^*) is satisfied, implying that $\int_{-\infty}^0 \mathscr{E}(U)_{u-}^{-1}\, \mathrm{d}\eta_u$ exists.

Conversely, assuming that $\int_{-\infty}^0 \mathscr{E}(U)_{u-}^{-1}\, \mathrm{d}\eta_u$ exists and defining V by (29) it is easily seen that (24) is satisfied. Moreover, since V_t is given a.s. as

$$V_t = \lim_{h\to\infty} \mathscr{E}(U)_t \int_{t-h}^{t} \mathscr{E}(U)_{u-}^{-1}\, \mathrm{d}\eta_u = \lim_{h\to\infty} \frac{\mathscr{E}(U)_t}{\mathscr{E}(U)_{t-h}} \int_{t-h}^{t} \frac{\mathscr{E}(U)_{t-h}}{\mathscr{E}(U)_{u-}}\, \mathrm{d}\eta_u \qquad (32)$$

and the distribution of the right-hand side does not depend on t, the distribution of V_t is also independent of t. The variable V_t is moreover determined by $(\eta_t - \eta_u, U_t - U_u)_{u \leq t}$ and is hence in particular independent of $(\eta_{u+t} - \eta_t, U_{u+t} - U_t)_{u\geq 0}$. From (24) it follows that V is stationary.

The proof of (2) is similar except that in (24) and (25) we fix s, rearrange terms to isolate V_s, and let $t \to \infty$. □

In the next theorem we study integrability properties and the so-called Wold-Karhunen representation of generalized Ornstein–Uhlenbeck processes. As above we consider the bivariate Lévy process (U, η) as well as ξ and L^* defined in (18) and (22). From now on we assume $m_U((-\infty, -1]) = 0$, that is, U has no jumps of size -1 or smaller. We will thus not need the process N. Whenever U is integrable let $\lambda := E[-U_1]$ and define $\overline{U} = (\overline{U}_t)_{t\in\mathbb{R}}$ as $\overline{U}_0 = 0$ and $\overline{U}_t - \overline{U}_s = U_t - U_s + \lambda(t-s)$ for $s < t$.

Theorem 3. *Assume* $m_U((-\infty, -1]) = 0$.

(1) For $r > 0$ *we have* $U_1 \in L^r(P)$ *if and only if* $E[e^{-r\xi_1}] < \infty$. *When* $U_1 \in L^1(P)$ *we have*

$$\exp(E[U_1]) = E[e^{-\xi_1}] = \exp\Big[-\gamma_\xi + \tfrac{1}{2}\sigma_\xi^2 + \int_{\mathbb{R}} (e^{-x} - 1 + x1_{\{|x|\leq 1\}})\, m_\xi(\mathrm{d}x)\Big]. \qquad (33)$$

(2) Assume $L_1^* \in L^2(P)$ *and* $E[e^{-2\xi_1}] < 1$. *Then* $\xi_t \to \infty$ *as* $t \to \infty$ *a.s and the integral* $\int_0^\infty e^{-\xi_t}\, \mathrm{d}L_t^*$ *exists and is in* $L^2(P)$. *Moreover,* λ *is strictly positive, the generalized Ornstein–Uhlenbeck process satisfying* (27) *exists and is square integrable, and*

$$V_t = \int_{-\infty}^{t} e^{-\lambda(t-s)}\,\mathrm{d}L_s^* + \int_{-\infty}^{t} e^{-\lambda(t-s)} V_{s-}\,\mathrm{d}\overline{U}_s, \quad t \in \mathbb{R}. \tag{34}$$

Remark 3. Assume $L_1^* \in L^2(P)$, $E[e^{-2\xi_1}] < 1$ and $E[L_1^*] = 0$. By (34),

$$V_t = \int_{-\infty}^{t} e^{-\lambda(t-s)}\,\mathrm{d}\Xi_s, \quad t \in \mathbb{R}, \tag{35}$$

where, for $s < t$, $\Xi_t - \Xi_s = L_t^* - L_s^* - \int_s^t V_{u-}\,\mathrm{d}\overline{U}_u$ and $\Xi_0 = 0$. The process Ξ is square integrable with zero mean and stationary orthogonal increments, implying that (35) is the Wold-Karhunen representation of V. To verify this fix $s \in \mathbb{R}$. It was noted in the proof of Theorem 2 that V_s is independent of $(L_{s+t}^* - L_s^*, U_{s+t} - U_s)_{t\geq 0}$. Using that the distribution of V_s as well as of $(L_{s+t}^* - L_s^*, U_{t+s} - U_s)_{t\geq 0}$ does not depend on s, it follows from (27) that Ξ has stationary increments. Since L^* as well as $\overline{U}$ are zero mean square integrable Lévy processes and V is square integrable and stationary it follows by definition of Ξ that $(\Xi_{t+s} - \Xi_s)_{t\geq 0}$ is a square integrable martingale in the filtration generated by V_s and $(L_{s+t}^* - L_s^*, U_{t+s} - U_s)_{t\geq 0}$. In particular this implies that Ξ has zero mean and orthogonal increments.

For results related to some of the integrability properties above see Proposition 3.1 and Theorem 3.3 in [3].

Proof. (1) Since $U_t = \mathscr{L}\mathrm{og}(-\xi_t)$ for $t > 0$, [12], Corollary II.8.16, shows that

$$\int_{u>1} u^r\, m_U(\mathrm{d}u) = \int_{e^{rx}-1>1} (e^{rx} - 1)\, m_{-\xi}(\mathrm{d}x) = \int_{x>\frac{\log 2}{r}} (e^{rx} - 1)\, m_{-\xi}(\mathrm{d}x).$$

Thus, the left-hand side is finite if and only if $\int_{|x|>1} e^{rx}\, m_{-\xi}(\mathrm{d}x)$ is finite. On the other hand, since $m_U((-\infty, -1]) = 0$, [17], Corollary 25.8, shows that finiteness of the left-hand side is equivalent to $U_1 \in L^r(P)$, and similarly finiteness of $\int_{|x|>1} e^{rx}\, m_{-\xi}(\mathrm{d}x)$ is equivalent to $E[e^{-r\xi_1}] < \infty$.

Now assume $E[e^{-\xi_1}] < \infty$. The second equality in (33) follows from [17], Theorem 25.17. Moreover, since $U_t = \mathscr{L}\mathrm{og}(-\xi_t)$ for $t > 0$, [12], Theorem II.8.10, implies that

$$U_t = -\xi_t + \tfrac{t}{2}\sigma_\xi^2 + [(e^{-x} - 1 + x) * \mu^\xi]_t, \quad t \geq 0.$$

Recalling the Lévy-Itô decomposition of ξ:

$$\xi_t = x1_{\{|x|\leq 1\}} * (\mu^\xi - \nu^\xi)_t + [x1_{\{|x|>1\}} * \mu^\xi]_t + t\gamma_\xi + G_t, \quad t \geq 0,$$

where G denotes the (mean zero) Gaussian component of ξ and $\nu^\xi(\mathrm{d}t \times \mathrm{d}x) = m_\xi(\mathrm{d}x)\,\mathrm{d}t$, it follows that

$$U_t = -x1_{\{|x|\leq 1\}} * (\mu^\xi - \nu^\xi)_t - t\gamma_\xi - G_t + \tfrac{t}{2}\sigma_\xi^2 + [(e^{-x} - 1 + x1_{\{|x|\leq 1\}}) * \mu^\xi]_t, \quad t \geq 0.$$

All terms on the right-hand side have finite mean, and the first and third term have mean zero, implying

$$E[U_1] = -\gamma_\xi + \tfrac{1}{2}\sigma_\xi^2 + \int_{\mathbb{R}} (e^{-x} - 1 + x1_{\{|x|\le 1\}})\, m_\xi(\mathrm{d}x).$$

In particular this gives (33).

(2) Since $E[e^{-\xi_1}] \le \sqrt{E[e^{-2\xi_1}]} < 1$ it follows from (33) that $\lambda > 0$. It is shown in [14], Proposition 4.1, that $\xi_t \to \infty$ a.s. Recall ([17], Theorem 25.17) that $E[e^{-2\xi_t}] = (E[e^{-2\xi_1}])^t$ for $t \ge 0$. Using that L_1^* is square integrable we can decompose $(L_t^*)_{t\ge 0}$ as $L_t^* = M_t + c_1 t$ where c_1 is a constant and $(M_t)_{t\ge 0}$ is a square integrable martingale with $\langle M\rangle_t = c_2 t$ for some $c_2 \ge 0$. Since

$$E\Big[\int_0^\infty e^{-2\xi_t}\,\mathrm{d}\langle M\rangle_t\Big] = \int_0^\infty (E[e^{-2\xi_1}])^t\,\mathrm{d}(c_2 t) < \infty$$

the integral $\int_0^\infty e^{-\xi_t}\,\mathrm{d}M_t$ exists and is square integrable. Moreover, since ξ_s is independent of $\xi_u - \xi_s$ for all $0 \le s < u$ and $E[e^{-\xi_1}], E[e^{-2\xi_1}] < 1$, we get

$$\begin{aligned} E\Big[\Big(\int_0^\infty e^{-\xi_s}\,\mathrm{d}s\Big)^2\Big] &= 2\int_0^\infty \int_s^\infty E[e^{-\xi_s-\xi_u}]\,\mathrm{d}u\,\mathrm{d}s \\ &= 2\int_0^\infty \int_s^\infty E[e^{-2\xi_s-(\xi_u-\xi_s)}]\,\mathrm{d}u\,\mathrm{d}s \\ &= 2\int_0^\infty \int_s^\infty (E[e^{-2\xi_1}])^s (E[e^{-\xi_1}])^{u-s}\,\mathrm{d}u\,\mathrm{d}s < \infty \end{aligned}$$

and hence the integrals $\int_0^\infty e^{-\xi_s}\,\mathrm{d}(c_1 s)$ and $\int_0^\infty e^{-\xi_s}\,\mathrm{d}L_s^*$ exist and are square integrable. By Theorem 2 and the remarks following it the generalized Ornstein–Uhlenbeck process $V = (V_t)_{t\in\mathbb{R}}$ exists and is square integrable. From (26) we have

$$V_t = V_s + [(L_t^* - L_s^*) + \int_s^t V_{u-}\,\mathrm{d}\overline{U}_u] - \lambda \int_s^t V_u\,\mathrm{d}u \quad \text{for } s < t.$$

Thus, by [10], Theoreme VI(6.8),

$$V_t = e^{-\lambda(t-s)} V_s + \int_s^t e^{-\lambda(t-u)}\,\mathrm{d}L_u^* + \int_s^t e^{-\lambda(t-u)} V_{u-}\,\mathrm{d}\overline{U}_u \quad \text{for } s < t. \tag{36}$$

Since V is a stationary square integrable process, λ is positive and $\overline{U}$ and L^* are square integrable, it follows from Remark 1 that $(V_{u-}e^{\lambda u}1_{\{u\le t\}})_{u\in\mathbb{R}}$ is integrable with respect to $\overline{U}$ and $(e^{\lambda u}1_{\{u\le t\}})_{u\in\mathbb{R}}$ is integrable with respect to L^*. Letting $s \to -\infty$ in (36) it follows that

$$V_t = \int_{-\infty}^t e^{-\lambda(t-u)}\,\mathrm{d}L_u^* + \int_{-\infty}^t e^{-\lambda(t-u)} V_{u-}\,\mathrm{d}\overline{U}_u \quad \text{for } t \in \mathbb{R}.$$

□

References

1. O.E. Barndorff-Nielsen, N. Shephard, Non-Gaussian Ornstein-Uhlenbeck-based models and some of their uses in financial economics. J. R. Stat. Soc. Ser. B Stat. Methodol. **63**(2), 167–241 (2001)
2. A. Basse-O'Connor, S.-E. Graversen, J. Pedersen, A unified approach to stochastic integration on the real line. Thiele Research report 2010-08 http://www.imf.au.dk/publication/publid/880, 2010
3. A. Behme, Distributional properties of solutions of $dV_t = V_{t-}dU_t + dL_t$ with Lévy Noise. Adv. Appl. Prob. **43**, 688–711 (2011)
4. A. Behme, A. Lindner, R.A. Maller, Stationary solutions of the stochastic differential equation $dV_t = V_{t-}dU_t + dL_t$ with Lévy Noise. Stochast. Process. Appl. **121**(1), 91–108 (2011)
5. J. Bertoin, A. Lindner, R. Maller, On continuity properties of the law of integrals of Lévy processes. In *Séminaire de Probabilités XLI*. Lecture Notes in Mathematics, vol. 1934 (Springer, Berlin, 2008), pp. 137–159
6. P. Carmona, F. Petit, M. Yor, Exponential functionals of Lévy processes. In *Lévy Processes* (Birkhäuser Boston, Boston, MA, 2001), pp. 41–55
7. A. Cherny A. Shiryaev, On stochastic integrals up to infinity and predictable criteria for integrability. In *Séminaire de Probabilités XXXVIII*. Lecture Notes in Mathematics, vol. 1857 (Springer, Berlin, 2005), pp. 165–185
8. R.A. Doney, R.A. Maller, Stability and attraction to normality for Lévy processes at zero and at infinity. J. Theor. Probab. **15**(3), 751–792 (2002)
9. K.B. Erickson, R.A. Maller, Generalised Ornstein-Uhlenbeck processes and the convergence of Lévy integrals. In *Séminaire de Probabilités XXXVIII*. Lecture Notes in Mathematics, vol. 1857 (Springer, Berlin, 2005), pp. 70–94
10. J. Jacod, *Calcul Stochastique et Problèmes de Martingales*. Lecture Notes in Mathematics, vol. 714 (Springer, Berlin, 1979)
11. J. Jacod, P. Protter, Time reversal on Lévy processes. Ann. Probab. **16**(2), 620–641 (1988)
12. J. Jacod, A.N. Shiryaev, *Limit Theorems for Stochastic Processes, vol. 288 of Grundlehren der Mathematischen Wissenschaften [Fundamental Principles of Mathematical Sciences]*, 2nd edn. (Springer, Berlin (2003)
13. H. Kondo, M. Maejima, K. Sato, Some properties of exponential integrals of Lévy processes and examples. Electron. Commun. Probab. **11**, 291–303 (electronic) (2006)
14. A. Lindner, R. Maller, Lévy integrals and the stationarity of generalised Ornstein-Uhlenbeck processes. Stochast. Process. Appl. **115**(10), 1701–1722 (2005)
15. A. Lindner, K. Sato, Continuity properties and infinite divisibility of stationary distributions of some generalized Ornstein-Uhlenbeck processes. Ann. Probab. **37**(1), 250–274 (2009)
16. A. Rocha-Arteaga, K. Sato, *Topics in Infinitely Divisible Distributions and Lévy Processes, vol. 17 of Aportaciones Matemáticas: Investigación [Mathematical Contributions: Research]* (Sociedad Matemática Mexicana, México, 2003)
17. K. Sato, *Lévy Processes and Infinitely Divisible Distributions, vol. 68 of Cambridge Studies in Advanced Mathematics* (Cambridge University Press, Cambridge, 1999). Translated from the 1990 Japanese original, Revised by the author
18. M. Yor, *Exponential Functionals of Brownian Motion and Related Processes* (Springer Finance. Springer, Berlin, 2001). With an introductory chapter by Hélyette Geman, Chapters 1, 3, 4, 8 translated from the French by Stephen S. Wilson

Martingale Representations for Diffusion Processes and Backward Stochastic Differential Equations

Zhongmin Qian and Jiangang Ying

Abstract In this paper we explain that the natural filtration of a continuous Hunt process is continuous, and show that martingales over such a filtration are continuous. We further establish a martingale representation theorem for a class of continuous Hunt processes under certain technical conditions. In particular we establish the martingale representation theorem for the martingale parts of (reflecting) symmetric diffusion in a bounded domain with a continuous boundary. Together with an approach put forward in (Liang et al., Ann. Probab.), our martingale representation theorem is then applied to the study of initial and boundary problems for quasi-linear parabolic equations by using solutions to backward stochastic differential equations over the filtered probability space determined by reflecting diffusions in a bounded domain with only continuous boundary.

Keywords Backward SDE • Dirichlet form • Hunt process • Martingale • Natural filtration • Non-linear equations

AMS Classification: 60H10, 60H30, 60J45

Z. Qian (✉)
Mathematical Institute, University of Oxford, Oxford OX1 3LB, England
e-mail: qianz@maths.ox.ac.uk

J. Ying
Institute of Mathematics, Fudan University, Shanghai, China
e-mail: jgying@fudan.edu.cn

C. Donati-Martin et al. (eds.), *Séminaire de Probabilités XLIV*, Lecture Notes in Mathematics 2046, DOI 10.1007/978-3-642-27461-9_4,

1 Introduction

Let $B = (B_t)_{t\geq 0}$ be the Brownian motion in $\mathbb{R}^d$ started with an initial distribution μ. The natural filtration $(\mathscr{F}_t^{\mu})_{t\geq 0}$ (called the Brownian filtration, for a definition, see [3]) is continuous. More importantly the Brownian motion has the *martingale representation property*: any martingale over $(\Omega, \mathscr{F}, \mathscr{F}_t^{\mu}, P^{\mu})$ can be expressed as an Itô integral against Brownian motion. In particular all martingales on the filtered probability space $(\Omega, \mathscr{F}, \mathscr{F}_t^{\mu}, P^{\mu})$ are continuous.

The (predictable) martingale representation property of a family of martingales has been studied in a more extended setting, and several general results have been obtained. For example, Jacod and Yor [16] have discovered the equivalence between the martingale representation property and the extremal property of martingale measures. Jacod [15] also obtained the martingale representation property in terms of the uniqueness of some martingale problems, and further present criteria in terms of predictable characteristics. These results have greatly illuminated the subject matter. When applied to specific situations, further work and indeed hard estimates are often required. As a matter of fact, we still have very limited examples of martingales and filtrations which possess martingale representation property (see, e.g. [1, 15, 26, 33]).

The renewed interest in recent years in the martingale presentation property has been motivated not only by its own right, but also by its important applications in the mathematical finance [11, 31], backward stochastic differential equations and their applications in some non-linear partial differential equations, see also for example [1, 5, 17, 23, 27, 32] and the reference therein.

An intimate question is the continuity of natural filtrations generated by semi-martingales and diffusion processes. A great knowledge about them has been obtained in the past. For example, a complete characterization of the natural filtrations generated by simple jump processes and Lévy processes in terms of their sample paths is known. Much information has been obtained for a class of Markov processes. We know, from the fundamental work by Blumenthal, Chung, Dynkin, Getoor, Hunt, Meyer etc., (see for example [4, 8, 25], in particular Hunt [14]), that the natural filtration of a Feller process is right continuous and quasi-left continuous. On the other hand, to the best knowledge of the present authors, there are no general conditions in literature to guarantee the martingales over the natural filtration of a Markov process to be continuous. A reasonable conjecture is that the natural filtration of a diffusion process (a continuous strong Markov process) should be continuous, so are the martingales over the natural filtration. Such a result is plausible but remains to prove.

In this paper, we show that all martingales over the natural filtration of a *continuous* Hunt process are continuous. Our proof follows a key idea originated from Blumenthal [3], formulated carefully in Meyer [25].

The main result of the present article is a martingale representation theorem for a class of continuous Hunt processes which satisfy a technical condition called *the Fukushima representation property* (FRP). The term FRP is so named, because for symmetric diffusions on a finite dimensional space, under a few regularity conditions, due to Fukushima's chain rule for energy measures, the FRP will be satisfied automatically. We are therefore able to provide a large class of diffusion processes which have the martingale representation property.

As a consequence of our main result, we establish the martingale representation theorem for symmetric diffusion processes on a domain, with Dirichlet or Neumann boundary condition. More precisely, let $D \subset \mathbb{R}^d$ be a bounded domain with a continuous boundary ∂D. Consider the symmetric diffusion in D with a formal infinitesimal generator

$$L = \frac{1}{2} \sum_{i,j=1}^{d} \frac{\partial}{\partial x^j} a^{ij}(x) \frac{\partial}{\partial x^i} \quad \text{in } D$$

subject to the Dirichlet or Neumann boundary condition (for precise meaning, see Sects. 4 and 5 below), where (a^{ij}) are only Borel measurable and satisfies the uniform elliptic condition. To make it clear, let

$$X = (\Omega, \mathscr{F}, \mathscr{F}_t, X_t, \theta_t, P^x).$$

be the symmetric diffusion associated with the Dirichlet form $(\mathscr{E}, \mathscr{F})$ where

$$\mathscr{E}(u,v) = \frac{1}{2} \int_D \sum_{i,j=1}^{d} a^{ij} \frac{\partial u}{\partial x^j} \frac{\partial v}{\partial x^i} dx$$

and the Dirichlet space $\mathscr{F} = H_0^1(D)$ or $H^1(D)$ depending on the Dirichlet or Neumann boundary condition. Here we use the same letter $\mathscr{F}$ to denote the filtration as well as the Dirichlet space: we hope it should be clear from the context which one $\mathscr{F}$ stands for. $\mathscr{F}$, $\mathscr{F}_t$ are the natural filtrations generated by the symmetric diffusion $(X_t)_{t \geq 0}$. It happens in this case that the coordinate functions $u^j(x) = x^j$ belong to the local Dirichlet space $\mathscr{F}_{\text{loc}}$, and $X_t = (X_t^1, \cdots, X_t^d)$ has the Fukushima's decomposition

$$X_t^j - X_0^j = M_t^j + A_t^j \quad P^x\text{-a.e. } j = 1, \cdots, d$$

for all $x \in D$ (or $\overline{D}$ in the Neumann boundary condition case) except for a zero capacity set with respect to the Dirichlet form $(\mathscr{E}, \mathscr{F})$, where $M^1, \cdots, M^d$ are continuous martingales additive functionals and $A^1, \cdots, A^d$ are continuous additive functionals with zero energy. The following martingale representation theorem follows from our main result.

Theorem 1. *Under the above assumptions, for any initial distribution μ which has no charge on capacity zero sets, the family $(M^1, \cdots, M^d)$ of martingales over $(\Omega, \mathscr{F}^\mu, \mathscr{F}_t^\mu, P^\mu)$ has the martingale representation property: for any square integrable martingale $N = (N_t)_{t \geq 0}$ over $(\Omega, \mathscr{F}^\mu, \mathscr{F}_t^\mu, P^\mu)$ there are unique*

$(\mathscr{F}_t^\mu)$-predictable processes $F^1, \cdots, F^d$ such that

$$N_t - N_0 = \sum_{j=1}^{d} \int_0^t F_s^j dM_s^j.$$

We must point out a subtle difference for the martingale representation property used in this article to that used in the semimartingale theory, such as Jacod and Yor [16]. Here the filtration $\mathscr{F}^\mu$ is generated by the diffusion process, not necessarily coincide with the filtration generated by the family of martingale parts M^j, and therefore the integrands F^j are predictable processes with respect to the diffusion process.

Theorem 1 may be applied to the symmetric diffusions in $\mathbb{R}^d$ with Dirichlet form $(\mathscr{E}, \mathscr{F})$ where $\mathscr{F} = H_0^1(\mathbb{R}^d)$. This special case has been proved in [33] and [1].

In [33], Zheng has pointed out that the martingale part of symmetric diffusion $(X_t)_{t\geq 0}$ in $\mathbb{R}^d$ with infinitesimal generator being a uniform second order elliptic operator in divergence form has the martingale predictable representation property and described a proof based on the results on the Dirichlet process $p(t, X_t)$ obtained in Lyons and Zheng [22] and [21], where $p(t, x)$ is the probability density function of X_t under the stationary distribution. More precisely, Lyons and Zheng [21] have extended Fukushima's representation theorem for martingale additive functionals to a class of processes which has a form $f(t, X_t)$, where f has finite space-time energy. Their results in particular yield that $p(t, X_t)$ is a Dirichlet process in the sense of Föllmer [12], and its martingale part can be expressed as an Itô integral against $(M^1, \cdots, M^d)$, which, together with the Markov property, allows to show that for $\xi = f_1(X_{t_1}) \cdots f_n(X_{t_n})$ the conditional expectation $E(\xi|\mathscr{F}_t)$ is again a Dirichlet process which can be expressed as an Itô's integral against $(M^1, \cdots, M^d)$, where the expectation is taken against the stationary distribution $P^m(\cdot) = \int_{\mathbb{R}^d} P^x(\cdot)dx$. A routine procedure based on the Doob's maximal inequality allows to prove the martingale representation theorem for the symmetric diffusion in $\mathbb{R}^d$ with the generator L an elliptic operator of second order. Apparently not knowing the work [33], in an independent work Bally, Pardoux and Stoica [1], among other things, a detailed proof has been provided.

The technical difficulty with the proof described above lies in the fact that even for a smooth function f, $f(X_t)$ may not be a semimartingale, so that Itô's calculus can not be applied. Instead of considering random variables with product form such as $\xi = f_1(X_{t_1}) \cdots f_n(X_{t_n})$ which linearly span a vector space dense in $L^p(\Omega, \mathscr{F}, P^\mu)$, we utilize a linear vector $\mathscr{C}$ spanned by those random variables which have a product form $\xi = \xi_1 \cdots \xi_n$ with

$$\xi_j = \int_0^\infty e^{-\alpha_j t} f_j(X_t)dt$$

where f_j are bounded Borel measurable functions, and $\alpha_j > 0$ for $j = 1, \cdots, n$. According to Meyer [25], $\mathscr{C}$ is dense in $L^p(\Omega, \mathscr{F}, P^\mu)$. The important feature is that $U^{\alpha}f(X_t)$ is a semimartingale for any $\alpha > 0$ and a bounded Borel function f, where U^{α} is the resolvent of the transition semigroup. Moreover, in the symmetric case, $U^{\alpha}f$ always belongs to the Dirichlet space (when f is bounded and square integrable) and thus Fukushima's representation theorem for *martingale additive functionals* can be applied to extend the representation to any martingales.

The martingale representation theorem for the symmetric diffusion process (X_t) in a domain D allows us to study the following type of backward stochastic differential equation (BSDE)

$$dY_t = -f(t, Y_t, Z_t)dt + \sum_{j=1}^{d} Z_t^j dM_t^j$$

with a terminal condition that $Y_T = \xi \in L^2(\Omega, \mathscr{F}_T^{\mu}, P^{\mu})$, and thus gives a probability representation for weak solutions to the initial and boundary value problem for non-linear parabolic equations. The existence and uniqueness of solutions to the BSDE follows from exactly the same approach as for the Brownian motion case, which is the pioneering work in BSDE done by Pardoux and Peng in [27]. We however describe an approach put forward in [20], which allows us to devise an alternative probability representation for the initial and boundary problem of the corresponding semi-linear parabolic equation

$$\frac{\partial}{\partial t}u - \frac{1}{2}\sum_{i,j=1}^{d} \frac{\partial}{\partial x^j}a^{ij}\frac{\partial}{\partial x^i}u + f(t, u, \nabla u) = 0$$

subject to $\frac{\partial}{\partial \nu}\big|_{\partial D} u(t,\cdot) = 0$ in a bounded domain with only continuous boundary. For related topics of BSDEs associated with general elliptic operators, we mention in particular the papers Lejay [19], Bally, Pardoux and Stoica [1], Rozkosz [30] and [28] and the references therein.

The paper is organized as follows. In Sect. 2, we develop further an idea put forward in Meyer [25], and show both the natural filtrations and martingales over the natural filtrations for a continuous Hunt process are continuous. This result is hardly new but it seems not appear in the literature yet. In order to prove this result, we devise an important while elementary formula for $E^{\mu}(\xi|\mathscr{F}_t^{\mu})$, which may be considered as a refined version of a classical formula devised firstly by Hunt and Blumenthal for potentials and multiple potential case by Meyer. In Sect. 3, we establish the main result of the paper: a martingale representation theorem for a continuous Hunt process under technical assumptions called the Fukushima representation property, and give some examples in which our result may apply. In Sect. 4, we outline the existence and uniqueness of solutions to backward stochastic differential equations over the natural filtered probability space over a reflecting symmetric diffusion in a bounded domain with non-smooth diffusion coefficients and non-smooth boundary, and finally we apply the theory of BSDE to the study of the initial and (Neumann) boundary problem of a non-linear parabolic equation in a bounded domain with only continuous boundary. We believe these results are new even for reflecting Brownian motion in a domain with non-smooth boundary.

The first version of this paper was posted on ArXiv a year ago. Subsequently many colleagues have kindly informed the authors that it is known to experts that the subspace of potential martingales is dense in the martingale space over a Hunt process, a result goes back to Kunita and Watanabe [18]. A proof for symmetric diffusions may be found in Fukushima et al. [13]. By using this fact, we can prove Theorem 2 without using the multiple potential formula (6) which is the main technical step in our proof. The first author would like to thank Professor Le Jan and Professor Emery for the discussions on this topic and the literature about filtrations and martingale representation property.

2 Martingales over the Filtrations of Continuous Hunt Processes

Consider a Markov process $(\Omega, \mathscr{F}^0, \mathscr{F}_t^0, X_t, \theta_t, P^x)$ in a state space $E' = E \cup \{\partial\}$, where E is a locally compact separable metric space E, with transition probability function $\{P_t(x,\cdot) : t \geq 0\}$, i.e.,

$$E^x\{f(X_{t+s})|\mathscr{F}_s^0\} = \int_E f(z)P_t(X_s, dz)\,. \tag{1}$$

In (1), E^x on the left-hand side stands for the (conditional) expectation with respect to the probability measure P^x, and the right-hand side may be abbreviated as $P_t f(X_s)$ where

$$P_t f(x) = \int_E f(z)P_t(x, dz)$$

which is well defined for a bounded or non-negative Borel measurable function f. The family of kernels $(P_t)_{t>0}$ is called the transition semigroup associated with the Markov process $(X_t)_{t\geq 0}$. Without specification, $(\mathscr{F}_t^0)_{t\geq 0}$ is the filtration generated by $(X_t)_{t\geq 0}$, that is $\mathscr{F}_t^0 = \sigma\{X_s : s \leq t\}$ for $t \geq 0$ and $\mathscr{F}^0 = \sigma\{X_s : s \geq 0\}$.

For a σ-finite measure μ on $(E, \mathscr{B}(E))$ (where $\mathscr{B}(M)$ always represents the Borel σ-algebra on a topological space M)

$$P^\mu(\Lambda) = \int_E P^x(\Lambda)\mu(dx), \quad \Lambda \in \mathscr{F}^0,$$

defines a measure on $(\Omega, \mathscr{F}^0)$. If μ is a probability, then $(X_t)_{t\geq 0}$ is Markovian under P^μ with transition semigroup $(P_t)_{t>0}$ and initial distribution μ, in the sense that

$$E^\mu\{f(X_{t+s})|\mathscr{F}_s^0\} = \int_E f(z)P_t(X_s, dz) \quad P^\mu\text{-a.e.}$$

and $P^{\mu}\{X_0 \in A\} = \mu(A)$ for any $A \in \mathscr{B}(E)$, where E^{μ} is the (conditional) expectation against the probability P^{μ}.

Denote by $\mathscr{P}(E)$ the space of all probability measures on $(E, \mathscr{B}(E))$. If $\mu \in \mathscr{P}(E)$, $\mathscr{F}^{\mu}$ denotes the completion of $\mathscr{F}^0$ under P^{μ}, and $\mathscr{F}_t^{\mu}$ is the smallest σ-algebra containing $\mathscr{F}_t^0$ and all sets in $\mathscr{F}^{\mu}$ with zero probability. $(\mathscr{F}_t^{\mu})_{t\geq 0}$ is called the natural filtration of the Markov process $(X_t)_{t\geq 0}$ with initial distribution μ. Let

$$\mathscr{F}_t = \bigcap_{\mu \in \mathscr{P}(E)} \mathscr{F}_t^{\mu}$$

and $(\mathscr{F}_t)_{t\geq 0}$ is called the natural filtration determined by the Markov process

$$X = (\Omega, \mathscr{F}^0, \mathscr{F}_t^0, X_t, \theta_t, P^x).$$

If $(\mathscr{G}_t)_{t\geq 0}$ is a filtration, i.e. an increasing family of σ-algebras on a common sample space, then $\mathscr{G}_{t+} = \bigcap_{s>t} \mathscr{G}_s$ for $t \geq 0$ and $\mathscr{G}_{t-} = \sigma\{\mathscr{G}_s : s < t\}$ for $t > 0$. The filtration is called right (resp. left) continuous if $\mathscr{G}_{t+} = \mathscr{G}_t$ for all $t \geq 0$ (resp. $\mathscr{G}_{t-} = \mathscr{G}_t$ for all $t > 0$). The sample function properties of a Markov process $(X_t)_{t\geq 0}$ and the continuity properties of its natural filtration had been studied by Blumenthal, Dynkin, Getoor, Hunt, Meyer etc. The fundamental results have been established via the regularity of the transition probability function $\{P_t(x, \cdot) : t > 0\}$. Their work achieved the climax for Markov processes with Feller transition semigroups.

As matter of fact, the continuity of the filtration $(\mathscr{F}_t^{\mu})$ (or $(\mathscr{F}_t)$) does not follow that of sample function $(X_t)_{t\geq 0}$. For example, a right continuous Markov process does not necessarily lead to the right continuity of its natural filtration $(\mathscr{F}_t^{\mu})$ (or $(\mathscr{F}_t)$). The same claim applies to the left continuity. In fact, the regularity of natural filtrations has much to do with the nature of the Markov property, such as strong Markov property.

Let $C_\infty(E)$ (resp. $C_0(E)$) denote the space of all continuous functions f on E which vanish at infinity ∂, i.e. $\lim_{x\to\partial} f(x) = 0$ (resp. with compact support). Recall that a transition semigroup $(P_t)_{t>0}$ on $(E, \mathscr{B}(E))$ is Feller, if for each $t > 0$, P_t preserves $C_\infty(E)$ and $\lim_{t\downarrow 0} P_t f(x) = f(x)$ for each $x \in E$ and $f \in C_\infty(E)$.

For a given Feller semigroup $(P_t)_{t>0}$ on $(E, \mathscr{B}(E))$, there is a Markov process

$$X = (\Omega, \mathscr{F}^0, \mathscr{F}_t^0, X_t, \theta_t, P^x)$$

with the Feller transition semigroup $(P_t)_{t>0}$ such that the sample function $t \to X_t$ is right continuous on $[0, \infty)$ with left hand limits on $(0, \infty)$. In this case, we call $(\Omega, \mathscr{F}^0, \mathscr{F}_t^0, X_t, \theta_t, P^x)$ a Feller process on E.

For a Feller process $(\Omega, \mathscr{F}^0, \mathscr{F}_t^0, X_t, \theta_t, P^x)$, the natural filtration $(\mathscr{F}_t^{\mu})_{t\geq 0}$ for any $\mu \in \mathscr{P}(E)$ and as well as $(\mathscr{F}_t)_{t\geq 0}$ are right continuous. $(X_t)_{t\geq 0}$ and $(\mathscr{F}_t^{\mu})_{t\geq 0}$ are also quasi-left continuous, that is, if T_n is an increasing family of $(\mathscr{F}_t^{\mu})$-stopping times, and $T_n \uparrow T$, then $\lim_{n\to\infty} X_{T_n} = X_T$ on $\{T < \infty\}$ and $\mathscr{F}_T^{\mu} = \sigma\{\mathscr{F}_{T_n}^{\mu} : n \in \mathbb{N}\}$. Therefore accessible $(\mathscr{F}_t^{\mu})$-stopping times are predictable.

An $(\mathscr{F}_t^\mu)$-stopping time T is totally inaccessible if and only if $P^\mu\{T < \infty\} > 0$ and $X_T \neq X_{T-}$ on $\{T < \infty\}$ P^μ-a.e. Similarly, T is accessible if and only if $X_T = X_{T-}$ P^μ-a.e. on $\{T < \infty\}$. Hence X has only inaccessible jump times.

What we are mainly concerned in this article is Hunt processes. Hunt processes are right continuous, strong Markov processes which are quasi-left continuous. These processes are defined in terms of sample functions, rather than transition semigroups, see [4] and [8] for details. It is well-known that Feller processes are stereotype of Hunt processes or the latter is an abstraction of the former.

We are interested in the martingales over the filtered probability space $(\Omega, \mathscr{F}^\mu, \mathscr{F}_t^\mu, P^x)$, and we are going to show that, if $(\Omega, \mathscr{F}^0, \mathscr{F}_t^0, X_t, \theta_t, P^x)$ is a Hunt process which has continuous sample function, then any martingale on this filtered probability space is continuous, a result one could expect for the natural filtration of a diffusion process. Indeed, this result was proved more or less by Meyer in his Lecture Notes in Mathematics 26, "Processus de Markov". Meyer himself credited his proof to Blumenthal and Getoor, more precisely a calculation done by Blumenthal [3]. However it is surprising that the full computation, which yields more information about martingales over the natural filtration of a Hunt process, was not reproduced either in the new edition of Meyer's "Probabilités et Potentiels" or Chung's "Lectures from Markov Processes to Brownian Motion", although it was mentioned in [9] where Chung and Walsh gave an alternative proof of Meyer's predictability result, so that Blumenthal's computation is no longer needed. However, it is fortunate that Blumenthal's calculation indeed leads to a proof of a martingale representation theorem we are going to establish for certain Hunt processes, see Sect. 3.

Let us first describe an elementary calculation, originally according to Meyer [25] due to Blumenthal. Let

$$X = (\Omega, \mathscr{F}^0, \mathscr{F}_t^0, X_t, \theta_t, P^x)$$

be a Hunt process in a state space $E' = E \cup \{\partial\}$ with the transition semigroup $(P_t)_{t\geq 0}$, where ∂ plays a role of cemetery. Let $\{U^\alpha : \alpha > 0\}$ be the resolvent of the transition semigroup $(P_t)_{t\geq 0}$:

$$U^\alpha(x, A) = \int_0^\infty e^{-\alpha t} P_t(x, A) dt$$

and $(U^\alpha)_{\alpha>0}$ the corresponding resolvent (operators), i.e.

$$\begin{aligned} U^\alpha f(x) &= \int_E f(z) U^\alpha(x, dz) \\ &= \int_0^\infty e^{-\alpha t} P_t f(x) dt \end{aligned}$$

for bounded or nonnegative Borel measurable function f on E. To save words, we use $\mathscr{B}_b(E)$ to denote the algebra of all bounded Borel measurable functions on E. Obviously, $C_\infty(E) \subset \mathscr{B}_b(E)$.

Let $K(E) \subset \mathscr{B}_b(E)$ be a vector space which generates the Borel σ-algebra $\mathscr{B}(E)$. Let $\mathscr{C} \subset L^1(\Omega, \mathscr{F}^\mu, P^\mu)$ (for any initial distribution μ) be the vector space spanned by all $\xi = \xi_1 \cdots \xi_n$ for some $n \in \mathbb{N}$,

$$\xi_j = \int_0^\infty e^{-\alpha_j t} f_j(X_t)dt$$

where α_j are positive numbers, $f_j \in K(E)$, $j = 1, \cdots, n$. Meyer [25] proved that $\mathscr{C}$ is dense in $L^1(\Omega, \mathscr{F}^\mu, P^\mu)$ for a Hunt process. Since this density result will play a crucial role in what follows, we include Meyer's a proof for completeness and for the convenience of the reader. The key observation in the proof is the following result from real analysis.

Lemma 1. *Let $T > 0$. Let $\mathbb{K}$ denote the vector space spanned by all functions $e_\alpha(t) = e^{-\alpha t}$, where $\alpha > 0$, then $\mathbb{K}$ is dense in $C[0, T]$ equipped with the uniform norm.*

The lemma follows from Stone–Weierstrass' theorem.

Lemma 2 (P. A. Meyer). *For any initial distribution μ and $p \in [1, \infty)$, $\mathscr{C}$ is dense in $L^p(\Omega, \mathscr{F}^\mu, P^\mu)$.*

Proof. First, by utilizing Doob's martingale convergence theorem, it is easy to show that the collection $\mathscr{A}$ of all random variables which have the following form

$$g_1(X_{t_1}) \cdots g_n(X_{t_n}),$$

where $n \in \mathbb{N}$, $0 < t_1 < \cdots < t_n < \infty$ and $g_j \in K(E)$, is dense in $L^p(\Omega, \mathscr{F}^\mu, P^\mu)$. Let $\mathscr{H}$ be the linear space spanned by all $\xi = \eta_1 \cdots \eta_n$, where

$$\eta_j = \int_0^\infty g_j(X_t)\varphi_j(t)dt,$$

where $g_j \in K(E)$ and $\varphi_j \in C[0, \infty)$ with compact supports. According to the previous lemma, for every $\varepsilon > 0$ we may choose $\psi_j \in \mathbb{K}$ such that

$$|\varphi_j(t) - \psi_j(t)| < \varepsilon e^{-\lambda t} \quad \text{for all } t \geq 0$$

for some $\lambda > 0$. Let

$$\xi_j = \int_0^\infty g_j(X_t)\psi_j(t)\,\mathrm{d}t.$$

Then $\tilde{\xi} = \xi_1 \cdots \xi_n \in \mathscr{C}$, and

$$|\eta_j - \xi_j| \leq \frac{1}{\lambda}||g_j||_\infty \varepsilon$$

where $||\cdot||_\infty$ is the supermum norm. It follows that

$$E|\xi-\tilde{\xi}|^p \le \frac{n^p}{\lambda^p}\max_j ||g_j||_\infty^{np}\varepsilon^p$$

and thus ξ belongs to the closure of $\mathscr{C}$. Finally it is clear that any element

$$g_1(X_{t_1})\cdots g_n(X_{t_n}) = \lim_{k\to\infty}\eta_1^k\cdots\eta_n^k$$

where

$$\eta_j^k = \int_0^\infty g_j(X_t)\varphi_j^k(t)dt$$

and φ_j^k has compact support and $\varphi_j^k \to \delta_{t_j}$ weakly. We thus have completed the proof. □

Let μ be any fixed initial distribution. If f is a bounded Borel measurable function on E and $\alpha > 0$, then

$$\xi = \int_0^\infty e^{-\alpha t}f(X_t)dt \in L^1(\Omega, \mathscr{F}^\mu, P^\mu).$$

Consider the martingale $M_t = E^\mu\left\{\xi|\mathscr{F}_t^\mu\right\}$ where $t \ge 0$. According to an elementary computation in the theory of Markov processes,

$$\begin{aligned}
M_t &= E^\mu\left\{\int_0^\infty e^{-\alpha s}f(X_s)ds|\mathscr{F}_t^\mu\right\}\\
&= \int_0^t e^{-\alpha s}f(X_s)ds + E^\mu\left\{\int_t^\infty e^{-\alpha s}f(X_s)ds|\mathscr{F}_t^\mu\right\}\\
&= \int_0^t e^{-\alpha s}f(X_s)ds + e^{-\alpha t}\int_0^\infty e^{-\alpha s}P_s f(X_t)ds\\
&= \int_0^t e^{-\alpha s}f(X_s)ds + e^{-\alpha t}U^\alpha f(X_t).
\end{aligned}$$

It is known that if $X=(X_t)_{t\ge 0}$ is a Hunt process,then for any $\alpha > 0$ and bounded Borel measurable function f, $U^\alpha f$ is finely continuous, i.e., $t \to U^\alpha f(X_t)$ is right continuous. Moreover if X is a continuous Hunt process, it follows from a result proved by Meyer that $t \to U^\alpha f(X_t)$ is continuous, and therefore, the martingale $M_t = E^\mu\{\xi|\mathscr{F}_t^\mu\}$ is continuous. We record Meyer's result as a lemma here. This result was proved in [25] for Hunt processes (see T15 THEOREME, page 89, [25]). A simpler proof for Feller processes may be found on page 168, [10].

Lemma 3 (Meyer). *Let $(X_t)_{t\ge 0}$ be a Hunt process, $f \in \mathscr{B}_b(E)$, $\alpha > 0$ and $h = U^\alpha f$ be a potential. Then*

$$h(X_{t-}) = h(X)_{t-} \qquad \forall t > 0 \quad P^{\mu}\text{-}a.e.$$

for any initial distribution μ.

P. A. Meyer pointed out that the previous computation can be carried out equally for random variables on $(\Omega, \mathscr{F}^0)$ which have a product form $\xi = \xi_1 \cdots \xi_n$ where each $\xi_j = \int_0^\infty e^{-\alpha_j s} f_j\,(X_s)ds$. Let π_n denote the permutation group of $\{1, \cdots, n\}$.

Lemma 4 (Blumenthal and Meyer). *Let* $\xi = \xi_1 \cdots \xi_n$ *where* $\xi_j = \int_0^\infty e^{-\alpha_j s} f_j(X_s)ds$, $\alpha_j > 0$ *and* $f_j \in \mathscr{B}_b(E)$, *and* $M_t = E^{\mu}\{\xi|\mathscr{F}_t^{\mu}\}$. *Then*

$$M_t = \sum_{k=0}^{n} \sum_{(j_1,\cdots,j_k,\cdots,j_n)\in\pi_n} \left(\prod_{i=1}^{k} \int_0^t e^{-\alpha_{j_i} s} f_{j_i}(X_s)ds\right) \cdot F_{(j_1,\cdots,j_k,\cdots,j_n)}(X_t) \tag{2}$$

where

$$F_{(j_1,\cdots,j_k,\cdots,j_n)}(x) = E^x\left\{\left(\prod_{l=k+1}^{n} e^{-\alpha_{j_l} t} \int_0^\infty e^{-\alpha_{j_l} s} f_{j_l}(X_s)ds\right)\right\}. \tag{3}$$

Proof. The task is to calculate the conditional expectation $M_t = E^{\mu}\{\xi|\mathscr{F}_t^{\mu}\}$. The idea is very simple: splitting each ξ_j into

$$\xi_j = \int_0^t e^{-\alpha_j s} f_j(X_s)ds + e^{-\alpha_j t} \int_0^\infty e^{-\alpha_j s} f_j(X_s \circ \theta_t)ds$$

so that

$$\xi = \sum_{k=0}^{n} \sum_{(j_1,\cdots,j_k,\cdots,j_n)\in\pi\{1,\cdots,n\}} \left(\prod_{i=1}^{k} \int_0^t e^{-\alpha_{j_i} s} f_{j_i}(X_s)ds\right)$$
$$\cdot\left(\prod_{l=k+1}^{n} e^{-\alpha_{j_l} t} \int_0^\infty e^{-\alpha_{j_l} s} f_{j_l}(X_s \circ \theta_t)ds\right).$$

By using the Markov property one thus obtains

$$\begin{aligned} M_t &= E^{\mu}\{\xi|\mathscr{F}_t^{\mu}\} \\ &= \sum_{k=1}^{n} \sum_{(j_1,\cdots,j_k,\cdots,j_n)\in\pi\{1,\cdots,n\}} \left(\prod_{i=1}^{k} \int_0^t e^{-\alpha_{j_i} s} f_{j_i}(X_s)ds\right) \\ &\quad \cdot E^{\mu}\left\{\left(\prod_{l=k+1}^{n} e^{-\alpha_{j_l} t} \int_0^\infty e^{-\alpha_{j_l} s} f_{j_l}(X_s \circ \theta_t)ds\right) |\mathscr{F}_t^{\mu}\right\} \end{aligned}$$

$$= \sum_{k=0}^{n} \sum_{(j_1,\cdots,j_k,\cdots,j_n)\in\pi\{1,\cdots,n\}} \left(\prod_{i=1}^{k} \int_0^t e^{-\alpha_{j_i} s} f_{j_i}(X_s) ds \right)$$

$$\cdot E^{X_t} \left\{ \left(\prod_{l=k+1}^{n} e^{-\alpha_{j_l} t} \int_0^\infty e^{-\alpha_{j_l} s} f_{j_l}(X_s) ds \right) \right\} . \tag{4}$$

□

Our only contribution in this aspect is the following formula, which allows to prove not only that all martingales over the natural filtration of a continuous Hunt process are continuous, but also a martingale representation theorem in the next section.

Lemma 5. *Let α_j be positive numbers and $f_j \in \mathscr{B}_b(E)$ for $j = 1, \cdots, k$. Consider*

$$F(x) = \int \cdots \int_{0<s_1<\cdots<s_k<\infty} e^{-\sum_{j=1}^{k} \alpha_j s_j} \int_{E^{\otimes k}} f_1(z_1) \cdots f_k(z_k) P_{s_1}(x, dz_1)$$

$$\cdot P_{s_2-s_1}(z_1, dz_2) \cdots P_{s_k-s_{k-1}}(z_{k-1}, dz_k) ds_1 \cdots ds_k.$$

Then

$$F = U^{\alpha_1+\cdots+\alpha_k} \left(f_1 (U^{\alpha_2+\cdots+\alpha_k} f_2 \cdots (U^{\alpha_k} f_k) \cdots) \right). \tag{5}$$

Proof. To see why it is true, we begin with the case that $k = 1$. In this case $F = \int_0^\infty e^{-\alpha s} P_s f ds = U^\alpha f$. If $k = 2$, then

$$\begin{aligned} F &= \iint_{0<s_1<s_2<\infty} e^{-\alpha_2 s_2} e^{-\alpha_1 s_1} P_{s_1} (f_1 P_{s_2-s_1} f_2) \, ds_1 ds_2 \\ &= \int_0^\infty \int_t^\infty e^{-\alpha_1 t} e^{-\alpha_2 s} P_t (f_1 P_{s-t} f_2) \, ds dt \\ &= \int_0^\infty e^{-\alpha_1 t} P_t \left(\int_t^\infty e^{-\alpha_2 s} f_1 P_{s-t} f_2 ds \right) dt \\ &= \int_0^\infty e^{-\alpha_1 t} e^{-\alpha_2 t} P_t \left(f_1 \int_0^\infty e^{-\alpha_2 s} P_s f_2 ds \right) dt \\ &= \int_0^\infty e^{-(\alpha_1+\alpha_2)t} P_t (f_1 U^{\alpha_2} f_2) \, dt \\ &= U^{\alpha_1+\alpha_2} (f_1 U^{\alpha_2} f_2) \end{aligned}$$

and by an induction argument, for a general case. Indeed, if $k > 2$, then

$$
\begin{aligned}
F(x) &= \int \cdots \int_{0<s_1<\cdots<s_{k+1}<\infty} e^{-\sum_{j=1}^k \alpha_j s_j} e^{-\alpha_{k+1} s_{k+1}} \\
&\quad \times \int_{E^{\otimes k}} f_1(z_1) \cdots \left(f_k(z_k) P_{s_{k+1}-s_k} f_{k+1}(z_k)\right) \\
&\quad \times P_{s_1}(x, dz_1) P_{s_2-s_1}(z_1, dz_2) \cdots P_{s_k-s_{k-1}}(, z_{k-1}, dz_k) ds_1 \cdots ds_k ds_{k+1} \\
&= \int \cdots \int_{0<s_1<\cdots<s_k<\infty} e^{-\sum_{j=1}^k \alpha_j s_j} \int_{E^{\otimes k}} f_1(z_1) \cdots \\
&\quad \times \left(f_k(z_k) \int_{s_k}^\infty e^{-\alpha_{k+1} s_{k+1}} P_{s_{k+1}-s_k} f_{k+1}(z_k) ds_{k+1} \right) \\
&\quad \times P_{s_1}(x, dz_1) P_{s_2-s_1}(z_1, dz_2) \cdots P_{s_k-s_{k-1}}(, z_{k-1}, dz_k) ds_1 \cdots ds_k \\
&= \int \cdots \int_{0<s_1<\cdots<s_k<\infty} e^{-\sum_{j=1}^k \alpha_j s_j} \int_{E^{\otimes k}} f_1(z_1) \cdots \\
&\quad \times \left(f_k(z_k) e^{-\alpha_{k+1} s_k} \int_0^\infty e^{-\alpha_{k+1} t} P_t f_{k+1}(z_k) dt \right) \\
&\quad \times P_{s_1}(x, dz_1) P_{s_2-s_1}(z_1, dz_2) \cdots P_{s_k-s_{k-1}}(, z_{k-1}, dz_k) ds_1 \cdots ds_k \\
&= \int \cdots \int_{0<s_1<\cdots<s_k<\infty} e^{-\sum_{j=1}^{k-1} \alpha_j s_j - (\alpha_{k+1}+\alpha_k) s_k} \\
&\quad \times \int_{E^{\otimes k}} f_1(z_1) \cdots f_k(z_k) U^{\alpha_{k+1}} f_{k+1}(z_k) dt
\end{aligned}
$$

and the formula follows the induction assumption. □

Lemma 6. *Let $f_1, \cdots, f_k \in \mathscr{B}_b(E)$, α_j positive numbers, and*

$$
F(x) = E^x \left\{ \left(\prod_{j=1}^k \int_0^\infty e^{-\alpha_j s} f_j(X_s) ds \right) \right\}.
$$

Then

$$
F = \sum_{\{j_1, \cdots, j_k\} \in \pi_k} U^{\alpha_{j_1} + \cdots + \alpha_{j_k}} \left(f_{j_1} (U^{\alpha_{j_2} + \cdots + \alpha_{j_k}} f_{j_2} \cdots (U^{\alpha_{j_k}} f_{j_k}) \cdots \right) \tag{6}
$$

where π_k is the permutation group of $\{1, \cdots, k\}$.

Proof. We have

$$
\begin{aligned}
F(x) &= E^x \left\{ \int_0^\infty \cdots \int_0^\infty e^{-\alpha_1 s_1} \cdots e^{-\alpha_k s_k} f_1(X_{s_1}) \cdots f_k(X_{s_k}) ds_1 \cdots ds_k \right\} \\
&= \int_0^\infty \cdots \int_0^\infty e^{-\alpha_1 s_1} \cdots e^{-\alpha_k s_k} E^x \{ f_1(X_{s_1}) \cdots f_k(X_{s_k}) \} ds_1 \cdots ds_k
\end{aligned}
$$

$$= \sum_{\{j_1,\cdots,j_k\}\in\pi_k} \int\cdots\int_{0<s_{j_1}<\cdots<s_{j_k}<\infty} e^{-\alpha_1 s_1}\cdots e^{-\alpha_k s_k}$$

$$\times E^x\{f_1(X_{s_1})\cdots f_k(X_{s_k})\}\,ds_1\cdots ds_k$$

$$= \sum_{\{j_1,\cdots,j_k\}\in\pi_k} \int\cdots\int_{0<s_1<\cdots<s_k<\infty} e^{-\alpha_{j_1} s_1}\cdots e^{-\alpha_{j_k} s_k}$$

$$\times E^x\left\{f_{j_1}(X_{s_1})\cdots f_{j_k}(X_{s_k})\right\} ds_1\cdots ds_k$$

$$= \sum_{\{j_1,\cdots,j_k\}\in\pi_k} \int\cdots\int_{0<s_1<\cdots<s_k<\infty} \int_{E^{\otimes k}} e^{-\alpha_{j_1} s_1}\cdots e^{-\alpha_{j_k} s_k} f_{j_1}(z_1)\cdots f_{j_k}(z_k)$$

$$\times P_{s_1}(x,dz_1)P_{s_2-s_1}(z_1,dz_2)\cdots P_{s_k-s_{k-1}}(z_{k-1},dz_k)ds_1\cdots ds_k$$

and (6) follows from Lemma 5. □

From now on, we assume that

$$X = (\Omega, \mathscr{F}^0, \mathscr{F}_t^0, X_t, \theta_t, P^x)$$

is a *continuous Hunt process* in $E' = E \cup \{\partial\}$ with the transition semigroup $(P_t)_{t\geq 0}$. In other words, it is a Hunt process with continuous sample paths. Therefore, $(X_t)_{t\geq 0}$ is a diffusion process in E, i.e. $(X_t)_{t\geq 0}$ possesses the strong Markov property with continuous sample function. Under our assumptions, any finite $(\mathscr{F}_t^\mu)$-stopping time is accessible and thus predictable, and therefore $\mathscr{F}_T^\mu = \mathscr{F}_{T-}^\mu$. In particular, $(\mathscr{F}_t^\mu)$ is left continuous, and thus the filtration $(\mathscr{F}_t^\mu)$ is continuous for any initial distribution μ.

Since any martingale on $(\Omega, \mathscr{F}^\mu, \mathscr{F}_t^\mu, P^\mu)$ has a right continuous modification, by a martingale we always mean a martingale with right continuous sample function.

Lemma 7. *Suppose $\xi = \xi_1\cdots\xi_n$ where each ξ_j has the following form*

$$\xi_j = \int_0^\infty e^{-\alpha_j s} f_j(X_s)ds$$

where $\alpha_j > 0$ and $f_j \in \mathscr{B}_b(E)$. Let $M_t = E^\mu\left\{\xi|\mathscr{F}_t^\mu\right\}$. Then $(M_t)_{t\geq 0}$ is a bounded continuous martingale on $(\Omega, \mathscr{F}^\mu, \mathscr{F}_t^\mu, P^\mu)$.

Proof. According to Lemma 4, we need only to show that for function of the following type

$$F(x) = E^x\left\{\left(\prod_{l=k+1}^n e^{-\alpha_{j_l} t}\int_0^\infty e^{-\alpha_{j_l} s} f_{j_l}(X_s)ds\right)\right\},$$

$t \mapsto F(X_t)$ is continuous. By Lemma 6, F is an α-potential, so that it is finely continuous, and together with Lemma 3, it implies that $t \to F(X_t)$ is continuous, which completes the proof. □

We now state the main result of this section. For simplicity, a square integrable martingale $(M_t)_{t\geq 0}$ over $(\Omega, \mathscr{M}, \mathscr{M}_t, P)$ means $M_t = E(\xi|\mathscr{M}_t)$ with $\xi \in L^2(\Omega, \mathscr{M}, P)$. This is equivalent to say $\sup_{t>0} E[M_t^2] < \infty$.

Theorem 2. *Let*

$$X = (\Omega, \mathscr{F}^0, \mathscr{F}_t^0, X_t, \theta_t, P^x)$$

be a continuous Hunt process in E, and $\mu \in \mathscr{P}(E)$. If $\xi \in L^2(\Omega, \mathscr{F}^\mu, P^\mu)$, then the martingale $M_t = E^\mu\{\xi|\mathscr{F}_t^\mu\}$ is continuous, that is, square-integrable martingales on $(\Omega, \mathscr{F}^\mu, \mathscr{F}_t^\mu, P^\mu)$ are continuous. Therefore local martingales over $(\Omega, \mathscr{F}^\mu, \mathscr{F}_t^\mu, P^\mu)$ are continuous.

Proof. We can choose a sequence $\xi_n \in \mathscr{C}$ such that $\xi_n \to \xi$ in L^2. Doob's maximal inequality implies that, if necessary by considering a subsequence, the martingales $\{E^\mu(\xi_n|\mathscr{F}_t^\mu) : t \geq 0\}$ converges (almost surely at least along a subsequence) to $\{E^\mu(\xi|\mathscr{F}_t^\mu) : t \geq 0\}$ uniformly on any finite interval of $t \geq 0$. It is shown in Lemma 7 that for each n, the martingale $E^\mu(\xi_n|\mathscr{F}_t^\mu)$ is continuous and thus the square integrable martingale $\{E^\mu(\xi|\mathscr{F}_t^\mu) : t \geq 0\}$ must be continuous.

By the localization technique, it follows thus that local martingales on $(\Omega, \mathscr{F}^\mu, \mathscr{F}_t^\mu, P^\mu)$ are continuous. □

3 Martingale Representation for Continuous Hunt Process

In this section we assume that

$$X = (\Omega, \mathscr{F}^0, \mathscr{F}_t^0, X_t, \theta_t, P^x)$$

is a *continuous* Hunt process in the state space $E' = E \cup \{\partial\}$ with transition semigroup $\{P_t(x, dy) : t \geq 0\}$, where E is a locally compact separable metric space. Let $\mu \in \mathscr{P}(E)$ be an initial distribution.

If $\alpha > 0$ and $f \in \mathscr{B}_b(E)$ then $M^{\alpha,f}$ denotes the continuous martingale

$$M_t^{\alpha,f} = E^\mu\left\{\int_0^\infty e^{-\alpha s} f(X_s) ds|\mathscr{F}_t^\mu\right\}.$$

Recall that, if u is an α-potential, i.e., $u = U^\alpha f$ where $f \in \mathscr{B}_b(E)$, then $u(X_t) - u(X_0)$ is a continuous semimartingale on $(\mathscr{F}^\mu, \mathscr{F}_t^\mu, P^\mu)$. and possesses Doob–Meyer's decomposition

$$u(X_t) - u(X_0) = M_t^{[u]} + A_t^{[u]}$$

where

$$M_t^{[u]} = \int_0^t e^{\alpha s} dM_s^{\alpha,f}, \ A_t^{[u]} = \int_0^t Lu(X_s) ds$$

and $Lu = \alpha u - f$.

We make the following assumptions on a continuous Hunt process X with an initial distribution $\mu \in \mathscr{P}(E)$.

Assumptions. There is an algebra (a vector space which is closed under the multiplication of functions) $K(E) \subset \mathscr{B}_b(E)$ which generates the Borel σ-algebra $\mathscr{B}(E)$ and is invariant under U^α for $\alpha > 0$, and there are finite many continuous martingales $M^1, \cdots, M^d$ over $(\Omega, \mathscr{F}^\mu, \mathscr{F}_t^\mu, P^\mu)$ such that the following conditions are satisfied:

(1) For any potential $u = U^\alpha f$ where $\alpha > 0$ and $f \in K(E)$, the martingale part $M^{[u]}$ of the semimartingale $u(X_t) - u(X_0)$ has the martingale representation in terms of $(M^1, \cdots, M^d)$, that is, there are predictable processes $F_1, \cdots, F_d$ on $(\Omega, \mathscr{F}^\mu, \mathscr{F}_t^\mu)$ such that

$$M_t^{[u]} = \sum_{j=1}^d \int_0^t F_s^j dM_s^j \quad P^\mu\text{-a.e.} \tag{7}$$

(2) $(\langle M^j, M^i \rangle_t)$ is strictly positive definite.

The first assumption means that the martingale $M^{[u]}$ with u being a potential may be represented. The second condition ensures that the representation (7) is unique. In this case we say that Fukushima representation property holds for X with initial law μ and martingales $\{(M^{(1)}, \cdots, M^{(d)})\}$.

The Fukushima representation property is mainly an abstraction of the chain rule for the martingale part of $u(X_t)$. Indeed, if $X_t = (X_t^1, \cdots, X_t^d)$ is a d-dimensional Brownian motion and u is an α-potential with $\alpha > 0$, then u is smooth and by Itô's formula

$$u(X_t) - u(X_0) = \sum_{j=1}^d \int_0^t \frac{\partial u}{\partial x^j}(X_s) dX_s^j + \int_0^t \frac{1}{2} \Delta u(X_s) ds$$

so that

$$M_t^{[u]} = \sum_{j=1}^d \int_0^t \frac{\partial u}{\partial x^j}(X_s) dX_s^j.$$

One can easily see that the Brownian motion satisfies the Fukushima representation property.

Theorem 3 (Martingale representation). *Let $\mu \in \mathscr{P}(E)$. If Fukushima representation property holds for X with initial law μ and a finite set of martingales $(M^1, \cdots, M^d)$, then for any square-integrable martingale $N = (N_t)_{t \geq 0}$ on $(\Omega, \mathscr{F}^\mu, \mathscr{F}_t^\mu, P^\mu)$, there are unique predictable processes (F_t^i) such that*

$$N_t - N_0 = \sum_{i=1}^d \int_0^t F_s^i dM_s^i \quad P^\mu\text{-a.e.}$$

Proof. The uniqueness follows from condition 2) in the Fukushima representation. We prove the existence. Take $\xi \in L^2(\Omega, \mathscr{F}^\mu, P^\mu)$ such that $N_t = E^\mu\{\xi|\mathscr{F}_t^\mu\}$. Since $\mathscr{C}$ is dense in $L^2(\Omega, \mathscr{F}^\mu, P^\mu)$, so we first prove the martingale representation for $\xi \in \mathscr{C}$. By the linearity, we only need to consider the case that $\xi = \xi_1 \cdots \xi_n$ where $\xi_j = \int_0^\infty e^{-\alpha_j s} f_j(X_s)ds$ for $\alpha_j > 0$ and $f_j \in K(E)$. In this case, according to 4, Lemmas 5 and 6

$$N_t = E^\mu\{\xi|\mathscr{F}_t^\mu\} = \sum_m Z_t^m$$

where the sum is a finite one, and for each m, $Z^m = Z_t$ has the following form

$$Z_t^m = V_t^m u^m(X_t)$$

(the superscript m will be dropped if no confusion may arise), where

$$V_t = \prod_{i=1}^{k'} \int_0^t e^{-\beta_i s} g_i(X_s)ds$$

and

$$u(x) = \int \cdots \int_{0<s_1<\cdots<s_k<\infty} e^{-\sum_{j=1}^k \beta_j s_j} \int_{E^{\otimes k}} h_1(z_1)\cdots h_k(z_k)P_{s_1}(x, dz_1) \\ \times P_{s_2-s_1}(z_1, dz_2)\cdots P_{s_k-s_{k-1}}(z_{k-1}, dz_k)ds_1\cdots ds_k$$

for some k' and k, $\beta_i > 0$ and functions g_i, h_j are bounded and continuous. According to Lemma 5

$$u = U^{\beta_1+\cdots+\beta_k}\left(h_1(U^{\beta_2+\cdots+\beta_k}h_2\cdots(U^{\beta_k}h_k)\cdots\right).$$

In particular, u is again a potential which has a form $u = U^\alpha g$ for

$$g = h_1(U^{\beta_2+\cdots+\beta_k}h_2\cdots(U^{\beta_k}h_k)\cdots) \in K(E)$$

and $\alpha = \beta_1 + \cdots + \beta_k$. Hence $u(X_t)$ is a continuous semimartingale with decomposition

$$u(X_t) - u(X_0) = M_t^{[u]} + A_t^{[u]}$$

where $A^{[u]}$ is continuous with finite variation, and due to the Fukushima representation property

$$M_t^{[u]} = \sum_{j=1}^d \int_0^t G_s^j dM_s^j$$

for some predictable processes G^j. In particular, each Z^m is a continuous semimartingale. Since, by Theorem 2, N is a continuous martingale, so that

$$N_t = \sum_m \text{the continuous martingale part of } V_t^m u^m(X_t).$$

Therefore we are interested in the martingale part of $Z_t = V_t u(X_t)$. Since V is a finite variation process, so according to Itô's formula

$$\begin{aligned} Z_t &= Z_0 + \int_0^t u(X_s)dV_s + \int_0^t V_s du(X_s) \\ &= Z_0 + \int_0^t u(X_s)dV_s + \int_0^t V_s dA_t^{[u]} + \int_0^t V_s dM_t^{[u]} \\ &= Z_0 + \int_0^t u(X_s)dV_s + \int_0^t V_s dA_t^{[u]} + \sum_{i=1}^d \int_0^t V_s \cdot G_s^i dM_s^i \end{aligned}$$

so that the martingale part of Z_t is

$$\sum_{i=1}^d \int_0^t V_s \cdot G_s^i dM_s^i.$$

Therefore

$$N_t = E^\mu\{\xi|\mathscr{F}_t^\mu\} = \sum_{i=1}^d \int_0^t \sum_m V_s^m \cdot G_s^{m,i} dM_s^i$$

which shows the martingale representation.

Suppose now $\xi \in L^2(\Omega, \mathscr{F}^\mu, P^\mu)$. Choose a sequence $\xi_n \in \mathscr{C}$ such that $\xi_n \to \xi$ in $L^2(\Omega, \mathscr{F}^\mu, P^\mu)$. Let $N_t^{(n)} = E^\mu(\xi_n|\mathscr{F}_t^\mu)$ and $N_t = E^\mu(\xi|\mathscr{F}_t^\mu)$. According to Doob's maximal inequality, if necessary by passing to a subsequence, we can assume that $N_t^{(n)}$ converges to N_t uniformly on any finite interval. $N_t^{(n)}$ has the martingale representation

$$N_t^{(n)} - N_0^{(n)} = \sum_{j=1}^d \int_0^t F(n)_s^j dM_s^j$$

so that

$$\begin{aligned} &\langle N_t^{(n)} - N_t^{(m)}, N_t^{(n)} - N_t^{(m)}\rangle \\ &= \sum_{i,j=1}^d \int_0^t (F(n)_s^i - F(m)_s^i)(F(n)_s^j - F(m)_s^j) d\langle M^i, M^j\rangle_s. \end{aligned}$$

Since $(\langle M^i, M^j\rangle_t)$ is positive, it follows that $(F(n)^1, \cdots, F(n)^d)$ converges to predictable processes $(F^1, \cdots, F^d)$ under the norm

$$\|(F^1, \cdots, F^d)\| = \sum_{N=1}^\infty \frac{1}{2^N} \sum_{i,j=1}^d E^\mu \left[\int_0^N F_s^i F_s^j d\langle M^i, M^j\rangle_s\right]. \tag{8}$$

Then

$$N_t - N_0 = \sum_{j=1}^{d} \int_0^t F_s^j \, dM_s^j .$$

□

This theorem claims that as long as every martingale of resolvent type is representable, so is any martingale. When is the Fukushima representation property satisfied? There are many examples. In the remain of this section, we shall give three interesting examples in symmetric situation.

Brownian motion with any initial distribution is certainly an example. Indeed, for Brownian motion in $\mathbb{R}^d$, we may choose $K(E) = C_\infty^\infty(R^d)$ (the space of smooth functions which vanish at infinity), then for $f \in K(E)$, $U^\alpha f$ is smooth, and (7) follows from Itô's formula applying to $U^\alpha f$. Theorem 3 gives a new proof for classical martingale representation theorem.

The second example is the reflecting Brownian motion. As Example 1.6.1 in [13], we consider Dirichlet form $(\frac{1}{2}\mathbf{D}, H^1(D))$ on $L^2(D)$ where $\mathbf{D}$ is the classical Dirichlet integral and D is a bounded domain on $\mathbb{R}^d$. We further assume that any $x \in \partial D$ has a neighborhood U such that

$$D \cap U = \{(x_i) \in \mathbb{R}^d : x_d > F(x_1, \cdots, x_{d-1})\} \cap U$$

for some continuous function F. Then $C_0^\infty(\overline{D})$ (the space of restriction to $\overline{D}$ of functions in $C_0^\infty(\mathbb{R}^d)$) is dense in $H^1(D)$ (see [24] for details), i.e., $(\frac{1}{2}\mathbf{D}, H^1(D))$ is a regular Dirichlet form on $L^2(\overline{D})$. The corresponding continuous Hunt process $X = (X_t, P^x)$ is called the reflecting Brownian motion. For $x = (x^i) \in \mathbb{R}^d$, we use $u_i(x) = x^i$, $1 \le i \le d$, to denote the coordinate functions. Then $u_i \in \mathscr{F}$ and we denote by $M^i = M^{[u_i]}$ the martingale part in Fukushima's decomposition. It can be seen from Corollary 5.6.2 [13] that for any $u \in C_0^\infty(\overline{D})$,

$$M_t^{[u]} = \sum_{i=1}^{d} \int_0^t \frac{\partial u}{\partial x_i}(X_s) dM_s^i, \ P^x\text{-a.s. for q.e. } x \in \overline{D},$$

where q.e. means "quasi-everywhere", i.e., except a set of zero-capacity. Then a routine approximation procedure shows that for any $u \in H^1(D)$, there exist Borel measurable functions $\{f_i : 1 \le i \le d\}$ on $\overline{D}$ such that

$$M_t^{[u]} = \sum_{i=1}^{d} \int_0^t f_i(X_s) dM_s^i, \ P^x\text{-a.s. for q.e. } x \in \mathbb{R}^d.$$

Therefore the reflecting Brownian motion has Fukushima representation property, by choosing $K(\overline{D})$ to be the space of bounded measurable functions and any initial distribution μ charging no set of zero capacity, i.e., a smooth distribution, because an exceptional set exists in above representation as is always when the process is constructed through a Dirichlet form. In particular, if the boundary is Lipschitz, then the transition function has density [2] and in this case, the exceptional set may be erased and Fukushima representation property holds for X with any initial law or starting at each point. Notice that under the current condition, the reflecting Brownian motion X itself is not necessarily a semimartingale. The readers who are interested may refer to [2, 6, 7, 29] about when a reflecting BM is a semimartingale and the corresponding Skorohod decomposition. It should be pointed out that, although the martingale part of the reflected Brownian motion is a Brownian motion, but the martingale representation property does not follow from the classical representation property for Brownian motion. The reason is that, as long as the boundary is not sufficiently smooth, the natural filtration $(\mathscr{F}_t^{\mu})_{t\geq 0}$ is much bigger in general than the natural filtration generated by the martingale part $(M^1, \cdots, M^d)$ of X.

Another example our main result may apply is symmetric diffusions in a domain killed at boundary. Actually Theorem 6.2.2 in [13] tells us that every continuous symmetric Hunt process with a smooth core enjoys the Fukushima representation property. More precisely let D be a domain of $\mathbb{R}^d$ with continuous boundary ∂D and m a Radon measure on D. Let X be a continuous Hunt process which is symmetric with respect to m and $(\mathscr{E}, \mathscr{F})$ the associated Dirichlet form on $L^2(D, m)$, which has $C_0^1(D)$ as a core. For $x = (x^i) \in \mathbb{R}^d$, we use $u_i(x) = x^i$, $1 \leq i \leq d$, to denote the coordinate functions. Then $u_i \in \mathscr{F}_{\mathrm{loc}}$ and we denote by $M^i = M^{[u_i]}$ the martingale part in Fukushima's decomposition. Let

$$\mu_{i,j} = \mu_{\langle M^i, M^j\rangle}, \ 1 \leq i, j \leq d,$$

the smooth measure associated with CAF $\langle M^i, M^j\rangle$. Then $\mathscr{E}$ is expressed as

$$\mathscr{E}(u, v) = \sum_{i,j=1}^{d} \int_D \frac{\partial u}{\partial x^i}\frac{\partial u}{\partial x^j} d\mu_{i,j}(x), \ u, v \in C_0^1(D).$$

As asserted in Theorem 6.2.2 [13], for any initial smooth distribution μ (i.e. a probability on $(D, \mathscr{B}(D))$ having no charge on capacity zero sets) and $u \in \mathscr{F}$, the martingale part $M^{[u]}$ in Fukushima's decomposition of u may be represented as

$$M_t^{[u]} = \sum_{i=1}^{d} \int_0^t f_i(X_s) dM_s^i \qquad P^{\mu}\text{-a.e.}$$

where $f_1, \cdots, f_d \in \mathscr{B}(D)$. If we take $K(E) = L^2(E, m) \cap \mathscr{B}_b(D)$, X satisfies the Fukushima representation property. In these examples, $\{M^i\}$ are the martingales corresponding to coordinate functions so we call them coordinate martingales.

To have the uniqueness, some kind of non-degenerateness is needed. We say that X is non-degenerate if the condition (2) in Fukushima representation property is satisfied: $(\langle M^i, M^j\rangle)_{1\leq i,j\leq d}$ is positive.

Corollary 1. *Assume that X is either the reflecting Brownian motion on a bounded domain or a non-degenerate symmetric Hunt diffusion on a domain $D \subset \mathbb{R}^d$ as stated above. Then the Fukushima representation property is satisfied and therefore the martingale representation holds in the sense of Theorem 3 with coordinate martingales and for a given initial distribution μ charging no sets of zero capacity.*

From this result, we may recover the martingale representation established in [1] and [33], where X is a diffusion process corresponding to non-degenerate symmetric elliptic operator on $\mathbb{R}^d$.

Without essential difference, the conclusion holds also for reflecting diffusions on such domain with generator being a symmetric uniformly elliptic differential operator of second order as introduced in the beginning of next section.

4 Backward Stochastic Differential Equations

In this section we consider backward stochastic differential equations which can be used to provide probability representations for weak solutions of the initial and boundary value problem of a quasi-linear parabolic equation.

Let $D \subset \mathbb{R}^d$ be a bounded domain with a continuous boundary ∂D, $\overline{D} = D \cup \partial D$ the closure of D. Let

$$L = \frac{1}{2} \sum_{i,j=1}^{d} \frac{\partial}{\partial x^j} a^{ij} \frac{\partial}{\partial x^i}$$

be an elliptic differential operator of second order, where $a = (a^{ij})$ is a positive-definite, symmetric, matrix-valued function on D, $a = (a^{ij})$ is Borel measurable, and satisfies the elliptic condition:

$$\lambda |\xi|^2 \leq \sum_{i,j=1}^{d} a^{ij}(x)\xi_i \xi_j \leq \lambda^{-1} |\xi|^2 \qquad \forall \xi = (\xi_i) \in \mathbb{R}^d$$

for all $x \in D$ for some constant $\lambda > 0$. Consider the Dirichlet form $(\mathscr{E}, \mathscr{F})$ on $L^2(D, dx)$, where

$$\mathscr{E}(u, v) = \frac{1}{2} \int_D \sum_{i,j=1}^{d} a^{ij} \frac{\partial u}{\partial x^j} \frac{\partial v}{\partial x^i} \mathrm{d}x \tag{9}$$

and $\mathscr{F} = H^1(D)$.

Let Ω be a space of all continuous paths in $\overline{D}$, $(X_t)_{t \geq 0}$ the coordinate process on Ω, $\mathscr{F}^0 = \sigma\{X_s : s \geq 0\}$, $\mathscr{F}_t^0 = \sigma\{X_s : s \leq t\}$ for each $t \geq 0$, and $(\theta_t)_{t \geq 0}$ shift operators on Ω. Let

$$X = (\Omega, \mathscr{F}^0, \mathscr{F}_t^0, X_t, \theta_t, P^x)$$

be the canonical realization of the symmetric diffusion process in the state space $\overline{D}$ associated with the Dirichlet space $(\mathscr{E}, \mathscr{F})$, which is called a reflecting symmetric diffusion in D.

The coordinate functions $u_j(x) = x^j$ $(j = 1, \cdots, d)$ belong to the local Dirichlet space $\mathscr{F}_{\text{loc}}$, so that

$$X_t^j - X_0^j = M_t^j + A_t^j \quad P^x\text{-a.e. } j = 1, \cdots, d \tag{10}$$

for all $x \in \overline{D}$ except for a capacity zero set, where $M^j = M^{[f_j]}$ etc.

Let $\mathscr{S}_1(\overline{D})$ denote the space of all probability $\mu \in \mathscr{P}(\overline{D})$ which has no charge on zero capacity sets (with respect to the Dirichlet form $(\mathscr{E}, H^1(D))$ defined by (9). According to Theorem 3, for any initial distribution $\mu \in \mathscr{S}_1(\overline{D})$, the family of martingales $\{M^j : j = 1, \cdots, d\}$ over $(\Omega, \mathscr{F}^\mu, \mathscr{F}_t^\mu, P^\mu)$ has the martingale representation property: for any square-integrable martingale $N = (N_t)_{t \geq 0}$ on $(\Omega, \mathscr{F}^\mu, \mathscr{F}_t^\mu, P^\mu)$, there are unique predictable processes (F_t^i) such that

$$N_t - N_0 = \sum_{i=1}^{d} \int_0^t F_s^i dM_s^i \quad P^\mu\text{-a.e.}$$

Let us work with a fixed smooth initial distribution $\mu \in \mathscr{S}_1(D)$ and the filtered probability space $(\Omega, \mathscr{F}^\mu, \mathscr{F}_t^\mu, P^\mu)$.

Consider the following backward stochastic differential equation

$$dY_t^i = -f^i(t, Y_t, Z_t)dt + \sum_{i,j=1}^{d} Z_t^{ij} dM_t^j, \; Y_T^i = \xi^i \tag{11}$$

$i = 1, \cdots, d'$, where $T > 0$, $\xi^i \in L^2(\Omega, \mathscr{F}_T^\mu, P^\mu)$ are given terminal values, and f^i are Lipschitz functions: there is a constant $C_1 \geq 0$

$$|f^i(t, y, z)| \leq C_1(1 + t + |y| + |z|)$$

and

$$|f^i(t, y, z) - f^i(t, \tilde{y}, \tilde{z})| \leq C_1 \left(|y - \tilde{y}| + |z - \tilde{z}|\right)$$

for all $t \geq 0$, $y, \tilde{y} \in \mathbb{R}^{d'}$, $z, \tilde{z} \in \mathbb{R}^{d' \times d}$. One seeks for a solution pair (Y, Z) which solves the following integral equation

$$Y_t^i - \xi^i = \int_t^T f^i(s, Y_s, Z_s)ds - \sum_{j=1}^{d} \int_t^T Z_s^{ij} dM_s^j \tag{12}$$

for $t \in [0, T]$. The integral equation (12) has a unique solution pair (Y, Z) such that Y^i is a continuous semimartingale, and Z^{ij} are predictable processes satisfying

$$E^\mu \int_0^T \sum_{k,l=1}^{d} a^{kl}(X_s) Z_s^{il} Z_s^{ki} ds < \infty.$$

This can be demonstrated by employing the Picard iteration for (Y, Z) as in the case of Brownian motion (see [27]). Another approach, proposed in a paper by Lyons et al. [20] which applies to a general filtered probability space, may be described as follows. The idea is to rewrite the integral equation (12) into a functional differential equation for the variation process part V of Y. Let $Y = N - V$ where V is a finite variation process, and

$$N_t^i - N_0^i = \sum_{j=1}^{d} \int_0^t Z_s^{ij} dM_s^j.$$

On the other hand

$$N_t = E^\mu\{\xi + V_T | \mathscr{F}_t^\mu\}.$$

Since Y is a continuous semimartingale, its decomposition is unique up to an initial value. The integral equation (12) leads to that

$$V_t = -\int_t^T f^i(s, Y_s, Z_s)ds + N_T - \xi$$

conditioned on $\mathscr{F}_t^\mu$ and we obtain

$$\begin{aligned} V_t &= -E^\mu\left\{\int_t^T f^i(s, Y_s, Z_s)ds | \mathscr{F}_t^\mu\right\} + N_t - E^\mu\left\{\xi | \mathscr{F}_t^\mu\right\} \\ &= -E^\mu\left\{\int_t^T f^i(s, Y_s, Z_s)ds | \mathscr{F}_t^\mu\right\} + E^\mu\left\{V_T | \mathscr{F}_t^\mu\right\} \\ &= -E^\mu\left\{\int_0^T f^i(s, Y_s, Z_s)ds - V_T | \mathscr{F}_t^\mu\right\} + \int_0^t f^i(s, Y_s, Z_s)ds. \end{aligned}$$

Therefore the integral equation (12) is equivalent to

$$V_t - V_0 = \int_0^t f^i(s, Y_s, Z_s)ds \tag{13}$$

where

$$Y_t = Y(V)_t = N(V)_t - V_t, \quad N(V)_t = E^\mu\left\{\xi + V_T | \mathscr{F}_t^\mu\right\}$$

and $Z_t = Z(V)_t$ is determined by the martingale representation theorem

$$N(V)_t^i - N(V)_0^i = \sum_{j=1}^{d} \int_0^t Z(V)_s^{ij} dM_s^j.$$

Equation (13) thus may be written as a functional equation

$$V_t - V_0 = \int_0^t f^i(s, Y(V)_s, Z(V)_s)ds \tag{14}$$

where $Y(V)$ and $Z(V)$ are considered as functionals of V. The Picard iteration applies to (14) we have

Theorem 4. *If $\xi \in L^2(\Omega, \mathscr{F}_T^{\mu}, P^{\mu})$ and f^i are Lipschitz continuous, then there is a unique pair (Y, Z) such that Y is a continuous semimartingale which solves BSDE (11).*

For a complete proof of Theorem 4, the reader may refer to [20].

5 Non-linear Parabolic Equations

We are under the same setting as in the previous section, and use the notations established therein.

To motivate our approach, let us begin with the case that a is smooth, and D is bounded domain with a smooth boundary.

In this case $X^j = (X_t^j)_{t\geq 0}$ in (10), Sect. 3, are continuous semimartingales, thus A^j are finite variation processes. For any $h \in C_b^{1,2}([0,\infty) \times \overline{D})$ satisfying the Neumann boundary condition that $\frac{\partial h}{\partial \nu}|_{\partial D} = 0$, where $\frac{\partial}{\partial \nu}$ denotes the normal derivative with respect to the Riemann metric $(a^{ij}) = (a_{ij})^{-1}$, we have

$$h(t, X_t) - h(0, X_0) = M_t^h + A_t^h$$

where

$$M_t^h = h(t, X_t) - h(0, X_0) - \int_0^t \left(\frac{\partial}{\partial s} + L\right) h(s, X_s) ds \tag{15}$$

is a martingale under P^x, and

$$A_t^h = \int_0^t \left(\frac{\partial}{\partial s} + L\right) h(s, X_s) ds. \tag{16}$$

On the other hand, applying Itô's formula

$$\begin{aligned} h(t, X_t) - h(0, X_0) &= \int_0^t \left(\frac{\partial}{\partial s} + \frac{1}{2}\sum_{i,j=1}^d a^{ij}\frac{\partial^2}{\partial x^i \partial x^j}\right) h(s, X_s) ds \\ &\quad + \int_0^t \sum_{j=1}^d \frac{\partial}{\partial x^j} h(s, X_s) d(M_s^j + A_s^j) \end{aligned}$$

it thus follows that

$$M_t^h = \sum_{j=1}^d \int_0^t \frac{\partial}{\partial x^j} h(s, X_s) dM_s^j \tag{17}$$

and

$$A_t^i = \frac{1}{2}\int_0^t \sum_{j=1}^d \frac{\partial}{\partial x^j} a^{ij}(X_s)ds. \tag{18}$$

Consider a solution $u(x,t)$ to the initial boundary problem to the non-linear parabolic equation

$$\begin{cases} \left(\frac{\partial}{\partial t} - L\right) u + f(t,u,\nabla u) = 0, & \\ u(0,x) = \varphi(x), & x \in \mathbb{R}^d, \\ \left.\frac{\partial u(t,\cdot)}{\partial \nu}\right|_{\partial D} = 0, & t > 0 \end{cases} \tag{19}$$

Then, by (15) and (17)

$$\begin{aligned} h(T,X_T) = h(t,X_t) &+ \int_t^T \left(\frac{\partial}{\partial s} + L\right) h(s,X_s)ds \\ &+ \sum_{j=1}^d \int_t^T \frac{\partial}{\partial x^j} h(s,X_s)dM_s^j \end{aligned}$$

together with the PDE (19) we deduce that

$$\begin{aligned} h(t,X_t) - h(T,X_T) = &\int_t^T f(T-s, h(s,X_s), \nabla h(s,X_s))ds \\ &- \sum_{j=1}^d \int_t^T \frac{\partial}{\partial x^j} h(s,X_s)dM_s^j. \end{aligned} \tag{20}$$

Let $Y_t = u(T-t, X_t)$ and $Z_t^j = \frac{\partial}{\partial x^j} h(t,X_t)$. Then the previous equation may be written as

$$Y_t - Y_T = \int_t^T f(T-s, Y_s, Z_s)ds - \sum_{j=1}^d \int_t^T Z_s^j dM_s^j \tag{21}$$

and $Y_T = u(0,X_T) = \varphi(X_T)$. That is to say that $Y_t = u(T-t, X_t)$ solves the scalar BSDE

$$dY_t = -f_T(t, Y_t, Z_t)dt + \sum_{j=1}^d Z_t^j dM_t^j, \quad Y_T = \varphi(X_T) \tag{22}$$

where $f_T = f(T-t, y, z)$.

For any fixed $T > 0$, let $Y^T = \{Y_t^T : t \in [0, T]\}$ be the unique solution to the BSDE (22) on $(\Omega, \mathscr{F}^\mu, \mathscr{F}_t^\mu, P^\mu)$. Since the solution to BSDE is unique, $Y_t^T = u(T-t, X_t)$. In particular $u(T, X_0) = Y_0^T$, and therefore

$$\int_{R^d} u(T, x)\mu(dx) = E^\mu\left(Y_0^T\right). \tag{23}$$

The above argument leading to the probabilistic representation (23) can not be justified in the case that $a = (a^{ij})$ is only Borel measurable or the boundary ∂D is only continuous, as in this case, $(X_t)_{t\geq 0}$ is no longer a semimartingale, both (16) and (18) no longer make sense. While, in this case, boundary problem (strong or weak solutions) to the non-linear PDE (19) also need to be interpreted. On the other hand, the BSDE (22), which relies on only the martingale representation, still make sense, thus the representation theorems stated in Sect. 3 can be made as the definition of a solution to (19). This is the approach we will carry out.

Consider the initial value problem of the following non-linear parabolic equation in a bounded domain D with a continuous boundary ∂D

$$\left(\frac{\partial}{\partial t} - \frac{1}{2}\sum_{i,j=1}^{d} \frac{\partial}{\partial x^j} a^{ij}(x) \frac{\partial}{\partial x^i}\right) u + f(t, u, \nabla u) = 0 \tag{24}$$

subject to the initial and boundary conditions

$$u(x, 0) = \varphi(x), \qquad \left.\frac{\partial}{\partial \nu} u(t, \cdot)\right|_{\partial D} = 0 \text{ for } t > 0$$

where $a = (a^{ij})$ is Borel measurable, satisfying the uniform ellipticity condition:

$$\lambda \sum_{i=1}^{d} |\xi^i|^2 \leq \sum_{i,j}^{d} \xi^i \xi^j a^{ij}(x) \leq \lambda^{-1} \sum_{i=1}^{d} |\xi^i|^2 \qquad \forall (\xi^i) \in \mathbb{R}^d,$$

for some constant $\lambda > 0$.

Definition 1. The functional on $\mathscr{S}_1(\overline{D})$ defined by $\mu \to E^\mu\{Y(t, \mu)_0\}$, denoted by $u(t, \mu)$, is called the stochastic solution of the initial and boundary problem of (24), where for each $t > 0$ and $\mu \in \mathscr{S}_1(\overline{D})$, $Y(t, \mu) = (Y_s)_{s\leq t}$ is the unique solution to the BSDE

$$\begin{cases} dY_s = -f(t-s, Y_s, Z_s)ds + \sum_{j=1}^{d} Z_s^j dM_s^j, \\ Y_t = \varphi(X_t) \end{cases} \tag{25}$$

on $(\Omega, \mathscr{F}^\mu, \mathscr{F}_t^\mu, X_t, \theta_t, P^\mu)$.

As a consequence we have

Theorem 5. *If φ is bounded and Borel measurable on $\overline{D}$, and f is Lipschitz continuous, then there is a unique stochastic solution to the non-linear parabolic equation (24)*

We will study the regularity theory of the stochastic solutions in a separate paper. On the other hand we would like to derive an alternative probability representation of the stochastic solution.

Let us apply the approach outlined in [20]. Let $Y_s = N_s - V_s$ where

$$N_s - N_0 = \sum_{j=1}^{d} \int_0^s Z_r^j \, dM_r^j .$$

Then $V = (V_s)_{s \in [0,t]}$ is the unique solution to the functional differential equation

$$V_s = \int_0^s f(t-r, Y(V)_r, Z(V)_r) dr, \;\; V_0 = 0 \tag{26}$$

where $N(V)_s = -E^{\mu}\{\varphi(X_t) + V_t | \mathscr{F}_s^{\mu}\}$,

$$Y(V)_s = E^{\mu}\{\varphi(X_t) + V_t | \mathscr{F}_s^{\mu}\} - V_s$$

for $s \in [0,t]$, and $Z(V)$ is given as the density process of $N(V)$ in the martingale representation. In particular

$$Y(V)_0 = E^{\mu}\{\varphi(X_t) + V_t | \mathscr{F}_0^{\mu}\}.$$

We therefore have the following

Theorem 6. *Let φ be bounded and measurable. For $t > 0$, let $V(t)$ be the unique solution to the functional differential equation (26). Then the stochastic solution to the Neumann boundary problem of the non-linear PDE (24) is given by*

$$u(t,\mu) = E^{\mu}\{\varphi(X_t) + V(t)_t\} \qquad \forall \mu \in \mathscr{S}_1(\overline{D}). \tag{27}$$

Acknowledgements The research of the first author was supported in part by EPSRC grant EP/F029578/1, and by the Oxford-Man Institute. The second author's research was supported in part by the National Basic Research Program of China (973 Program) under grant No. 2007CB814904, and a Royal Society Visiting grant.

References

1. V. Bally, E. Pardoux, L. Stoica, Backward stochastic differential equations associated to a symmetric Markov process. Potential Anal. **22**(1), 17–60 (2005)
2. R.F. Bass, P. Hsu, Some potential theory for reflecting Brownian motion in Hölder and Lipschitz domains. Ann. Probab. **19**(2), 486–508 (1991)

3. R.M. Blumenthal, An extended Markov property. Trans. Am. Math. Soc. **85**, 52–72 (1957)
4. R.M. Blumenthal, R.K. Getoor, Markov Processes and Potential Theory. *Pure and Applied Mathematics*, vol 29 (Academic Press, New York, 1968)
5. P. Briand, Y. Hu, BSDE with quadratic growth and unbounded terminal value. Probab. Theory Relat. Fields **136**(4), 604–618 (2006)
6. Z.Q. Chen, Pseudo Jordan domains and reflecting Brownian motions. Probab. Theory Relat. Fields **94**(2), 271–280 (1992)
7. Z.Q. Chen, P.J. Fitzsimmons, R.J. Williams, Reflecting Brownian motions: quasimartingales and strong Caccioppoli sets. Potential Anal. **2**(3), 219–243 (1993)
8. K.L. Chung, Lectures from Markov processes to Brownian motion, Grundlehren der Mathematischen Wissenschaften [Fundamental Principles of Mathematical Science], vol. 249 (Springer, Berlin, Heidelberg, 1982)
9. K.L. Chung, J.B. Walsh, Meyer's theorem on predictability, Z. Wahrscheinlichkeitstheorie und Verw. Gebiete **29**, 253–256 (1974)
10. C. Dellacherie, P.A. Meyer, *Probabilités et potentiel*, Hermann, Paris, 1975, Chapitres I à IV, Édition entièrement refondue, Publications de l'Institut de Mathématique de l'Université de Strasbourg, No. XV, Actualités Scientifiques et Industrielles, No. 1372.
11. N. El Karoui, S. Peng, M.C. Quenez, Backward stochastic differential equations in finance. Math. Finance **7**(1), 1–71 (1997)
12. H. Föllmer, *Dirichlet Processes*, Stochastic integrals (Proc. Sympos., Univ. Durham, Durham, 1980), Lecture Notes in Mathematics, vol. 851 (Springer, Berlin, Heidelberg, 1981), pp. 476–478
13. M. Fukushima, Y. Ōshima, M. Takeda, *Dirichlet Forms and Symmetric Markov Processes*, de Gruyter Studies in Mathematics, vol. 19 (Walter de Gruyter & Co., Berlin, 1994)
14. G.A. Hunt, Markoff processes and potentials. I, II. Illinois J. Math. **1**, 44–93, 316–369 (1957)
15. J. Jacod, in *Calcul Stochastique et Problèmes de Martingales*. Lecture Notes in Mathematics, vol. 714 (Springer, Berlin, Heidelberg, 1979)
16. J. Jacod and M. Yor, Étude des solutions extrémales et représentation intégrale des solutions pour certains problèmes de martingales. Z. Wahrscheinlichkeitstheorie und Verw. Gebiete **38**(2), 83–125 (1977)
17. M. Kobylanski, Backward stochastic differential equations and partial differential equations with quadratic growth. Ann. Probab. **28**(2), 558–602 (2000)
18. H. Kunita, S. Watanabe, On square integrable martingales. Nagoya Math. J. **30**, 209–245 (1967)
19. A. Lejay, BSDE driven by Dirichlet process and semi-linear parabolic PDE. Application to homogenization. Stochastic Process. Appl. **97**(1), 1–39 (2002)
20. G.C. Liang, T. Lyons, Z.M. Qian, Backward stochastic dynamics on a filtered probability space. Ann. Probab. **39**(4), 1422–1448 (2011)
21. T.J. Lyons, W.A. Zheng, Diffusion processes with nonsmooth diffusion coefficients and their density functions. Proc. R. Soc. Edinburgh Sect. A **115**(3–4), 231–242 (1990)
22. T.J. Lyons, W.A. Zheng, A crossing estimate for the canonical process on a Dirichlet space and a tightness result. Astérisque, 157–158, 249–271 (1988), Colloque Paul Lévy sur les Processus Stochastiques (Palaiseau, 1987)
23. J. Ma, P. Protter, J.M. Yong, Solving forward-backward stochastic differential equations explicitly—a four step scheme. Probab. Theory Relat. Fields **98**(3), 339–359 (1994)
24. V.G. Maz'ja, in *Sobolev spaces*, Springer Series in Soviet Mathematics (Springer, Berlin, Heidelberg, 1985), Translated from the Russian by T. O. Shaposhnikova
25. P.A. Meyer, *Processus de Markov*. Lecture Notes in Mathematics, No. 26 (Springer, Berlin, Heidelberg, 1967)
26. D. Nualart, W. Schoutens, Chaotic and predictable representations for Lévy processes. Stochast. Process. Appl. **90**(1), 109–122 (2000)
27. É. Pardoux, S.G. Peng, Adapted solution of a backward stochastic differential equation. Syst. Control Lett. **14**(1), 55–61 (1990)
28. A. Rozkosz, Backward SDEs and Cauchy problem for semilinear equations in divergence form. Probab. Theory Relat. Fields **125**(3), 393–407 (2003a)

29. A. Rozkosz, On a decomposition of symmetric diffusions with reflecting boundary conditions. Stochast. Process. Appl. **103**(1), 101–122 (2003b)
30. A. Rozkosz, BSDEs with random terminal time and semilinear elliptic PDEs in divergence form. Studia Math. **170**(1), 1–21 (2005)
31. J.M. Yong, X.Y. Zhou, Stochastic controls. *Applications of Mathematics (New York)*, vol. 43 (Springer, Berlin, Heidelberg, 1999), Hamiltonian systems and HJB equations
32. Y.N. Zhang, W.A. Zheng, Discretizing a backward stochastic differential equation. Int. J. Math. Math. Sci. **32**(2), 103–116 (2002)
33. W.A. Zheng, *On Symmetric Diffusion Processes*, Probability theory and its applications in China, Contemp. Math., vol. 118, Am. Math. Soc., Providence, RI, 1991, pp. 329–333

Quadratic Semimartingale BSDEs Under an Exponential Moments Condition

Markus Mocha and Nicholas Westray

Abstract In the present article we provide existence, uniqueness and stability results under an exponential moments condition for quadratic semimartingale backward stochastic differential equations (BSDEs) having convex generators. We show that the martingale part of the BSDE solution defines a true change of measure and provide an example which demonstrates that pointwise convergence of the drivers is not sufficient to guarantee a stability result within our framework.

Keywords Quadratic semimartingale BSDEs • Convex generators • Exponential moments

AMS Classification: 60H10

1 Introduction

Since their introduction by Bismut [3] within the Pontryagin maximum principle, backward stochastic differential equations (BSDEs) have attracted much attention in the mathematical literature. In a Brownian framework such equations are usually

---

This chapter was originally published within the PhD thesis "Utility Maximization and Quadratic BSDEs under Exponential Moments" by Markus Mocha, Humboldt University Berlin, 2012. With kind permission of the author.

M. Mocha (✉)
Institut für Mathematik, Humboldt-Universität zu Berlin, Unter den Linden 6, 10099 Berlin, Germany
e-mail: mocha@math.hu-berlin.de

N. Westray
Department of Mathematics, Imperial College, London SW7 2AZ, UK
e-mail: n.westray@imperial.ac.uk

C. Donati-Martin et al. (eds.), *Séminaire de Probabilités XLIV*, Lecture Notes in Mathematics 2046, DOI 10.1007/978-3-642-27461-9_5,

written

$$dY_t = Z_t\, dW_t - F(t, Y_t, Z_t)\, dt, \quad Y_T = \xi, \tag{1}$$

where ξ is an $\mathscr{F}_T$-measurable random variable, the *terminal value*, and F is the so called *driver* or *generator*. Here $(\mathscr{F}_t)_{t\in[0,T]}$ denotes the filtration generated by the one-dimensional Brownian motion W. Solving such a BSDE corresponds to finding a *pair* of adapted processes (Y, Z) such that the integrated version of (1) holds. The presence of the *control process* Z stems from the requirement of adaptedness for Y together with the fact that Y must be driven into the random variable ξ at time T. One may think of Z as arising from the martingale representation theorem.

In the more general semimartingale framework, where the main source of randomness is encoded in a given continuous local martingale M on a filtration $(\mathscr{F}_t)_{t\in[0,T]}$ that is not necessarily generated by M, we have to add an extra orthogonal component N. The corresponding BSDE then takes the form

$$dY_t = Z_t\, dM_t + dN_t - f(t, Y_t, Z_t)\, d\langle M\rangle_t - g_t\, d\langle N\rangle_t, \quad Y_T = \xi. \tag{2}$$

Solving (2) now corresponds to finding an adapted *triple* (Y, Z, N) of processes satisfying the integrated version of (2), where N is a (continuous) local martingale orthogonal to M. We refer to $Z \cdot M + N$ as the *martingale part* of a solution to the BSDE (2).

BSDEs of type (1) and (2) have found many fields of application in mathematical finance. The first problems to be attacked by means of such equations included pricing and hedging, superreplication and recursive utility, for the latter see for instance Schroder and Skiadas [30] and Skiadas [32]. For a general overview the reader is directed to the survey articles El Karoui, Peng and Quenez [12] and El Karoui, Hamadéne and Matoussi [10] and the references therein. A second large focus has been on their use in constrained utility maximization. In a Brownian setting Hu, Imkeller and Müller [16] used the martingale optimality principle to derive a BSDE for the value process characterizing the optimal wealth and investment strategy. Their article can be regarded as an extension of earlier work by Rouge and El Karoui [29] as well as Sekine [31]. In related work in a semimartingale setting Mania and Schweizer [22] used a BSDE to describe the dynamic indifference price for exponential utility. Their stochastic control approach was extended to robust utility in Bordigoni, Matoussi and Schweizer [4] and to an infinite time horizon in the recent article by Hu and Schweizer [17]. We also mention Becherer [2] for further extensions to BSDEs with jumps and Mania and Tevzadze [23] to backward stochastic partial differential equations. This list is by far not exhaustive and additional references can be found in the stated papers.

With regards to the theory of BSDEs, existence and uniqueness results were first provided in a Brownian setting by Pardoux and Peng [27] and in a semimartingale setting by El Karoui and Huang [11] both under Lipschitz conditions. These were extended (in the Brownian case) by Lepeltier and San Martín [21] to continuous drivers with linear growth and by Kobylanski [20] to generators which are quadratic as a function of the control variable Z. The corresponding results for the semimartingale case may be found in Morlais [25] and Tevzadze [33], where in the former the main theorems of [16] are extended. In addition a stability result for quadratic BSDEs may also be found in the recent article by Frei [13]. In the situation when the generator has superquadratic growth, Delbaen, Hu and Bao [8] show that such BSDEs are essentially ill-posed.

A strong requirement present in the articles [20, 25, 33] is that the terminal condition be bounded. In a Brownian setting Briand and Hu [5,6] have replaced this by the assumption that it need only have exponential moments but in addition the driver is convex in the Z variable. By interpreting the Y component as the solution to a stochastic control problem, Delbaen, Hu and Richou [9] extend their results and show that one can reduce the order of exponential moments required. Let us finally mention a paper of Barrieu and El Karoui [1], prepared independently of the present article. They prove stability theorems for a class of continuous semimartingales whose finite variation processes satisfy a quadratic growth condition. Viewing BSDEs as a subclass they then derive existence and monotone stability results under a slightly weaker exponential moments condition than that given here, however there is no uniqueness.

The present article has two main contributions, the first is to extend the existence, uniqueness and stability theorems of [6] and [25] to the unbounded semimartingale case. The motivation here is predominantly mathematical, having results in greater generality increases the range of applications for BSDEs. We remark however, that there are additional practical applications for the results derived here, e.g. related to utility maximization with an unbounded mean variance tradeoff, see Nutz [26] and Mocha and Westray [24], which provides a second motivation for the present work.

In order to prove the respective results in the unbounded semimartingale framework technical difficulties related to an a priori estimate must be overcome. This requires an additional assumption when compared to [6] and [25]. As a biproduct of establishing our results we are able to show via an example that the stability theorem as stated in [6] Proposition 7 needs a minor amendment to the mode of convergence assumed on the drivers and we include the appropriate formulation.

Our second contribution is to address the question of measure change. It is a classical result that when the generator has quadratic growth in z then the solution processes Y is bounded if and only if the martingale part $Z \cdot M + N$ is a BMO martingale. In the present setting, where Y is assumed to satisfy an exponential moments condition only, such a correspondence is lost. However, we are able to show that whilst $Z \cdot M + N$ need not be a BMO martingale, see Frei, Mocha and Westray [14] for further discussion and some examples, the stochastic exponential $\mathscr{E}\big(q(Z \cdot M + N)\big)$ is still a true martingale for q valid in some half-line. However, it is not only mathematically interesting to be able to describe the properties of the martingale part of the BSDE but also relevant for applications, notably in an unbounded setting. For instance, the above result can be used to extend the results of [16] and [25] on utility maximization. Moreover such a theorem may be used in the partial equilibrium framework of Horst, Pirvu and dos Reis [15] where the market price of external risk is given by equilibrium considerations and is typically unbounded.

The paper is organized as follows. In the next section we lay out the notation and the assumptions and state the main results. The subsequent sections contain the proofs. Section 3 gives the a priori estimates together with some remarks on the necessity of an additional assumption, Sect. 4 deals with existence and Sect. 5 includes the comparison and uniqueness results. In Sects. 6 and 7 we prove the stability property as well as providing an interesting counterexample. In Sect. 8, we turn our attention to the measure change problem and finally, in Sect. 9, we give interesting applications of our results to constrained utility maximization and partial equilibrium models.

2 Model Formulation and Statement of Results

We work on a filtered probability space $(\Omega, \mathscr{F}, (\mathscr{F}_t)_{0\leq t\leq T}, \mathbb{P})$ satisfying the usual conditions of right-continuity and completeness. We also assume that $\mathscr{F}_0$ is the completion of the trivial σ-algebra. The time horizon T is a finite number in $(0,\infty)$ and all semimartingales are considered equal to their càdlàg modification.

Throughout this paper $M = (M^1, \dots, M^d)^\top$ stands for a continuous d-dimensional local martingale, where $^\top$ denotes transposition. We refer the reader to Jacod and Shiryaev [18] and Protter [28] for further details on the general theory of stochastic integration.

The objects of study in the present paper will be semimartingale BSDEs considered on $[0, T]$. In the d-dimensional case such a BSDE may be written

$$dY_t = Z_t^\top dM_t + dN_t - \mathbf{1}^\top d\langle M\rangle_t f(t, Y_t, Z_t) - g_t\, d\langle N\rangle_t, \quad Y_T = \xi. \tag{3}$$

Here ξ is an $\mathbb{R}$-valued $\mathscr{F}_T$-measurable random variable and f and g are random predictable functions $[0,T]\times\Omega\times\mathbb{R}\times\mathbb{R}^d \to \mathbb{R}^d$ and $[0,T]\times\Omega \to \mathbb{R}$, respectively. We set $\mathbf{1} := (1,\dots,1)^\top \in \mathbb{R}^d$.

The format in which the BSDE (3) encodes its finite variation parts is not so tractable from the point of view of analysis. Therefore we write semimartingale BSDEs by factorizing the matrix-valued process $\langle M\rangle = \langle M^i, M^j\rangle_{i,j=1,\dots,d}$. This separates its matrix property from its nature as measure.

For $i, j \in \{1,\dots,d\}$ we may write $\langle M^i, M^j\rangle = C^{ij}\cdot A$ where C^{ij} are the components of a predictable process C valued in the space of symmetric positive semidefinite $d\times d$ matrices and A is a predictable increasing process. There are many such factorizations (cf. [18] Sect. III.4a). We may choose $A := \arctan\left(\sum_{i=1}^d \langle M^i\rangle\right)$ so that A is uniformly bounded by $K_A = \pi/2$ and derive the absolute continuity of all the $\langle M^i, M^j\rangle$ with respect to A from the Kunita–Watanabe inequality. This together with the Radon-Nikodým theorem provides C. Furthermore, we can factorize C as $C = B^\top B$ for a predictable process B valued in the space of $d\times d$ matrices. We note that all the results below do not rely on the specific choice of A, but only on its *boundedness*. In particular, if $M = W$ is a d-dimensional Brownian motion we may choose $A_t = t$, $t \in [0, T]$, and B the identity matrix. Then A is bounded by $K_A = T$.

We let $\mathscr{P}$ denote the predictable σ-algebra on $[0, T] \times \Omega$ generated by all the left-continuous adapted processes. The process A induces a measure μ^A on $\mathscr{P}$, the *Doléans measure*, defined for $E \in \mathscr{P}$ by

$$\mu^A(E) := \mathbb{E}\left[\int_0^T \mathbf{1}_E(t)\, dA_t\right].$$

Given the above discussion (3) may be rewritten as

$$dY_t = Z_t^{\mathsf{T}}\, dM_t + dN_t - F(t, Y_t, Z_t)\, dA_t - g_t\, d\langle N\rangle_t, \quad Y_T = \xi, \tag{4}$$

where again ξ is an $\mathbb{R}$-valued $\mathscr{F}_T$-measurable random variable, the *terminal condition*, and F and g are random predictable functions $[0, T] \times \Omega \times \mathbb{R} \times \mathbb{R}^d \to \mathbb{R}$ and $[0, T] \times \Omega \to \mathbb{R}$ respectively, called *generators* or *drivers*. This formulation of the BSDE is very flexible, allowing for various applications and being amenable to analysis.

Starting with (3) and setting $F(t, y, z) := \mathbf{1}^{\mathsf{T}} C_t f(t, y, z) = \mathbf{1}^{\mathsf{T}} B_t^{\mathsf{T}} B_t f(t, y, z)$ we get (4). The reversion of this procedure is not so clear, however is not relevant in applications.

Under boundedness assumptions, existence of solutions to (4) is provided in [25] via an exponential transformation that makes the $d\langle N\rangle$ term disappear. A necessary condition for this kind of transformation to work properly is $dg = 0$. In the sequel we thus consider the above BSDE to be given in the form

$$dY_t = Z_t^{\mathsf{T}}\, dM_t + dN_t - F(t, Y_t, Z_t)\, dA_t - \frac{1}{2}\, d\langle N\rangle_t, \quad Y_T = \xi, \tag{5}$$

except in specific situations where a solution is assumed to exist.

Definition 1. A *solution* to the BSDE (4), or (5), is a triple (Y, Z, N) of processes valued in $\mathbb{R} \times \mathbb{R}^d \times \mathbb{R}$ satisfying (4), or (5), $\mathbb{P}$-a.s. such that:

(i) The function $t \mapsto Y_t$ is continuous $\mathbb{P}$-a.s.
(ii) The process Z is predictable and M-integrable, in particular $\int_0^T Z_t^{\mathsf{T}}\, d\langle M\rangle_t Z_t < +\infty$ $\mathbb{P}$-a.s.
(iii) The local martingale N is continuous and orthogonal to each component of M, i.e. $\langle N, M^i\rangle = 0$ for all $i = 1, \ldots, d$.
(iv) We have that $\mathbb{P}$-a.s.

$$\int_0^T |F(t, Y_t, Z_t)|\, dA_t + \langle N\rangle_T < +\infty.$$

As in the introduction we call $Z \cdot M + N$ the *martingale part* of a solution.

In what follows we collect together the assumptions that allow for all the assertions of this paper to hold simultaneously. However we want to point out that not all of our results require that every item of Assumption 1 be satisfied, as will be indicated in appropriate remarks.

Assumption 1 *There exist nonnegative constants β and $\overline{\beta}$, positive numbers β_f and $\gamma \geq \max(1, \beta)$ together with an M-integrable (predictable) $\mathbb{R}^d$-valued process λ so that writing*

$$\alpha := \|B\lambda\|^2 \text{ and } |\alpha|_1 := \int_0^T \alpha_t \, dA_t = \int_0^T \lambda_t^T d\langle M \rangle_t \lambda_t$$

we have $\mathbb{P}$-a.s.

(i) The random variable $|\xi| + |\alpha|_1$ has exponential moments of all orders, i.e. for all $p > 1$

$$\mathbb{E}\Big[\exp\Big(p\,[|\xi| + |\alpha|_1]\Big)\Big] < +\infty. \tag{6}$$

(ii) For all $t \in [0, T]$ the driver $(y, z) \mapsto F(t, y, z)$ is continuous in (y, z), convex in z and Lipschitz continuous in y with Lipschitz constant $\overline{\beta}$, i.e. for all y_1, y_2 and z we have

$$|F(t, y_1, z) - F(t, y_2, z)| \leq \overline{\beta}\, |y_1 - y_2|. \tag{7}$$

(iii) The generator F satisfies a quadratic growth condition in z, i.e. for all t, y and z we have

$$|F(t, y, z)| \leq \alpha_t + \alpha_t \beta |y| + \frac{\gamma}{2} \|B_t z\|^2. \tag{8}$$

(iv) The function F is locally Lipschitz in z, i.e. for all t, y, z_1 and z_2

$$|F(t, y, z_1) - F(t, y, z_2)| \leq \beta_f \Big(\|B_t \lambda_t\| + \|B_t z_1\| + \|B_t z_2\|\Big) \|B_t (z_1 - z_2)\|.$$

(v) The constant β in (iii) equals zero and then we set $c_A := 0$. Alternatively, $\beta > 0$, but additionally assume that for all t, y and z we have

$$|F(t, y, z) - F(t, 0, z)| \leq \overline{\beta}\, |y| \quad \text{and} \quad A_t \leq c_A \cdot t$$

for a positive constant c_A.

If this assumption is satisfied we refer to (5) as BSDE(F, ξ) with the set of parameters $(\alpha, \beta, \overline{\beta}, \beta_f, \gamma)$.

Remark 1. The above items (i)–(iv) correspond to the assumptions made in [6] and [25]. In particular, the BSDEs under consideration are of *quadratic type* (in the control variable z) and of Lipschitz type in y. Item (v) is new and arises from the fact that the methods used in [25] to derive an a priori estimate may no longer be directly applied so that an additional assumption is required. We elaborate further on this topic in Sect. 3. Observe that in the key application of utility maximization, cf. [24], the associated driver is independent of y and hence $\beta = 0$ applies. In particular, we need not assume that the mean-variance tradeoff of the underlying market be bounded.

Notice that items (ii) and (iii) from above provide

$$|F(t, y, z)| \leq \alpha_t + \overline{\beta}|y| + \frac{\gamma}{2} \|B_t z\|^2, \tag{9}$$

for all t, y and z, $\mathbb{P}$-a.s. This is an inequality which does not involve α in the $|y|$ term on the right hand side and which is used repeatedly throughout the proofs. We also define the constant

$$\beta^* := c_A \cdot \overline{\beta}. \tag{10}$$

Before giving the main results of the paper let us introduce some notation. For $p \geq 1$, $\mathscr{S}^p$ denotes the set of $\mathbb{R}$-valued, adapted and continuous processes Y on $[0, T]$ such that

$$\mathbb{E}\left[\sup_{0 \leq t \leq T} |Y_t|^p\right]^{1/p} < +\infty.$$

The space $\mathscr{S}^\infty$ consists of the continuous bounded processes. An $\mathbb{R}$-valued, adapted and continuous process Y belongs to $\mathfrak{E}$ if the random variable

$$Y^* := \sup_{t \in [0,T]} |Y_t|$$

has exponential moments of all orders. We also recall that Y is called *of class D* if the family $\{Y_\tau | \tau \in [0, T] \text{ stopping time}\}$ is uniformly integrable. The set of (equivalence classes of) $\mathbb{R}^d$-valued predictable processes Z on $[0, T] \times \Omega$ satisfying

$$\mathbb{E}\left[\left(\int_0^T Z_t^\top d\langle M\rangle_t Z_t\right)^{p/2}\right]^{1/p} < +\infty$$

is denoted by $\mathfrak{M}^p$. Finally, $\mathscr{M}^p$ stands for the set of $\mathbb{R}$-valued martingales N on $[0, T]$, such that

$$\|N\|_{\mathscr{M}^p} := \mathbb{E}\left[\langle N\rangle_T^{p/2}\right]^{1/p} < +\infty.$$

Notice that if the following assumption on the filtration holds the elements of $\mathscr{M}^p$ are continuous.

Assumption 2 *The filtration $(\mathscr{F}_t)_{t\in[0,T]}$ is a continuous filtration, in the sense that all local $(\mathscr{F}_t)_{t\in[0,T]}$-martingales are continuous.*

The following four theorems constitute the main results of the paper. We mention that only the existence result requires the assumption of the continuity of the filtration.

Theorem 1 (Existence). *If Assumptions 1 and 2 hold there exists a solution (Y, Z, N) to the BSDE (5) such that $Y \in \mathfrak{E}$ and $Z \cdot M + N \in \mathscr{M}^p$ for all $p \geq 1$.*

Theorem 2 (Uniqueness). *Suppose that Assumption 1 holds. Then any two solutions (Y, Z, N) and (Y', Z', N') in $\mathfrak{E} \times \mathfrak{M}^2 \times \mathscr{M}^2$ to the BSDE (5) coincide in the sense that Y and Y', $Z \cdot M$ and $Z' \cdot M$, and N and N' are indistinguishable.*

Theorem 3 (Stability). *Consider a family of BSDEs(F^n, ξ^n) indexed by the extended natural numbers $n \geq 0$ for which Assumption 1 holds true with parameters $(\alpha^n, \beta^n, \overline{\beta}, \beta_f, \gamma)$. Assume that the exponential moments assumption (6) holds uniformly in n, i.e. for all $p > 1$,*

$$\sup_{n \geq 0} \mathbb{E}\left[e^{p\,(|\xi^n| + |\alpha^n|_1)}\right] < +\infty.$$

If for $n \geq 0$ (Y^n, Z^n, N^n) is the solution in $\mathfrak{E} \times \mathfrak{M}^2 \times \mathscr{M}^2$ to the BSDE(F^n, ξ^n) and if

$$|\xi^n - \xi^0| + \int_0^T \left|F^n - F^0\right|(s, Y_s^0, Z_s^0)\, dA_s \longrightarrow 0 \quad \text{in probability, as } n \to +\infty, \tag{11}$$

then for each $p \geq 1$ as $n \to +\infty$

$$\mathbb{E}\left[\left(\exp\left(\sup_{0 \leq t \leq T} \left|Y_t^n - Y_t^0\right|\right)\right)^p\right] \longrightarrow 1 \quad \text{and}$$

$$Z^n \cdot M + N^n \longrightarrow Z^0 \cdot M + N^0 \text{ in } \mathscr{M}^p.$$

Theorem 4 (Exponential Martingales). *Suppose that Assumption 1 holds, let $|q| > \gamma/2$ and let $(Y, Z, N) \in \mathfrak{E} \times \mathfrak{M}^2 \times \mathscr{M}^2$ be a solution to the BSDE (5). Then the local martingale $\mathscr{E}\big(q\,(Z \cdot M + N)\big)$ is a true martingale on $[0, T]$.*

Remark 2. The preceding theorems generalize the results of [6] and [25]. For their proofs we combine the localization and θ-technique from [6] together with the existence and stability results for BSDEs with bounded solutions found in [25]. Similar ideas are used in [17] on a specific quadratic BSDE arising in a robust utility maximization problem where the authors also investigate the measure change problem for their special BSDE, however here we pursue the general theory. We point out that when the BSDE is of quadratic type and $|\xi| + |\alpha|_1$ does not have sufficiently large exponential moments there are examples where the BSDE admits no solution, see [14]. In particular, we present all the theoretical background for the study of utility maximization under exponential moments, see [24], as well as partial equilibrium, see [15].

3 A Priori Estimates

In this section we show that, under appropriate conditions, solutions to the BSDE (4) satisfy some a priori norm bounds. After giving an important result used in the subsequent sections we motivate Assumption 1 (v) by showing that without such an assumption the method utilized in [25] for the purpose of deriving appropriate a priori bounds fails in the present unbounded case.

Let (Y, Z, N) be a solution to (4), suppose that Assumption 1 (iii) and (v) hold and that g is uniformly bounded by $\gamma/2$. Recall β^* from (10), fix $s \in [0, T]$ and set, for $t \in [s, T]$,

$$\widetilde{H}_t := \exp\left(\gamma e^{\beta^*(t-s)}|Y_t| + \gamma \int_s^t e^{\beta^*(r-s)}\, d\langle \lambda \cdot M\rangle_r\right).$$

where we have written $\langle \lambda \cdot M\rangle_t := \int_0^t \lambda_r^\top \, d\langle M\rangle_r \lambda_r = \int_0^t \alpha_r \, dA_r$. First we show that $\widetilde{H}$ is, up to integrability, a local submartingale.

From Tanaka's formula,

$$d|Y_t| = \operatorname{sgn}(Y_t)(Z_t^\top dM_t + dN_t) - \operatorname{sgn}(Y_t)\big(F(t, Y_t, Z_t)\, dA_t + g_t \, d\langle N\rangle_t\big) + dL_t, \tag{12}$$

where L is the local time of Y at 0. Itô's formula then yields

$$\begin{aligned} d\widetilde{H}_t = \gamma \widetilde{H}_t e^{\beta^*(t-s)}\Bigg[& \operatorname{sgn}(Y_t)(Z_t^\top dM_t + dN_t) + \overline{\beta}|Y_t|(c_A\, dt - dA_t) \\ & + \left(-\operatorname{sgn}(Y_t)F(t, Y_t, Z_t) + \alpha_t + \overline{\beta}|Y_t| + \frac{\gamma}{2} e^{\beta^*(t-s)}\|B_t Z_t\|^2\right) dA_t \\ & + \left(-\operatorname{sgn}(Y_t)\, g_t + \frac{\gamma}{2} e^{\beta^*(t-s)}\right) d\langle N\rangle_t + dL_t\Bigg]. \end{aligned} \tag{13}$$

An inspection of the finite variation parts shows that under the present assumptions they are nonnegative. In particular, the semimartingale $\widetilde{H}$ is a local submartingale, which leads to the following result.

Proposition 1 (A Priori Estimate). *Suppose Assumption 1 (iii) and (v) hold and assume that the function g is uniformly bounded by $\gamma/2$, $\mathbb{P}$-a.s. Let (Y, Z, N) be a solution to the BSDE* (4) *and let the process*

$$\exp\left(\gamma e^{\beta^* T}|Y| + \gamma \int_0^T e^{\beta^* r}\, d\langle \lambda \cdot M\rangle_r\right)$$

be of class D. Then $\mathbb{P}$-a.s. for all $s \in [0, T]$,

$$|Y_s| \leq \frac{1}{\gamma} \log \mathbb{E}\left[\exp\left(\gamma e^{\beta^*(T-s)} |\xi| + \gamma \int_s^T e^{\beta^*(r-s)}\, d\langle \lambda \cdot M \rangle_r \right) \middle| \mathscr{F}_s \right]. \tag{14}$$

Proof. Fix $s \in [0, T]$ and set $\widetilde{H}$ as above. Since $\widetilde{H}$ is a local submartingale there exists a sequence of stopping times $(\tau_n)_{n \geq 1}$ valued in $[s, T]$, which converges $\mathbb{P}$-a.s. to T, such that $\widetilde{H}^{\tau_n}$ is a submartingale for each $n \geq 1$. We then derive

$$\begin{aligned} \exp(\gamma |Y_s|) &\leq \mathbb{E}[\widetilde{H}_{T \wedge \tau_n} \,|\, \mathscr{F}_s] \\ &\leq \mathbb{E}\left[\exp\left(\gamma e^{\beta^*(T-s)} |Y_{T \wedge \tau_n}| + \gamma \int_s^T e^{\beta^*(r-s)}\, d\langle \lambda \cdot M \rangle_r \right) \middle| \mathscr{F}_s \right]. \end{aligned}$$

Letting $n \to +\infty$ the claim follows from the class D assumption. □

Proposition 1 provides the appropriate a priori estimate, indeed suppose that $|\xi|$ and $|\alpha|_1$ are bounded random variables and (Y, Z, N) is a solution to (5). If the current assumptions hold and $\exp(\gamma e^{\beta^* T} |Y|)$ is of class D, then Y satisfies

$$|Y| \leq \left\| e^{\beta^* T} (|\xi| + |\alpha|_1) \right\|_\infty . \tag{15}$$

Comparing with (14) this indicates that the inclusion of Assumption 1 (v) allows us to prove similar estimates to the bounded case which enables us to establish existence for the BSDE (5) when $|\xi| + |\alpha|_1$ has exponential moments of all orders, to be more precise, an order of at least $\gamma e^{\beta^* T}$.

Contrary to the above let us investigate the method utilized in [25] under Assumption 1 (iii) only, supposing that g be bounded by $\gamma/2$. We set

$$H_t := \exp\left(\gamma e^{\beta \langle \lambda \cdot M \rangle_{s,t}} |Y_t| + \gamma \int_s^t e^{\beta \langle \lambda \cdot M \rangle_{s,r}}\, d\langle \lambda \cdot M \rangle_r \right), \tag{16}$$

where $\langle \lambda \cdot M \rangle_{s,t} := \langle \lambda \cdot M \rangle_t - \langle \lambda \cdot M \rangle_s = \int_s^t \alpha_r\, dA_r$. We derive from Itô's formula

$$\begin{aligned} dH_t = \gamma H_t\, e^{\beta \langle \lambda \cdot M \rangle_{s,t}} \Bigg[& \operatorname{sgn}(Y_t)(Z_t^\top dM_t + dN_t) \\ & + \left(-\operatorname{sgn}(Y_t) F(t, Y_t, Z_t) + \alpha_t + \alpha_t \beta |Y_t| + \frac{\gamma}{2} e^{\beta \langle \lambda \cdot M \rangle_{s,t}} \| B_t Z_t \|^2 \right) dA_t \\ & + \left(-\operatorname{sgn}(Y_t)\, g_t + \frac{\gamma}{2} e^{\beta \langle \lambda \cdot M \rangle_{s,t}} \right) d\langle N \rangle_t + dL_t \Bigg]. \end{aligned}$$

Once again, the finite variation parts are nonnegative. We conclude in the same way as for Proposition 1 that the corresponding a priori result holds for H as well. To sum up, we have that under a similar class D assumption, now on

$$\exp\left(\gamma e^{\beta\langle\lambda\cdot M\rangle_T}|Y|+\gamma\int_0^T e^{\beta\langle\lambda\cdot M\rangle_r}\,d\langle\lambda\cdot M\rangle_r\right),$$

$\mathbb{P}$-a.s. for all $s\in[0,T]$,

$$|Y_s|\leq\frac{1}{\gamma}\log\mathbb{E}\left[\exp\left(\gamma e^{\beta\langle\lambda\cdot M\rangle_{s,T}}|\xi|+\gamma\int_s^T e^{\beta\langle\lambda\cdot M\rangle_{s,r}}\,d\langle\lambda\cdot M\rangle_r\right)\Bigg|\mathscr{F}_s\right].\tag{17}$$

If $\beta=0$, then $\widetilde{H}$ from above equals H and there is no difference with the statement of Proposition 1. However when $\beta>0$ the estimate (17) is not sufficient for our purposes. We aim at using the a priori estimate to show the existence of solutions to the BSDE (5) in $\mathfrak{E}\times\mathfrak{M}^2\times\mathscr{M}^2$ using an appropriate approximating procedure. If $|\xi|$ and $|\alpha|_1$ are bounded random variables there exists a solution (Y,Z,N) to (5) with Y bounded, cf. [25]. With (17) at our disposal we then have the estimate

$$|Y|\leq\left\|e^{\beta|\alpha|_1}\left(|\xi|+|\alpha|_1\right)\right\|_\infty.\tag{18}$$

Our goal is to remove the boundedness assumption and to replace it with the assumption on the existence of exponential moments of $|\xi|+|\alpha|_1$ in the spirit of [6]. However a closer inspection of the a priori estimate from (17) together with (18) already indicates that more restrictive assumptions are necessary. More specifically, when $\beta>0$ we cannot deduce any integrability of $\exp\big(\gamma e^{\beta|\alpha|_1}\left(|\xi|+|\alpha|_1\right)\big)$ when $|\xi|$ and $|\alpha|_1$ have only exponential moments. This motivates Assumption 1 (v). Note that we could opt for deriving the existence result under the weaker assumption that the above random variable be integrable. In this case, describing the space in which a solution to the BSDE exists is more technical, as would be a statement of uniqueness. See also [1].

4 Existence

In the present section we establish Theorem 1 together with some related results on norm bounds of the solution. The proof of existence follows the following recipe. First we truncate $\langle\lambda\cdot M\rangle$ to get approximate solutions. Then by using the estimate from Proposition 1 we localize and work on a random time interval so that the approximations are uniformly bounded and we can apply a stability result. Finally we glue together on $[0,T]$ to construct a solution. The a priori estimates ensure that we may take all limits in the described procedure.

Theorem 5 (Existence). *Let Assumptions 1 (ii)–(v) and 2 hold and $|\xi|+|\alpha|_1$ have an exponential moment of order $\gamma e^{\beta^* T}$. Then the BSDE (5) has a solution (Y,Z,N) such that*

$$|Y_t| \le \frac{1}{\gamma}\log \mathbb{E}\left[\exp\left(\gamma\, e^{\beta^*(T-t)}|\xi| + \gamma \int_t^T e^{\beta^*(r-t)}\, d\langle \lambda \cdot M\rangle_r\right)\middle|\, \mathscr{F}_t\right]. \tag{19}$$

Proof. Exactly as in [6] we first assume that F and ξ are nonnegative. For each integer $n \ge 1$, set

$$\sigma_n := \inf\left\{t \in [0,T]\,\middle|\, \langle \lambda \cdot M\rangle_t := \int_0^t \alpha_s\, dA_s \ge n\right\} \wedge T,$$

$\xi^n := \xi \wedge n$, $\lambda_t^n := \mathbf{1}_{\{t\le\sigma_n\}}\lambda_t$ and $F^n(t,y,z) := \mathbf{1}_{\{t\le\sigma_n\}}F(t,y,z)$. Then F^n satisfies Assumption 1 (ii)–(v) with the same constants, but with the processes λ^n and α^n where

$$\alpha_t^n := \|B_t\lambda_t^n\|^2 = \mathbf{1}_{\{t\le\sigma_n\}}\|B_t\lambda_t\|^2 = \mathbf{1}_{\{t\le\sigma_n\}}\alpha_t.$$

In particular, $|\alpha^n|_1 = \int_0^{\sigma_n}\alpha_s\, dA_s \le n$ and

$$\int_0^T (\lambda_t^n)^\top d\langle M\rangle_t \lambda_t^n = \int_0^T \|B_t\lambda_t^n\|^2\, dA_t = |\alpha^n|_1 \le n,$$

so we may apply [25] Theorems 2.5 and 2.6 to conclude that there exists a unique solution $(Y^n, Z^n, N^n) \in \mathscr{S}^\infty \times \mathfrak{M}^2 \times \mathscr{M}^2$ to the BSDE (5), where F is replaced by F^n and ξ by ξ^n. From Proposition 1 we derive

$$\begin{aligned}
|Y_t^n| &\le \frac{1}{\gamma}\log \mathbb{E}\left[\exp\left(\gamma e^{\beta^*(T-t)}|\xi^n| + \gamma \int_t^T e^{\beta^*(r-t)}\, d\langle \lambda^n \cdot M\rangle_r\right)\middle|\, \mathscr{F}_t\right]\\
&\le \frac{1}{\gamma}\log \mathbb{E}\left[\exp\left(\gamma e^{\beta^*(T-t)}|\xi| + \gamma \int_t^T e^{\beta^*(r-t)}\, d\langle \lambda \cdot M\rangle_r\right)\middle|\, \mathscr{F}_t\right]\\
&\le \frac{1}{\gamma}\log \mathbb{E}\left[\exp\left(\gamma e^{\beta^* T}\left(|\xi| + |\alpha|_1\right)\right)\middle|\, \mathscr{F}_t\right] =: X_t.
\end{aligned} \tag{20}$$

Let $n \le m$ so that then we have $\sigma_n \le \sigma_m$ and $\mathbf{1}_{\{t\le\sigma_n\}} \le \mathbf{1}_{\{t\le\sigma_m\}}$. In particular, $\xi^n \le \xi^m$ and $F^n \le F^m$, from which we deduce that the Assumptions 1 (ii)–(v), hence the corresponding assumptions in [25], hold for both F^n and F^m with the same set of parameters $(\alpha^m, \beta, \overline{\beta}, \beta_f, \gamma)$ where the additional c_θ in [25] is equal to m. An application of Theorem 2.7 therein now shows that $Y^n \le Y^m$ so that $(Y^n)_{n\ge1}$ is an increasing sequence of bounded continuous processes.

The next step would be to send n to infinity, however, we do not dispose of a suitable stability result. Indeed we have only [25] Lemma 3.3 which applies for bounded processes under uniform growth assumptions on the drivers, hence we introduce an additional truncation. Let $k \ge 1$ be a fixed integer and

$$\tau_k := \inf\left\{t \in [0,T]\,\middle|\, X_t \ge k \text{ or } \langle \lambda \cdot M\rangle_t \ge k\right\} \wedge T.$$

Thanks to the continuity of the filtration the martingale $\exp(\gamma X)$ is continuous so that the random variable

$$V := \max_{t\in[0,T]}(X_t) \vee \langle \lambda \cdot M\rangle_T$$

is finite $\mathbb{P}$-a.s. We derive that $\mathbb{P}$-a.s. $\tau_k = T$ for large k. Due to (20) the sequence $(Y^{n,k})_{n\geq 1}$ given by

$$Y_t^{n,k} := Y_{t\wedge\tau_k}^n,$$

is uniformly bounded by k. For the martingale parts we define

$$Z_t^{n,k} := \mathbf{1}_{\{t\leq\tau_k\}} Z_t^n \text{ and } N_t^{n,k} := \mathbf{1}_{\{t\leq\tau_k\}} N_t^n.$$

An inspection of the respective cases shows that

$$\begin{aligned} Y_t^{n,k} &= Y_{\tau_k}^n - \int_t^T (Z_s^{n,k})^\top dM_s - \int_t^T dN_s^{n,k} \\ &\quad + \int_t^T \mathbf{1}_{\{s\leq\tau_k\wedge\sigma_n\}} F(s, Y_s^{n,k}, Z_s^{n,k})\, dA_s + \frac{1}{2}\int_t^T d\langle N^{n,k}\rangle_s. \end{aligned}$$

Moreover, $Y_{\tau_k}^n \xrightarrow{n\uparrow+\infty} \sup_{n\geq 1} Y_{\tau_k}^n =: \xi_k$, where ξ_k is bounded by k. Next we appeal to the stability result stated in [25] Lemma 3.3, noting Remark 3.4 therein. Note that this result requires estimates that are uniform in n which is accomplished by the specific choice of the stopping time τ_k. Hence $(Y^{n,k}, Z^{n,k}, N^{n,k})$ converges to $(Y^{\infty,k}, Z^{\infty,k}, N^{\infty,k})$ in the sense that

$$\lim_{n\to+\infty} \mathbb{E}\left(\sup_{0\leq t\leq T}\left|Y_t^{n,k} - Y_t^{\infty,k}\right|\right) = 0,$$

$$\lim_{n\to+\infty} \mathbb{E}\left(\int_0^T \left(Z_s^{n,k} - Z_s^{\infty,k}\right)^\top d\langle M\rangle_s \left(Z_s^{n,k} - Z_s^{\infty,k}\right)\right) = 0$$

and

$$\lim_{n\to+\infty} \mathbb{E}\left(\left|N_T^{n,k} - N_T^{\infty,k}\right|^2\right) = 0,$$

where the triples $(Y^{\infty,k}, Z^{\infty,k}, N^{\infty,k})$ solve the BSDE

$$\begin{aligned} dY_t^{\infty,k} &= \left(Z_t^{\infty,k}\right)^\top dM_t + dN_t^{\infty,k} \\ &\quad - \mathbf{1}_{\{t\leq\tau_k\}} F(t, Y_t^{\infty,k}, Z_t^{\infty,k})\, dA_t - \frac{1}{2} d\langle N^{\infty,k}\rangle_t, \quad Y_{\tau_k}^{\infty,k} = \xi_k, \end{aligned}$$

on the random horizon $[\![0, \tau_k]\!] \subset [0, T]$. The stopping times τ_k are monotone in k and therefore it follows that

$$Y^{n,k+1}_{t\wedge\tau_k} = Y^{n,k}_t, \quad \mathbf{1}_{\{t\le\tau_k\}} Z^{n,k+1}_t = Z^{n,k}_t \quad \text{and} \quad \mathbf{1}_{\{t\le\tau_k\}} N^{n,k+1}_t = N^{n,k}_t,$$

so that the above convergence yields (for the two last objects in $\mathscr{M}^2$)

$$Y^{\infty,k+1}_{t\wedge\tau_k} = Y^{\infty,k}_t, \quad \left(\left(\mathbf{1}_{\{t\le\tau_k\}} Z^{\infty,k+1}\right)\cdot M\right)_t = \left(Z^{\infty,k}\cdot M\right)_t$$
$$\text{and } \mathbf{1}_{\{t\le\tau_k\}} N^{\infty,k+1}_t = N^{\infty,k}_t.$$

To finish the proof, we define the processes

$$Y_t := \mathbf{1}_{\{t\le\tau_1\}} Y^{\infty,1}_t + \sum_{k\ge 2} \mathbf{1}_{\{t\in]\!]\tau_{k-1},\tau_k]\!]\}} Y^{\infty,k}_t,$$
$$Z_t := \mathbf{1}_{\{t\le\tau_1\}} Z^{\infty,1}_t + \sum_{k\ge 2} \mathbf{1}_{\{t\in]\!]\tau_{k-1},\tau_k]\!]\}} Z^{\infty,k}_t$$
$$\text{and } N_t := \mathbf{1}_{\{t\le\tau_1\}} N^{\infty,1}_t + \sum_{k\ge 2} \mathbf{1}_{\{t\in]\!]\tau_{k-1},\tau_k]\!]\}} N^{\infty,k}_t.$$

By construction this gives a solution to the BSDE

$$dY_t = Z^\top_t \, dM_t + dN_t - F(t, Y_t, Z_t)\, dA_t - \frac{1}{2}\, d\langle N\rangle_t, \quad Y_T = \xi,$$

since $\mathbf{1}_{\{t\le\tau_1\}} + \sum_{k\ge 2} \mathbf{1}_{\{t\in]\!]\tau_{k-1},\tau_k]\!]\}} = \mathbf{1}_{\{t\in[0,T]\!]}$ $\mathbb{P}$-a.s. More precisely, there is a $\mathbb{P}$-null set $\mathfrak{N}$ such that for all $\omega \in \mathfrak{N}^c$ there is a minimal $k_0(\omega)$ with $\tau_{k_0(\omega)}(\omega) = T$ and such that $Y^{\infty,k}_{\tau_k(\omega)}(\omega) = \xi_k(\omega)$ for all k, which yields that (possibly after another modification of $\mathfrak{N}$)

$$Y_T(\omega) = Y^{\infty,k_0(\omega)}_T(\omega) = \xi_{k_0(\omega)}(\omega) = \sup_{n\ge 1} Y^n_T(\omega) = \xi(\omega).$$

The bound in (19) holds as we have it for all n and k from (20).
In the case when ξ and f are not necessarily nonnegative, we proceed as in [6] by using a double truncation defined by $\xi^{n,m} := \xi^+ \wedge n - \xi^- \wedge m$, $\lambda^{n,m} := \mathbf{1}_{\{t\le\sigma_n\}}\lambda^+ - \mathbf{1}_{\{t\le\sigma_m\}}\lambda^-$ and $F^{n,m} := \mathbf{1}_{\{t\le\sigma_n\}} F^+ - \mathbf{1}_{\{t\le\sigma_m\}} F^-$. □

As an immediate corollary we deduce

Corollary 1 (Norm Bounds).

(i) Let Assumptions 1 (ii)–(v) and 2 hold and $|\xi| + |\alpha|_1$ have an exponential moment of order $\delta^* > \gamma e^{\beta^* T}$. Then the BSDE (5) has a solution (Y, Z, N) such that $e^{\gamma Y} \in \mathscr{S}^{p^*}$ for $p^* := \frac{\delta^*}{\gamma e^{\beta^* T}} > 1$.

When additionally $|\xi| + |\alpha|_1$ *has exponential moments of all orders, i.e. Assumption 1 (i) holds, this solution is such that* $Y \in \mathfrak{E}$. *In particular, for each* $p > 1$ *we have the estimate*

$$\mathbb{E}\Big[e^{p\gamma Y^*}\Big] \leq \left(\frac{p}{p-1}\right)^p \mathbb{E}\Big[\exp\Big(p\gamma e^{\beta^* T}\big(|\xi| + |\alpha|_1\big)\Big)\Big]. \tag{21}$$

(ii) Let Assumption 1 (i)–(iii) and (v) hold and suppose there exists a solution (Y, Z, N) *to the BSDE* (5) *such that* $Y \in \mathfrak{E}$. *Then* $(Z, N) \in \mathfrak{M}^p \times \mathscr{M}^p$ *for all* $p \geq 1$, *more precisely*

$$\mathbb{E}\left[\left(\int_0^T Z_s^\top d\langle M\rangle_s Z_s + d\langle N\rangle_s\right)^{p/2}\right] \leq c_{p,\gamma}\, \mathbb{E}\Big[\exp\Big(4p\gamma e^{\beta^* T}\big(|\xi| + |\alpha|_1\big)\Big)\Big], \tag{22}$$

where $c_{p,\gamma}$ *is a positive constant depending on* p *and* γ. *The estimate* (21) *then holds as well.*

Proof. (i) Let (Y, Z, N) be the solution to (5) obtained in Theorem 5. As in the previous section set

$$\widetilde{H}_t := \exp\left(\gamma e^{\beta^* t}|Y_t| + \gamma \int_0^t e^{\beta^* r}\, d\langle \lambda \cdot M\rangle_r\right), \tag{23}$$

which is a local submartingale. Moreover, from the estimate (19), Jensen's inequality and the adaptedness of $\int_0^\cdot e^{\beta^* r}\, d\langle\lambda \cdot M\rangle_r$ we deduce that

$$\begin{aligned}
\widetilde{H}_t &= \Big[\exp(\gamma|Y_t|)\Big]^{e^{\beta^* t}} \exp\left(\gamma \int_0^t e^{\beta^* r}\, d\langle \lambda \cdot M\rangle_r\right) \\
&\leq \mathbb{E}\left[\exp\left(\gamma e^{\beta^*(T-t)}|\xi| + \gamma \int_t^T e^{\beta^*(r-t)}\, d\langle \lambda \cdot M\rangle_r\right)\Bigg| \mathscr{F}_t\right]^{e^{\beta^* t}} \\
&\quad \times \exp\left(\gamma \int_0^t e^{\beta^* r}\, d\langle \lambda \cdot M\rangle_r\right) \\
&\leq \mathbb{E}\left[\exp\Big(\gamma e^{\beta^* T}\big(|\xi| + |\alpha|_1\big)\Big)\Bigg| \mathscr{F}_t\right].
\end{aligned}$$

Observe that this upper estimate is a uniformly integrable martingale, in particular it is of class D and therefore $\widetilde{H}$ is a true submartingale. Then, via the Doob maximal inequality, we find that for $p > 1$

$$\mathbb{E}\Big[e^{p\gamma Y^*}\Big] \le \mathbb{E}\Big[\sup_{0\le t\le T} \widetilde{H}_t^p\Big] \le \Big(\frac{p}{p-1}\Big)^p \mathbb{E}[\widetilde{H}_T^p]$$
$$\le \Big(\frac{p}{p-1}\Big)^p \mathbb{E}\Big[\exp\Big(p\gamma e^{\beta^* T}\big(|\xi| + |\alpha|_1\big)\Big)\Big], \quad (24)$$

provided the right hand side is finite. In particular, $e^{\gamma Y} \in \mathscr{S}^{p^*}$ and $Y \in \mathfrak{E}$ as soon as $|\xi| + |\alpha|_1$ has exponential moments of all orders, in which case (21) holds.

(ii) We first verify that (21) continues to hold when (Y, Z, N) is a solution to (5) with $Y \in \mathfrak{E}$. Observe that we may reformulate the result of Proposition 1 under the condition that

$$\exp\Big(\gamma e^{\beta^* T}|Y_\cdot| + \gamma \int_0^\cdot e^{\beta^* r}\, d\langle \lambda \cdot M\rangle_r\Big)$$

be of class D. Repeating the argument from (i) using (14) instead of (19) leads to the same conclusion, since we have the relation

$$\exp\Big(\gamma e^{\beta^* t}|Y_t| + \gamma \int_0^t e^{\beta^* r}\, d\langle \lambda \cdot M\rangle_r\Big) \le \mathbb{E}\Big[\exp\Big(\gamma e^{\beta^* T}\big(Y^* + |\alpha|_1\big)\Big)\Big|\, \mathscr{F}_t\Big],$$

where the right hand side is indeed a process of class D. For the remaining claim, relation (22), define the functions $u, v : \mathbb{R} \to \mathbb{R}_+$ via $u(x) := \frac{1}{\gamma^2}(e^{\gamma x} - 1 - \gamma x)$ and $v(x) := u(|x|)$. We have that v is a $\mathscr{C}^2$ function, so we use Itô's formula to see that for a stopping time τ (to be chosen later)

$$\begin{aligned}
v(Y_0) = v(Y_{t\wedge\tau}) &- \int_0^{t\wedge\tau} u'(|Y_s|)\operatorname{sgn}^*(Y_s)(Z_s^\top dM_s + dN_s)\\
&+ \int_0^{t\wedge\tau} u'(|Y_s|)\operatorname{sgn}^*(Y_s)\Big(F(s, Y_s, Z_s)\, dA_s + \frac{1}{2}\, d\langle N\rangle_s\Big)\\
&- \frac{1}{2}\int_0^{t\wedge\tau} u''(|Y_s|)\Big(Z_s^\top d\langle M\rangle_s Z_s + d\langle N\rangle_s\Big),
\end{aligned}$$

where use the notation $\operatorname{sgn}^*(x) := -\mathbf{1}_{\{x\le 0\}} + \mathbf{1}_{\{x>0\}}$ and observe that $u'(0) = 0$. Assumption 1 (iii) yields

$$\begin{aligned}
v(Y_0) \le v(Y_{t\wedge\tau}) &- \int_0^{t\wedge\tau} u'(|Y_s|)\operatorname{sgn}^*(Y_s)(Z_s^\top dM_s + dN_s)\\
&+ \int_0^{t\wedge\tau} u'(|Y_s|)\Big(\alpha_s + \alpha_s\beta|Y_s|\Big)dA_s\\
&+ \frac{1}{2}\int_0^{t\wedge\tau} \Big(\gamma u'(|Y_s|) - u''(|Y_s|)\Big)Z_s^\top d\langle M\rangle_s Z_s
\end{aligned}$$

$$+\frac{1}{2}\int_0^{t\wedge\tau}\Big(u'(|Y_s|)\,\mathrm{sgn}^*(Y_s)-u''(|Y_s|)\Big)\,d\langle N\rangle_s,$$

since $u'(x)=\frac{1}{\gamma}(e^{\gamma x}-1)\geq 0$ for $x\geq 0$. Using the relation $\gamma u'(x)-u''(x)=-1$ together with $\gamma\geq 1$ it follows that

$$\begin{aligned}0\leq v(Y_0)\leq v(Y_{t\wedge\tau})&-\int_0^{t\wedge\tau}u'(|Y_s|)\,\mathrm{sgn}^*(Y_s)(Z_s^\top\,dM_s+dN_s)\\&+\int_0^{t\wedge\tau}u'(|Y_s|)\Big(\alpha_s+\alpha_s\beta|Y_s|\Big)\,dA_s-\frac{1}{2}\int_0^{t\wedge\tau}Z_s^\top\,d\langle M\rangle_sZ_s+d\langle N\rangle_s.\end{aligned}\tag{25}$$

Suppose first that $p\geq 2$. Then (22) can be proved using the Burkholder–Davis–Gundy inequalities as follows. From (25) we deduce that

$$\begin{aligned}\frac{1}{2}\int_0^{\tau}Z_s^\top\,d\langle M\rangle_sZ_s+d\langle N\rangle_s\leq\frac{1}{\gamma^2}e^{\gamma Y^*}&+\frac{1}{\gamma}\int_0^{T}e^{\gamma|Y_s|}\Big(\alpha_s+\alpha_s\beta|Y_s|\Big)\,dA_s\\&+\sup_{0\leq t\leq T}\left|\int_0^{t\wedge\tau}u'(|Y_s|)\,\mathrm{sgn}^*(Y_s)(Z_s^\top\,dM_s+dN_s)\right|,\end{aligned}$$

where we used the estimates $u'(x)\leq e^{\gamma x}/\gamma$ and $v(x)\leq e^{\gamma x}/\gamma^2$, valid for $x\geq 0$. From the inequalities $y\leq e^y-1$ and $\beta\leq\gamma$ we derive

$$\begin{aligned}\left(\int_0^{\tau}Z_s^\top\,d\langle M\rangle_sZ_s+d\langle N\rangle_s\right)^{p/2}\leq 2^{3p/2-2}\Bigg(&\frac{1}{\gamma^p}e^{p/2\gamma Y^*}+\frac{1}{\gamma^{p/2}}e^{p\gamma Y^*}|\alpha|_1^{p/2}\\&+\sup_{0\leq t\leq T}\left|\int_0^{t\wedge\tau}u'(|Y_s|)\mathrm{sgn}^*(Y_s)(Z_s^\top\,dM_s+dN_s)\right|^{p/2}\Bigg),\end{aligned}$$

which yields, after taking expectation and applying the estimate $|x|^{p/2}<e^{p/2|x|}$ and the Burkholder–Davis–Gundy inequality,

$$\begin{aligned}\mathbb{E}\left[\left(\int_0^{\tau}Z_s^\top\,d\langle M\rangle_sZ_s+d\langle N\rangle_s\right)^{p/2}\right]&\leq c_{p,\gamma}\,\mathbb{E}\left[e^{p/2\gamma Y^*}+e^{p\gamma Y^*}e^{p/2\gamma|\alpha|_1}\right]\\&+c_{p,\gamma}\,\mathbb{E}\left[\left(\int_0^{\tau}e^{2\gamma|Y_s|}\Big(Z_s^\top\,d\langle M\rangle_sZ_s+d\langle N\rangle_s\Big)\right)^{p/4}\right],\end{aligned}$$

where we used the estimate $u'(x)\leq e^{\gamma x}/\gamma$ for $x\geq 0$. Note that in the above and in what follows $c_{p,\gamma}>0$ is a generic constant depending on p and γ that may change from line to line. We apply the generalized Young inequality, $|ab| \leq \frac{\varepsilon}{2} a^2 + \frac{b^2}{2\varepsilon}$, for $\varepsilon := 1$ and for $\varepsilon := c_{p,\gamma}$. Then, after an adjustment of $c_{p,\gamma}$,

$$\begin{aligned}
&\mathbb{E}\Big[\Big(\int_0^\tau Z_s^\top d\langle M\rangle_s Z_s + d\langle N\rangle_s\Big)^{p/2}\Big] \\
&\quad \leq c_{p,\gamma}\Big(\mathbb{E}\Big[e^{p/2\gamma Y^*}\Big] + \frac{1}{2}\mathbb{E}\Big[e^{2p\gamma Y^*}\Big] + \frac{1}{2}\mathbb{E}\Big[e^{p\gamma|\alpha|_1}\Big]\Big) \\
&\qquad + c_{p,\gamma}\,\mathbb{E}\Big[e^{p\gamma Y^*}\Big] + \frac{1}{2}\mathbb{E}\Big[\Big(\int_0^\tau Z_s^\top d\langle M\rangle_s Z_s + d\langle N\rangle_s\Big)^{p/2}\Big] \\
&\quad \leq c_{p,\gamma}\Big(\mathbb{E}\Big[e^{2p\gamma Y^*}\Big] + \mathbb{E}\Big[e^{2p\gamma|\alpha|_1}\Big]\Big) \\
&\qquad + \frac{1}{2}\mathbb{E}\Big[\Big(\int_0^\tau Z_s^\top d\langle M\rangle_s Z_s + d\langle N\rangle_s\Big)^{p/2}\Big].
\end{aligned}$$

Next define, for each integer $n \geq 1$, the stopping time

$$\tau_n := \inf\Big\{t \in [0,T]\,\Big|\,\int_0^t e^{2\gamma|Y_s|}\Big(Z_s^\top d\langle M\rangle_s Z_s + d\langle N\rangle_s\Big) \geq n\Big\} \wedge T.$$

Inserting τ_n into the above calculation and using $e^a + e^b \leq 2e^{a+b}$ for $a, b \geq 0$ together with (21), we may rewrite the last estimate as

$$\mathbb{E}\Big[\Big(\int_0^{\tau_n} Z_s^\top d\langle M\rangle_s Z_s + d\langle N\rangle_s\Big)^{p/2}\Big] \leq c_{p,\gamma}\,\mathbb{E}\Big[\exp\Big(2p\gamma e^{\beta^* T}\Big(|\xi| + |\alpha|_1\Big)\Big)\Big]. \tag{26}$$

By Fatou's lemma, since $\tau_n \to T$ as $n \to +\infty$,

$$\mathbb{E}\Big[\Big(\int_0^T Z_s^\top d\langle M\rangle_s Z_s + d\langle N\rangle_s\Big)^{p/2}\Big] \leq c_{p,\gamma}\,\mathbb{E}\Big[\exp\Big(2p\gamma e^{\beta^* T}\Big(|\xi| + |\alpha|_1\Big)\Big)\Big]$$

and (22) follows. In the situation where $p < 2$, $q := 2/p > 1$ and we may combine Jensen's inequality with (26), which is valid for $p = 2$, to get

$$\begin{aligned}
&\mathbb{E}\Big[\Big(\int_0^{\tau_n} Z_s^\top d\langle M\rangle_s Z_s + d\langle N\rangle_s\Big)^{p/2}\Big]^q \\
&\quad \leq \mathbb{E}\Big[\int_0^{\tau_n} Z_s^\top d\langle M\rangle_s Z_s + d\langle N\rangle_s\Big] \leq c_{2,\gamma}\,\mathbb{E}\Big[\exp\Big(4\gamma e^{\beta^* T}\Big(|\xi| + |\alpha|_1\Big)\Big)\Big]
\end{aligned}$$

from which (22) follows after another application of Fatou's lemma together with the fact that the right hand side in the inequality above is greater or equal one while $1/q = p/2 < 1$. □

Remark 3. We point out that the results of this section do not require that F be convex in z, but only that F be continuous in (y, z). The reader may have noticed that the continuity of F is not used directly in the proofs. However in Theorem 5 we rely on the results of [25] where continuity is a technical condition needed for an application of Dini's theorem. In addition our results also apply to the BSDE (4) if g is identically equal to a nonzero constant $\gamma_g/2$, in which case we assume without loss of generality that $\gamma \geq |\gamma_g|$.

5 Uniqueness

We now provide a comparison theorem that yields uniqueness of the BSDE triple. The proof makes use of the θ-technique applied in the context of second order Bellman-Isaacs equations by Da Lio and Ley [7] and subsequently adapted to the framework of Brownian BSDEs in [6]. We extend these ideas to take into account the orthogonal part of the BSDE solution.

Theorem 6 (Comparison Principle). *Let (Y, Z, N) and (Y', Z', N') be solutions to the BSDE* (5) *with drivers F and F' and terminal conditions ξ and ξ', respectively. Suppose in addition that $Y \in \mathfrak{E}$ and $Y' \in \mathfrak{E}$. If $\mathbb{P}$-a.s. for all $t \in [0, T]$,*

$$\xi \leq \xi' \quad \textit{and} \quad F(t, Y_t', Z_t') \leq F'(t, Y_t', Z_t'),$$

and if (F, ξ) satisfies Assumption 1 (i)–(iii) then $\mathbb{P}$-a.s. for each $t \in [0, T]$

$$Y_t \leq Y_t'.$$

Proof. Let θ be a real number in $(0, 1)$ and set $U := Y - \theta Y'$, $V := Z - \theta Z'$ and $W := N - \theta N'$. Consider the process

$$\rho_s := \begin{cases} \frac{F(s,Y_s,Z_s) - F(s,\theta Y_s',Z_s)}{U_s} & \text{if } U_s \neq 0, \\ \overline{\beta} & \text{if } U_s = 0. \end{cases}$$

By Assumption 1 (ii), ρ is bounded by $\overline{\beta}$ and we define $R_s := \int_0^s \rho_r \, dA_r$. Notice that by the boundedness of A we have that $|R| \leq \overline{\beta} A_T \leq \overline{\beta} K_A$. From Itô's formula we deduce

$$\begin{aligned} e^{R_t} U_t = e^{R_T} U_T &- \int_t^T e^{R_s} (V_s^{\top} \, dM_s + dW_s) \\ &+ \int_t^T e^{R_s} \Big(F_s^{\theta} \, dA_s + \frac{1}{2} \big(d\langle N \rangle_s - \theta \, d\langle N' \rangle_s \big) \Big), \end{aligned}$$

where we define $F_s^\theta := F(s, Y_s, Z_s) - \theta F'(s, Y_s', Z_s') - \rho_s U_s$. We also set

$$\Delta F(s) := (F - F')(s, Y_s', Z_s') \leq 0$$

and observe that from the convexity of F in z together with (9) we get

$$\begin{aligned} &F(s, Y_s', Z_s) - \theta F(s, Y_s', Z_s') \\ &\quad = F\left(s, Y_s', \theta Z_s' + (1-\theta)\frac{Z_s - \theta Z_s'}{1-\theta}\right) - \theta F(s, Y_s', Z_s') \\ &\quad \leq (1-\theta)\, F\left(s, Y_s', \frac{Z_s - \theta Z_s'}{1-\theta}\right) \\ &\quad \leq (1-\theta)\alpha_s + (1-\theta)\overline{\beta}|Y_s'| + \frac{\gamma}{2(1-\theta)} \|B_s V_s\|^2. \end{aligned} \tag{27}$$

Another application of the Lipschitz Assumption 1 (ii), yields

$$\begin{aligned} F(s, Y_s, Z_s) - F(s, Y_s', Z_s) &= F(s, Y_s, Z_s) - F(s, \theta Y_s', Z_s) \\ &\quad + F(s, \theta Y_s', Z_s) - F(s; Y_s', Z_s) \\ &= \rho_s U_s + F(s, \theta Y_s', Z_s) - F(s, Y_s', Z_s) \\ &\leq \rho_s U_s + (1-\theta)\overline{\beta}|Y_s'|. \end{aligned} \tag{28}$$

Combining (27) and (28) we see that

$$\begin{aligned} F_s^\theta &= F(s, Y_s, Z_s) - \theta F(s, Y_s', Z_s') + \theta\, \Delta F(s) - \rho_s U_s \\ &= [F(s, Y_s, Z_s) - F(s, Y_s', Z_s)] + [F(s, Y_s', Z_s) - \theta F(s, Y_s', Z_s')] \\ &\quad + \theta\, \Delta F(s) - \rho_s U_s \\ &\leq (1-\theta)\left(\alpha_s + 2\overline{\beta}|Y_s'|\right) + \frac{\gamma}{2(1-\theta)} \|B_s V_s\|^2 + \theta\, \Delta F(s). \end{aligned} \tag{29}$$

Let $\kappa := \frac{\gamma \exp(\overline{\beta} K_A)}{1-\theta} > 0$ and $P_t := \exp(\kappa e^{R_t} U_t) > 0$. The logic is now similar to how we derived the a priori estimates, namely to show that, by removing an appropriate drift, P is a (local) submartingale. By Itô's formula, for $t \in [0, T]$,

$$\begin{aligned} P_t = P_T - \int_t^T \kappa P_s e^{R_s} (V_s^\top dM_s + dW_s) + \int_t^T \kappa P_s e^{R_s} \left(F_s^\theta - \frac{\kappa e^{R_s}}{2} \|B_s V_s\|^2\right) dA_s \\ + \int_t^T \kappa P_s e^{R_s} \Big(-\frac{\kappa e^{R_s}}{2} d\langle W\rangle_s + \frac{1}{2}\Big(d\langle N\rangle_s - \theta\, d\langle N'\rangle_s\Big)\Big). \end{aligned} \tag{30}$$

To simplify notation set

$$G := \kappa P e^{R}\left(F^{\theta} - \frac{\kappa e^{R}}{2}\,\|BV\|^{2}\right) \quad \text{and} \tag{31}$$

$$H := \int_0^{\cdot} \kappa P_s e^{R_s}\left(-\frac{\kappa e^{R_s}}{2}\,d\langle W\rangle_s + \frac{1}{2}\Big(d\langle N\rangle_s - \theta\, d\langle N'\rangle_s\Big)\right). \tag{32}$$

Let us first investigate the finite variation process H. We claim that H is decreasing, indeed for all $r, u \in [0,T]$, $r \leq u$, we have

$$\int_r^u d\langle W\rangle_s = \int_r^u d\langle N\rangle_s - 2\theta\, d\langle N, N'\rangle_s + \theta^2\, d\langle N'\rangle_s.$$

Applying the Kunita–Watanabe and Young inequalities,

$$\begin{aligned}
\int_r^u d\langle W\rangle_s &\geq \int_r^u d\langle N\rangle_s + \int_r^u \theta^2\, d\langle N'\rangle_s - 2\theta\left(\int_r^u d\langle N\rangle_s\right)^{1/2}\left(\int_r^u d\langle N'\rangle_s\right)^{1/2}\\
&\geq \int_r^u d\langle N\rangle_s + \int_r^u \theta^2\, d\langle N'\rangle_s - \theta\left(\int_r^u d\langle N\rangle_s + \int_r^u d\langle N'\rangle_s\right)\\
&= (1-\theta)\left(\int_r^u d\langle N\rangle_s - \theta\, d\langle N'\rangle_s\right).
\end{aligned}$$

In particular, since $\gamma \geq 1$ and $|R| \leq \overline{\beta} K_A$ we have,

$$\int_r^u \kappa e^{R_s} d\langle W\rangle_s \geq \frac{\gamma}{1-\theta}\int_r^u d\langle W\rangle_s \geq \int_r^u d\langle N\rangle_s - \theta\, d\langle N'\rangle_s,$$

which shows that the process H is decreasing and hence the third integral in (30) is nonpositive.

Next we consider the other finite variation integral in (30). Combining (29), $\Delta F \leq 0$ and the boundedness of R we have

$$G = \kappa P e^{R}\left(F^{\theta} - \frac{\kappa e^{R}}{2}\,\|BV\|^{2}\right) \leq \kappa P e^{R}\Big((1-\theta)\big(\alpha + 2\overline{\beta}|Y'|\big)\Big) \leq PJ, \tag{33}$$

where

$$J := \gamma e^{2\overline{\beta} K_A}\big(\alpha + 2\overline{\beta}|Y'|\big) \geq 0.$$

We set

$$D_t := \exp\left(\int_0^t J_s \, dA_s\right) \quad \text{and} \quad \widetilde{P}_t := D_t P_t.$$

Partial integration yields

$$\begin{aligned} d\widetilde{P}_t &= D_t\Big(-G_t \, dA_t - dH_t + \kappa P_t e^{R_t}\big(V_t^{\top} dM_t + dW_t\big)\Big) + P_t D_t J_t \, dA_t \\ &= D_t\Big((P_t J_t - G_t) \, dA_t - dH_t + \kappa P_t e^{R_t}\big(V_t^{\top} dM_t + dW_t\big)\Big) \end{aligned} \tag{34}$$

and we conclude that the finite variation parts in the latter expression are nonnegative. We can now use the following stopping time argument to derive

$$P_t \le \mathbb{E}\left[\frac{D_T}{D_t} P_T \,\middle|\, \mathscr{F}_t\right]. \tag{35}$$

Namely, consider the stopping time

$$\tau_n := \inf\left\{u \in [t, T] \,\middle|\, \int_t^u \kappa^2 \widetilde{P}_s^2 e^{2R_s}\left(V_s^{\top} d\langle M\rangle_s V_s + d\langle W\rangle_s\right) \ge n\right\} \wedge T,$$

where $n \ge 1$ is an integer. Observe that $\tau_n \to T$ as $n \to +\infty$ due to the integrability assumptions on α, Y and Y', as well as the boundedness of A. Then (34) provides the estimate

$$\begin{aligned} P_t &\le \mathbb{E}\left[\exp\left(\int_t^{\tau_n} J_s \, dA_s\right) P_{\tau_n} \,\middle|\, \mathscr{F}_t\right] \\ &= \mathbb{E}\left[\exp\left(\int_t^{\tau_n} \gamma e^{2\overline{\beta} K_A}\left(\alpha_s + 2\overline{\beta}|Y_s'|\right) dA_s\right) P_{\tau_n} \,\middle|\, \mathscr{F}_t\right]. \end{aligned}$$

In view of the current integrability and boundedness assumptions we can send n to infinity and deduce (35).

Notice that we also have $\xi - \theta\xi' \le (1-\theta)|\xi| + \theta\Delta\xi$, where $\Delta\xi := \xi - \xi' \le 0$. Then together with the definition of P the inequality (35) shows that

$$\begin{aligned} &\exp\left(\frac{\gamma e^{\overline{\beta} K_A + R_t}}{1-\theta}\left(Y_t - \theta Y_t'\right)\right) \\ &\quad \le \mathbb{E}\left[\exp\left(\int_t^T \gamma e^{2\overline{\beta} K_A}\left(\alpha_s + 2\overline{\beta}|Y_s'|\right) dA_s\right) \exp\left(\kappa e^{R_T}(\xi - \theta\xi')\right) \,\middle|\, \mathscr{F}_t\right] \\ &\quad \le \mathbb{E}\left[\exp\left(\gamma e^{2\overline{\beta} K_A} \int_t^T \left(\alpha_s + 2\overline{\beta}|Y_s'|\right) dA_s\right) \exp\left(\gamma e^{2\overline{\beta} K_A}|\xi|\right) \,\middle|\, \mathscr{F}_t\right]. \end{aligned}$$

Thus, we can derive the estimate

$$Y_t - \theta Y_t' \leq \frac{1-\theta}{\gamma} \log \mathbb{E}\left[\exp\left(\gamma e^{2\overline{\beta} K_A} \left(|\xi| + \int_t^T \left(\alpha_s + 2\overline{\beta} |Y_s'| \right) dA_s \right) \right) \middle| \mathscr{F}_t \right],$$

which follows from the above by checking the cases $Y_t - \theta Y_t' \geq 0$ and $Y_t - \theta Y_t' < 0$ separately, noting that $R + \overline{\beta} K_A \geq 0$. Once again, by the integrability assumptions on ξ, α and Y' and the boundedness of A, the conditional expectation on the right hand side is finite, $\mathbb{P}$-a.s. Taking $\theta \uparrow 1$ then gives $Y_t \leq Y_t'$ and the continuity of Y and Y' yields the claim. □

The following corollary is then immediate.

Corollary 2 (Uniqueness). *Let Assumption 1 (i)–(iii) hold and let* (Y, Z, N) *and* (Y', Z', N') *be two solutions to the BSDE* (5) *with* $Y \in \mathfrak{E}$ *and* $Y' \in \mathfrak{E}$. *Then* Y *and* Y', $Z \cdot M$ *and* $Z' \cdot M$, *and* N *and* N' *are indistinguishable. In addition* $(Z \cdot M, N)$ *and* $(Z' \cdot M, N')$ *both belong to* $\mathscr{M}^p \times \mathscr{M}^p$ *for all* $p \geq 1$.

Proof. By Theorem 6 and Corollary 1 (ii) only the assertion regarding the indistinguishability of the martingale part remains. Itô's formula gives $\mathbb{P}$-a.s.

$$\begin{aligned} 0 = (Y_T - Y_T')^2 &= (Y_0 - Y_0')^2 + 2 \int_0^T (Y_t - Y_t')\, d(Y_t - Y_t') \\ &\quad + \int_0^T (Z_t - Z_t')^\top d\langle M \rangle_t (Z_t - Z_t') + d\langle N - N' \rangle_t \\ &= \int_0^T (Z_t - Z_t')^\top d\langle M \rangle_t (Z_t - Z_t') + d\langle N - N' \rangle_t, \end{aligned}$$

from which $Z \cdot M \equiv Z' \cdot M$ and $N \equiv N'$. □

6 Stability

It follows from the previous results that the BSDE (5) has a unique solution in $\mathfrak{E}$ under appropriate Lipschitz and convexity assumptions on the driver F and an exponential moments condition on the terminal value ξ and process α. In the present section we show that a stability result for such BSDEs also holds. More precisely, given a sequence of terminal values and a sequence of drivers such that the exponential moments condition is fulfilled uniformly, and such that they both converge to a fixed terminal value and a fixed generator in a suitable sense, then we gain convergence on the level of the respective BSDE solutions exactly as in [6].

Theorem 7 (Stability). *Let* $(F^n)_{n \geq 0}$ *be a sequence of generators for the BSDE* (5) *such that Assumption 1 (ii)–(iii) and (v) hold for each* F^n *with the set of parameters* $(\alpha^n, \beta^n, \overline{\beta}, \beta_f, \gamma)$. *If* $(\xi^n)_{n \geq 0}$ *are the associated random terminal values then suppose that, for each* $p > 0$,

$$\sup_{n\geq 0}\mathbb{E}\Big[e^{p\,(|\xi^n|+|\alpha^n|_1)}\Big] < +\infty. \tag{36}$$

Let (Y^n, Z^n, N^n) *be the solution to the BSDE* (5) *with driver* F^n *and terminal condition* ξ^n *such that* $Y^n \in \mathfrak{E}$ *for all* $n \geq 0$*. If*

$$|\xi^n - \xi^0| + \int_0^T \big|F^n - F^0\big|\,(s, Y^0_s, Z^0_s)\,dA_s \longrightarrow 0 \quad \textit{in probability, as } n \to +\infty, \tag{37}$$

then for each $p > 0$*,*

$$\lim_{n\to+\infty}\mathbb{E}\bigg[\bigg(\exp\Big(\sup_{0\leq t\leq T}\big|Y^n_t - Y^0_t\big|\Big)\bigg)^p\bigg] = 1 \quad \textit{and}$$

$$\lim_{n\to+\infty}\mathbb{E}\bigg[\bigg(\int_0^T (Z^n_s - Z^0_s)^\top d\langle M\rangle_s(Z^n_s - Z^0_s) + d\langle N^n - N^0\rangle_s\bigg)^{p/2}\bigg] = 0.$$

Remark 4. Let us briefly indicate how the above stability theorem differs from those in the literature, [13] Theorem 2.1 and [25] Lemma 3.3. The key points are that first in our conditions the parameters α^n and β^n are allowed to depend on n, whereas in [13] and [25] they are assumed independent of n. Second, we assume a uniform exponential moments condition, as opposed to a uniform boundedness condition in the cited references. Finally, in the unbounded setting we require the mode of convergence assumed above, this is in contrast to the setting of [13] Theorem 2.1 where the weaker notion of pointwise convergence is sufficient for a stability result to hold.

Proof. Note that Assumption 1 (i) holds for each n thanks to (36). Exactly as in the statement of Corollary 1 we deduce that for each $p \geq 1$

$$\sup_{n\geq 0}\mathbb{E}\bigg[\bigg(\exp\Big(\sup_{0\leq t\leq T}\big|Y^n_t\big|\Big)\bigg)^p + \bigg(\int_0^T (Z^n_s)^\top d\langle M\rangle_s Z^n_s + d\langle N^n\rangle_s\bigg)^{p/2}\bigg] < +\infty.$$

Hence the sequences in n of random variables

$$\bigg(\exp\Big(\sup_{0\leq t\leq T}\big|Y^n_t\big|\Big)\bigg)^p \quad \text{and} \quad \bigg(\int_0^T (Z^n_s)^\top d\langle M\rangle_s Z^n_s + d\langle N^n\rangle_s\bigg)^{p/2}$$

are uniformly integrable for all $p \geq 1$. By the Vitali convergence theorem, it is thus sufficient to prove that

$$\sup_{0\leq t\leq T}\big|Y^n_t - Y^0_t\big| + \int_0^T (Z^n_s - Z^0_s)^\top d\langle M\rangle_s(Z^n_s - Z^0_s) + d\langle N^n - N^0\rangle_s \longrightarrow 0$$

in probability when n tends to infinity.

We split the proof into four steps. The first two steps construct one-sided estimates for the difference of Y^n and Y^0 proceeding very similarly to the proof of the comparison result. In the third step we combine the aforementioned estimates to show that $Y^n - Y^0$ converges to zero uniformly on $[0, T]$ in probability, i.e. in *ucp*. Finally, we use this result to show the required convergence of the martingale parts.

Step 1. First fix $\theta \in (0, 1)$ and $n \geq 1$ and proceed in the same way as in the proof of Theorem 6 by defining the same objects U, V, W, ρ, R, F^θ, κ, P, G, and H, subject to the following modification. All the objects X', $X \in \{Y, Z, N, F\}$, with a prime $'$ are replaced by the respective object X^0 with a superscript 0. All the objects $X \in \{Y, Z, N, F, \alpha\}$ without a prime are replaced by the respective object X^n with a superscript n, e.g. $U := Y^n - \theta Y^0$. We observe that the above objects depend on n however we omit this dependence for brevity. In addition set $\Delta^n F(s) = (F^n - F^0)(s, Y_s^0, Z_s^0)$. From (29) and (31),

$$G \leq \kappa P e^R \Big((1-\theta)\Big(\alpha^n + 2\overline{\beta}|Y^0|\Big) + \theta\, \Delta^n F\Big)$$
$$\leq \gamma e^{2\overline{\beta} K_A} P \left(\alpha^n + 2\overline{\beta}|Y^0| + \frac{|\Delta^n F|}{1-\theta}\right) = P J^n + \gamma e^{2\overline{\beta} K_A} P \frac{|\Delta^n F|}{1-\theta},$$

where, consistent with our modification,

$$J^n := \gamma e^{2\overline{\beta} K_A}\Big(\alpha^n + 2\overline{\beta}|Y^0|\Big) \geq 0.$$

Observe that in contrast to the proof of the comparison theorem, the difference $\Delta^n F$ of the drivers cannot be bounded above by zero. Considering

$$D_t^n := \exp\left(\int_0^t J_s^n \, dA_s\right) \quad \text{and} \quad \widetilde{P}_t^n := D_t^n P_t$$

and applying partial integration yields

$$d\widetilde{P}_t^n + \gamma e^{2\overline{\beta} K_A} \widetilde{P}_t^n \frac{|\Delta^n F|}{1-\theta} \, dA_t = D_t^n \, dP_t + P_t \, dD_t^n + \gamma e^{2\overline{\beta} K_A} \widetilde{P}_t^n \frac{|\Delta^n F|}{1-\theta} \, dA_t$$
$$= D_t^n \left[\left(P_t J_t^n + \gamma e^{2\overline{\beta} K_A} P_t \frac{|\Delta^n F|}{1-\theta} - G_t\right) \times dA_t - dH_t + \kappa P_t e^{R_t}\left(V_t^\top dM_t + dW_t\right)\right].$$

We again conclude that the finite variation parts in the last expression are nonnegative. We now use the stopping time argument from the proof of Theorem 6 to derive

$$P_t \leq D_t^n P_t \leq \mathbb{E}\left[D_T^n P_T + \frac{\gamma e^{2\overline{\beta} K_A}}{1-\theta} \int_t^T D_s^n P_s |\Delta^n F(s)| \, dA_s \,\middle|\, \mathscr{F}_t\right]. \tag{38}$$

From the boundedness of ρ and the definition $P = \exp\big(\kappa e^R U\big)$ we derive, for $s \in [0,T]$,

$$P_s \leq \sup_{0\leq t\leq T}\left[\exp\left(\frac{\gamma e^{2\bar\beta K_A}}{1-\theta}\Big(|Y_t^0| + |Y_t^n|\Big)\right)\right] =: \Upsilon^n(\theta) \quad \text{and}$$

$$P_T \leq \exp\Big(\tfrac{\gamma e^{2\bar\beta K_A}}{1-\theta}\big|\xi^n - \theta\xi^0\big|\Big) \leq \exp\Big(\tfrac{\gamma e^{2\bar\beta K_A}}{1-\theta}\Big(\big|\xi^n - \theta\xi^0\big| \vee \big|\xi^0 - \theta\xi^n\big|\Big)\Big) =: \chi^n(\theta).$$

Using the boundedness of ρ, the inequalities $\log(x) \leq x$, (38) and $1 \leq D_s^n \leq D_T^n$ we then find

$$\begin{aligned} Y_t^n - Y_t^0 &\leq (1-\theta)|Y_t^0| + Y_t^n - \theta Y_t^0 = (1-\theta)|Y_t^0| + U_t \\ &= (1-\theta)|Y_t^0| + \frac{1-\theta}{\gamma}\exp\Big(-\bar\beta K_A - R_t\Big)\log(P_t) \\ &\leq (1-\theta)|Y_t^0| \\ &\quad + \frac{1-\theta}{\gamma}\mathbb{E}\left[D_T^n\chi^n(\theta) + \frac{\gamma e^{2\bar\beta K_A}}{1-\theta}D_T^n\Upsilon^n(\theta)\int_t^T |\Delta^n F(s)|\,dA_s\,\middle|\,\mathscr{F}_t\right]. \end{aligned} \tag{39}$$

Step 2. With regards to the converse inequality we proceed as in the proof of [6] Proposition 7 so that finally

$$\begin{aligned} Y_t^0 - Y_t^n &\leq (1-\theta)|Y_t^n| \\ &\quad + \frac{1-\theta}{\gamma}\mathbb{E}\left[D_T^n\chi^n(\theta) + \frac{\gamma e^{2\bar\beta K_A}}{1-\theta}D_T^n\Upsilon^n(\theta)\int_t^T |\Delta^n F(s)|\,dA_s\,\middle|\,\mathscr{F}_t\right], \end{aligned} \tag{40}$$

where J^n and thus D^n, Υ^n and $\chi^n(\theta)$ are as in Step 1.

Step 3. Let us now prove that $\big(\sup_{0\leq t\leq T}\big|Y_t^n - Y_t^0\big|\big)_{n\geq 1}$ converges to zero in probability. Summing up (39) and (40) we deduce

$$\begin{aligned} \big|Y_t^n - Y_t^0\big| \leq (1-\theta)\left(|Y_t^0| + |Y_t^n|\right) + \frac{1-\theta}{\gamma}\mathbb{E}\left[D_T^n\chi^n(\theta)\,\middle|\,\mathscr{F}_t\right] \\ + e^{2\bar\beta K_A}\mathbb{E}\left[D_T^n\Upsilon^n(\theta)\int_t^T |\Delta^n F(s)|\,dA_s\,\middle|\,\mathscr{F}_t\right]. \end{aligned}$$

We note that by the usual assumptions on the filtration and by continuity of Y^n and Y^0 this holds for all t, $\mathbb{P}$-a.s. Applying the Doob, Markov and Hölder inequalities, we deduce the existence of some nonnegative constants c_1, c_2, independent of θ, as well as $c_3(\theta)$ such that, for $\varepsilon > 0$,

$$\mathbb{P}\left(\sup_{0\le t\le T}\left|Y_t^n - Y_t^0\right| \ge \varepsilon\right) \le \frac{c_1(1-\theta)}{\varepsilon} + \frac{c_2(1-\theta)}{\varepsilon}\mathbb{E}\left[\chi^n(\theta)^2\right]^{1/2} + \frac{c_3(\theta)}{\varepsilon}\mathbb{E}\left[\left(\int_0^T |\Delta^n F(s)|\,dA_s\right)^2\right]^{1/2}, \tag{41}$$

where the latter inequality is due to the fact that, by our assumptions, the sequences $\left(\sup_{0\le t\le T}\left(|Y_t^0| + |Y_t^n|\right)\right)_{n\ge 1}$, $(D_T^n)_{n\ge 1}$ and $(\Upsilon^n(\theta))_{n\ge 1}$ are bounded in all $L^p(\mathbb{P})$ spaces, $p \ge 1$. In addition, for the application of Doob's inequality, we used that

$$\begin{aligned}|\Delta^n F(s)| &\le |F^n(s, Y_s^0, Z_s^0) - F^n(s,0,Z_s^0)| + |F^n(s,0,Z_s^0)| + |F^0(s,0,Z_s^0)| \\ &\quad + |F^0(s, Y_s^0, Z_s^0) - F^0(s,0,Z_s^0)| \\ &\le 2\overline{\beta}|Y_s^0| + \alpha_s^n + \alpha_s^0 + \gamma\|B_s Z_s^0\|^2, \end{aligned} \tag{42}$$

which in turn also implies that for all $p \ge 1$ the family $\left(\left(\int_0^T |\Delta^n F(s)|\,dA_s\right)^p\right)_{n\ge 1}$ is uniformly integrable due to Corollary 1 and (36). Here, for reasons explained in Sect. 7 we deviate from [6]. The Vitali convergence theorem and (37) imply that $\int_0^T |\Delta^n F(s)|\,dA_s \to 0$ in all $L^p(\mathbb{P})$ spaces. Observe that $\chi^n(\theta)$ converges in probability to $\exp\left(\gamma e^{2\overline{\beta}K_A}|\xi^0|\right)$ as n goes to infinity. This convergence is also in all $L^p(\mathbb{P})$ spaces because of the uniform integrability assumption on $(\xi^n)_{n\ge 1}$. More precisely, for all $p \ge 1$, we have

$$\begin{aligned}\sup_{n\ge 1}\mathbb{E}[\chi^n(\theta)^p] &\le \sup_{n\ge 1}\mathbb{E}\left[\exp\left(\frac{p\gamma e^{2\overline{\beta}K_A}}{1-\theta}\left(|\xi^n| + |\xi^0|\right)\right)\right] \\ &\le \sup_{n\ge 1}\mathbb{E}\left[\exp\left(\frac{2p\gamma e^{2\overline{\beta}K_A}}{1-\theta}|\xi^n|\right)\right]^{1/2}\mathbb{E}\left[\exp\left(\frac{2p\gamma e^{2\overline{\beta}K_A}}{1-\theta}|\xi^0|\right)\right]^{1/2} < +\infty.\end{aligned}$$

From (41) we then deduce that for all $\theta \in (0,1)$

$$\limsup_{n\to+\infty}\mathbb{P}\left(\sup_{0\le t\le T}\left|Y_t^n - Y_t^0\right| \ge \varepsilon\right) \le \frac{c_1(1-\theta)}{\varepsilon} + \frac{c_2(1-\theta)}{\varepsilon}\mathbb{E}\left[\exp\left(2\gamma e^{2\overline{\beta}K_A}|\xi^0|\right)\right]^{\frac{1}{2}}.$$

We then send θ to 1 to conclude that

$$\lim_{n\to+\infty}\mathbb{P}\left(\sup_{0\le t\le T}\left|Y_t^n - Y_t^0\right| \ge \varepsilon\right) = 0.$$

Step 4. Let us now turn to the last assertion of the theorem. We derive from Itô's formula that

$$\mathbb{E}\Big[\int_0^T (Z_s^n - Z_s^0)^\top d\langle M\rangle_s (Z_s^n - Z_s^0) + d\langle N^n - N^0\rangle_s\Big]$$

$$\leq \mathbb{E}\Big[(\xi^n - \xi^0)^2 + 2\sup_{0\leq t\leq T}|Y_t^n - Y_t^0| \cdot \int_0^T \big|F^n(s, Y_s^n, Z_s^n) - F^0(s, Y_s^0, Z_s^0)\big|\, dA_s\Big]$$

$$+\,\mathbb{E}\Big[\sup_{0\leq t\leq T}|Y_t^n - Y_t^0| \cdot \Big|\int_0^T d\langle N^n\rangle_s - d\langle N^0\rangle_s\Big|\Big]$$

after observing that the local martingale arising therein is in fact a true martingale thanks to the present integrability assumptions, cf. Corollary 1. By (9),

$$\big|F^n(s, Y_s^n, Z_s^n) - F^0(s, Y_s^0, Z_s^0)\big| \leq \alpha_s^n + \alpha_s^0 + \overline{\beta}|Y_s^n| + \overline{\beta}|Y_s^0|$$

$$+ \frac{\gamma}{2}\|B_s Z_s^n\|^2 + \frac{\gamma}{2}\|B_s Z_s^0\|^2.$$

Applying Hölder's inequality, the formula (22) and the condition (36) we recognize (the expectation of the squares of) the integrals from the right hand side above as uniformly bounded (in n). The result now follows from the fact that $\xi^n \to \xi^0$ and that by Steps 1–3 $\big(\sup_{0\leq t\leq T}|Y_t^n - Y_t^0|\big) \to 0$ in $L^2(\mathbb{P})$. To sum up, we conclude in particular that

$$\int_0^T (Z_s^n - Z_s^0)^\top d\langle M\rangle_s (Z_s^n - Z_s^0) + d\langle N^n - N^0\rangle_s \xrightarrow{\mathbb{P}} 0 \quad \text{as} \quad n \to +\infty,$$

which completes the proof. □

Remark 5. As previously discussed the sense of convergence given here differs from that in [6] where the pointwise convergence of the drivers is assumed, namely

$$\mu^A\text{-a.e. for all } y \text{ and } z \text{ we have } \lim_{n\to+\infty} F^n(\cdot, y, z) = F^0(\cdot, y, z). \qquad (43)$$

We provide an example in the next section showing that this condition is not sufficient in the present setting so that the statement of [6] Proposition 7 needs a small modification.

7 Stability Counterexample

Suppose our filtration is the augmentation of the filtration generated by a one dimensional Brownian motion W so that we may set $A_t = t$ and $B \equiv 1$. The measure μ^A now becomes the product of $\mathbb{P}$ and the Lebesgue measure on $[0, T]$. In this setting BSDEs take the following form

$$dY_t = Z_t\, dW_t - F(t, Y_t, Z_t)\, dt, \quad Y_T = \xi. \tag{44}$$

A solution now consists of a *pair* (Y, Z) such that Y has continuous paths, Z is a predictable process with $\int_0^T Z_t^2\, dt < +\infty$ $\mathbb{P}$-a.s., $\int_0^T |F(t, Y_t, Z_t)|\, dt < +\infty$ $\mathbb{P}$-a.s. and such that (44) holds, $\mathbb{P}$-a.s.

Suppose our condition $\int_0^T |F^n - F^0|\,(s, Y_s^0, Z_s^0)\, ds \xrightarrow{\mathbb{P}} 0$ is replaced by (43), i.e. F^n converges to F^0 pointwise (t, ω)-almost everywhere on $[0, T] \times \Omega$, where the μ^A-null set does not depend on (y, z). One may ask whether this is sufficient for Theorem 7 to hold, in particular if

$$\sup_{0 \leq t \leq T} |Y_t^n - Y_t^0| \xrightarrow{\mathbb{P}} 0. \tag{45}$$

We now present an example to show that this is in fact not the case. The example resembles the standard counterexample to the dominated convergence theorem and shows that such a stability statement (under the present assumptions) already fails to hold in an essentially deterministic situation.

Consider $T > 1$ together with parameters $F^0 \equiv \alpha^0 \equiv \xi^0 \equiv 0$. Then all the assumptions in [6] and in the present paper are satisfied and the unique solution to the BSDE (44) with parameters (F^0, ξ^0) is given by $(Y^0, Z^0) \equiv 0$, up to appropriate null sets.

Furthermore, for integers $n \geq 1$, define the terminal values $\xi^n \equiv 0$ and drivers

$$F^n \equiv \alpha^n \equiv n \cdot \mathbf{1}_{[0, \frac{1}{n}] \times \Omega} \geq 0.$$

Observe that F^n does not depend on y or on z and is constant in ω, hence deterministic. In particular $|\alpha^n|_1 = \int_0^T \alpha_s^n\, ds = 1$, independently of ω and n, which shows that again all the assumptions in [6] and in the present paper are satisfied by each pair (F^n, ξ^n).

The unique solution to the BSDE (44) with parameters (F^n, ξ^n) is given $\mathbb{P}$-a.s. by $Z^n \equiv 0$, more precisely the zero element in $L^2([0, T] \times \Omega)$, and

$$Y_t^n = (1 - nt) \cdot \mathbf{1}_{[0, \frac{1}{n}] \times \Omega}(t, \cdot).$$

We deduce that Y^n is nonnegative, nonincreasing and that $Y_0^n = 1$, independent of n, $\mathbb{P}$-a.s. It now follows that, $\mathbb{P}$-a.s. for all $n \geq 1$, $\sup_{0 \leq t \leq T} |Y_t^n - Y_t^0| = Y_0^n = 1$, from which

$$\lim_{n\to+\infty}\left(\sup_{0\le t\le T}|Y_t^n - Y_t^0|\right) = 1 \quad \mathbb{P}\text{-a.s.} \tag{46}$$

However, by construction, $\lim_{n\to+\infty} F^n = 0 = F$ on $(0,T]\times\Omega$, hence μ^A-a.e. independently of y and z, so that (43) holds. Since (45) and (46) cannot hold simultaneously, the condition in (43) is not sufficient for a stability theorem to hold under the present assumptions. We remark that this phenomenon is not dependent on the non differentiability of the paths of the Y^n as one can choose F^n to be arbitrarily smooth in t. Indeed, independent of ω, take a smooth nonnegative function on $[0,T]$ that is identically zero on $(\frac{1}{n},T]$ and integrates to one over $[0,T]$. The corresponding Y^n in the BSDE solution are smooth in t and lead to the same contradiction.

The problem arising in the proof of [6] Proposition 7 can be observed from (41) and (42). More specifically, the authors require $L^2(\mathbb{P})$-convergence of the random variables $\int_0^T |\Delta^n F(s)|\,ds$ however they only dispose of an estimate on the product space $[0,T]\times\Omega$ of the form

$$|\Delta^n F| \le 2\overline{\beta}|Y^0| + \alpha^n + \alpha^0 + \gamma\|Z^0\|^2,$$

together with uniform integrability assumptions that are on the level of Ω, *with the t-component integrated away*. There is no guarantee that the pointwise convergence of $|\Delta^n F|$ on the product space $[0,T]\times\Omega$ will transform to pointwise convergence of the integrals $\int_0^T |\Delta^n F(s)|\,ds$ on Ω, which is necessary to utilize the uniform integrability assumptions. This is the insight behind the present example and motivates the modified condition.

We now move on to look at whether the martingale part of our BSDE solution determines a change of measure.

8 Change of Measure

In this section we show that under the exponential moments assumption the martingale part of a solution (Y,Z,N) to the BSDE (4) defines a measure change. In particular, we do not need to show that $Z\cdot M + N$ is a BMO martingale, which is a stronger statement that may indeed not hold, see [14] for some examples and related discussion. Here we do not require that the driver F be convex in z. Our proof is based upon Kazamaki [19] Lemmas 1.6 and 1.7 which we state here for martingales on compact time intervals.

Lemma 1 ([19] Lemmas 1.6 and 1.7). *If $\widetilde{M}$ is a martingale on $[0,T]$ such that*

$$\sup_{\substack{\tau \text{ stopping time}\\ \text{valued in } [0,T]}} \mathbb{E}\left[\exp\left(\eta\widetilde{M}_\tau + \left(\frac{1}{2}-\eta\right)\langle\widetilde{M}\rangle_\tau\right)\right] < +\infty, \tag{47}$$

for a real number $\eta \ne 1$, then $\mathcal{E}(\eta\widetilde{M})$ is a true martingale on $[0,T]$. Moreover, if condition (47) *holds for* some $\eta^* > 1$ *then it holds for* all $\eta \in (1,\eta^*)$.

We deduce the following result.

Theorem 8. *Let Assumption 1 (iii) hold, $|\alpha|_1$ have all exponential moments, q be a real number with $|q| > \gamma/2$ and (Y, Z, N) a solution to the BSDE* (4) *such that g is bounded by $\gamma/2$, $Y \in \mathfrak{E}$ and $Z \cdot M + N$ is a martingale. If $\beta > 0$ we also require that $Y^*|\alpha|_1$ has exponential moments of all orders or that* (7) *holds with fixed $y_2 = 0$. Then $\mathscr{E}\Big(q\,(Z \cdot M + N)\Big)$ is a true martingale on $[0, T]$. In particular, when $\gamma < 2$, $\mathscr{E}(Z \cdot M + N)$ is a true martingale.*

Remark 6. In [22] Proposition 7 the authors show that the martingale part of solutions to the BSDE (3) with bounded first component and $\lambda \cdot M$ a BMO martingale also belongs to the class of BMO martingales so that it yields a measure change. Our theorem may thus be seen as a generalization of this result to the case in which Y is not necessarily bounded. We mention that it follows from the proof of this theorem that the assumption of all exponential moments may be weakened to requiring exponential moments of some specific order.

Proof. We apply Lemma 1 with $\widetilde{M} := \tilde{q}(Z \cdot M + N)$ for some fixed $|\tilde{q}| > \gamma/2$. First, we assume that $\beta > 0$ and that $Y^*|\alpha|_1$ has exponential moments of all orders. Considering

$$\log G_\eta(t) := \tilde{q}\eta\left[(Z \cdot M)_t + N_t\right] + \tilde{q}^2\left(\frac{1}{2} - \eta\right)\langle Z \cdot M + N\rangle_t$$

for $\eta > 0$ we get from the BSDE (4) and the growth condition in (8),

$$\begin{aligned}\log G_\eta(t) &= \tilde{q}\eta\Big(Y_t - Y_0 + \int_0^t F(s, Y_s, Z_s)\, dA_s + \int_0^t g_s\, d\langle N\rangle_s\Big)\\ &\qquad\qquad + \tilde{q}^2\Big(\frac{1}{2} - \eta\Big)\langle Z \cdot M + N\rangle_t\\ &\le |\tilde{q}|\eta(Y^* + |Y_0|) + |\tilde{q}|\eta|\alpha|_1 + |\tilde{q}|\eta\beta\, Y^*|\alpha|_1\\ &\qquad\qquad + |\tilde{q}|\eta\left(\frac{\gamma}{2} + \frac{|\tilde{q}|}{\eta}\left(\frac{1}{2} - \eta\right)\right)\langle Z \cdot M + N\rangle_t.\end{aligned}$$

Noting that

$$\frac{\gamma}{2} + \frac{|\tilde{q}|}{\eta}\left(\frac{1}{2} - \eta\right) \le 0 \iff \eta \ge \frac{|\tilde{q}|}{2|\tilde{q}| - \gamma} =: q_0,$$

we have that $\mathbb{P}$-a.s. for all $t \in [0, T]$,

$$G_\eta(t) \le \exp\Big(|\tilde{q}|\eta(Y^* + |Y_0|)\Big)\exp\Big(|\tilde{q}|\eta|\alpha|_1 + |\tilde{q}|\eta\beta\, Y^*|\alpha|_1\Big), \tag{48}$$

for all $\eta \ge q_0$. By the exponential moments assumption on Y^*, $|\alpha|_1$ and $Y^*|\alpha|_1$, we conclude from Hölder's inequality that

$$\sup_{\substack{\tau \text{ stopping time}\\ \text{valued in } [0,T]}} \mathbb{E}\big[G_\eta(\tau)\big] < +\infty \tag{49}$$

for all $\eta \geq q_0 > 1/2$. It now follows from Lemma 1 that $\mathscr{E}(\tilde{q}\eta(Z \cdot M + N))$ is a true martingale for all $\eta \in [q_0, \infty)\backslash\{1\}$. The second part of this lemma ensures that in fact $\mathscr{E}(\tilde{q}\eta(Z \cdot M + N))$ is a true martingale for all $\eta > 1$. Thus, if $|q| > \gamma/2$ we apply this result for some fixed $|\tilde{q}| \in (\gamma/2, |q|)$ and $\eta := q/\tilde{q} = |q/\tilde{q}| > 1$ to conclude that indeed $\mathscr{E}(q(Z \cdot M + N))$ is a true martingale.

Now if $\beta > 0$ and (7) holds with fixed $y_2 = 0$, we use (9) to derive, similarly to the above,

$$\log G_\eta(t) \leq |\tilde{q}|\eta(Y^* + |Y_0|) + |\tilde{q}|\eta|\alpha|_1 + |\tilde{q}|\eta\overline{\beta}\, Y^* A_T$$
$$+ |\tilde{q}|\eta\left(\frac{\gamma}{2} + \frac{|\tilde{q}|}{\eta}\left(\frac{1}{2} - \eta\right)\right)\langle Z \cdot M + N\rangle_t$$

so that the claim follows from the boundedness of A_T using exactly the same arguments. The reasoning from above also applies when Assumption 1 (iii) holds with $\beta = 0$, without any further conditions. □

9 Applications

In the final section we explore two applications of our results, specifically focusing on utility maximization and partial equilibrium. Our contribution is to show that the standard results continue to hold when the usual boundedness assumptions are replaced by appropriate exponential moments conditions, allowing for more generality.

9.1 *Constrained Utility Maximization Under Exponential Moments*

In the context of the constrained utility maximization problem with power utility the following BSDE appears, cf. [25] Sect. 4.2.1,

$$dY_t = Z_t^\top dM_t + dN_t - F(t, Z_t)\, dA_t - \frac{1}{2}\, d\langle N\rangle_t, \quad Y_T = 0,$$

where the generator is given by

$$F(t,z) = \frac{p(p-1)}{2} \inf_{\nu \in \mathscr{C}} \left\| B_t\left(\nu - \frac{z - \lambda_t}{1-p}\right)\right\|^2 + \frac{p(1-p)}{2}\left\| B_t\left(\frac{z-\lambda_t}{1-p}\right)\right\|^2$$
$$+ \frac{1}{2}\|B_t z\|^2.$$

In the above $1-p \in (0,+\infty)$ is the investor's relative risk aversion and ν refers to investment strategies in a stock whose returns are driven by the continuous local martingale M under the market price of risk process λ and must be valued in the closed constraint set $\mathscr{C}$. Writing the infimum in terms of the distance function, which is Lipschitz continuous, one can show that the driver F satisfies Assumption 1 (ii)–(v), so that there exist constants c_λ and c_z such that

$$|F(t,z)| \leq c_\lambda \|B_t\lambda_t\|^2 + c_z\|B_t z\|^2.$$

When we enforce that the *mean-variance trade off* $\int_0^T \lambda_t^\top \langle M\rangle_t \lambda_t$ has all exponential moments, an assumption weaker than that of boundedness given in the cited literature, we are in the current framework and see that the BSDE admits a unique solution in $\mathfrak{E}\times\mathfrak{M}^2\times\mathscr{M}^2$. The crucial step in [16] and [25] is, given a solution triple (Y,Z,N), to construct the relevant optimizers; this is the process of verification. Using the theorems of the present paper, it is possible to show that one can repeat the reasoning of [16] and [25] and that similar results continue to hold for more general classes of market price of risk processes under appropriate trading constraints such as bounded short-selling and borrowing.

A biproduct of the analysis described here is a direct link between solutions to the utility maximization problem and solutions to the BSDE in an exponential moments setting, building on [26]. This allows for a detailed study of the stability of the utility maximization problem, undertaken in [24], which is important in many applications.

9.2 *Partial Equilibrium and Market Completion Under Exponential Moments*

We now briefly describe the partial equilibrium framework of [15] in which structured securities that are written on nontradable assets are priced via a market clearing condition.

The agents in this economy have preferences which are given by dynamic convex risk measures. The risk they are exposed to is given by two sources. The first is encoded in a financial market in which frictionless trading in a stock S is possible. The second is a non-financial risk factor R that can only be dealt with via a derivative written on this external factor. It is assumed that this derivative completes the market, in fact it is shown that in equilibrium the market is complete.

More specifically, while the market price λ^S of financial risk is given exogenously the market price λ^R of external risk is determined via an equilibrium condition. This states that when the derivative is priced according to the pricing rule arising from (λ^S, λ^R) the agents' aggregated demand matches the fixed supply. The demand is in this setting given by the solutions to the agents' individual risk minimization problems and is a function of λ^R.

To ease the exposition we put ourselves in a representative agent setting where the agent's preferences are of entropic type, i.e. their utility function is exponential. Then the following BSDE for the agent's dynamic risk Y appears

$$dY_t = Z_t^{\top} dW_t - \frac{1}{2}\left((\lambda_t^S)^2 - 2\lambda_t^S Z_t^1 - (Z_t^2)^2\right) dt, \quad Y_T = H,$$

where W is a two dimensional Brownian motion representing the two sources of risk, $Z = (Z^1, Z^2)$ is the corresponding control process to Y and H is the agent's endowment. Under suitable exponential moments assumptions the present article provides the existence of a unique solution $(\hat{Y}, \hat{Z})$ to the above BSDE. Once we check that $\hat{Z}^2$ defines a valid pricing rule, i.e. that $\mathscr{E}\big((\lambda^S, \hat{Z}^2) \cdot W\big)$ is a true martingale, we know that the equilibrium market price λ^R of external risk is given by $\lambda^R := \hat{Z}^2$. To conclude we can generalize [15], full details will appear elsewhere.

References

1. P. Barrieu, N. El Karoui, Monotone stability of quadratic semimartingales with applications to general quadratic BSDEs and unbounded existence result. arXiv:1101.5282v2, 2012
2. D. Becherer, Bounded solutions to backward SDE's with jumps for utility optimization and indifference hedging. Ann. Appl. Probab. **16**(4), 2027–2054 (2006)
3. J.-M. Bismut, Conjugate convex functions in optimal stochastic control. J. Math. Anal. Appl. **44**, 384–404 (1973)
4. G. Bordigoni, A. Matoussi, M. Schweizer, A stochastic control approach to a robust utility maximization problem. In *Stochastic Analysis and Applications*, vol. 2 of *Abel Symposium* (Springer, Berlin, 2007), pp. 125–151
5. P. Briand, Y. Hu, BSDE with quadratic growth and unbounded terminal value. Probab. Theory Relat. Fields **136**(4), 604–618 (2006)
6. P. Briand, Y. Hu, Quadratic BSDEs with convex generators and unbounded terminal conditions. Probab. Theory Relat. Fields **141**(3–4), 543–567 (2008)
7. F. Da Lio, O. Ley, Uniqueness results for second-order Bellman-Isaacs equations under quadratic growth assumptions and applications. SIAM J. Control Optim. **45**(1), 74–106 (electronic) (2006)
8. F. Delbaen, Y. Hu, X. Bao, Backward SDEs with superquadratic growth. Probab. Theory Relat. Fields (2010) DOI 10.1007/s00440-010-0271-1
9. F. Delbaen, Y. Hu, A. Richou, On the uniqueness of solutions to quadratic BSDEs with convex generators and unbounded terminal conditions. Ann. Inst. Henri Poincaré Probab. Stat. **47**(2), 559–574, (2011)
10. N. El Karoui, S. Hamadène, A. Matoussi, *Backward stochastic differential equations and applications.* In R. Carmona (ed.), Indifference Pricing: Theory and Applications, Princeton Series in Financial Engineering (Princeton University Press, Princeton, NJ, 2009), pp. 267–320
11. N. El Karoui, S.-J. Huang, A general result of existence and uniqueness of backward stochastic differential equations. In *Backward Stochastic Differential Equations (Paris, 1995–1996)*, vol. 364 of *Pitman Res. Notes Math. Ser.* (Longman, Harlow, 1997), pp. 27–36
12. N. El Karoui, S. Peng, M.C. Quenez, Backward stochastic differential equations in finance. Math. Finance **7**(1), 1–71 (1997)
13. C. Frei, Convergence results for the indifference value based on the stability of BSDEs, Working Paper (2011)

14. C. Frei, M. Mocha, N. Westray, BSDEs in utility maximization with BMO price of risk, arXiv:1107.0183 (2011)
15. U. Horst, T. Pirvu, G. dos Reis, On securitization, market completion and equilibrium risk transfer. Math. Finance Econ. **2**(4), 211–252 (2010)
16. Y. Hu, P. Imkeller, M. Müller, Utility maximization in incomplete markets. Ann. Appl. Probab. **15**(3), 1691–1712 (2005)
17. Y. Hu, M. Schweizer, Some new BSDE results for an infinite-horizon stochastic control problem. In G.D. Nunno, B. Øksendal (eds), *Advanced Mathematical Methods for Finance* (Springer, Berlin, 2011), pp. 367–395
18. J. Jacod, A.N. Shiryaev, *Limit Theorems for Stochastic Processes*, vol. 288 of *Grundlehren der Mathematischen Wissenschaften [Fundamental Principles of Mathematical Sciences]*, 2nd edn. (Springer, Berlin, 2003)
19. N. Kazamaki, *Continuous Exponential Martingales and BMO*. Lecture Notes in Mathematics, vol. 1579 (Springer, Berlin, Heidelberg, 1994)
20. M. Kobylanski, Backward stochastic differential equations and partial differential equations with quadratic growth. Ann. Probab. **28**(2), 558–602 (2000)
21. J.P. Lepeltier, J. San Martín, Backward stochastic differential equations with continuous coefficient. Statist. Probab. Lett. **32**(4), 425–430 (1997)
22. M. Mania, M. Schweizer, Dynamic exponential utility indifference valuation. Ann. Appl. Probab. **15**(3), 2113–2143 (2005)
23. M. Mania, R. Tevzadze, Backward stochastic partial differential equations related to utility maximization and hedging. J. Math. Sci. **153**, 291–380 (2008)
24. M. Mocha, N. Westray, The stability of the constrained utility maximization problem – a BSDE approach arXiv:1107.0190 (2011)
25. M.-A. Morlais, Quadratic BSDEs driven by a continuous martingale and applications to the utility maximization problem. Finance Stoch. **13**(1), 121–150 (2009)
26. M. Nutz, The Bellman equation for power utility maximization with semimartingales. Ann. Appl. Probab. arXiv:0912.1883 (2012 in press)
27. É. Pardoux, S.G. Peng, Adapted solution of a backward stochastic differential equation. Syst. Control Lett. **14**(1), 55–61 (1990)
28. P.E. Protter, *Stochastic Integration and Differential Equations*, vol. 21 of *Applications of Mathematics (New York)*, 2nd edn. Stochastic Modelling and Applied Probability (Springer, Berlin, 2004)
29. R. Rouge, N. El Karoui, Pricing via utility maximization and entropy. Math. Finance **10**(2), 259–276 (2000) INFORMS Applied Probability Conference (Ulm, 1999)
30. M. Schroder, C. Skiadas, Optimal consumption and portfolio selection with stochastic differential utility. J. Econ. Theory **89**(1), 68–126 (1999)
31. J. Sekine, On exponential hedging and related quadratic backward stochastic differential equations. Appl. Math. Optim. **54**, 131–158 (2006)
32. C. Skiadas, Robust control and recursive utility. Finance Stoch. **7**(4), 475–489 (2003)
33. R. Tevzadze, Solvability of backward stochastic differential equations with quadratic growth. Stochast. Process. Appl. **118**(3), 503–515 (2008)

The Derivative of the Intersection Local Time of Brownian Motion Through Wiener Chaos

Greg Markowsky

Abstract Rosen (Séminaire de Probabilités XXXVIII, 2005) proved the existence of a process known as the derivative of the intersection local time of Brownian motion in one dimension. The purpose of this paper is to use the methods developed in Nualart and Vives (Publicacions Matematiques 36(2):827–836, 1992) in order to give a simple new proof of the existence of this process. Some related theorems and conjectures are discussed.

1 Introduction

The study of the derivative of the intersection local time of Brownian motion essentially began with the work of Rogers and Walsh in [8–10]. There, the following functional was studied

$$A(t, B_t) = \int_0^t 1_{[0,\infty)}(B_t - B_s)ds.$$

A formal application of Itō's Lemma to $A(t, B_t)$ gives rise to the following expression:

$$\int_0^1 \int_0^t \delta_0'(B_t - B_s)dsdt. \tag{1}$$

Let $p_\varepsilon(x) = \frac{1}{(\varepsilon)^{d/2}\sqrt{2\pi}} e^{\frac{x^2}{2\varepsilon}}$ be the Gaussian density with mean 0 and variance ε in any dimension d. In [11] Rosen proved the following, among other things.

G. Markowsky (✉)
Monash University, VIC 3800, Australia
e-mail: gmarkowsky@gmail.com

C. Donati-Martin et al. (eds.), *Séminaire de Probabilités XLIV*, Lecture Notes in Mathematics 2046, DOI 10.1007/978-3-642-27461-9_6,

Theorem 1. *Let B_t be 1-dimensional Brownian motion. Then*

$$\alpha'_\varepsilon = \int_0^1 \int_0^t p'_\varepsilon(B_t - B_s)dsdt$$

converges in L^2 to a random variable α' as $\varepsilon \longrightarrow 0$.

Note that ρ'_ε converges weakly to δ'_0, so α' can be interpreted as a realization of the formal expression (1). α' has attracted some attention in the time since Rosen's paper. In [4], a Tanaka formula for α' was proved, and in the process another proof of Theorem 1 was obtained. Yet another proof of Theorem 1, together with an interesting stochastic integral representation of α', was given in [3]. In [15], a study was begun into A and α' driven by fractional Brownian motion.

In a different research thread, in [7] the Wiener chaos expansion was used in the study of various local times of Brownian motion. In particular, a new proof of the following well-known theorem, known as *Varadhan's renormalization* (see [13]), was given.

Theorem 2. *Let B_t be 2-dimensional Brownian motion. Then*

$$L_\varepsilon = \int_{0<s<t<1} (p_\varepsilon(B_t - B_s) - E[p_\varepsilon(B_t - B_s)])dsdt$$

converges in L^2 to a random variable as $\varepsilon \longrightarrow 0$.

The primary purpose of this note is to show how the technique developed in [7] can be used to give a simple proof of Theorem 1. The next section is devoted to this, and gives the explicit Wiener chaos expansion of α' as well. The final section is devoted to the discussion of related questions involving higher derivatives and dimensions.

2 The Wiener Chaos Expansion of α'

For more information on the Wiener chaos expansion, see [6] or [14]. Given symmetric $f \in L^2([0,\infty)^n)$ let

$$I_n(f) = n! \int_0^\infty \int_0^{t_n} \cdots \int_0^{t_2} f(t_1, t_2, \ldots, t_n) dW_{t_1} \ldots dW_{t_n}$$

and for $f \in L^2([0,\infty))$ let $I_n(f^{\otimes n})$ denote the multiple stochastic integral I_n applied to the tensor product $f^{\otimes n} = f(t_1)f(t_2)\ldots f(t_n)$. Following [7], we can deduce the following Wiener chaos expansion for α', in the process obtaining another proof of Theorem 1.

Theorem 3.

$$\alpha' = \sum_{m=1}^{\infty} c_m I_{2m-1}\Big(1 + \frac{1}{(\overline{v} - \underline{v})^{m-3/2}} - \frac{1}{(1-\underline{v})^{m-3/2}} - \frac{1}{(\overline{v})^{m-3/2}}\Big),$$

where

$$c_m = \frac{(-1)^m}{\sqrt{2\pi}\,(m-1)!\,2^{m-1}\,(m-1/2)\,(m-3/2)}$$

and

$$\overline{v} = v_1 \vee \ldots \vee v_{2m-1}$$

$$\underline{v} = v_1 \wedge \ldots \wedge v_{2m-1}.$$

Proof (Theorems 1 and 3). By Stroock's formula, given in [12] and [6, p. 35],

$$p_\varepsilon'(W_t) = \sum_{n=0}^{\infty} \frac{1}{n!} E[p_\varepsilon^{(n+1)}(W_t)] I_n(1_{[0,t]}{}^{\otimes n}). \tag{2}$$

The n-th (probabilists') Hermite polynomial is given by

$$H_n(x) = (-1)^n e^{x^2/2} \frac{d^n}{dx^n}(e^{-x^2/2}).$$

As a consequence of this, and as is shown in [7],

$$p_\varepsilon^{(n)}(x) = (-1)^n\, \varepsilon^{-n/2} p_\varepsilon(x) H_n\Big(\frac{x}{\sqrt{\varepsilon}}\Big). \tag{3}$$

Plugging (3) into (2) gives

$$p_\varepsilon'(W_t) = \sum_{n=0}^{\infty} \frac{(-1)^n}{n!\,\varepsilon^{(n+1)/2}} E[p_\varepsilon(W_t) H_{n+1}\Big(\frac{W_t}{\sqrt{\varepsilon}}\Big)] I_n(1_{[0,t]}{}^{\otimes n}). \tag{4}$$

The expectation in the last expression of (4) is calculated in [7]. It is equal to 0 when n is even, and equal to

$$\frac{(2m)!}{\sqrt{2\pi}\,2^m\,m!} \frac{(-\varepsilon)^m}{(t+\varepsilon)^{m+1/2}}$$

when $n = 2m-1$ is odd. We arrive at

$$p_\varepsilon'(W_t) = \sum_{m=1}^{\infty} \frac{(-1)^{m+1}}{\sqrt{2\pi}\,(m-1)!\,2^{m-1}\,(t+\varepsilon)^{m+1/2}} I_{2m-1}(1_{[0,t]}{}^{\otimes 2m-1}).$$

This leads to

$$\alpha'_\varepsilon = \sum_{m=1}^{\infty} \frac{(-1)^{m+1}}{\sqrt{2\pi}\,(m-1)!\,2^{m-1}} \int_0^1 \int_0^t \frac{I_{2m-1}(1_{(s,t]}{}^{\otimes 2m-1})}{(t-s+\varepsilon)^{m+1/2}} dsdt. \tag{5}$$

We want to let $\varepsilon \longrightarrow 0$ in this expression. In order to do so, we will apply the following lemma from [7], which is a consequence of the dominated convergence theorem.

Lemma 1. *Let F_ε be a family of square integrable random variables with chaos expansions $F_\varepsilon = \sum_{n=0}^{\infty} I_n(f_n^\varepsilon)$. If f_n^ε converges in L^2 to some function f_n as $\varepsilon \longrightarrow 0$ for each n, and if*

$$\sum_{n=0}^{\infty} \sup_\varepsilon \{n! || f_n^\varepsilon ||_2^2\} < \infty, \tag{6}$$

then F_ε converges in L^2 to $F = \sum_{n=0}^{\infty} I_n(f_n)$ as $\varepsilon \longrightarrow 0$.

We must therefore show that

$$\sum_{n=0}^{\infty} \sup_\varepsilon \{(2m-1)! || f_{2m-1}^\varepsilon ||_2^2\} < \infty, \tag{7}$$

with f_{2m-1}^ε determined by (5). For fixed m and ε, $(2m-1)!|| f_{2m-1}^\varepsilon ||_2^2$ is given by

$$\frac{(2m-1)!}{2\pi((m-1)!)^2 2^{2m-2}} \int_{s<t,u<v} \frac{\langle 1_{(s,t]}, 1_{(u,v]}\rangle^{2m-1}}{(t-s+\varepsilon)^{m+1/2}(v-u+\varepsilon)^{m+1/2}} dsdtdudv$$
$$= \frac{(2m-1)!}{2\pi((m-1)!)^2 2^{2m-2}} \int_{s<t,u<v} \frac{|(s,t] \cap (u,v]|^{2m-1}}{(t-s+\varepsilon)^{m+1/2}(v-u+\varepsilon)^{m+1/2}} dsdtdudv. \tag{8}$$

Using Stirling's approximation,

$$\frac{(2m-1)!}{2\pi((m-1)!)^2 2^{2m-2}} = O(\sqrt{m}) \tag{9}$$

To estimate the integral, we break it into two regions as

$$2 \int_{0<u<s<v<t<1} \frac{(v-s)^{2m-1}}{(\varepsilon+t-s)^{m+1/2}(\varepsilon+v-u)^{m+1/2}} dsdtdudv$$
$$+ 2 \int_{0<u<s<t<v<1} \frac{(t-s)^{m-3/2}}{(\varepsilon+v-u)^{m+1/2}} dsdtdudv. \tag{10}$$

We set $\varepsilon = 0$ in order to obtain an upper bound for this expression. To bound the second integral for large m, perform the linear transformation $(u, a, b, c) = (u, s-u, t-s, v-t)$ to get

$$\int_0^1 \int_0^{1-u} \int_0^{1-u-a} \int_0^{1-u-a-b} \frac{b^{m-3/2}}{(a+b+c)^{m+1/2}} dc\,db\,da\,du \qquad (11)$$
$$\leq \int_0^1 \int_0^1 \int_0^1 \frac{b^{m-3/2}}{(a+b+c)^{m+1/2}} dc\,da\,db$$
$$\leq \frac{k}{(m-1/2)(m-3/2)}$$

with k a constant. To obtain the final inequality in (11) we twice employed the trivial bound

$$\int_0^1 \frac{1}{(x+h)^n} dx \leq \frac{1}{(n-1)h^{n-1}} \qquad (12)$$

valid for $n > 1$. To estimate the first integral in (10), substitute $(u, a, b, c) = (u, s-u, v-s, t-v)$ to get

$$\int_0^1 \int_0^{1-u} \int_0^{1-u-a} \int_0^{1-u-a-b} \frac{b^{2m-1}}{(a+b+c)^{m+1/2}(a+b)^{m+1/2}} \qquad (13)$$
$$\leq \int_0^1 \int_0^1 \int_0^1 \frac{b^{2m-1}}{(a+b+c)^{m+1/2}(a+b)^{m+1/2}} dc\,da\,db$$
$$\leq \frac{k}{(m-1/2)(2m-1)}$$

for large m, again with k a fixed constant. Combining (11) and (13) we see that (10) is $O(m^{-2})$. Taking into account (9), we see that $(2m-1)!\|f_{2m-1}^{\varepsilon}\|_2^2$ is $O(m^{-3/2})$. This is summable, so we can indeed apply Lemma 1 to let $\varepsilon \longrightarrow 0$ in (5). This shows that α'_ε converges in L^2, and to arrive at the form for the Wiener chaos decomposition given in the statement of the theorem, note that

$$\alpha' = \sum_{m=1}^{\infty} \frac{(-1)^{m+1}}{\sqrt{2\pi}\,(m-1)!\,2^{m-1}} \int_0^1 \int_0^t \frac{I_{2m-1}(1_{(s,t]}{}^{\otimes 2m-1})}{(t-s)^{m+1/2}} ds\,dt$$
$$= \sum_{m=1}^{\infty} \frac{(-1)^{m+1}}{\sqrt{2\pi}\,(m-1)!\,2^{m-1}} I_{2m-1}\Big(\int_0^1 \int_0^t \frac{1_{(s,t]}{}^{\otimes 2m-1}}{(t-s)^{m+1/2}} ds\,dt\Big). \qquad (14)$$

Now, for $0 < s < t < 1$,

$$1_{(s,t]}{}^{\otimes 2m-1}(v_1, \ldots, v_{2m-1}) = 1_{(v_1 \vee \ldots \vee v_{2m-1}, 1]}(t) 1_{[0, v_1 \wedge \ldots \wedge v_{2m-1}]}(s).$$

Using the notation given in (2), we therefore have

$$\int_0^1 \int_0^t \frac{1_{(s,t]}{}^{\otimes 2m-1}}{(t-s)^{m+1/2}} ds dt$$

$$= \int_0^1 1_{(\overline{v},1]}(t) \int_0^{\underline{v}} \frac{1}{(t-s)^{m+1/2}} ds dt$$

$$= \frac{1}{m-1/2} \int_{\overline{v}}^1 \Big(\frac{1}{(t-\underline{v})^{m-1/2}} - \frac{1}{(t)^{m-1/2}} \Big) dt$$

$$= \frac{1}{(m-1/2)(m-3/2)} \Big(1 + \frac{1}{(\overline{v}-\underline{v})^{m-3/2}} - \frac{1}{(1-\underline{v})^{m-3/2}} - \frac{1}{(\overline{v})^{m-3/2}} \Big).$$

Plugging this into (14) gives

$$\alpha' = \sum_{m=1}^{\infty} \frac{(-1)^{m+1}}{\sqrt{2\pi}\,(m-1)!\,2^{m-1}\,(m-1/2)\,(m-3/2)}$$

$$I_{2m-1} \Big(1 + \frac{1}{(\overline{v}-\underline{v})^{m-3/2}} - \frac{1}{(1-\underline{v})^{m-3/2}} - \frac{1}{(\overline{v})^{m-3/2}} \Big).$$

This completes the proof of Theorems 1 and 3. □

3 Higher Dimensions and Derivatives

In [2], these methods were extended considerably in order to prove a number of results concerning fractional Brownian motion. As relates to Brownian motion, a proof was given for the following central limit theorem, which was originally proved by Yor in [16].

Theorem 4. *Let B_t be 3-dimensional Brownian motion. Then*

$$\frac{1}{\sqrt{\log(1/\varepsilon)}} \int_{0<s<t<1} (p_\varepsilon(B_t - B_s) - E[p_\varepsilon(B_t - B_s)]) ds dt$$

converges in law to a normal law as $\varepsilon \longrightarrow 0$.

A similar central limit theorem has been proved in [5] concerning the derivative of an intersection local time.

Theorem 5. *Let B_t be 2-dimensional Brownian motion. Then*

$$\frac{1}{\log(1/\varepsilon)} \int_{0<s<t<1} \frac{\delta p_\varepsilon}{\delta x}(B_t - B_s) ds dt$$

converges in law to a normal law as $\varepsilon \longrightarrow 0$.

Note that $\frac{\delta p_\varepsilon}{\delta x}$ is the partial derivative with respect to one of the space variables, x. It seems likely that the methods from [2] can be adapted to yield a proof of Theorem 5, as well as of the following conjecture.

Conjecture 1. (a) Let B_t be 1-dimensional Brownian motion. Then, for some $\gamma > 0$,

$$\frac{1}{\log(1/\varepsilon)^\gamma}\int_{0<s<t<1}(p_\varepsilon''(B_t - B_s) - E[p_\varepsilon''(B_t - B_s)])dsdt$$

converges in law to a normal law as $\varepsilon \longrightarrow 0$, where p_ε'' is the second derivative in the space variable of p_ε.

(b) Let B_t be a d-dimensional Brownian motion. Then, if D is a (possibly mixed) partial derivative of order $n \geq 4 - d$,

$$\varepsilon^{(n+d)/2-3/2}\int_{0<s<t<1}(Dp_\varepsilon(B_t - B_s) - E[Dp_\varepsilon(B_t - B_s)])dsdt$$

converges in law to a normal law as $\varepsilon \longrightarrow 0$.

We should mention that part (b) of the conjecture has already been proved for $n = 0$ in [1] and [2]. The remaining cases of the conjecture come from examining the scaling properties of the integrals which result from expanding in Wiener chaos. Unfortunately, however, the details seem to become a bit more onerous with the presence of the derivatives, and handling the relevant integrals which result would seem to require some new techniques.

Acknowledgements I would like to thank Paul Jung, David Nualart, and Jay Rosen for helpful conversations. I would also like to thank the referee for comments which improved the exposition. I am grateful for support from the Priority Research Centers Program through the National Research Foundation of Korea (NRF) funded by the Ministry of Education, Science, and Technology (Grant #2009-0094070) and from Australian Research Council Grant DP0988483.

References

1. J.Y. Calais, M. Yor, Renormalization et convergence en loi pour certaines intégrales multiples associées au mouvement Brownien dans $\mathbb{R}^d$. In *Séminaire de Probabilités XXI*. Lecture Notes in Mathematics, vol. 1247 (Springer, Berlin, Heidelberg, 1987)
2. Y. Hu, D. Nualart, Renormalized self-intersection local time for fractional Brownian motion. Ann. Probab. **33**(3), 948–983 (2005)
3. Y. Hu, D. Nualart, Central limit theorem for the third moment in space of the Brownian local time increments. Electron. Commun. Prob. **15**, 396–410 (2009)
4. G. Markowsky, Proof of a Tanaka-like formula stated by J. Rosen in Séminaire XXXVIII. *Séminaire de Probabilités XLI*, 2008, pp. 199–202
5. G. Markowsky, Renormalization and convergence in law for the derivative of intersection local time in $\mathbb{R}^2$. Stochast. Proces. Appl. **118**(9), 1552–1585 (2008)
6. D. Nualart, *The Malliavin Calculus and Related Topics* (Springer, Berlin, Heidelberg, 1995)

7. D. Nualart, J. Vives, Chaos expansions and local times. Publicacions Matematiques **36**(2), 827–836 (1992)
8. L.C.G. Rogers, J.B. Walsh, $A(t, B_t)$ is not a semimartingale. *Seminar on Stochastic Process*, Progr. Probab., 24, Birkhäuser Boston, Boston, MA, 1991, pp. 275–283
9. L.C.G. Rogers, J.B. Walsh, The intrinsic local time sheet of Brownian motion. Prob. Theory Relat. Fields **88**(3), 363–379 (1991)
10. L.C.G. Rogers, J.B. Walsh, Local time and stochastic area integrals. Ann. Probab. **19**(2), 457–482 (1991)
11. J. Rosen, Derivatives of self-intersection local times. *Séminaire de Probabilités XXXVIII*. Lecture Notes in Mathematics, vol. 1857, 2005, pp. 263–281
12. D. Stroock, Homogeneous chaos revisited. *Séminaire de Probabilités XXI*. Lecture Notes in Mathematics, vol. 1247, 1987, pp. 1–7
13. S.R.S. Varadhan, Appendix to "Euclidean quantum field theory" by K. Symanzik. Local Quant. Theory (1969)
14. N. Wiener, The homogeneous chaos. Am. J. Math. **60**(4), 897–936 (1938)
15. L. Yan, X. Yang, Y. Lu, p-variation of an integral functional driven by fractional Brownian motion. Stat. Prob. Lett. **78**(9), 1148–1157 (2008)
16. M. Yor, Renormalisation et convergence en loi pour les temps locaux d'intersection du mouvement Brownien dans $\mathbb{R}^3$. *Séminaire de Probabilités XIX*. Lecture Notes in Mathematics, vol. 1123, 1985, pp. 350–365

On the Occupation Times of Brownian Excursions and Brownian Loops

Hao Wu

Abstract We study properties of occupation times by Brownian excursions and Brownian loops in two-dimensional domains. This allows for instance to interpret some Gaussian fields, such as the Gaussian Free Fields as (properly normalized) fluctuations of the total occupation time of a Poisson cloud of Brownian excursions when the intensity of the cloud goes to infinity.

Keywords Conformal invariance • Brownian excursion measure • Brownian loop measure • Green's function

1 Introduction

Conformal invariance of planar Brownian motion has been derived and exploited long ago by Paul Lévy [8]. See also B. Davis (Annals of Proba 1979) in particular his derivation of Picard's big theorem. More recently, conformal invariance turned out to be an instrumental idea in the study of various critical models from statistical physics in the plane (see for instance [4, 16] and the references therein). Two basic important conformally invariant measures on random geometric objects are the Brownian excursion measure and the Brownian loop measure. Let us now very briefly describe these measures and the meaning of conformal invariance relatively to these measures. For each open domain D with non-polar boundary in the plane, one can define these two measures in D respectively denoted by μ_D and λ_D. These are infinite but σ-finite measures on Brownian-type paths with particular properties:

H. Wu (✉)
Département de Mathématiques, Université Paris-Sud, 91405 Orsay Cedex, France
e-mail: hao.wu@math.u-psud.fr

C. Donati-Martin et al. (eds.), *Séminaire de Probabilités XLIV*, Lecture Notes in Mathematics 2046, DOI 10.1007/978-3-642-27461-9_7,

- μ_D is supported on the set of Brownian excursions $(B_t, t \leq \tau)$ in D i.e. Brownian paths such that B_0 and B_τ are in ∂D, while $B(0,\tau) \subset D$.
- λ_D is supported on the set of Brownian loops $(B_t, t \leq \tau)$ i.e. Brownian paths in D such that $B_0 = B_\tau$.

In fact, in both cases, it is useful to view these paths up to monotone reparametrization (in the loop-case, one views the time-set modulo τ i.e., there is no "starting point" on the loop). Then, it turns out (see [5, 15] for details) that for any conformal map Φ from D onto $\Phi(D)$, the image measures of μ_D and λ_D under Φ are exactly $\mu_{\Phi(D)}$ and $\lambda_{\Phi(D)}$.

These two measures on loops and on excursions allow in some sense to get rid of the dependence of the measure on Brownian paths with respect to its starting point, see for instance the discussion in [16].

In the present text, we shall focus on the following type of results (here and in the sequel, dx or dy will denote the area measure, and x or y will always denote points in the plane):

Proposition 1. *Suppose that D is a simply connected domain and that A and B are two open proper subsets of D. Then,*

$$\mu_D\Big(\int_0^\tau ds 1_A(\gamma_s) \int_0^\tau ds 1_B(\gamma_s)\Big) = 4 \int_{A\times B} dx\, dy\, G_D(x,y) \tag{1}$$

and

$$\lambda_D\Big(\int_0^\tau ds 1_A(\gamma_s) \int_0^\tau ds 1_B(\gamma_s)\Big) = \int_{A\times B} dx\, dy\, (G_D(x,y))^2, \tag{2}$$

where $(\gamma_s, 0 \leq s \leq \tau)$ is a Brownian excursion in (1) and a Brownian loop in (2), $G_D(x,y)$ denotes the usual Green's function in D (with Dirichlet boundary conditions).

The Brownian excursion measure and the loop measure are infinite measures, but they can be used to define random conformally invariant collections of excursions and loops (i.e. under a probability measure) by a Poissonization procedure. As explained in [16], both these Poissonian clouds are of interest and useful in the context of random planar conformally invariant curves of SLE-type: The "excursion clouds" give rise to the restriction measures [15], while the "loop-soups (loop clouds)" are related to Conformal Loop Ensembles (see [14]).

It is natural to study the cumulative occupation time of these random collections of Brownian paths. The previous proposition can then be viewed as a description of the covariance structure of these cumulative occupation times (even if as we shall explain later, things are slightly more complicated in the case of the loop measure because cumulative occupation times are infinite, so that a renormalization procedure is needed). By the classical central limit theorem, in the asymptotic regime where the intensity of these clouds goes to infinity, the fluctuations of these occupation times converge (if properly normalized of course) to a Gaussian process with the same covariance structure. This will in particular enable us to interpret the

Gaussian Free Field in terms of fluctuations of occupation times of high-intensity clouds of Brownian excursions.

Note that in [7], a different and more direct (as it involves no asymptotic) relation between the loop-soup occupation times and the Gaussian Free Field (or rather its square) is pointed out.

Here is how the present paper is structured: In Sect. 2, we review various very elementary facts concerning Green's functions, their conformal invariance and their relation to Brownian motion and the Gaussian Free Field. In Sect. 3, we recall the definition of the Brownian excursion measure, we derive (1) and deduce from it the interpretation of the Gaussian Free Field as asymptotic fluctuations of the Excursions occupation time measure. In passing, we note a representation of the solution to the standard Dirichlet problem using Brownian excursions, that does not seem so well-known despite its simplicity. Section 4 is the counterpart of Sect. 3 for Brownian loops instead of Brownian excursions. Finally, in Sect. 5, we briefly mention a generalization of the previous results using some clouds of interacting pairs of excursions (via their intersection local-time) that exhibits some relations between loops and excursions.

We will focus on two-dimensional domains, but many of our statements (in particular those on Brownian excursions) are also valid in higher dimensions. However, as the reader will see, we choose to base our proofs on conformal invariance, so that another approach would be needed to derive the results in dimensions greater than two. We should also point out that the statements are in fact valid in non-simply connected domains, but again, some of our proofs, in particular those dealing with the loop-measure, would need to be changed in order to cover non-simply connected planar domains (as we will use explicit expressions for the unit disc).

2 Review of Basic Notions

2.1 *Generalities*

We first recall some classical facts about Brownian motion and its relation to harmonic functions, see for instance [1, 10, 11] for further details or background.

Suppose that D is a bounded planar domain, and that it has a smooth boundary. Then, for any point x in D, the distribution of the exit position from D by a Brownian motion started at x has a continuous density with respect to the surface measure $\sigma(dz)$ on ∂D, called the *Poisson kernel*, that we will denote by $h_D(x,z)$ for $z \in \partial D$. In other words, the exit distribution is $h_D(x,z)\sigma(dz)$.

This Poisson kernel is closely related to the solutions of the *Dirichlet problem* in D (i.e., to find a harmonic function u in D, that is continuous on $\overline{D}$ and equal to some prescribed continuous function f on the boundary of D). Indeed, the solution to the Dirichlet problem, if it exists, is given by

$$u(x) = \int \sigma(dz)h_D(x,z)f(z) = E_x(f(Z_\tau))$$

where Z is a planar Brownian motion started from x under the probability measure P_x and τ denotes its exit time from D.

The *Green's function* in D, is the unique function in $D \times D$, such that for each $x \in D$, $y \mapsto G_D(x,y)$ is harmonic, vanishes on ∂D, and satisfies $G_D(x,y) \sim \pi^{-1}\log(1/|x-y|)$ when $y \to x$.

Alternatively, one can think of $G_D(x,y)dy$ as the expected time spent by Z in the infinitesimal neighborhood of y before exiting D. More precisely, if A denotes an open set, the expected time spent by the Brownian motion Z (started from $Z_0 = x$) in A before exiting D is

$$E_x(\int_0^\tau dt 1_A(Z_t)) = \int_A dy G_D(x,y).$$

The Green's function is closely related to the *Poisson problem* (i.e. to find a C^2 function u in D such that $\Delta u = -2g$, where g is some given continuous function in D, with the property that u is continuous on $\overline{D}$ and equal to 0 on ∂D). Under mild assumptions on D, the solution to this problem exists, is unique, and

$$u(x) = \int_D dy G_D(x,y)g(y) = E_x(\int_0^\tau dt f(Z_t)).$$

Not surprisingly, the Poisson kernel is closely related to the Green's function. More precisely, if $n = n_{z,D}$ is the inwards pointing normal vector at $z \in \partial D$, then, as ϵ goes to 0,

$$G_D(x, z + \epsilon n) \sim 2\epsilon h_D(x,z).$$

In the case of the unit disc $\mathbb{U} := \{x : |x| < 1\}$ in the complex plane, the Poisson kernel and the Green's function can be explicitly computed:

$$h_{\mathbb{U}}(x,z) = \frac{1-|x|^2}{2\pi|x-z|^2}$$

and

$$G_{\mathbb{U}}(x,y) = \frac{-1}{\pi}\log\frac{|x-y|}{|1-x\bar{y}|}$$

for $x \in \mathbb{U}$, $y \in \mathbb{U}$, and $z \in \partial\mathbb{U}$.

2.2 Conformal Invariance

Conformal invariance of planar Brownian motion, first observed by Paul Lévy [8], can be described as follows: if one considers a planar Brownian motion Z started from x and stopped at its first exit time of a simply connected domain D, and if Φ denotes a conformal map from D onto some other domain D', then the law of $\Phi(Z)$ is that of a Brownian motion started from $\Phi(x)$ and stopped at its first exit time of D'. Actually, for this statement to be fully true, one has to reparametrize time of $\Phi(Z)$ in a proper way. The rigorous statement is that for all $t < \tau$,

$$\Phi(Z_t) = Z'_{H_t} \text{ with } H_t = \int_0^t ds|\Phi'(Z_s)|^2,$$

where Z' is a Brownian motion started from $\Phi(x)$, stopped at $\tau' = H_\tau$, which is its exit time of D'.

Conformal invariance of Brownian motion is closely related to the conformal invariance of the Green's function and of the Poisson kernel. Let us give a rather convoluted explanation of the conformal invariance of Green's functions using Brownian motion (a direct proof using the analytic characterization of the Green's function is much more straightforward) that will be helpful for what follows. Suppose that x and y are in D and that ϵ is very small. We have seen that the expected time spent in the ball $U(y, \epsilon)$, centered at y and of radius ϵ, by the Brownian motion Z started at x behaves like

$$\pi\epsilon^2 G_D(x, y)$$

when $\epsilon \to 0$. Equivalently, the expected time spent in the ball $U(\Phi(y), |\Phi'(y)|\epsilon)$ by the Brownian motion β started at $\Phi(x)$, behaves like

$$\pi|\Phi'(y)|^2\epsilon^2 G_{D'}(\Phi(x), \Phi(y))$$

as $\epsilon \to 0$. The process $\Phi(Z)$ can be viewed as a time-changed Brownian motion, and the time-change when Z is close to y is described via H_t. It follows easily that this expected time of $\Phi(Z)$ spent in the ball $U(\Phi(y), |\Phi'(y)|\epsilon)$ behaves like

$$\frac{\pi|\Phi'(y)|^2\epsilon^2 G_{D'}(\Phi(x), \Phi(y))}{|\Phi'(y)|^2} = \pi\epsilon^2 G_{D'}(\Phi(x), \Phi(y)).$$

As a result, we have indeed that

$$G_{\Phi(D)}(\Phi(x), \Phi(y)) = G_D(x, y). \tag{3}$$

For a more rigorous derivation along the same lines, we can use the integral representation of occupation times of domains: on the one hand,

$$\begin{aligned} E_{\Phi(x)}\Big(\int_0^{\tau_{D'}} dt f(Z'_t)\Big) &= \int_{D'} dy G_{D'}(\Phi(x), y) f(y) \\ &= \int_D |\Phi'(y)|^2 dy G_{D'}(\Phi(x), \Phi(y)) f(\Phi(y)) \end{aligned}$$

for indicator functions $f = 1_A$, and on the other hand,

$$E_{\Phi(x)}\left(\int_0^{\tau_{D'}} dt f(Z'_t)\right) = E_x\left(\int_0^{\tau_D} |\Phi'(Z_t)|^2 dt f(\Phi(Z_t))\right)$$
$$= \int_D dy G_D(x,y) f(\Phi(y)) |\Phi'(y)|^2.$$

Conformal invariance of planar Brownian motion can also be used in a similar way to see that

$$|\Phi'(z)|\, h_{\Phi(D)}(\Phi(x), \Phi(z)) = h_D(x,z) \tag{4}$$

for all $x \in D, z \in \partial D$ when ∂D is smooth. Let us stress again that these conformal invariance properties of the Green's functions and of the Poisson kernel can be derived much more directly without any reference to Brownian paths.

Note that $G_{\mathbb{U}}(0, y_0) = -\pi^{-1} \log |y_0|$ for all $y_0 \neq 0$. The formula for $G_{\mathbb{U}}(x,y)$ then follows immediately, using the Möbius transformation ϕ_x of $\mathbb{U}$ onto itself that maps x onto 0 and vice-versa (this is the map $z \mapsto (z-x)/(1-\bar{x}z)$) because then $G_{\mathbb{U}}(x,y) = G_{\mathbb{U}}(0, \phi_x(y))$. Note also that this conformal invariance also provides one possible explanation of the symmetry of the Green's function $G_{\mathbb{U}}(x,y) = G_{\mathbb{U}}(y,x)$ (because for any x and y, there exists a conformal map from D onto itself that maps x onto y and y onto x).

Similarly, since clearly $h_{\mathbb{U}}(0,z) = 1/(2\pi)$ for all $z \in \partial\mathbb{U}$, the formula for $h_{\mathbb{U}}(x,z)$ recalled at the end of the previous subsection follows using conformal invariance.

2.3 *The Gaussian Free Field*

In the present text, we will briefly relate our Brownian excursions to the Gaussian Free Field, which is a classical and basic building block in Field theory, see for instance [2, 9]. So we recall its definition, in the Gaussian Hilbert space framework (as in [12] for instance): Consider the space $H_s(D)$ of smooth, real-valued functions on $\mathbb{R}^2$ that are supported on a compact subset of a domain $D \subset \mathbb{R}^d$ (so that, in particular, their first derivatives are in $L^2(D)$). This space can be endowed with the *Dirichlet inner product* defined by

$$(f_1, f_2)_\nabla = \int_D dx (\nabla f_1 \cdot \nabla f_2)$$

It is immediate to see that this Dirichlet inner product is invariant under conformal transformation. Denote by $H(D)$ the Hilbert space completion of $H_s(D)$. The quantity $(f, f)_\nabla$ is called the *Dirichlet energy* of f.

A *Gaussian Free Field* is any Gaussian Hilbert space $\mathcal{G}(D)$ of random variables denoted by "$(h, f)_\nabla$"—one variable for each $f \in H(D)$—that inherits the Dirichlet inner product structure of $H(D)$, i.e.,

$$\mathbb{E}[(h,a)_\nabla (h,b)_\nabla] = (a,b)_\nabla.$$

In other words, the map from f to the random variable $(h, f)_\nabla$ is an inner product preserving map from $H(D)$ to $\mathscr{G}(D)$. The reason for this notation is that it is possible to view h as a random linear operator, but we will not need this approach. We also view (h,ρ) as being well defined for all $\rho \in (-\Delta)H(D)$ (if $\rho = -\Delta f$ for some $f \in H(D)$, then we denote $(h,\rho) = (h,f)_\nabla$).

When ρ_1 and ρ_2 are in $H_s(D)$, the covariance of (h,ρ_1) and (h,ρ_2) can be written as $(-\Delta^{-1}\rho_1, -\Delta^{-1}\rho_2)_\nabla = (\rho_1, -\Delta^{-1}\rho_2) = (-\Delta^{-1}\rho_1, \rho_2)$. From the Poisson problem that we discussed before, $-\Delta^{-1}\rho$ can be written using the Green's function as

$$[-\Delta^{-1}\rho](x) = \frac{1}{2}\int_D dy\ G_D(x,y)\rho(y),$$

we may also write:

$$\mathrm{Cov}[(h,\rho_1),(h,\rho_2)] = \frac{1}{2}\int dxdy\ G_D(x,y)\rho_1(x)\rho_2(y) \tag{5}$$

Both the Dirichlet inner product and the Gaussian Free Field inherit naturally conformal invariance properties from the conformal invariance of the Green's function. The 2-dimensional Gaussian free field (GFF) is a particular rich object, in which a number of geometric features can be detected, and that has been shown to play a central role in the theory of random surfaces and conformally invariant geometric structures, see [13] and the references therein.

3 Occupation Times of Brownian Excursions

Brownian excursion measure. Let us first very briefly recall the construction of Brownian excursion measures. For the unit disc $\mathbb{U}$, for each $\epsilon > 0$, let μ_ϵ denote the measure of total mass $1/\epsilon$ defined as $1/\epsilon$ times the law of a Brownian motion started uniformly on the circle of radius $(1-\epsilon)$, and stopped at its first hitting time of the unit circle. In some appropriate topology, the measures μ_ϵ converge when $\epsilon \to 0$ to an infinite measure μ on two-dimensional paths that start and end on the unit circle. For a general simply connected domain D, the excursion measure μ_D can either be defined as the image of μ by the conformal map Φ that maps $\mathbb{U}$ onto D, or alternatively in an analogous way as in the disc, by integrating over the choice of the starting point of the excursion on ∂D. The fact that these two definitions are equivalent is the conformal invariance property of the Brownian excursion measures. See e.g. [16] for details and references.

Note that μ is a measure on paths $(B_t, 0<t<\tau)$ that start and end on ∂D (i.e., $B_0 \in \partial D$ and $B_\tau \in \partial D$) that are "oriented", i.e. B_0 and B_τ do a priori not play the same role. However, it turns out that the Brownian excursions are reversible i.e., that $(B_t, 0<t<\tau)$ and $(B_{\tau-t}, 0<t<\tau)$ are defined under the same measure (this can for instance be easily seen using the definition in the case where D is the upper half-plane).

Brownian Excursion Occupation Times and the Dirichlet Problem. Let us first make a comment on the relation between the Brownian excursion measure and the Dirichlet problem. Let u be the solution to the Dirichlet problem, i.e. $\Delta u = 0$ in $\mathbb{U}$ and $u = f$ on $\partial\mathbb{U}$. For all $z \in \partial\mathbb{U}$ and all positive ϵ, we have that

$$\begin{aligned}
E_{(1-\epsilon)z}\Big(\int_0^\tau dt 1_A(\gamma_t) f(\gamma_\tau)\Big) &= E_{(1-\epsilon)z}\Big(\int_0^\infty dt 1_A(\gamma_t) 1_{t\le\tau} f(\gamma_\tau)\Big)\\
&= E_{(1-\epsilon)z}\Big(\int_0^\infty dt 1_A(\gamma_t) 1_{t\le\tau} E(f(\gamma_\tau)|\mathscr{F}_t)\Big)\\
&= E_{(1-\epsilon)z}\Big(\int_0^\tau dt 1_A(\gamma_t) E_{\gamma_t}(f(\gamma_\tau))\Big)\\
&= E_{(1-\epsilon)z}\Big(\int_0^\tau dt 1_A(\gamma_t) u(\gamma_t)\Big)\\
&= \int_A dy G_{\mathbb{U}}((1-\epsilon)z, y) u(y)
\end{aligned}$$

And for the Brownian excursion measure $\mu = \mu_{\mathbb{U}}$, we have that

$$\begin{aligned}
\mu\Big(\int_0^\tau dt 1_A(\gamma_t) f(\gamma_\tau)\Big) &= \lim_{\epsilon\to 0}\int_0^{2\pi}\frac{d\theta}{\epsilon} E_{(1-\epsilon)e^{i\theta}}\Big(\int_0^\tau dt 1_A(\gamma_t) f(\gamma_\tau)\Big)\\
&= \lim_{\epsilon\to 0}\int_0^{2\pi}\frac{d\theta}{\epsilon}\int_A dy G_{\mathbb{U}}((1-\epsilon)e^{i\theta}, y) u(y)\\
&= \int_0^{2\pi} 2d\theta \int_A dy\, h_{\mathbb{U}}(y, e^{i\theta}) u(y)\\
&= 2\int_A dy\, u(y)\int_0^{2\pi} d\theta h_{\mathbb{U}}(y, e^{i\theta})\\
&= 2\int_A dy\, u(y)
\end{aligned}$$

That is to say, we can represent the solution to the Dirichlet problem via the Brownian excursion measure by the formula

$$\mu\Big(\int_0^\tau dt 1_A(\gamma_t) f(\gamma_\tau)\Big) = 2\int_A dy\, u(y)$$

Since the Brownian excursion is reversible, we also have that

$$\mu\Big(f(\gamma_0)\int_0^\tau dt 1_A(\gamma_t)\Big) = 2\int_A dy\, u(y) \tag{6}$$

Hence, if we put a weight f on starting point of the excursion, then the mean occupation time spent in A is measured by the integral of u on A, where u is the solution to the corresponding Dirichlet problem. By conformal invariance, (6) also holds for any simply connected domain.

We would like to note that, if we set $f = 1$ in (6), we get that $\mu_D(\int_0^\tau dt 1_A(\gamma_t))$ is equal to twice the area of A. In particular, $\mu_D(\tau)$ is therefore just twice the area of D.

The Covariance Structure. We now turn our attention towards the proof of (1). This formula can be understood as follows: we can cut $A \times B$ into very small pieces, calculate on each small piece and then add all these pieces together. On each small piece $dx \times dy$, the Brownian excursion starts from the boundary, firstly it hits the small piece dx (with a small probability), after this time, it is a true Brownian motion starting nearby x, which is (almost) independent of the past and then the expected time of this new Brownian motion spent in the neighborhood of y before exiting D is close to $G_D(x, y)dy$. When we add up all these small pieces together and we obtain the right-hand side of the formula.

For a precise calculation, we first consider the case where $D = \mathbb{U}$ as the general case will then follow from conformal invariance. We also use the notation that $\mu = \mu_{\mathbb{U}}$. Let γ denote a Brownian excursion in $\mathbb{U}$. For all $z \in \partial\mathbb{U}$ and all positive ϵ,

$$\begin{aligned}
E_{(1-\epsilon)z}\Big(\int_0^\tau ds 1_A(\gamma_s)\int_s^\tau dt 1_B(\gamma_t)\Big) &= E_{(1-\epsilon)z}\Big(\int_0^\tau ds 1_A(\gamma_s) E\Big(\int_s^\tau dt 1_B(\gamma_t)|\mathscr{F}_s\Big)\Big)\\
&= E_{(1-\epsilon)z}\Big(\int_0^\tau ds 1_A(\gamma_s) E_{\gamma_s}\Big(\int_0^\tau dt 1_B(\gamma_t)\Big)\Big)\\
&= E_{(1-\epsilon)z}\Big(\int_0^\tau ds 1_A(\gamma_s) G_{\mathbb{U}}(\gamma_s, B)\Big)\\
&= \int_A dy G_{\mathbb{U}}((1-\epsilon)z, y) G_{\mathbb{U}}(y, B).
\end{aligned}$$

And for the Brownian excursion measure, we have that

$$\begin{aligned}
\mu\Big(\int_0^\tau ds 1_A(\gamma_s)\int_s^\tau dt 1_B(\gamma_t)\Big) &= \lim_{\epsilon\to 0}\int_0^{2\pi}\frac{d\theta}{\epsilon} E_{(1-\epsilon)e^{i\theta}}\Big(\int_0^\tau ds 1_A(\gamma_s)\int_s^\tau dt 1_B(\gamma_t)\Big)\\
&= \lim_{\epsilon\to 0}\int_0^{2\pi}\frac{d\theta}{\epsilon}\int_A dy G_{\mathbb{U}}((1-\epsilon)e^{i\theta}, y) G_{\mathbb{U}}(y, B)\\
&= \int_0^{2\pi} 2d\theta \int_A dy h_{\mathbb{U}}(y, e^{i\theta}) G_{\mathbb{U}}(y, B)\\
&= 2\int_A dy G_{\mathbb{U}}(y, B)\int_0^{2\pi} d\theta h_{\mathbb{U}}(y, e^{i\theta})\\
&= 2\int_A dy G_{\mathbb{U}}(y, B)
\end{aligned}$$

By symmetry of the Green's function ($G_{\mathbb{U}}(x,y) = G_{\mathbb{U}}(y,x)$), we have that

$$\mu(\int_0^\tau ds 1_A(\gamma_s) \int_0^\tau ds 1_B(\gamma_s)) = 4 \int_{A\times B} dxdy G_{\mathbb{U}}(x,y).$$

This concludes the proof of the (1), since we can use to conformal invariance to derive the formula for general simply connected domain D. More generally, we have that

$$\mu_D(\int_0^\tau ds f(\gamma_s) \int_0^\tau ds g(\gamma_s)) = 4 \int dx\, dy G_D(x,y) f(x) g(y) \qquad (7)$$

for all measurable bounded functions f and g.

Large Intensity Clouds of Excursions and GFF. Let us now use this formula to make a link between Brownian excursions and the GFF. For this we are going to use Poissonian cloud of excursions in D, as in [15]. Recall that a Poisson cloud of excursions with intensity $c\mu_D$ is a random countable family of Brownian excursions in D, that is defined as a Poisson point process with intensity $c\mu_D$.

In particular, the union of two independent Poissonian clouds of Brownian excursions in D with intensity $c_1\mu_D$ and $c_2\mu_D$ is a Poissonian cloud of excursions in D with intensity $(c_1 + c_2)\mu_D$.

Let us now consider an i.i.d. sequence $M^j, j \geq 1$ of Poissonian clouds of excursions in D with the common intensity μ_D. For each $j \geq 1$, and each $f \in (-\Delta)H(D)$, define the "cumulative occupation" time of M^j by

$$X_f^j = \sum_{\gamma \in M^j} \int_0^{\tau(\gamma)} ds f(\gamma_s).$$

The fact that $\mu(\tau)$ is finite (as soon as the area of D is finite) ensures that X_f^j is almost surely finite (as soon as f is bounded) because its expectation is bounded. We then define

$$\tilde{X}_f^j = X_f^j - E(X_f^j).$$

On an enlarged probability space, we can also define an i.i.d. family of random variable ϵ_γ indexed by the set of excursions in $\cup_j M^j$ such that $P(\epsilon_\gamma = 1) = P(\epsilon_\gamma = -1) = 1/2$. We can then define

$$Y_f^j = \sum_{\gamma \in M^j} \epsilon_\gamma \int_0^{\tau(\gamma)} ds f(\gamma_s).$$

It is easy to see that $Y_f^1, Y_f^2, Y_f^3, \ldots$ are i.i.d. centered random variables with common variance

$$\sigma_f^2 = \mu_D(\int_0^\tau ds f(\gamma_s) \int_0^\tau ds f(\gamma_s)) = 4 \int dxdy G_D(x,y) f(x) f(y).$$

The same is true for $\tilde{X}_f^1, \tilde{X}_f^2, \tilde{X}_f^3, \ldots$. By the Central Limit Theorem, we have that

$$\frac{1}{\sqrt{N}}(Y_f^1 + \ldots + Y_f^N)$$

converges in law as $N \to \infty$ to a centered Gaussian random variable Y_f with variance σ_f^2. The same holds for the sequence $(\tilde{X}_f^1 + \ldots + \tilde{X}_f^N)/\sqrt{N}$.

Hence, we see that the GFF can be viewed as the limit (in law, and in the sense of finite-dimensional distributions) of the occupation times fluctuations of a Poisson cloud of Brownian excursions, when the intensity tends to infinity.

Higher-Order "Moments". We just mention that our proof can be adapted directly in order to show that for all $p \geq 2$:

$$\mu_D\left(\int_{(0,\tau)^p} dt_1 \ldots dt_p 1_{t_1<\ldots<t_p} 1_{A_1}(\gamma_{t_1}) \cdots 1_{A_p}(\gamma_{t_p})\right)$$
$$= 2\int_{A_1\times\cdots\times A_p} dx_1 \cdots dx_p G_D(x_1, x_2) \times \cdots \times G_D(x_{p-1}, x_p)$$

which gives for instance (when one sums over all possible order of visits) a formula for $\mu_D((\int_0^\tau f(\gamma_s)ds)^p)$. We have chosen to focus on the case $p = 2$ because of the above-mentioned link with Gaussian fields.

Non-Simply-Connected Domains. Suppose that D is a finitely connected open domain in the plane. Then, by Koebe's uniformization Theorem (see [3]), it is possible to map it conformally onto a circular domain i.e., the unit disk U punctured by a finite number of disjoint closed disks. It is very easy to generalize the definition of the Brownian excursion measure in circular domains (adding the contributions corresponding to starting points in the neighborhood of each of the boundary disks), and to see that all our proofs go through without any real difficulty, so that all our statements are in fact valid also in circular domains. One can then *define* the excursion measure in D via conformal invariance starting from circular domains, and then, by conformal invariance of all the quantities involved, we easily see that all our statements are also valid in D.

4 Occupation Times of Brownian Loops

Brownian Loop Measure. We now briefly recall the construction of the Brownian loop measure [5]. As for the Brownian excursion measure, we can first define it in the unit disc, and then define it in any other simply connected domain using conformal invariance (and one then checks that this is indeed consistent with other possible constructions).

For any $r \in (0,1]$, define $U_r = r\mathbb{U}$. For any $x \in U_r$ and any $z \in \partial U_r$, one can define the Brownian motion started at x and conditioned to exit U_r at z (this can be rigorously defined as the limit when $\epsilon \to 0$ of the law of the Brownian motion conditioned to exit U_r in an ϵ-neighborhood of z). Let us denote this probability measure by $P^r_{x\to z}$. Then, as for the excursion measure, one can let $x \to z$, and renormalize it in order to get a measure on macroscopic sets i.e. define

$$m^r_z(\cdot) = \lim_{\epsilon\to 0} \epsilon^{-1} h_{U_r}(z+\epsilon n, z) P^r_{z+\epsilon n\to z}(\cdot)$$

where $n = n_{z,U_r}$ is the inwards pointing normal vector at $z \in \partial U_r$. Then, one can define the loop measure in $\mathbb{U}$ by integrating z on ∂U_r, and then integrating r from 0 to 1:

$$\lambda_{\mathbb{U}}(\cdot) = \int_0^1 r dr \int_0^{2\pi} d\theta \, m^r_{re^{i\theta}}(\cdot).$$

In fact, the above definition is not quite the loop measure because it defines a measure on parametrized loops. We will forget about the precise parametrization of the loop and view $\lambda_{\mathbb{U}}$ as a measure on loops defined modulo monotone reparametrization (where the time-parameter should be viewed as an element of the circle, because the end-point of the loop is the same as the starting point, this is possible). It turns out that this definition of $\lambda_{\mathbb{U}}$ is then invariant under the Moebius transformations that map the unit disc onto itself. Hence, it is possible to define, for a general simply connected domain D, the loop measure λ_D as the image of $\lambda_{\mathbb{U}}$ by any conformal map Φ that maps $\mathbb{U}$ onto D. And we usually denote $\lambda = \lambda_{\mathbb{U}}$.

Before going on, we would like to say a word on the value of $\lambda(\tau)$. In fact, by direct computation we have that $\lambda(\tau) = \infty$ which is very different from $\mu(\tau)$ mentioned before. A direct way to check that $\lambda(\tau) = \infty$ goes as follows. Consider D to be the square $[0,1]^2$. For any dyadic square d in D with sidelength 2^{-n}, a direct scaling argument shows that the mass (for λ) of the set of loops that stay in d and have a time-length in $[4^{-n}, 2\times 4^{-n})$ does not depend on d. Hence, if we sum this quantity over all dyadic squares d in D, and because $\sum_n 4^n 4^{-n} = \infty$, we readily see that $\lambda(\tau) = \infty$.

However, almost the same argument ensures that $\lambda(\tau^{1+\epsilon})$ is finite for $\epsilon > 0$ (and bounded D). Indeed, in the case of the unit square, we can decompose the set of loops with time-length in $[4^{-n}, 4^{1-n})$ according to the dyadic square in which its lowest point lies. This leads readily to the bound

$$\lambda(1_{\tau<1}\tau^{1+\epsilon}) \leq C \sum_{n\geq 1} 4^n (4^{1-n})^{1+\epsilon} < \infty$$

and one can see by other means that $\lambda(\tau > t)$ decays exponentially fast as $t \to \infty$. In particular, we get that $\lambda(\tau^2)$ is finite (as soon as D is bounded).

Covariance Structure. Our goal is now to prove (2). As before, we are going to derive the result first in the case where $D = \mathbb{U}$, and the general result will then follow using conformal invariance. Again, it will be convenient to (loosely speaking) divide $A \times B$ into infinitesimal pieces $dx \times dy$, make the computation on each piece, and then add all these pieces together. Clearly, this will give a formula of the type

$$\lambda_D(\int_0^\tau ds 1_A(\gamma_s) \int_0^\tau ds 1_B(\gamma_s)) = \int_{A\times B} dxdy F_D(x,y)$$

where $F_D(x, y)$ is the "covariance" function between x and y determined by the Brownian loop measure. Just as what we have done to derive the conformal invariance of the Green's function in the (3), we can also derive the conformal invariance of F:

$$F_{\Phi(D)}(\Phi(x), \Phi(y)) = F_D(x, y).$$

To determine $F_D(x, y)$, it is enough to describe $F_{\mathbb{U}}(0, y_0)$ for $y_0 \in (0, 1)$, because there exists a y_0 and a conformal map Φ from D onto $\mathbb{U}$ such that $\Phi(x) = 0, \Phi(y) = y_0$.

And now begin our computation. For $r \in (0, 1), z \in \partial U_r$, we can write

$$\begin{aligned}
&E^r_{z+\epsilon n\to z}(\int_0^\tau ds 1_A(\gamma_s) \int_s^\tau dt 1_B(\gamma_t)) \\
&= \lim_{\epsilon'\to 0}\Big(\frac{1}{P^r_{z+\epsilon n}(\gamma_\tau \in U(z,\epsilon') \cap \partial U_r)} \\
&\quad \times E^r_{z+\epsilon n}(\int_0^\tau ds 1_A(\gamma_s) \int_s^\tau dt 1_B(\gamma_t) 1_{\gamma_\tau \in U(z,\epsilon')\cap\partial U_r})\Big) \\
&= \lim_{\epsilon'\to 0}\Big(\frac{1}{h_{U_r}(z+\epsilon n, U(z,\epsilon') \cap \partial U_r)} \\
&\quad \times E^r_{z+\epsilon n}(\int_0^\tau ds 1_A(\gamma_s) \int_s^\tau dt 1_B(\gamma_t) h_{U_r}(\gamma_t, U(z,\epsilon') \cap \partial U_r))\Big) \\
&= \frac{1}{h_{U_r}(z+\epsilon n, z)} E^r_{z+\epsilon n}(\int_0^\tau ds 1_A(\gamma_s) \int_s^\tau dt 1_B(\gamma_t) h_{U_r}(\gamma_t, z)).
\end{aligned}$$

Hence,

$$\begin{aligned}
&h_{U_r}(z+\epsilon n, z) E^r_{z+\epsilon n\to z}(\int_0^\tau ds 1_A(\gamma_s) \int_s^\tau dt 1_B(\gamma_t)) \\
&= E^r_{z+\epsilon n}(\int_0^\tau ds 1_A(\gamma_s) \int_s^\tau dt 1_B(\gamma_t) h_{U_r}(\gamma_t, z)) \\
&= \int_A dx\, G_{U_r}(z+\epsilon n, x) \int_B dy\, G_{U_r}(x,y) h_{U_r}(y,z)
\end{aligned}$$

and letting $\epsilon \to 0$, we get

$$\begin{aligned} m_z^r(\int_0^\tau ds 1_A(\gamma_s) \int_0^\tau ds 1_B(\gamma_s)) &= \lim_{\epsilon\to 0} \frac{2}{\epsilon} \int_A dx\, G_{U_r}(z+\epsilon n, x) \int_B dy\, G_{U_r}(x,y) h_{U_r}(y,z) \\ &= 4\int_{A\times B} dxdy G_{U_r}(x,y) h_{U_r}(x,z) h_{U_r}(y,z). \end{aligned}$$

For simplicity, we define a new kernel

$$K_{U_r}(x,y) = 4\int_0^{2\pi} d\theta h_{U_r}(x, re^{i\theta}) h_{U_r}(y, re^{i\theta})$$

and then we have that

$$\lambda_{\mathbb{U}}(\int_0^\tau ds 1_A(\gamma_s) \int_0^\tau ds 1_B(\gamma_s)) = \int_0^1 rdr \int_{A\times B} dxdy G_{U_r}(x,y) K_{U_r}(x,y).$$

Note that

$$K_{U_r}(rx, ry) = \frac{1}{r^2} K_{\mathbb{U}}(x,y)$$

and

$$G_{U_r}(rx, ry) = G_{\mathbb{U}}(x,y).$$

Furthermore, $K_{\mathbb{U}}(0,y) = 2/\pi$.

Suppose that $A = U(0,\epsilon)$ and $B = U(y_0, \delta)$ where ϵ and δ are both small. In our decomposition of $\lambda_{\mathbb{U}}$, the loop can visit B only if it started on a circle of radius $r > y_0$. Hence, on the one hand, as ϵ and δ tend to 0,

$$\begin{aligned} \lambda_{\mathbb{U}}(\int_0^\tau ds 1_A(\gamma_s) \int_0^\tau ds 1_B(\gamma_s)) &= \int_{y_0}^1 rdr \int_{(A\bigcap U_r)\times(B\bigcap U_r)} dxdy G_{U_r}(x,y) K_{U_r}(x,y) \\ &\sim \int_{y_0}^1 rdr (\pi\epsilon^2\pi\delta^2) G_{U_r}(0,y_0) K_{U_r}(0,y_0) \\ &= (\pi\epsilon^2\pi\delta^2) \int_{y_0}^1 \frac{1}{r} dr G_{\mathbb{U}}(0, \frac{y_0}{r}) K_{\mathbb{U}}(0, \frac{y_0}{r}) \\ &= (\pi\epsilon^2\pi\delta^2) \frac{2}{\pi^2} \int_{y_0}^1 dr (-\frac{1}{r}\log(\frac{y_0}{r})) \\ &= (\pi\epsilon^2\pi\delta^2) \frac{1}{\pi^2} (\log y_0)^2 \end{aligned}$$

On the other hand, this quantity is precisely behaving as $(\pi\epsilon^2\pi\delta^2)F_{\mathbb{U}}(0, y_0)$ and as a result, we get that

$$F_{\mathbb{U}}(0, y_0) = \frac{1}{\pi^2}(\log y_0)^2 = (G_{\mathbb{U}}(0, y_0))^2.$$

We can then conclude that (2) holds in $\mathbb{U}$, and then also in D by conformal invariance. More generally, we have that

$$\lambda_D(\int_0^\tau dsf(\gamma_s)\int_0^\tau dsg(\gamma_s)) = \int_{A\times B} dxdy(G_D(x, y))^2 f(x)g(y).$$

for all measurable bounded functions f and g.

Brownian Loop-Soups and Fields. Just as in the case of Brownian excursion measure, we can use this formula to make a link between Brownian loops and some Gaussian Fields. Let $M^j, j \geq 1$ be a sequence of i.i.d Poissonian clouds of loops in D with the common intensity λ_D. We can try to give the same definitions of the quantities $\tilde{X}_f^j, Y_f^j, j \geq 1$. However, things are a little more complicated, due to the fact that the same scaling argument that showed that $\lambda(\tau) = \infty$ implies that

$$\sum_{\gamma\in M^j} \tau(\gamma) = \infty$$

almost surely, so that some care is needed.

The definition of Y_f^j is however not a big problem. Recall that on an enlarged probability space, one associates to each loop γ a random variable ϵ_γ with $E(\epsilon_\gamma) = 0$ and $E((\epsilon_\gamma)^2) = 1$. But (2) precisely ensures that the sum

$$\sum_{\gamma\in M^j} \epsilon_\gamma \int_0^{\tau(\gamma)} f(\gamma_s)ds$$

makes sense in L^2, and that its second moment is equal to

$$\sigma_f^2 = \lambda_D(\int_0^\tau dsf(\gamma_s)\int_0^\tau dsf(\gamma_s)) = \int dxdy(G_D(x, y))^2 f(x)f(y)$$

which is finite.

Then, just as in the case of the clouds of excursions, the sequence $Y_f^1, Y_f^2, \ldots$ is made of i.i.d centered random variables with common variance σ_f^2. By the Central Limit Theorem,

$$\frac{1}{\sqrt{N}}(Y_f^1 + \ldots + Y_f^N)$$

converges in law as $N \to \infty$ to a centered Gaussian random variable with variance σ_f^2. Hence, we obtain another Gaussian Field, characterized by this new covariance structure.

It is also still possible to make sense of $\tilde{X}_f^j$ even though it is not possible to define X_f^j. It suffices to partition the set of loops (in D) into a countable set of loops $A_k, k \geq 1$ such that for each k, $\lambda(\tau 1_{\gamma \in A_k})$ is finite (for instance, one can take $A_k = \{\gamma : \tau(\gamma) > 1/k\} \setminus (A_1 \cup \ldots \cup A_{k-1})\}$. Then, one can define

$$\tilde{X}_f^j = \sum_{k \geq 1} \left(\sum_{\gamma \in A_k \cap M^j} \int_0^{\tau(\gamma)} f(\gamma_s) ds - E(\sum_{\gamma \in A_k \cap M^j} \int_0^{\tau(\gamma)} f(\gamma_s) ds) \right)$$

and check that this sum with respect to k converges in L^2, and that its second moment is the same as that of Y_f. The rest of the argument is again the same.

5 Intersections of Brownian Excursions

In this section, we try to find the relation between intersections of Brownian excursion "occupations times" and Brownian loop occupation times, the former being defined via the intersection local time.

Let us first recall some features of Brownian intersection local times. Let $p \geq 2$ be an integer, and let $Z^1, ..., Z^p$ denote p independent Brownian motions in $\mathbb{R}^2$, started at $x^1, ..., x^p$ respectively. The intersection local time of $Z^1, ..., Z^p$ is a random measure $\alpha(ds_1 ... ds_p)$ on $\mathbb{R}_+^p$, supported on

$$\{(s_1, ..., s_p) \in \mathbb{R}_+^p : Z_{s_1}^1 = ... = Z_{s_p}^p\}.$$

The basic description concerning the intersection local time that we will use goes as follows (see [6] for details):

Proposition 2. *Almost surely, one can define a (random) measure $\alpha(ds_1 ... ds_p)$ on $\mathbb{R}_+^p$ such that, for any $A^1, ..., A^p$ bounded Borel subsets of $\mathbb{R}_+$,*

$$\alpha(A^1 \times ... \times A^p) = \lim_{\epsilon \to 0} \alpha_\epsilon(A^1 \times ... \times A^p)$$

in the L^n-norm, for any $n < \infty$, where

$$\alpha_\epsilon(ds_1 ... ds_p) = ds_1 ... ds_p \int_{\mathbb{R}^2} dy \delta_y^\epsilon(Z_{s_1}^1) ... \delta_y^\epsilon(Z_{s_p}^p)$$

with $\delta_y^\epsilon(z) = \frac{1}{\pi \epsilon^2} 1_{U(y,\epsilon)}(z)$.

Let us use this in the context of the Brownian excursion measure. This time we shall consider two Brownian excursions γ and γ' defined under the (infinite) measure $\mu_D \otimes \mu_D$, and study the behavior of their intersection local time that spent in two disjoint sets A and B, as before:

$$\begin{aligned}
&\mu_D \otimes \mu_D\Big(\int_0^\tau \int_0^{\tau'} \alpha(dtdt')1_{(\gamma_t=\gamma'_{t'}\in A)} \int_0^\tau \int_0^{\tau'} \alpha(dsds')1_{(\gamma_s=\gamma'_{s'}\in B)}\Big) \\
&= \lim_{\epsilon\to 0} \mu_D \otimes \mu_D\Big(\int_0^\tau \int_0^{\tau'} \alpha_\epsilon(dtdt')1_{(\gamma_t\in A)}1_{(\gamma'_{t'}\in A)} \\
&\quad \times \int_0^\tau \int_0^{\tau'} \alpha_\epsilon(dsds')1_{(\gamma_s\in B)}1_{(\gamma'_{s'}\in B)}\Big) \\
&= \lim_{\epsilon\to 0} \mu_D \otimes \mu_D\Big(\int_0^\tau \int_0^{\tau'} dtdt' \int dx\delta_x^\epsilon(\gamma_t)\delta_x^\epsilon(\gamma'_{t'})1_{(\gamma_t\in A)}1_{(\gamma'_{t'}\in A)} \\
&\quad \times \int_0^\tau \int_0^{\tau'} dsds' \int dy\delta_y^\epsilon(\gamma_s)\delta_y^\epsilon(\gamma'_{s'})1_{(\gamma_s\in B)}1_{(\gamma'_{s'}\in B)}\Big) \\
&= \lim_{\epsilon\to 0} \int dx \int dy \quad \mu_D \otimes \mu_D\Big(\int_0^\tau dt\delta_x^\epsilon(\gamma_t)1_{(\gamma_t\in A)} \int_0^\tau ds\delta_y^\epsilon(\gamma_s)1_{(\gamma_s\in B)} \\
&\quad \times \int_0^{\tau'} dt'\delta_x^\epsilon(\gamma'_{t'})1_{(\gamma'_{t'}\in A)} \int_0^{\tau'} ds'\delta_y^\epsilon(\gamma'_{s'})1_{(\gamma'_{s'}\in B)}\Big) \\
&= \lim_{\epsilon\to 0} \int dx \int dy (4 \int_{A\times B} dadb\delta_x^\epsilon(a)\delta_y^\epsilon(b)G_D(a,b))^2 \\
&= 16 \int_{A\times B} dxdy(G_D(x,y))^2
\end{aligned}$$

Hence, we see that pairs of Brownian excursions give rise to the same covariance structure as the Brownian loops. In a way, this is not too surprising, as for two points x and y that are both visited by γ and by γ', one sees in a way a loop structure (the part of γ from x to y, and then the part of γ' back from y to x).

Note that by a similar calculation, one gets that for any $p \geq 3$, if one defines for any A,

$$T_p(A;\gamma^1,\dots,\gamma^p) = \int_0^{\tau_1} \dots \int_0^{\tau_p} \alpha(dt_1 \dots dt_p)1_{(\gamma^1_{t_1}=\dots=\gamma^p_{t_p}\in A)},$$

then

$$\mu_D^{\otimes p}(T_p(A)T_p(B)) = 4^p \int_{A\times B} dxdy(G_D(x,y))^p.$$

Acknowledgements This paper is based on my Master's thesis and was completed under the guidance of my supervisor Professor Wendelin Werner. The author acknowledges the support from a Fondation CFM-JP Aguilar grant.

References

1. R. Durrett, *Brownian Motion and Martingales in Analysis* (Wadsworth Mathematics Series, 1984)
2. K. Gawędzki, *Lectures on Conformal Field Theory*, Quantum fields and strings: a course for mathematicians, Vol. 1, 2 (Princeton, NJ, 1996/1997), Amer. Math. Soc., (Providence, RI, 1999), pp. 727–805
3. P. Koebe, Abhandlungen zur Theorie der konformen Abbildung VI. Abbildung mehrfach zusammenhängender schlichter Bereiche auf Kreisbereiche, Math. Z. **7** 235–301 (1920)
4. G.F. Lawler, *Conformally Invariant Processes in the Plane*, (Mathematical Surveys and Monographs, AMS, 2005), no. 114, xii+242
5. G.F. Lawler, W. Werner, The brownian loop soup, Probab. Theor. Relat. Field. **128**(4), 565–588 (2004)
6. J.-F. Le Gall, *Some properties of Planar Brownian Motion*, École d'Été de Probabilités de Saint-Flour XX—1990, Lecture Notes in Mathematics, vol. 1527, (Springer, Berlin, 1992), pp. 111–235
7. Y. Le Jan, *Markov Paths, Loops and Fields*, École d'Été de Probabilités de Saint-Flour XXXVIII—2008, Lecture Notes in Mathematics, vol. 2026, (Springer, Berlin, 2011), p. 134
8. P. Lévy, *Processus Stochastiques et Mouvement Brownien*, Les Grands Classiques Gauthier-Villars. Éditions (Jacques Gabay, Sceaux, 1992); Followed by a note by M. Loève, Reprint of the second (1965) edition.
9. E. Nelson, The free markoff field, J. Funct. Anal. **12**, 211–227 (1973)
10. S.C. Port, C.J. Stone, *Brownian Motion and Classical Potential Theory*, (Academic Press, New York, 1978), p. 236
11. M. Rao, *Brownian Motion and Classical Potential Theory*, Lecture Notes Series, No. 47. Matematisk Institut, (Aarhus University, Aarhus, 1977)
12. S. Sheffield, Gaussian free fields for mathematicians, Probab. Theor. Relat. Field. **139**(3–4), 521–541 (2007)
13. S. Sheffield, *Conformal Weldings of Random Surfaces: SLE and the Quantum Gravity Zipper*, preprint, arXiv:1012.4797 (2010)
14. S. Sheffield, W. Werner, *Conformal Loop Ensembles: The Markovian Characterization and the Construction Via Loop-Soups*, Ann. Math. (to appear), arXiv:1006.2374v3 (2011)
15. W. Werner, Conformal restriction and related questions, Probab. Surv. **2**, 145–190 (2005)
16. W. Werner, *Some Recent Aspects of Random Conformally Invariant Systems*, (Mathematical statistical physics, Elsevier B. V., Amsterdam, 2006), pp. 57–99

Discrete Approximations to Solution Flows of Tanaka's SDE Related to Walsh Brownian Motion

Hatem Hajri

Abstract In a previous work, we have defined a Tanaka's SDE related to Walsh Brownian motion which depends on kernels. It was shown that there are only one Wiener solution and only one flow of mappings solving this equation. In the terminology of Le Jan and Raimond, these are respectively the stronger and the weaker among all solutions. In this paper, we obtain these solutions as limits of discrete models.

1 Introduction and Main Results

Consider Tanaka's equation:

$$\varphi_{s,t}(x) = x + \int_s^t \operatorname{sgn}(\varphi_{s,u}(x))dW_u, \quad s \le t, x \in \mathbb{R}, \tag{1}$$

where $\operatorname{sgn}(x) = 1_{\{x>0\}} - 1_{\{x\le 0\}}$, $W_t = W_{0,t}1_{\{t>0\}} - W_{t,0}1_{\{t\le 0\}}$ and $(W_{s,t}, s \le t)$ is a real white noise on a probability space $(\Omega, \mathscr{A}, \mathbb{P})$ (see Definition 1.10 [6]). This is an example of a stochastic differential equation which admits a weak solution but has no strong solution. If K is a stochastic flow of kernels (see Sect. 2.1 [5]) and W is a real white noise, then by definition, (K, W) is a solution of Tanaka's SDE if for all $s \le t, x \in \mathbb{R}$, $f \in C_b^2(\mathbb{R})$ (f is C^2 on $\mathbb{R}$ and f', f'' are bounded),

$$K_{s,t}f(x) = f(x) + \int_s^t K_{s,u}(f'\operatorname{sgn})(x)W(du) + \frac{1}{2}\int_s^t K_{s,u}f''(x)du \ a.s. \tag{2}$$

H. Hajri (✉)
Département de Mathématiques, Université Paris-Sud, 91405 Orsay Cedex, France
e-mail: Hatem.Hajri@math.u-psud.fr

C. Donati-Martin et al. (eds.), *Séminaire de Probabilités XLIV*, Lecture Notes in Mathematics 2046, DOI 10.1007/978-3-642-27461-9_8,

When $K = \delta_\varphi$ is a flow of mappings, K solves (2) if and only if φ solves (1) by Itô's formula. In [7], Le Jan and Raimond have constructed the unique flow of mappings associated to (1). It was shown also that

$$K^W_{s,t}(x) = \delta_{x+\mathrm{sgn}(x)W_{s,t}} 1_{\{t\leq\tau_{s,x}\}} + \frac{1}{2}(\delta_{W^+_{s,t}} + \delta_{-W^+_{s,t}})1_{\{t>\tau_{s,x}\}}, \quad s \leq t, x \in \mathbb{R},$$

is the unique $\mathscr{F}^W$ adapted solution (Wiener flow) of (2) where

$$\tau_{s,x} = \inf\{r \geq s : W_{s,r} = -|x|\}, \ W^+_{s,t} := W_{s,t} - \inf_{u\in[s,t]} W_{s,u}.$$

In [5], an extension of (2) in the case of Walsh Brownian motion was defined as follows

Definition 1. Fix $N \in \mathbb{N}^*$, $\alpha_1, \cdots, \alpha_N > 0$ such that $\sum_{i=1}^{N} \alpha_i = 1$ and consider the graph G consisting of N half lines $(D_i)_{1\leq i\leq N}$ emanating from 0 (see Fig. 1).

Let $\mathbf{e}_i$ be a vector of modulus 1 such that $D_i = \{h\mathbf{e}_i, h \geqslant 0\}$ and define for all $z \in G$, $\mathbf{e}(z) = \mathbf{e}_i$ if $z \in D_i, z \neq 0$ (convention $\mathbf{e}(0) = \mathbf{e}_N$). Define the following distance on G:

$$d(h\mathbf{e}_i, h'\mathbf{e}_j) = \begin{cases} h + h' & \text{if } i \neq j, (h, h') \in \mathbb{R}^2_+, \\ |h - h'| & \text{if } i = j, (h, h') \in \mathbb{R}^2_+. \end{cases}$$

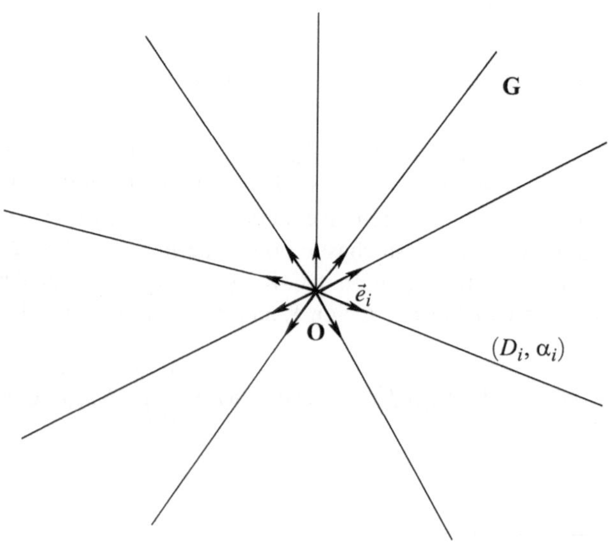

Fig. 1 Graph G

For $x \in G$, we will use the simplified notation $|x| := d(x, 0)$.
We equip G with its Borel σ-field $\mathscr{B}(G)$ and set $G^* = G \setminus \{0\}$. Let $C_b^2(G^*)$ be the space of all $f : G \longrightarrow \mathbb{R}$ such that f is continuous on G and has bounded first and second derivatives (f' and f'') on G^* (here $f'(z)$ is the derivative of f at z in the direction $\mathbf{e}(z)$ for all $z \neq 0$), both $\lim_{z \to 0, z \in D_i, z \neq 0} f'(z)$ and $\lim_{z \to 0, z \in D_i, z \neq 0} f''(z)$ exist for all $i \in [1, N]$. Define

$$D(\alpha_1, \cdots, \alpha_N) = \left\{ f \in C_b^2(G^*) : \sum_{i=1}^{N} \alpha_i \lim_{z \to 0, z \in D_i, z \neq 0} f'(z) = 0 \right\}.$$

Now, Tanaka's SDE on G extended to kernels is the following (see Remarks 3 (1) in [5] for a discussion of its origin).

Tanaka's equation on G. On a probability space $(\Omega, \mathscr{A}, \mathbb{P})$, let W be a real white noise and K be a stochastic flow of kernels on G. We say that (K, W) solves (T) if for all $s \leq t$, $f \in D(\alpha_1, \cdots, \alpha_N)$, $x \in G$,

$$K_{s,t} f(x) = f(x) + \int_s^t K_{s,u} f'(x) W(du) + \frac{1}{2} \int_s^t K_{s,u} f''(x) du \quad a.s.$$

If $K = \delta_\varphi$ is a solution of (T), we just say that (φ, W) solves (T).

Equation (T) is a particular case of an equation (E) studied in [5] (it corresponds to $\varepsilon = 1$ with the notations of [5]). It was shown (see Corollary 2 [5]) that if (K, W) solves (T), then $\sigma(W) \subset \sigma(K)$ and therefore one can just say that K solves (T). We also recall

Theorem 1. *[5] There exists a unique Wiener flow K^W (resp. flow of mappings φ) which solves (T).*

As described in Theorem 1 [5], the unique Wiener solution of (T) is simply

$$K_{s,t}^W(x) = \delta_{x + \mathbf{e}(x) W_{s,t}} 1_{\{t \leq \tau_{s,x}\}} + \sum_{i=1}^{N} \alpha_i \delta_{\mathbf{e}_i W_{s,t}^+} 1_{\{t > \tau_{s,x}\}}. \tag{3}$$

where

$$\tau_{s,x} = \inf\{r \geq s : x + \mathbf{e}(x) W_{s,r} = 0\} = \inf\{r \geq s : W_{s,r} = -|x|\}. \tag{4}$$

However, the construction of the unique flow of mappings φ associated to (T) relies on flipping Brownian excursions and is more complicated. Another construction of φ using Kolmogorov extension theorem can be derived from Sect. 4.1 [7] similarly to Tanaka's equation. Here, we restrict our attention to discrete models.

The one point motion associated to any solution of (T) is the Walsh Brownian motion $W(\alpha_1, \cdots, \alpha_N)$ on G (see Proposition 3 [5]) which we define as a strong Markov process with càdlàg paths, state space G and Feller semigroup $(P_t)_{t \geq 0}$ as given in Sect. 2.2 [5]. When $N = 2$, it corresponds to the famous skew Brownian motion [4].

Our first result is the following Donsker approximation of $W(\alpha_1, \cdots, \alpha_N)$

Proposition 1. *Let* $M = (M_n)_{n\geq 0}$ *be a Markov chain on* G *started at* 0 *with stochastic matrix* Q *given by:*

$$Q(0, \mathbf{e}_i) = \alpha_i, \quad Q(n\mathbf{e}_i, (n+1)\mathbf{e}_i) = Q(n\mathbf{e}_i, (n-1)\mathbf{e}_i) = \frac{1}{2} \;\; \forall i \in [1, N], n \in \mathbb{N}^*. \tag{5}$$

Let $t \longmapsto M(t)$ *be the linear interpolation of* $(M_n)_{n\geq 0}$ *and* $M_t^n = \frac{1}{\sqrt{n}} M(nt), n \geq 1$. *Then*

$$(M_t^n)_{t\geq 0} \xrightarrow[n \to +\infty]{law} (Z_t)_{t\geq 0}$$

in $C([0, +\infty[, G)$ *where* Z *is an* $W(\alpha_1, \cdots, \alpha_N)$ *started at* 0.

This result extends that of [2] who treated the case $\alpha_1 = \cdots = \alpha_N = \frac{1}{N}$ and of course the Donsker theorem for the skew Brownian motion (see [1] for example). We show in fact that Proposition 1 can be deduced immediately from the case $N = 2$.

In this paper we study the approximation of flows associated to (T). Among recent papers on the approximation of flows, let us mention [8] where the author construct an approximation for the Harris flow and the Arratia flow.

Let $G_{\mathbb{N}} = \{x \in G; |x| \in \mathbb{N}\}$ and $\mathscr{P}(G)$ (resp. $\mathscr{P}(G_{\mathbb{N}})$) be the space of all probability measures on G (resp. $G_{\mathbb{N}}$). We now come to the discrete description of (φ, K^W) and introduce

Definition 2 (Discrete flows). We say that a process $\psi_{p,q}(x)$ (resp. $N_{p,q}(x)$) indexed by $\{p \leq q \in \mathbb{Z}, x \in G_{\mathbb{N}}\}$ with values in $G_{\mathbb{N}}$ (resp. $\mathscr{P}(G_{\mathbb{N}})$) is a discrete flow of mappings (resp. kernels) on $G_{\mathbb{N}}$ if:

(i) The family $\{\psi_{i,i+1}; i \in \mathbb{Z}\}$ (resp. $\{N_{i,i+1}; i \in \mathbb{Z}\}$) is independent.
(ii) $\forall p \in \mathbb{Z}, x \in G_{\mathbb{N}}, \psi_{p,p+2}(x) = \psi_{p+1,p+2}(\psi_{p,p+1}(x))$
(resp. $N_{p,p+2}(x) = N_{p,p+1}N_{p+1,p+2}(x)$) a.s. where

$$N_{p,p+1}N_{p+1,p+2}(x, A) := \sum_{y \in G_{\mathbb{N}}} N_{p+1,p+2}(y, A) N_{p,p+1}(x, \{y\}) \text{ for all } A \subset G_{\mathbb{N}}.$$

We call (ii), the cocycle or flow property.

The main difficulty in the construction of the flow φ associated to (1) [7] is that it has to keep the consistency of the flow. This problem does not arise in discrete time. Starting from the following two remarks:

(1) $\varphi_{s,t}(x) = x + \text{sgn}(x) W_{s,t}$ if $s \leq t \leq \tau_{s,x}$,
(2) $|\varphi_{s,t}(0)| = W_{s,t}^+$ and $\text{sgn}(\varphi_{s,t}(0))$ is independent of W for all $s \leq t$,

one can easily expect the discrete analogous of φ as follows: consider an original random walk S and a family of signs (η_i) which are independent. Then:

(1) A particle at time k and position $n \neq 0$, just follows what the $S_{k+1} - S_k$ tells him (goes to $n+1$ if $S_{k+1} - S_k = 1$ and to $n-1$ if $S_{k+1} - S_k = -1$).
(2) A particle at 0 at time k does not move if $S_{k+1} - S_k = -1$, and moves according to η_k if $S_{k+1} - S_k = 1$.

The situation on a finite half-lines is very close. Let $S = (S_n)_{n\in\mathbb{Z}}$ be a simple random walk on $\mathbb{Z}$, that is $(S_n)_{n\in\mathbb{N}}$ and $(S_{-n})_{n\in\mathbb{N}}$ are two independent simple random walks on $\mathbb{Z}$ and $(\eta_i)_{i\in\mathbb{Z}}$ be a sequence of i.i.d random variables with law $\sum_{i=1}^{N} \alpha_i \delta_{\mathbf{e}_i}$ which is independent of S. For $p \leq n$, set

$$S_{p,n} = S_n - S_p,\ S^+_{p,n} = S_n - \min_{h\in[p,n]} S_h = S_{p,n} - \min_{h\in[p,n]} S_{p,h}.$$

and for $p \in \mathbb{Z}, x \in G_{\mathbb{N}}$, define

$$\Psi_{p,p+1}(x) = x + \mathbf{e}(x) S_{p,p+1} \text{ if } x \neq 0,\ \Psi_{p,p+1}(0) = \eta_p S^+_{p,p+1}.$$

$$K_{p,p+1}(x) = \delta_{x+\mathbf{e}(x)S_{p,p+1}} \text{ if } x \neq 0,\ K_{p,p+1}(0) = \sum_{i=1}^{N} \alpha_i \delta_{S^+_{p,p+1}\mathbf{e}_i}.$$

In particular, we have $K_{p,p+1}(x) = E[\delta_{\Psi_{p,p+1}(x)} | \sigma(S)]$. Now we extend this definition for all $p \leq n \in \mathbb{Z}, x \in G_{\mathbb{N}}$ by setting

$$\Psi_{p,n}(x) = x 1_{\{p=n\}} + \Psi_{n-1,n} \circ \Psi_{n-2,n-1} \circ \cdots \circ \Psi_{p,p+1}(x) 1_{\{p>n\}},$$
$$K_{p,n}(x) = \delta_x 1_{\{p=n\}} + K_{p,p+1} \cdots K_{n-2,n-1} K_{n-1,n}(x) 1_{\{p>n\}}.$$

We equip $\mathscr{P}(G)$ with the following topology of weak convergence:

$$\beta(P,Q) = \sup\left\{ \left| \int g dP - \int g dQ \right|, \|g\|_\infty + \sup_{x\neq y} \frac{|g(x)-g(y)|}{|x-y|} \leq 1, g(0)=0 \right\}.$$

In this paper, starting from (Ψ, K), we construct (φ, K^W) and in particular show the following

Theorem 2. *(1) Ψ (resp. K) is a discrete flow of mappings (resp. kernels) on $G_{\mathbb{N}}$.*
(2) There exists a joint realization (ψ, N, φ, K^W) on a common probability space $(\Omega, \mathscr{A}, \mathbb{P})$ such that

(i) $(\psi, N) \overset{law}{=} (\Psi, K)$.
(ii) (φ, W) *(resp.* (K^W, W)*) is the unique flow of mappings (resp. Wiener flow) which solves* (T).
(iii) For all $s \in \mathbb{R}, T > 0, x \in G, x_n \in \frac{1}{\sqrt{n}} G_{\mathbb{N}}$ *such that* $\lim_{n\to\infty} x_n = x$*, we have*

$$\lim_{n\to\infty}\sup_{s\le t\le s+T}|\frac{1}{\sqrt{n}}\psi_{\lfloor ns\rfloor,\lfloor nt\rfloor}(\sqrt{n}x_n)-\varphi_{s,t}(x)|=0\ a.s.$$

and

$$\lim_{n\to\infty}\sup_{s\le t\le s+T}\beta(K_{\lfloor ns\rfloor,\lfloor nt\rfloor}(\sqrt{n}x_n)(\sqrt{n}.),K^W_{s,t}(x))=0\ a.s. \tag{6}$$

This theorem implies also the following

Corollary 1. *For all $s\in\mathbb{R}, x\in G_{\mathbb{N}}$, let $t\longmapsto\Psi(t)$ be the linear interpolation of $\left(\Psi_{\lfloor ns\rfloor,k}(x),k\ge\lfloor ns\rfloor\right)$ and $\Psi^n_{s,t}(x):=\frac{1}{\sqrt{n}}\Psi(nt)$, $K^n_{s,t}(x)=E[\delta_{\Psi^n_{s,t}(x)}|\sigma(S)], t\ge s$, $n\ge 1$. For all $1\le p\le q$, $(x_i)_{1\le i\le q}\subset G$, let $x^n_i\in\frac{1}{\sqrt{n}}G_{\mathbb{N}}$ such that $\lim_{n\to\infty}x^n_i=x_i$. Define*

$$Y^n=\left(\Psi^n_{s_1,\cdot}(\sqrt{n}x^n_1),\cdots,\Psi^n_{s_p,\cdot}(\sqrt{n}x^n_p),K^n_{s_{p+1},\cdot}(\sqrt{n}x^n_{p+1}),\cdots,K^n_{s_q,\cdot}(\sqrt{n}x^n_q)\right).$$

Then

$$Y^n\xrightarrow[n\to+\infty]{law}Y\ \ in\ \prod_{i=1}^{p}C([s_i,+\infty[,G)\times\prod_{j=p+1}^{q}C([s_j,+\infty[,\mathscr{P}(G))$$

where

$$Y=\left(\varphi_{s_1,\cdot}(x_1),\cdots,\varphi_{s_p,\cdot}(x_p),K^W_{s_{p+1},\cdot}(x_{p+1}),\cdots,K^W_{s_q,\cdot}(x_q)\right).$$

Our proof of Theorem 2 is based on a remarkable transformation introduced by Csaki and Vincze [9] which is strongly linked with Tanaka's SDE. Let S be a simple random walk on $\mathbb{Z}$ (SRW) and ε be a Bernoulli random variable independent of S (just one!). Then there exists a SRW M such that

$$\sigma(M)=\sigma(\varepsilon,S)$$

and moreover

$$(\frac{1}{\sqrt{n}}S(nt),\frac{1}{\sqrt{n}}M(nt))_{t\ge0}\xrightarrow[n\to+\infty]{\text{law}}(B_t,W_t)_{t\ge0}\ \text{in}\ C([0,\infty[,\mathbb{R}^2).$$

where $t\longmapsto S(t)$ (resp. $M(t)$) is the linear interpolation of S (resp. M) and B,W are two Brownian motions satisfying Tanaka's equation

$$dW_t=\text{sgn}(W_t)dB_t.$$

We will study this transformation with more details in Sect. 2 and then extend the result of Csaki and Vincze to Walsh Brownian motion (Proposition 2); Let

$S = (S_n)_{n\in\mathbb{N}}$ be a SRW and associate to S the process $Y_n := S_n - \min_{k\le n} S_k$, flip independently every "excursion " of Y to each ray D_i with probability α_i, then the resulting process is not far from a random walk on G whose law is given by (5). In Sect. 3, we prove Proposition 1 and study the scaling limits of Ψ, K.

2 Csaki-Vincze Transformation and Consequences

In this section, we review a relevant result of Csaki and Vincze and then derive some useful consequences offering a better understanding of Tanaka's equation.

2.1 *Csaki-Vincze Transformation*

Theorem 3. *([9] page 109) Let $S = (S_n)_{n\ge0}$ be a SRW. Then, there exists a SRW $\overline{S} = (\overline{S}_n)_{n\ge0}$ such that:*

$$\overline{Y}_n := \max_{k\le n}\overline{S}_k - \overline{S}_n \Rightarrow |\overline{Y}_n - |S_n|| \le 2 \ \ \forall n \in \mathbb{N}.$$

Sketch of the proof. Here, we just give the expression of $\overline{S}$ with some useful comments (see also the figures below). We *insist* that a careful reading of the pages 109 and 110 [9] is recommended for the sequel. Let $X_i = S_i - S_{i-1}, i \ge 1$ and define

$$\tau_1 = \min\{i > 0 : S_{i-1}S_{i+1} < 0\},\ \tau_{l+1} = \min\{i > \tau_l : S_{i-1}S_{i+1} < 0\}\ \forall l \ge 1.$$

For $j \ge 1$, set

$$\overline{X}_j = \sum_{l\ge0}(-1)^{l+1}X_1X_{j+1}1_{\{\tau_l+1\le j\le\tau_{l+1}\}}.$$

Let $\overline{S}_0 = 0,\ \overline{S}_j = \overline{X}_1 + \cdots + \overline{X}_j,\ j \ge 1$. Then, the theorem holds for $\overline{S}$. We call $T(S) = \overline{S}$ the Csaki-Vincze transformation of S (Fig. 2).

Note that T is an even function, that is $T(S) = T(-S)$. As a consequence of *(iii)* and *(iv)* [9] (page 110), we have

$$\tau_l = \min\{n \ge 0, \overline{S}_n = 2l\}\ \forall l \ge 1. \tag{7}$$

This entails the following

Corollary 2. *(1) Let S be a SRW and define $\overline{S} = T(S)$. Then:*

(i) For all $n \ge 0$, we have $\sigma(\overline{S}_j, j \le n) \vee \sigma(S_1) = \sigma(S_j, j \le n+1)$.

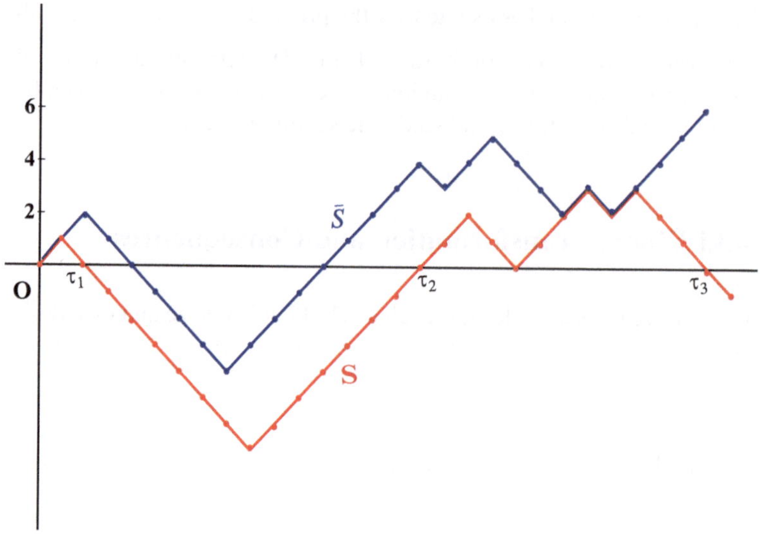

Fig. 2 S and $\overline{S}$

(ii) S_1 *is independent of* $\sigma(\overline{S})$.

(2) Let $\overline{S} = (\overline{S}_k)_{k\geq 0}$ *be a SRW. Then:*

(i) There exists a SRW S *such that:*

$$\overline{Y}_n := \max_{k\leq n}\overline{S}_k - \overline{S}_n \Rightarrow |\overline{Y}_n - |S_n|| \leq 2 \ \ \forall n \in \mathbb{N}.$$

(ii) $T^{-1}\{\overline{S}\}$ *is reduced to exactly two elements* S *and* $-S$ *where* S *is obtained by adding information to* $\overline{S}$.

Proof. (1) We retain the notations just before the corollary. (i) To prove the inclusion $\subset$, we only need to check that $\{\tau_l + 1 \leq j \leq \tau_{l+1}\} \in \sigma(S_h, h \leq n+1)$ for a fixed $j \leq n$. This is clear since $\{\tau_l = m\} \in \sigma(S_h, h \leq m+1)$ for all $l, m \in \mathbb{N}$. For all $1 \leq j \leq n$, we have $X_{j+1} = \sum_{l\geq 0}(-1)^{l+1}X_1\overline{X}_j 1_{\{\tau_l+1\leq j\leq\tau_{l+1}\}}$. By (7), $\{\tau_l + 1 \leq j \leq \tau_{l+1}\} \in \sigma(\overline{S}_h, h \leq j-1)$ and so the inclusion $\supset$ holds. (ii) We may write

$$\tau_1 = \min\{i > 1 : X_1 S_{i-1} X_1 S_{i+1} < 0\},$$

$$\tau_{l+1} = \min\{i > \tau_l : X_1 S_{i-1} X_1 S_{i+1} < 0\} \ \forall l \geq 1.$$

This shows that $\overline{S}$ is $\sigma(X_1 X_{j+1}, j \geq 0)$-measurable and (ii) is proved.
(2) (i) Set $\overline{X}_j = \overline{S}_j - \overline{S}_{j-1}, j \geq 1$ and $\tau_l = \min\{n \geq 0, \overline{S}_n = 2l\}$ for all $l \geq 1$. Let ε be a random variable independent of $\overline{S}$ such that:

$$\mathbb{P}(\varepsilon = 1) = \mathbb{P}(\varepsilon = -1) = \frac{1}{2}.$$

Define

$$X_{j+1} = \varepsilon 1_{\{j=0\}} + \left(\sum_{l \geq 0} (-1)^{l+1} \varepsilon \overline{X}_j 1_{\{\tau_l+1 \leq j \leq \tau_{l+1}\}} \right) 1_{\{j \geq 1\}}.$$

Then set $S_0 = 0, S_j = X_1 + \cdots X_j, j \geq 1$. It is not hard to see that the sequence of the random times $\tau_i(S), i \geq 1$ defined from S as in Theorem 3 is exactly $\tau_i, i \geq 1$ and therefore $T(S) = \overline{S}$. (ii) Let S such that $T(S) = \overline{S}$. By (1), $\sigma(\overline{S}) \vee \sigma(S_1) = \sigma(S)$ and S_1 is independent of $\overline{S}$ which proves (ii). □

2.2 The Link with Tanaka's Equation

Let S be a SRW, $\overline{S} = -T(S)$ and $t \longmapsto S(t)$ (resp. $\overline{S}(t)$) be the linear interpolation of S (resp. $\overline{S}$) on $\mathbb{R}$. Define for all $n \geq 1$, $S_t^{(n)} = \frac{1}{\sqrt{n}} S(nt), \overline{S}_t^{(n)} = \frac{1}{\sqrt{n}} \overline{S}(nt)$. Then, it can be easily checked (see Proposition 2.4 in [3] page 107) that

$$(\overline{S}_t^{(n)}, S_t^{(n)})_{t \geq 0} \xrightarrow[n \to +\infty]{\text{law}} (B_t, W_t)_{t \geq 0} \text{ in } C([0, \infty[, \mathbb{R}^2).$$

In particular B and W are two standard Brownian motions. On the other hand, $|Y_n^+ - |S_n|| \leq 2 \; \forall n \in \mathbb{N}$ with $Y_n^+ := \overline{S}_n - \min_{k \leq n} \overline{S}_k$ by Theorem 3 which implies $|W_t| = B_t - \min_{0 \leq u \leq t} B_u$. Tanaka's formula for local time gives

$$|W_t| = \int_0^t \text{sgn}(W_u) dW_u + L_t(W) = B_t - \min_{0 \leq u \leq t} B_u,$$

where $L_t(W)$ is the local time at 0 of W and so

$$dW_u = \text{sgn}(W_u) dB_u. \tag{8}$$

We deduce that for each SRW S the couple $(-T(S), S)$, suitably normalized and time scaled converges in law towards (B, W) satisfying (8). Finally, remark that $-T(S) = \overline{S} \Rightarrow -T(-S) = \overline{S}$ is the analogue of W solves (8) $\Rightarrow -W$ solves (8). We have seen how to construct solutions to (8) by means of T. In the sequel, we will use this approach to construct a stochastic flow of mappings which solves equation (T) in general.

2.3 Extensions

Let $S = (S_n)_{n\geq 0}$ be a SRW and set $Y_n := \max_{k\leq n} S_k - S_n$. For $0 \leq p < q$, we say that $E = [p, q]$ is an excursion for Y if the following conditions are satisfied (with the convention $Y_{-1} = 0$):

- $Y_p = Y_{p-1} = Y_q = Y_{q+1} = 0$.
- $\forall\, p \leq j < q, Y_j = 0 \Rightarrow Y_{j+1} = 1$.

For example in Fig. 3, [2, 14], [16, 18] are excursions for $\overline{Y}$. If $E = [p, q]$ is an excursion for Y, define $e(E) := p,\ f(E) := q$.
Let $(E_i)_{i\geq 1}$ be the random set of all excursions of Y ordered such that: $e(E_i) < e(E_j)\ \forall i < j$. From now on, we call E_i the ith excursion of Y. Then, we have

Proposition 2. *On a probability space $(\Omega, \mathscr{A}, P)$, consider the following jointly independent processes:*

- $\eta = (\eta_i)_{i\geq 1}$, *a sequence of i.i.d random variables distributed according to* $\sum_{i=1}^{N} \alpha_i \delta_{\mathbf{e}_i}$.
- $(\overline{S}_n)_{n\in\mathbb{N}}$ *a SRW.*

Then, on an extension of $(\Omega, \mathscr{A}, P)$, there exists a Markov chain $(M_n)_{n\in\mathbb{N}}$ started at 0 with stochastic matrix given by (5) such that:

$$\overline{Y}_n := \max_{k\leq n} \overline{S}_k - \overline{S}_n \Rightarrow |M_n - \eta_i \overline{Y}_n| \leq 2$$

on the ith excursion of $\overline{Y}$.

Proof. Fix $S \in T^{-1}\{\overline{S}\}$. Then, by Corollary 2, we have $|\overline{Y}_n - |S_n|| \leq 2\ \forall n \in \mathbb{N}$. Consider a sequence $(\boldsymbol{\beta}_i)_{i\geq 1}$ of i.i.d random variables distributed according to $\sum_{i=1}^{N} \alpha_i \delta_{\mathbf{e}_i}$ which is independent of $(\overline{S}, \boldsymbol{\eta})$. Denote by $(\tau_l)_{l\geq 1}$ the sequence of random times constructed in the proof of Theorem 3 from S. It is sufficient to look to what happens at each interval $[\tau_l, \tau_{l+1}]$ (with the convention $\tau_0 = 0$).
Using (7), we see that in $[\tau_l, \tau_{l+1}]$ there are two jumps of $\max_{k\leq n} \overline{S}_k$; from $2l$ to $2l + 1$ (J_1) and from $2l + 1$ to $2l + 2$ (J_2). The last jump (J_2) occurs always at τ_{l+1} by (7). Consequently there are only 3 possible cases:

(i) There is no excursion of $\overline{Y}$ (J_1 and J_2 occur respectively at $\tau_l + 1$ and $\tau_l + 2$, see $[0, \tau_1]$ in Fig. 3).
(ii) There is just one excursion of $\overline{Y}$ (see $[\tau_1, \tau_2]$ in Fig. 3).
(iii) There are 2 excursions of $\overline{Y}$ (see $[\tau_2, \tau_3]$ in Fig. 3).

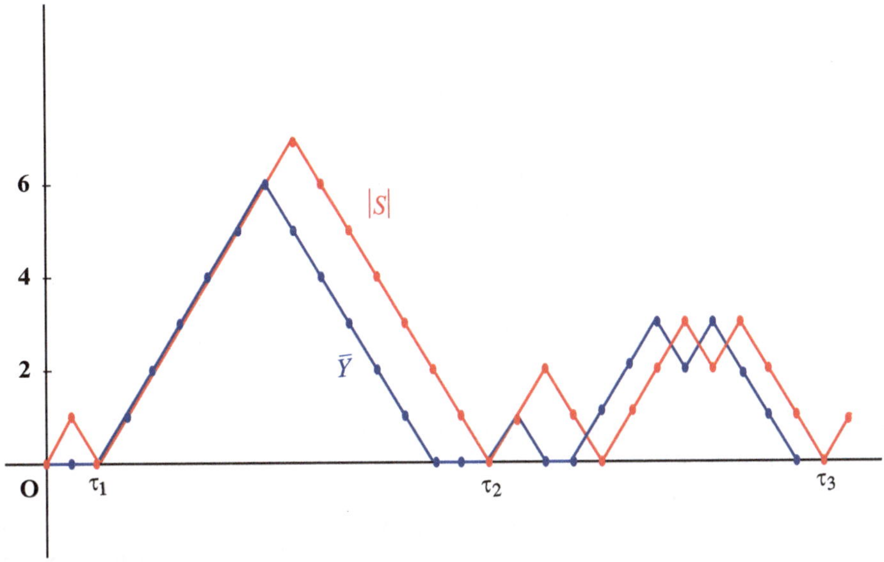

Fig. 3 $|S|$ and $\overline{Y}$

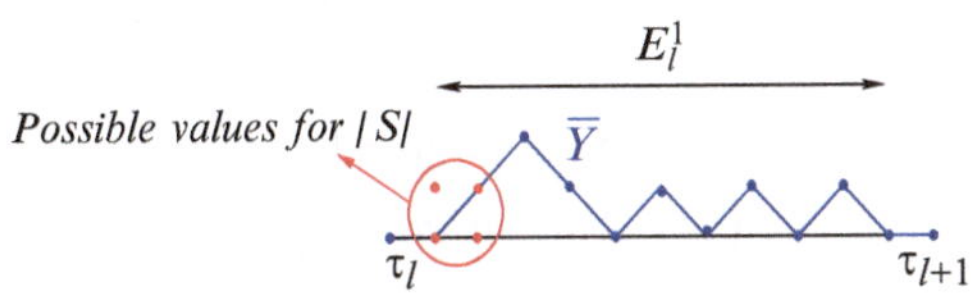

Fig. 4 The case (ii2)

Note that: $\overline{Y}_{\tau_l} = \overline{Y}_{\tau_{l+1}} = S_{\tau_l} = S_{\tau_{l+1}} = 0$. In the case (i), we have necessarily $\tau_{l+1} = \tau_l + 2$. Set $M_n = \boldsymbol{\beta}_l.|S_n| \quad \forall n \in [\tau_l, \tau_{l+1}]$.
To treat other cases, the following remarks may be useful: from the expression of $\overline{S}$, we have $\forall l \geq 0$:

(a) If $k \in [\tau_l + 2, \tau_{l+1}]$, $\overline{S}_{k-1} = 2l + 1 \iff S_k = 0$.
(b) If $k \in [\tau_l, \tau_{l+1}]$, $\overline{Y}_k = 0 \Rightarrow |S_{k+1}| \in \{0, 1\}$ and $S_{k+1} = 0 \Rightarrow \overline{Y}_k = 0$.

In the case (ii), let E_l^1 be the unique excursion of $\overline{Y}$ in the interval $[\tau_l, \tau_{l+1}]$. Then, we have two subcases:
(ii1) $f(E_l^1) = \tau_{l+1} - 2$ (J_1 occurs at $\tau_{l+1} - 1$).
If $\tau_l + 2 \leq k \leqslant f(E_l^1) + 1$, then $k - 1 \leqslant f(E_l^1)$, and so $\overline{S}_{k-1} \neq 2l + 1$. Using (a), we get: $S_k \neq 0$. Thus, in this case the first zero of S after τ_l is τ_{l+1}. Set: $M_n = \eta_{N(E_l^1)}.|S_n|$, where $N(E)$ is the number of the excursion E.
(ii2) $f(E_l^1) = \tau_{l+1} - 1$ (J_1 occurs at $\tau_l + 1$ and so $\overline{Y}_{\tau_l+1} = 0$)). In this case, using (b) and Fig. 4, we see that the first zero τ_l^* of S after τ_l is $e(E_l^1) + 1 = \tau_l + 2$ (Fig. 4).

Set

$$M_n = \begin{cases} \boldsymbol{\beta}_l.|S_n| & \text{if } n \in [\tau_l, \tau_l^* - 1] \\ \eta_{N(E_l^1)}.|S_n| & \text{if } n \in [\tau_l^*, \tau_{l+1}] \end{cases}$$

In the case (iii), let E_l^1 and E_l^2 denote respectively the first and 2nd excursion of $\overline{Y}$ in $[\tau_l, \tau_{l+1}]$. We have, $\tau_l + 2 \leq k \leq e(E_l^2) \Rightarrow k-1 \leq e(E_l^2) - 1 = f(E_l^1) \Rightarrow \overline{S}_{k-1} \neq 2l+1 \Rightarrow S_k \neq 0$ by (a). Hence, the first zero of S after τ_l is $\tau_l^* := e(E_l^2) + 1$ using $\overline{Y}_k = 0 \Rightarrow |S_{k+1}| \in \{0, 1\}$ in (b). Set:

$$M_n = \begin{cases} \eta_{N(E_l^1)}.|S_n| & \text{if } n \in [\tau_l, \tau_l^* - 1] \\ \eta_{N(E_l^2)}.|S_n| & \text{if } n \in [\tau_l^*, \tau_{l+1}] \end{cases}$$

Let $(M_n)_{n\in\mathbb{N}}$ be the process constructed above. Then clearly $|M_n - \eta_i \overline{Y}_n| \leq 2$ on the ith excursion of $\overline{Y}$.
To complete the proof, it suffices to show that the law of $(M_n)_{n\in\mathbb{N}}$ is given by (5). The only point to verify is $\mathbb{P}(M_{n+1} = \mathbf{e}_i | M_n = 0) = \alpha_i$. For this, consider on another probability space the jointly independent processes (S, γ, λ) such that S is a SRW and γ, λ have the same law as $\boldsymbol{\eta}$. Let $(\tau_l)_{l\geq 1}$ be the sequence of random times defined from S as in Theorem 3. For all $l \in \mathbb{N}$, denote by τ_l^* the first zero of S after τ_l and set

$$V_n = \begin{cases} \gamma_l.|S_n| & \text{if } n \in [\tau_l, \tau_l^* - 1] \\ \lambda_l.|S_n| & \text{if } n \in [\tau_l^*, \tau_{l+1}] \end{cases}$$

It is clear, by construction, that $M \stackrel{law}{=} V$. We can write:

$$\{\tau_0, \tau_0^*, \tau_1, \tau_1^*, \tau_2, \cdots\} = \{T_0, T_1, T_2, \cdots\} \text{ with } T_0 = 0 < T_1 < T_2 < \cdots.$$

For all $k \geq 0$, let $\boldsymbol{\zeta}_k := \sum_{j=0}^{N} \mathbf{e}_j 1_{\{V|_{[T_k, T_{k+1}]} \in D_j\}}$. Obviously, S and $\boldsymbol{\zeta}_k$ are independent and $\boldsymbol{\zeta}_k \stackrel{law}{=} \sum_{i=1}^{N} \alpha_i \delta_{\mathbf{e}_i}$. Furthermore

$$\begin{aligned} \mathbb{P}(V_{n+1} = \mathbf{e}_i | V_n = 0) &= \frac{1}{\mathbb{P}(S_n = 0)} \sum_{k=0}^{+\infty} \mathbb{P}(V_{n+1} = \mathbf{e}_i, S_n = 0, n \in [T_k, T_{k+1}[) \\ &= \frac{1}{\mathbb{P}(S_n = 0)} \sum_{k=0}^{+\infty} \mathbb{P}(\boldsymbol{\zeta}_k = \mathbf{e}_i, S_n = 0, n \in [T_k, T_{k+1}[) \\ &= \alpha_i \end{aligned}$$

This completes the proof of the proposition. □

Remark 1. With the notations of Proposition 2, let $(\eta.\overline{Y})$ be the Markov chain defined by $(\eta.\overline{Y})_n = \eta_i \overline{Y}_n$ on the ith excursion of $\overline{Y}$ and $(\eta.\overline{Y})_n = 0$ if $\overline{Y}_n = 0$. Then the stochastic Matrix of $(\eta.\overline{Y})$ is given by

$$M(0,0) = \frac{1}{2}, M(0, \mathbf{e}_i) = \frac{\alpha_i}{2}, M(n\mathbf{e}_i, (n+1)\mathbf{e}_i) \\ = M(n\mathbf{e}_i, (n-1)\mathbf{e}_i) = \frac{1}{2}, i \in [1, N],\ n \in \mathbb{N}^*. \tag{9}$$

3 Proof of Main Results

3.1 *Proof of Proposition 1*

Let $(Z_t)_{t\geq 0}$ be a $W(\alpha_1, \cdots, \alpha_N)$ on G started at 0. For all $i \in [1, N]$, define $Z_t^i = |Z_t| 1_{\{Z_t \in D_i\}} - |Z_t| 1_{\{Z_t \notin D_i\}}$. Then $Z_t^i = \Phi^i(Z_t)$ where $\Phi^i(x) = |x| 1_{\{x \in D_i\}} - |x| 1_{\{x \notin D_i\}}$. Let Q^i be the semigroup of the skew Brownian motion of parameter α_i ($SBM(\alpha_i)$) (see [10] page 87). Then the following relation is easy to check: $P_t(f \circ \Phi^i) = Q_t^i f \circ \Phi^i$ for all bounded measurable function f defined on $\mathbb{R}$. This shows that Z^i is a $SBM(\alpha_i)$ started at 0. For $n \geq 1, i \in [1, N]$, define

$$T_0^n = 0,\ \ T_{k+1}^n = \inf\{r \geq 0 : d(Z_r, Z_{T_k^n}) = \frac{1}{\sqrt{n}}\}, k \geq 0.$$

$$T_0^{n,i} = 0,\ \ T_{k+1}^{n,i} = \inf\{r \geq 0 : |Z_r^i - Z_{T_k^{n,i}}^i| = \frac{1}{\sqrt{n}}\}, k \geq 0.$$

Remark that $T_{k+1}^n = T_{k+1}^{n,i} = \inf\{r \geq 0 : ||Z_r| - |Z_{T_k^n}|| = \frac{1}{\sqrt{n}}\}$. Furthermore if $Z_t \in D_i$, then obviously $d(Z_t, Z_s) = |Z_t^i - Z_s^i|$ for all $s \geq 0$ and consequently

$$d(Z_t, Z_s) \leq \sum_{i=1}^{N} |Z_t^i - Z_s^i|. \tag{10}$$

Now define $Z_k^n = \sqrt{n} Z_{T_k^n}, Z_k^{n,i} = \sqrt{n} Z_{T_k^{n,i}}^i$. Then $(Z_k^n, k \geq 0) \overset{law}{=} M$ (see the proof of Proposition 2 in [5]). For all $T > 0$, we have

$$\sup_{t\in[0,T]} d(Z_t, \frac{1}{\sqrt{n}} Z_{\lfloor nt \rfloor}^n) \leq \sum_{i=1}^{N} \sup_{t\in[0,T]} |Z_t^i - \frac{1}{\sqrt{n}} Z_{\lfloor nt \rfloor}^{n,i}| \xrightarrow[n\to+\infty]{} 0 \text{ in probability}$$

by Lemma 4.4 [1] which proves our result.

Remark 2. (1) By (10), a.s. $t \mapsto Z_t$ is continuous. We will always suppose that Walsh Brownian motion is continuous.

(2) By combining the two propositions 1 and 2, we deduce that $(\eta.\overline{Y})$ rescales as Walsh Brownian motion in the space of continuous functions. It is also possible to prove this result by showing that the family of laws is tight and that any limit process along a subsequence is the Walsh Brownian motion.

3.2 Scaling Limits of (Ψ, K)

Set $\eta_{p,n} = \mathbf{e}(\Psi_{p,n})$ for all $p \le n$ where $\Psi_{p,n} = \Psi_{p,n}(0)$:

Proposition 3. *(i) For all* $p \le n$, $|\Psi_{p,n}| = S^+_{p,n}$.
(ii) For all $p < n < q$,

$$\mathbb{P}(\eta_{p,q} = \eta_{n,q} | min_{h\in[p,q]} S_h = min_{h\in[n,q]} S_h) = 1$$

and

$$\mathbb{P}(\eta_{p,n} = \eta_{p,q} | min_{h\in[p,n]} S_h = min_{h\in[p,q]} S_h, S^+_{p,j} > 0 \ \forall j \in [n,q]) = 1.$$

(iii) Set $T_{p,x} = \inf\{q \ge p : S_q - S_p = -|x|\}$. *Then for all* $p \le n$, $x \in G_{\mathbb{N}}$,

$$\Psi_{p,n}(x) = (x + \mathbf{e}(x)S_{p,n})1_{\{n \le T_{p,x}\}} + \Psi_{p,n}1_{\{n > T_{p,x}\}};$$

$$K_{p,n}(x) = E[\delta_{\Psi_{p,n}(x)}|\sigma(S)] = \delta_{x+\mathbf{e}(x)S_{p,n}}1_{\{n\le T_{p,x}\}} + \sum_{i=1}^{N} \alpha_i \delta_{S^+_{p,n}\mathbf{e}_i} 1_{\{n>T_{p,x}\}}.$$

Proof. (i) We take $p = 0$ and prove the result by induction on n. For $n = 0$, this is clear. Suppose the result holds for n. If $\Psi_{0,n} \in G^*$, then $S^+_{0,n} > 0$ and so $\min_{h\in[0,n]} S_h = \min_{h\in[0,n+1]} S_h$. Moreover $\Psi_{0,n+1} = \Psi_{0,n} + \eta_{0,n}S_{n,n+1} = (S_{n+1} - \min_{h\in[0,n]} S_h)\eta_{0,n} = S^+_{0,n+1}\eta_{0,n}$. If $\Psi_{0,n} = 0$, then $S^+_{0,n} = 0$ and $|\Psi_{0,n+1}| = S^+_{n,n+1}$. But $\min_{h\in[0,n+1]} S_h = \min(\min_{h\in[0,n]} S_h, S_{n+1}) = \min(S_n, S_{n+1})$ since $S^+_{0,n} = 0$ which proves (i).
(ii) Let $p < n < q$. If $\min_{h\in[p,q]} S_h = \min_{h\in[n,q]} S_h$, then $S^+_{p,q} = S^+_{n,q}$. When $S^+_{p,q} = 0$, we have $\eta_{p,q} = \eta_{n,q} = \mathbf{e}_N$ by convention. Suppose that $S^+_{p,q} > 0$, then clearly

$$J := \sup\{j < q : S^+_{p,j} = 0\} = \sup\{j < q : S^+_{n,j} = 0\}.$$

By the flow property of Ψ, we have $\Psi_{p,q} = \Psi_{n,q} = \Psi_{J,q}$. The second assertion of (ii) is also clear.
(iii) By (i), we have $\Psi_{p,n} = \Psi_{p,n}(x) = 0$ if $n = T_{p,x}$ and so $\Psi_{p,.}(x)$ is given by $\Psi_{p,.}$ after $T_{p,x}$ using the cocycle property. The last claim is easy to establish. □

For all $s \in \mathbb{R}$, let d_s (resp. d_∞) be the distance of uniform convergence on every compact subset of $C([s, +\infty[, G)$ (resp. $C(\mathbb{R}, \mathbb{R})$). Denote by $\mathbb{D} = \{s_n, n \in \mathbb{N}\}$

the set of all dyadic numbers of $\mathbb{R}$ and define $\widetilde{C} = C(\mathbb{R},\mathbb{R}) \times \prod_{n=0}^{+\infty} C([s_n,+\infty[,G)$ equipped with the metric:

$$d(x,y) = d_\infty(x',y') + \sum_{n=0}^{+\infty} \frac{1}{2^n} \inf(1, d_{s_n}(x_n,y_n)) \text{ where } x = (x', x_{s_0}, \cdots),$$
$$y = (y', y_{s_0}, \cdots).$$

Let $t \longmapsto S(t)$ be the linear interpolation of S on $\mathbb{R}$ and define $S_t^{(n)} = \frac{1}{\sqrt{n}} S(nt), n \geq 1$. If $u \leq 0$, we define $\lfloor u \rfloor = -\lfloor -u \rfloor$. Then, we have

$$S_t^{(n)} = S_t^n + o(\frac{1}{\sqrt{n}}), \text{ with } S_t^n := \frac{1}{\sqrt{n}} S_{\lfloor nt \rfloor}.$$

Let $\Psi_{s,t}^n = \Psi_{s,t}^n(0)$ (defined in Corollary 1). Then $\Psi_{s,t}^{(n)} := \frac{1}{\sqrt{n}} \Psi_{\lfloor ns \rfloor, \lfloor nt \rfloor} + o(\frac{1}{\sqrt{n}})$ and we have the following

Lemma 1. *Let $\mathbb{P}_n$ be the law of $Z^n = (S_.^{(n)}, (\Psi_{s_i,.}^{(n)})_{i\in\mathbb{N}})$ in $\widetilde{C}$. Then $(\mathbb{P}_n, n \geq 1)$ is tight.*

Proof. By Donsker theorem $\mathbb{P}_{S^{(n)}} \longrightarrow \mathbb{P}_W$ in $C(\mathbb{R},\mathbb{R})$ as $n \to \infty$ where $\mathbb{P}_W$ is the law of any Brownian motion on $\mathbb{R}$. Let $\mathbb{P}_{Z_{s_i}}$ be the law of any $W(\alpha_1, \cdots, \alpha_N)$ started at 0 at time s_i. Plainly, the law of $\Psi_{p,p+.}$ is given by (9) and so by Propositions 1 and 2, for all $i \in \mathbb{N}$, $\mathbb{P}_{\Psi_{s_i,.}^{(n)}} \longrightarrow \mathbb{P}_{Z_{s_i}}$ in $C([s_i,+\infty[,G)$ as $n \to \infty$. Now the lemma holds using Proposition 2.4 [3] (page 107). □

Fix a sequence $(n_k, k \in \mathbb{N})$ such that $Z^{n_k} \xrightarrow[k\to+\infty]{\text{law}} Z$ in $\widetilde{C}$. In the next paragraph, we will describe the law of Z. Notice that $(\Psi_{p,n})_{p\leq n}$ and S can be recovered from $(Z^{n_k})_{k\in\mathbb{N}}$. Using Skorokhod representation theorem, we may assume that Z is defined on the original probability space and the preceding convergence holds almost surely. Write $Z = (W, \psi_{s_1,.}, \psi_{s_2,.}, \cdots)$. Then, $(W_t)_{t\in\mathbb{R}}$ is a Brownian motion on $\mathbb{R}$ and $(\psi_{s,t})_{t\geq s}$ is an $W(\alpha_1, \cdots, \alpha_N)$ started at 0 for all $s \in \mathbb{D}$.

3.2.1 Description of the Limit Process

Set $\gamma_{s,t} = \mathbf{e}(\psi_{s,t}), s \in \mathbb{D}, s < t$ and define $\min_{u,v} = \min_{r\in[u,v]} W_r, u \leq v \in \mathbb{R}$. Then, we have:

Proposition 4. *(i) For all $s \leq t, s \in \mathbb{D}$, $|\psi_{s,t}| = W_{s,t}^+$.*
(ii) For all $s < t, u < v,\ s, u \in \mathbb{D}$,

$$\mathbb{P}(\gamma_{s,t} = \gamma_{u,v} | min_{s,t} = min_{u,v}) = 1 \textit{ if } \mathbb{P}(min_{s,t} = min_{u,v}) > 0.$$

Proof. (i) is immediate from the convergence of Z^{n_k} towards Z and Proposition 3 (i). (ii) We first prove that for all $s < t < u$,

$$\mathbb{P}(\gamma_{s,u} = \gamma_{t,u} | \min_{s,u} = \min_{t,u}) = 1 \text{ if } s, t \in \mathbb{D} \tag{11}$$

and

$$\mathbb{P}(\gamma_{s,t} = \gamma_{s,u} | \min_{s,t} = \min_{s,u}) = 1 \text{ if } s \in \mathbb{D}. \tag{12}$$

Fix $s < t < u$ with $s, t \in \mathbb{D}$ and let show that a.s.

$$\{\min_{s,u} = \min_{t,u}\} \subset \{\exists k_0, \ \boldsymbol{\eta}_{\lfloor n_k s \rfloor, \lfloor n_k u \rfloor} = \boldsymbol{\eta}_{\lfloor n_k t \rfloor, \lfloor n_k u \rfloor} \text{ for all } k \geq k_0\}. \tag{13}$$

We have $\{\min_{s,u} = \min_{t,u}\} = \{\min_{s,t} < \min_{t,u}\}$ $a.s.$ By uniform convergence the last set is contained in

$$\{\exists k_0, \min_{\lfloor n_k s \rfloor \leq j \leq \lfloor n_k t \rfloor} S_j < \min_{\lfloor n_k t \rfloor \leq j \leq \lfloor n_k u \rfloor} S_j \text{ for all } k \geq k_0\}$$

which is a subset of

$$\{\exists k_0, \min_{\lfloor n_k s \rfloor \leq j \leq \lfloor n_k u \rfloor} S_j = \min_{\lfloor n_k t \rfloor \leq j \leq \lfloor n_k u \rfloor} S_j \text{ for all } k \geq k_0\}.$$

This gives (13) using Proposition 3 (ii). Since $x \longrightarrow \mathbf{e}(x)$ is continuous on G^*, on $\{\min_{s,u} = \min_{t,u}\}$, we have

$$\gamma_{s,u} = \lim_{k \to \infty} \mathbf{e}(\frac{1}{\sqrt{n_k}} \Psi_{\lfloor n_k s \rfloor, \lfloor n_k u \rfloor}) = \lim_{k \to \infty} \mathbf{e}(\frac{1}{\sqrt{n_k}} \Psi_{\lfloor n_k t \rfloor, \lfloor n_k u \rfloor}) = \gamma_{t,u} \ a.s.$$

which proves (11). If $s \in \mathbb{D}, t > s$ and $\min_{s,t} = \min_{s,u}$, then s and t are in the same excursion interval of $W^+_{s,}$ and so $W^+_{s,r} > 0$ for all $r \in [t, u]$. As preceded, $\{\min_{s,t} = \min_{s,u}\}$ is a.s. included in

$$\{\exists k_0, \min_{\lfloor n_k s \rfloor \leq j \leq \lfloor n_k t \rfloor} S_j = \min_{\lfloor n_k s \rfloor \leq j \leq \lfloor n_k u \rfloor} S_j \, , S^+_{\lfloor n_k s \rfloor, j} > 0 \, \forall j \in [\lfloor n_k t \rfloor, \lfloor n_k u \rfloor], k \geq k_0\}.$$

Now it is easy to deduce (12) using Proposition 3 (ii). To prove (ii), suppose that $s \leq u, \min_{s,t} = \min_{u,v}$. There are two cases to discuss, (a) $s \leq u \leq v \leq t$, (b) $s \leq u \leq t \leq v$ (in any other case $\mathbb{P}(\min_{s,t} = \min_{u,v}) = 0$). In case (a), we have $\min_{s,t} = \min_{u,v} = \min_{u,t}$ and so $\gamma_{s,t} = \gamma_{u,t} = \gamma_{u,v}$ by (11) and (12). Similarly in case (b), we have $\gamma_{s,t} = \gamma_{u,t} = \gamma_{u,v}$. □

Proposition 5. *Fix* $s < t, s \in \mathbb{D}, n \geq 1$ *and* $\{(s_i, t_i); 1 \leq i \leq n\}$ *with* $s_i < t_i, s_i \in \mathbb{D}$. *Then:*

(i) $\gamma_{s,t}$ *is independent of* $\sigma(W)$.
(ii) For all $i \in [1, N]$, $h \in [1, n]$, *we have*

$$E[1_{\{\gamma_{s,t}=\mathbf{e}_i\}}|(\gamma_{s_i,t_i})_{1\le i\le n}, W] = 1_{\{\gamma_{s_h,t_h}=\mathbf{e}_i\}} \text{ on } \{min_{s,t} = min_{s_h,t_h}\}.$$

(iii) The law of $\gamma_{s,t}$ knowing $(\gamma_{s_i,t_i})_{1\le i\le n}$ and W is given by $\sum_{i=1}^{N} \alpha_i \delta_{\mathbf{e}_i}$ when $min_{s,t} \notin \{min_{s_i,t_i}; 1 \le i \le n\}$.

This entirely describes the law of $(W, \psi_{s,\cdot}, s \in \mathbb{D})$ in $\widetilde{C}$ independently of $(n_k, k \in \mathbb{N})$ and consequently $Z^n \xrightarrow[n\to+\infty]{law} Z$ in $\widetilde{C}$.

Proof. (i) is clear. (ii) is a consequence of Proposition 4 (ii). (iii) Write $\{s, t, s_i, t_i, 1 \le i \le n\} = \{r_k, 1 \le k \le m\}$ with $r_j < r_{j+1}$ for all $1 \le j \le m-1$. Suppose that $s = r_i, t = r_h$ with $i < h$. Then a.s. $\{\min_{r_j,r_{j+1}}, i \le j \le h-1\}$ are distinct and it will be sufficient to show that $\gamma_{s,t}$ is independent of $\sigma((\gamma_{s_i,t_i})_{1\le i\le n}, W)$ conditionally to $A = \{\min_{s,t} = \min_{r_j,r_{j+1}}, \min_{s,t} \ne \min_{s_i,t_i}$ for all $1 \le i \le n\}$ for $j \in [i, h-1]$. On A, we have $\gamma_{s,t} = \gamma_{r_j,r_{j+1}}$, $\{\min_{s_i,t_i}, 1 \le i \le n\} \subset \{\min_{r_k,r_{k+1}}, k \ne j\}$ and so $\{\gamma_{s_i,t_i}, 1 \le i \le n\} \subset \{\gamma_{r_k,r_{k+1}}, k \ne j\}$. Since $\gamma_{r_1,r_2}, \cdots, \gamma_{r_{m-1},r_m}, W$ are independent, it is now easy to conclude. □

In the sequel, we still assume that all processes are defined on the same probability space and that $Z^n \xrightarrow[n\to+\infty]{\text{a.s.}} Z$ in $\widetilde{C}$. In particular $\forall s \in \mathbb{D}, T > 0$,

$$\lim_{k\to+\infty} \sup_{s\le t\le s+T} \left|\frac{1}{\sqrt{k}}\Psi_{\lfloor ks\rfloor,\lfloor kt\rfloor} - \psi_{s,t}\right| = 0 \ a.s. \tag{14}$$

3.2.2 Extension of the Limit Process

For a fixed $s < t$, $\min_{s,t}$ is attained in $]s, t[$ a.s. By Proposition 4 (ii), on a measurable set $\Omega_{s,t}$ with probability 1, $\lim_{s'\to s+, s'\in\mathbb{D}} \gamma_{s',t}$ exists. Define $\varepsilon_{s,t} = \lim_{s'\to s+, s'\in\mathbb{D}} \gamma_{s',t}$ on $\Omega_{s,t}$ and give an arbitrary value to $\varepsilon_{s,t}$ on $\Omega_{s,t}^c$. Now, let $\varphi_{s,t} = \varepsilon_{s,t} W_{s,t}^{+}$. Then for all $s \in \mathbb{D}, t > s$, $(\varepsilon_{s,t}, \varphi_{s,t})$ is a modification of $(\gamma_{s,t}, \psi_{s,t})$. For all $s \in \mathbb{R}, t > s$, $\varphi_{s,t} = \lim_{n\to\infty} \varphi_{s_n,t}$ a.s., where $s_n = \frac{\lfloor 2^n s\rfloor+1}{2^n}$ and therefore $(\varphi_{s,t})_{t\ge s}$ is an $W(\alpha_1, \cdots, \alpha_N)$ started at 0. Again, Proposition 4 (ii) yields

$$\forall s < t, u < v,\ \mathbb{P}(\varepsilon_{s,t} = \varepsilon_{u,v}|\min_{s,t} = \min_{u,v}) = 1 \text{ if } \mathbb{P}(\min_{s,t} = \min_{u,v}) > 0. \tag{15}$$

Define:

$$\varphi_{s,t}(x) = (x + \mathbf{e}(x)W_{s,t})1_{\{t\le\tau_{s,x}\}} + \varphi_{s,t}1_{\{t>\tau_{s,x}\}}, s \le t, x \in G,$$

where $W_{s,t} = W_t - W_s$ and $\tau_{s,x}$ is given by (4).

Proposition 6. *Let $x \in G, x_n \in \frac{1}{\sqrt{n}}G_{\mathbb{N}}, \lim_{n\to\infty} x_n = x$, $s \in \mathbb{R}, T > 0$. Then, we have*

$$\lim_{n\to+\infty} \sup_{s\le t\le s+T} |\frac{1}{\sqrt{n}}\Psi_{\lfloor ns\rfloor,\lfloor nt\rfloor}(\sqrt{n}x_n) - \varphi_{s,t}(x)| = 0 \;\; a.s.$$

Proof. Let s' be a dyadic number such that $s < s' < s + T$. By (15), for $t > s'$:

$$\{\min_{s,t} = \min_{s',t}\} \subset \{\varphi_{s,t} = \varphi_{s',t}\} \;\; a.s.$$

and so, a.s.

$$\forall t > s', t \in \mathbb{D}; \; \{\min_{s,t} = \min_{s',t}\} \subset \{\varphi_{s,t} = \varphi_{s',t}\}.$$

If $t > s', \min_{s,t} = \min_{s',t}$ and $t_n \in \mathbb{D}, t_n \downarrow t$ as $n \to \infty$, then $\min_{s,t_n} = \min_{s',t_n}$ which entails that $\varphi_{s,t_n} = \varphi_{s',t_n}$ and therefore $\varphi_{s,t} = \varphi_{s',t}$ by letting $n \to \infty$. This shows that a.s.

$$\forall t > s'; \; \{\min_{s,t} = \min_{s',t}\} \subset \{\varphi_{s,t} = \varphi_{s',t}\}.$$

As a result a.s.

$$\forall s' \in \mathbb{D}\cap]s, s+T[, \forall t > s'; \; \{\min_{s,t} = \min_{s',t}\} \subset \{\varphi_{s,t} = \varphi_{s',t}\}. \tag{16}$$

By standard properties of Brownian paths, a.s. $\min_{s,s+T} \notin \{W_s, W_{s+T}\}$ and

$$\forall p \in \mathbb{N}^*; \min_{s,s+\frac{1}{p}} < W_s, \; \min_{s,s+\frac{1}{p}} \neq W_{s+\frac{1}{p}}, \exists! u_p \in]s, s+\frac{1}{p}[: \min_{s,s+\frac{1}{p}} = W_{u_p}.$$

The reasoning below holds almost surely: Take $p \ge 1, \min_{s,s+\frac{1}{p}} > \min_{s,s+T}$. Let $\mathscr{S}_p \in]s, s+\frac{1}{p}[: \min_{s,s+\frac{1}{p}} = W_{\mathscr{S}_p}$ and s' be a (random) dyadic number in $]s, \mathscr{S}_p[$. Then $\min_{s,s'} > \min_{s',t}$ for all $t \in [\mathscr{S}_p, s+T]$. By uniform convergence:

$$\exists n_0 \in \mathbb{N} : \forall n \ge n_0, \; \forall \mathscr{S}_p \le t \le s+T, \; \min_{u\in[s,s']} S_{\lfloor nu\rfloor} > \min_{u\in[s',t]} S_{\lfloor nu\rfloor}$$

and so

$$\Psi_{\lfloor ns'\rfloor,\lfloor nt\rfloor} = \Psi_{\lfloor ns\rfloor,\lfloor nt\rfloor}.$$

Therefore for $n \ge n_0$, we have

$$\sup_{\mathscr{S}_p\le t\le s+T} |\frac{1}{\sqrt{n}}\Psi_{\lfloor ns\rfloor,\lfloor nt\rfloor} - \varphi_{s,t}| = \sup_{\mathscr{S}_p\le t\le s+T} |\frac{1}{\sqrt{n}}\Psi_{\lfloor ns'\rfloor,\lfloor nt\rfloor} - \varphi_{s',t}| \text{ (using (16))}$$

and so

$$\sup_{s\le t\le s+T}|\frac{1}{\sqrt{n}}\Psi_{\lfloor ns\rfloor,\lfloor nt\rfloor}-\varphi_{s,t}|$$
$$\le \sup_{s\le t\le \mathscr{S}_p}|\frac{1}{\sqrt{n}}\Psi_{\lfloor ns\rfloor,\lfloor nt\rfloor}-\varphi_{s,t}|+\sup_{\mathscr{S}_p\le t\le s+T}|\frac{1}{\sqrt{n}}\Psi_{\lfloor ns\rfloor,\lfloor nt\rfloor}-\varphi_{s,t}|$$
$$\le \sup_{s\le t\le s+\frac{1}{p}}(\frac{1}{\sqrt{n}}S^+_{\lfloor ns\rfloor,\lfloor nt\rfloor}+W^+_{s,t})+\sup_{\mathscr{S}_p\le t\le s+T}|\frac{1}{\sqrt{n}}\Psi_{\lfloor ns'\rfloor,\lfloor nt\rfloor}-\varphi_{s',t}|$$
$$\le \sup_{s\le t\le s+\frac{1}{p}}(\frac{1}{\sqrt{n}}S^+_{\lfloor ns\rfloor,\lfloor nt\rfloor}+W^+_{s,t})+\sup_{s'\le t\le s'+T}|\frac{1}{\sqrt{n}}\Psi_{\lfloor ns'\rfloor,\lfloor nt\rfloor}-\varphi_{s',t}|.$$

From (14), a.s. $\forall u\in\mathbb{D}$, $\lim_{n\to+\infty}\sup_{u\le t\le u+T}|\frac{1}{\sqrt{n}}\Psi_{\lfloor nu\rfloor,\lfloor nt\rfloor}-\varphi_{u,t}|=0$. By letting n go to $+\infty$ and then p go to $+\infty$, we obtain

$$\lim_{n\to\infty}\sup_{s\le t\le s+T}|\frac{1}{\sqrt{n}}\Psi_{\lfloor ns\rfloor,\lfloor nt\rfloor}-\varphi_{s,t}|=0\ a.s. \tag{17}$$

We now show that

$$\lim_{n\to+\infty}\frac{1}{n}T_{\lfloor ns\rfloor,\sqrt{n}x_n}=\tau_{s,x}\ a.s. \tag{18}$$

We have

$$\frac{1}{n}T_{\lfloor ns\rfloor,\sqrt{n}x_n}=\inf\{r\ge\frac{\lfloor ns\rfloor}{n}:S^n_r-S^n_s=-|x_n|\}.$$

For $\epsilon>0$, from

$$\lim_{n\to\infty}\sup_{u\in[\tau_{s,x},\tau_{s,x}+\epsilon]}|(S^n_u-S^n_s+|x_n|)-(W_{s,u}+|x|)|=0,$$

we get

$$\lim_{n\to\infty}\inf_{u\in[\tau_{s,x},\tau_{s,x}+\epsilon]}(S^n_u-S^n_s+|x_n|)=\inf_{u\in[\tau_{s,x},\tau_{s,x}+\epsilon]}(W_{s,u}+|x|)<0$$

which implies $\frac{1}{n}T_{\lfloor ns\rfloor,\sqrt{n}x_n}<\tau_{s,x}+\epsilon$ for n large. If $x=0$, $\frac{1}{n}T_{\lfloor ns\rfloor,\sqrt{n}x_n}\ge\frac{\lfloor ns\rfloor}{n}$ entails obviously (18). If $x\neq 0$, then working in $[s,\tau_{s,x}-\epsilon]$ as before and using $\inf_{u\in[s,\tau_{s,x}-\epsilon]}(W_u-W_s+|x|)>0$, we prove that $\frac{1}{n}T_{\lfloor ns\rfloor,\sqrt{n}x_n}\le\tau_{s,x}-\epsilon$ for n large which establishes (18).
Now

$$\sup_{s\le t\le s+T}|\frac{1}{\sqrt{n}}\Psi_{\lfloor ns\rfloor,\lfloor nt\rfloor}(\sqrt{n}x_n)-\varphi_{s,t}(x)|\le\sup_{s\le t\le s+T}Q^{1,n}_{s,t}+\sup_{s\le t\le s+T}Q^{2,n}_{s,t} \tag{19}$$

where

$$Q_{s,t}^{1,n} = |(x_n + \mathbf{e}(x_n)(S_t^n - S_s^n))1_{\{\lfloor nt\rfloor \le T_{\lfloor ns\rfloor,\sqrt{n}x_n}\}} - (x + \mathbf{e}(x)W_{s,t})1_{\{t\le \tau_{s,x}\}}|,$$

$$Q_{s,t}^{2,n} = |\frac{1}{\sqrt{n}}\Psi_{\lfloor ns\rfloor,\lfloor nt\rfloor}1_{\{\lfloor nt\rfloor > T_{\lfloor ns\rfloor,\sqrt{n}x_n}\}} - \varphi_{s,t}1_{\{t>\tau_{s,x}\}}|.$$

By (17), (18) and the convergence of $\frac{1}{\sqrt{n}}S_{\lfloor n.\rfloor}$ towards W on compact sets, the right-hand side of (19) converges to 0 when $n \to +\infty$. □

Remark 3. From the definition of $\varepsilon_{s,t}$ (or Proposition 6), it is obvious that $\varepsilon_{r_1,r_2}, \cdots, \varepsilon_{r_{m-1},r_m}, W$ are independent for all $r_1 < \cdots < r_m$. Using (15), we easily check that (i), (ii) and (iii) of Proposition 5 are satisfied for all $s < t$, $n \ge 1, \{(s_i, t_i); 1 \le i \le n\}$ with $s_i < t_i$ (the proof remains the same as Proposition 5).

Proposition 7. *φ is the unique stochastic flow of mappings solution of (T).*

Proof. Fix $s < t < u, x \in G$ and let prove that $\varphi_{s,u}(x) = \varphi_{t,u} \circ \varphi_{s,t}(x)$ a.s. We follow Lemma 4.3 [7] and denote $\tau_{s,x}$ by $\tau_s(x)$. All the equalities below hold a.s. On the event $\{u < \tau_s(x)\}$, $\varphi_{s,t}(x) = x + \mathbf{e}(x)W_{s,t}$, $\tau_t(\varphi_{s,t}(x)) = \tau_s(x) < u$ and

$$\varphi_{t,u} \circ \varphi_{s,t}(x) = x + \mathbf{e}(x)(W_{s,t} + W_{t,u}) = x + \mathbf{e}(x)W_{s,u} = \varphi_{s,u}(x).$$

On the event $\{\tau_s(x) \in]t,u]\}$, we still have $\varphi_{s,t}(x) = x + \mathbf{e}(x)W_{s,t}$ and $\tau_t(\varphi_{s,t}(x)) = \tau_s(x) \le u$, thus

$$\varphi_{t,u} \circ \varphi_{s,t}(x) = \varepsilon_{t,u}W_{t,u}^+ = \varepsilon_{s,u}W_{s,u}^+ = \varphi_{s,u}(x).$$

since on the event $\{\tau_s(x) \in]t,u]\}$, $\min_{s,u} = \min_{t,u}$ and $W_{s,u}^+ = W_u - \min_{s,u} = W_{t,u}^+$.

On the event $\{\tau_s(x) \le t\} \cap \{\tau_t(\varphi_{s,t}(x)) \le u\}$, $\varphi_{s,t}(x) = \varepsilon_{s,t}W_{s,t}^+$ and

$$\varphi_{t,u} \circ \varphi_{s,t}(x) = \varphi_{t,u}(\varepsilon_{s,t}W_{s,t}^+) = \varepsilon_{t,u}W_{t,u}^+ = \varepsilon_{s,u}W_{s,u}^+ = \varphi_{s,u}(x)$$

since $W_{s,\tau_t(\varphi_{s,t}(x))}^+ = 0$ and thus $\min_{s,u} = \min_{t,u}$ which implies $\varepsilon_{s,u} = \varepsilon_{t,u}$ and $W_{s,u}^+ = W_{t,u}^+$.

On the event $\{\tau_s(x) \le t\} \cap \{\tau_t(\varphi_{s,t}(x)) > u\}$, $\varphi_{s,t}(x) = \varepsilon_{s,t}W_{s,t}^+$ and

$$\varphi_{t,u} \circ \varphi_{s,t}(x) = \varphi_{t,u}(\varepsilon_{s,t}W_{s,t}^+) = \varepsilon_{s,t}(W_{s,t}^+ + W_{t,u}) = \varepsilon_{s,u}W_{s,u}^+ = \varphi_{s,u}(x).$$

since in this case $\min_{s,u} = \min_{s,t}$ which implies $\varepsilon_{s,u} = \varepsilon_{s,t}$ and

$$\begin{aligned} W_{s,u}^+ &= W_u - \min_{s,u} \\ &= W_u - W_s + W_s - \min_{s,t} \\ &= W_{s,t}^+ + W_{t,u}. \end{aligned}$$

Thus we have, a.s. $\varphi_{s,u}(x) = \varphi_{t,u} \circ \varphi_{s,t}(x)$ which proves the cocycle property for φ. It is now easy to check that φ is a stochastic flow of mappings in the sense of Definition 4 [5].

Note that $(\varphi_{0,t}, t \geq 0)$ is an $W(\alpha_1, \cdots, \alpha_N)$ started at 0 and therefore satisfies Freidlin-Sheu formula (Theorem 3 [5]). Let $f \in D(\alpha_1, \cdots, \alpha_N)$, then for all $t \geq 0$,

$$f(\varphi_{0,t}) = f(0) + \int_0^t f'(\varphi_{0,u})dB_u + \frac{1}{2}\int_0^t f''(\varphi_{0,u})du \ a.s.$$

where $B_t = |\varphi_{0,t}| - \tilde{L}_t(|\varphi_{0,.}|)$ and $\tilde{L}_t(|\varphi_{0,.}|)$ is the symmetric local time at 0 of $|\varphi_{0,.}|$. Since $|\varphi_{0,t}| = W_t - \min_{0,t}$, we get $B_t = W_t$. Let $x \in D_i \setminus \{0\}$ and $f_i(r) = f(r\mathbf{e}_i), r \geq 0$. Since $\lim_{z\to 0, z\in D_i, z\neq 0} f'(z)$ and $\lim_{z\to 0, z\in D_i, z\neq 0} f''(z)$ exist, we can construct g which is C^2 on $\mathbb{R}$ and coincides with f_i on $\mathbb{R}_+$. By Itô's formula

$$g(|x| + W_t) = g(|x|) + \int_0^t g'(|x| + W_u)dW_u + \frac{1}{2}\int_0^t g''(|x| + W_u)du \ a.s.$$

and so for $t \leqslant \tau_0(x)$, we have

$$f(\varphi_{0,t}(x)) = f(x) + \int_0^t f'(\varphi_{0,u}(x))dW_u + \frac{1}{2}\int_0^t f''(\varphi_{0,u}(x))du \ a.s.$$

Set $\alpha = f(0) + \int_0^{\tau_0(x)} f'(\varphi_{0,u})dW_u + \frac{1}{2}\int_0^{\tau_0(x)} f''(\varphi_{0,u})du = f(\varphi_{0,\tau_0(x)}) = f(0)$ since $W^+_{0,\tau_0(x)} = 0$. Then for $t > \tau_0(x)$, write

$$\begin{aligned} f(\varphi_{0,t}(x)) = f(\varphi_{0,t}) &= \alpha + \int_{\tau_0(x)}^t f'(\varphi_{0,u})dW_u + \frac{1}{2}\int_{\tau_0(x)}^t f''(\varphi_{0,u})du \\ &= f(0) + \int_{\tau_0(x)}^t f'(\varphi_{0,u}(x))dW_u + \frac{1}{2}\int_{\tau_0(x)}^t f''(\varphi_{0,u}(x))du. \end{aligned}$$

But $f(x) + \int_0^{\tau_0(x)} f'(\varphi_{0,u}(x))dW_u + \frac{1}{2}\int_0^{\tau_0(x)} f''(\varphi_{0,u}(x))du = f(\varphi_{0,\tau_0(x)}(x)) = f(0)$ and so, for all $t \geq 0$, $f \in D(\alpha_1, \cdots, \alpha_N), x \in G$,

$$f(\varphi_{0,t}(x)) = f(x) + \int_0^t f'(\varphi_{0,u}(x))dW_u + \frac{1}{2}\int_0^t f''(\varphi_{0,u}(x))du \ a.s. \qquad (20)$$

Now, let (ψ, W) be a any flow of mappings solution of (T). Lemma 6 [5] implies

$$\psi_{0,t}(x) = x + \mathbf{e}(x)W_{0,t} \text{ for } 0 \leq t \leq \tau_{0,x} \text{ with } \tau_{0,x} \text{ given by (4).} \qquad (21)$$

By considering a sequence $(x_k)_{k\geq 0}$ converging to ∞, this shows that $\sigma(W_t) \subset \sigma(\psi_{0,t}(y), y \in G)$. Therefore, we can define a Wiener stochastic flow K^* obtained by filtering δ_ψ with respect to $\sigma(W)$ (Lemma 3-2 (ii) in [6]) satisfying: $\forall s \leq t$,

$x \in G, K^*_{s,t}(x) = E[\delta_{\psi_{s,t}(x)}|\sigma(W)]$ $a.s.$ In particular K^* solves (T) and since K^W given by (3) is the unique Wiener solution of (T), we get: $\forall s \le t, x \in G$, $K^W_{s,t}(x) = E[\delta_{\psi_{s,t}(x)}|\sigma(W)]$ $a.s.$ (see Proposition 8 [5]). As $K^W_{0,t}(0)$ is supported on $\{W^+_{0,t}\mathbf{e}_i,\ 1 \le i \le N\}$, we deduce that $|\psi_{0,t}(0)| = W^+_{0,t}$. Combining this with (21), we see that

$$\inf\{r \ge 0 : \psi_{0,r}(x) = \psi_{0,r}(0)\} = \tau_{0,x}.$$

This implies $\psi_{0,r}(x) = \psi_{0,r}(0)$ for all $r \ge \tau_{0,x}$ by applying the following

Lemma 2. *For all $(x_1, \cdots, x_n) \in G^n$, denote by $\mathbb{P}_{x_1,\cdots,x_n}$ the law of $(\psi_{0,.}(x_1), \cdots, \psi_{0,.}(x_n))$ in $C(\mathbb{R}_+, G^n)$. Let T be a finite $(\mathscr{F}_t)$ stopping time where $\mathscr{F}_t = \sigma(\psi_{0,u}, u \le t), t \ge 0$. Then the law of $(\psi_{0,T+.}(x_1), \cdots, \psi_{0,T+.}(x_n))$ knowing $\mathscr{F}_T$ is given by $\mathbb{P}_{\psi_{0,T}(x_1),\cdots,\psi_{0,T}(x_n)}$.*

Note that $W_{0,.}$ can be recovered out from $W^+_{0,.}$ and consequently $\psi_{0,.}(x)$ is a measurable function of $\psi_{0,.}(0)$ for all $x \in G$. Therefore, for all $(x_1, \cdots, x_n) \in G^n$, $(\psi_{0,\cdot}(x_1), \cdots, \psi_{0,\cdot}(x_n))$ is unique in law since $\psi_{0,\cdot}(0)$ is a Walsh Brownian motion. This completes the proof. □

3.2.3 The Wiener Flow

Remark that $K^W_{s,t}(x) = E[\delta_{\varphi_{s,t}(x)}|\sigma(W)]$ which entails that K^W is a stochastic flow of kernels. By conditioning with respect to $\sigma(W)$ in (20), we easily see that (K^W, W) solves (T). In order to finish the proof of Theorem 2 and Corollary 1, we need only check the following lemma (the proof of (6) is similar)

Lemma 3. *Under the hypothesis of Proposition 6, we have*

$$\sup_{t\in[s,s+T]} \beta(K^W_{s,t}(x), K^n_{s,t}(\sqrt{n}x_n)) \xrightarrow[n \to +\infty]{} 0 \ a.s.$$

Proof. Let $g : G \longrightarrow \mathbb{R}$ such that $\|g\|_\infty + \sup_{x\neq y} \dfrac{|g(x) - g(y)|}{|x - y|} \le 1, g(0) = 0$. Then,

$$\left| \int_G g(y) K^W_{s,t}(x)(dy) - \int_G g(y) K^n_{s,t}(\sqrt{n}x_n)(dy) \right| \le V^{1,n}_{s,t} + V^{2,n}_{s,t}$$

where

$$V^{1,n}_{s,t} = \left| g(x_n + \mathbf{e}(x_n)S^n_{s,t}) 1_{\{\lfloor nt\rfloor \le T_{\lfloor ns\rfloor,\sqrt{n}x_n}\}} - g(x + \mathbf{e}(x)W_{s,t}) 1_{\{t \le \tau_{s,x}\}} \right|,$$

$$V^{2,n}_{s,t} = \sum_{j=1}^{N} \alpha_j \left| g(\mathbf{e}_j W^+_{s,t}) 1_{\{t>\tau_{s,x}\}} - g(\mathbf{e}_j S^+_{n,s,t} + o_n) 1_{\{\lfloor nt\rfloor > T_{\lfloor ns\rfloor,\sqrt{n}x_n}\}} \right|$$

and $o_n \in G$ is a $\sigma(S)$ measurable random variable such that $|o_n| \le \frac{1}{\sqrt{n}}$, $S^n_{s,t} = S^n_t - S^n_s$, $S^+_{n,s,t} = \frac{1}{\sqrt{n}} S^+_{\lfloor ns\rfloor,\lfloor nt\rfloor}$. As $\lfloor x\rfloor - 1 \le x \le \lfloor x\rfloor + 1$ for all $x \in \mathbb{R}$, we get

$$V_{s,t}^{1,n} \leq \sup_{t \in I_{n,s,x}} |x_n + \mathbf{e}(x_n) S_{s,t}^{(n)} - x - \mathbf{e}(x) W_{s,t}|$$
$$+ \sup_{t \in J_{n,s,x}} |g(x_n + \mathbf{e}(x_n) S_{s,t}^{(n)})| + \sup_{t \in K_{n,s,x}} |g(x + \mathbf{e}(x) W_{s,t})| \tag{22}$$

with

$$I_{n,s,x} = [s, \tau_{s,x} \vee (\frac{1}{n} + \frac{1}{n} T_{\lfloor ns \rfloor, \sqrt{n} x_n})],$$
$$J_{n,s,x} = [\tau_{s,x}, (\frac{1}{n} T_{\lfloor ns \rfloor, \sqrt{n} x_n} + \frac{1}{n}) \vee \tau_{s,x}],$$
$$K_{n,s,x} = [\tau_{s,x} \wedge (\frac{1}{n} T_{\lfloor ns \rfloor, \sqrt{n} x_n} - \frac{1}{n}), \tau_{s,x}].$$

Using $|g(y)| \leq |y|$, we obtain

$$\sup_{t \in J_{n,s,x}} |g(x_n + \mathbf{e}(x_n) S_{s,t}^{(n)})| + \sup_{t \in K_{n,s,x}} |g(x + \mathbf{e}(x) W_{s,t})|$$
$$\leq \sup_{t \in J_{n,s,x}} ||x_n| + S_{s,t}^{(n)}| + \sup_{t \in K_{n,s,x}} ||x| + W_{s,t}|.$$

Since $\lim_{n \to +\infty} \frac{1}{n} T_{\lfloor ns \rfloor, \sqrt{n} x_n} = \tau_{s,x}$ a.s., the right-hand side converges to 0. By discussing the cases $x = 0, x \neq 0$, we easily see that $\lim_{n \to \infty} \sup_{t \in I_{n,s,x}} |x_n + \mathbf{e}(x_n) S_{s,t}^{(n)} - x - \mathbf{e}(x) W_{s,t}| = 0$ and therefore $\lim_{n \to \infty} \sup_{t \in [s,s+T]} V_{s,t}^{1,n} = 0$. By the same manner, we arrive at $\lim_{n \to \infty} \sup_{t \in [s,s+T]} V_{s,t}^{2,n} = 0$ which proves the lemma. □

Acknowledgements I sincerely thank Yves Le Jan, Olivier Raimond and Sophie Lemaire for very useful discussions. I am also grateful to the referee for his helpful comments.

References

1. A.S. Cherny, A.N. Shiryaev, M. Yor, Limit behaviour of the "horizontal-vertical" random walk and some extensions of the Donsker-Prokhorov invariance principle. Teor. Veroyatnost. i Primenen **47**(3) 498–517 (2002)
2. N. Enriquez, Y. Kifer, Markov chains on graphs and Brownian motion. J. Theoret. Probab. **14**(2) 495–510 (2001)
3. S. Ethier, T. Kurtz, *Markov Processes: Characterization and Convergence*, Wiley Series in Probability and Mathematical Statistics, (Wiley, New York, 1986)
4. J.M. Harrison, L.A. Shepp, On skew Brownian motion. Ann. Probab. **9**(2) 309–313 (1981)
5. H. Hajri, Stochastic flows related to Walsh Brownian motion. Electron. J. Probab. **16**, 1563–1599 (2011)

6. Y. Le Jan, O. Raimond, Flows, coalescence and noise. Ann. Probab. **32**(2), 1247–1315 (2004)
7. Y. Le Jan, O. Raimond, Flows associated to Tanaka's SDE. ALEA Lat. Am. J. Probab. Math. Stat. **1**, 21–34 (2006)
8. I.I. Nishchenko, Discrete time approximation of coalescing stochastic flows on the real line. Theory Stoch. Proc. **17(33)**(1), 70–78 (2011)
9. P. Révész, *Random Walk in Random and Non-random Environments*. 2nd ed. (World Scientific Publishing, 2005)
10. D. Revuz, M. Yor, *Continuous Martingales and Brownian Motion*, 3rd ed. (Grundlehren der Mathematischen Wissenschaften, 1999)
11. S. Watanabe, The stochastic flow and the noise associated to Tanaka stochastic differential equation. Ukraïn. Mat. Zh. **52**(9), 1176–1193 (2000)

Spectral Distribution of the Free Unitary Brownian Motion: Another Approach

Nizar Demni and Taoufik Hmidi

Abstract We revisit the description provided by Ph. Biane of the spectral measure of the free unitary Brownian motion. We actually construct for any $t \in (0,4)$ a Jordan curve γ_t around the origin, not intersecting the semi-axis $[1, \infty[$ and whose image under some meromorphic function h_t lies in the circle. Our construction is naturally suggested by a residue-type integral representation of the moments and h_t is up to a Möbius transformation the main ingredient used in the original proof. Once we did, the spectral measure is described as the push-forward of a complex measure under a local diffeomorphism yielding its absolute-continuity and its support. Our approach has the merit to be an easy yet technical exercise from real analysis.

1 Reminder and Motivation

In his pioneering paper [1], Ph. Biane defined and studied the so-called free unitary or multiplicative Brownian motion. It is a unitary operator-valued Lévy process with respect to the free multiplicative convolution of probability measures on the unit circle $\mathbb{T}$ (or equivalently the multiplication of unitary operators that are free in some non commutative probability space). Besides, the spectral distribution μ_t at any time $t \geq 0$ is characterized by its moments

$$m_n(t) := \int_{\mathbb{T}} z^n d\mu_t(z) = e^{-nt/2} \sum_{k=0}^{n-1} \frac{(-t)^k}{k!} n^{k-1} \binom{n}{k+1}, n \geq 1,$$

N. Demni (✉) · T. Hmidi
IRMAR, Université de Rennes 1, Campus de Beaulieu, 35042 Rennes cedex, France
e-mail: nizar.demni@univ-rennes1.fr; thmidi@univ-rennes1.fr

C. Donati-Martin et al. (eds.), *Séminaire de Probabilités XLIV*, Lecture Notes in Mathematics 2046, DOI 10.1007/978-3-642-27461-9_9,

and $m_{-n}(t) = m_n(t), n \geq 1$ since Y^{-1} defines a free unitary Brownian motion too. This alternate sum is not easy to handle analytically since for instance if we try to work out the moments generating function of $(m_n(t))_{n\geq 1}$

$$M_t(w) := \sum_{n\geq 1} m_n(t)w^n, \; |w| < 1,$$

then we are led to

$$M_t(w) = u\sum_{k\geq 0} \frac{(-ut)^k}{k!} S_k(u)$$

where

$$S_k(u) := \sum_{n\geq 0}(n+k+1)^{k-1}\binom{n+k+1}{k+1}u^n, \quad k \geq 0,$$

and $u = we^{-t/2}$. Nonetheless, the explicit inverse function of

$$\tau_t(w) := \int_{\mathbb{T}} \frac{z+w}{z-w}\mu_t(dz) = 1 + 2M_t(w)$$

in the open unit disc played a key role in the description of μ_t [2]. More precisely, it was proved there that μ_t is an absolutely continuous probability measure with respect to the normalized Haar measure on $\mathbb{T}$ and that its density is a real analytic function inside its support. The latter coincides with $\mathbb{T}$ when $t > 4$ while it is given by the angle

$$|\theta| \leq \beta(t) := (1/2)\sqrt{t(4-t)} + \arccos(1-(t/2))$$

when $t \leq 4$. When proving these important results, the author relied on free stochastic integration (Lemma 11 p. 266), Caratheodory's extension Theorem for Riemann maps (Lemma 12 p. 270) and a Poisson-type integral representation for this kind of maps (see the proof of Proposition 10 p. 270). In the present paper, we shall recover Biane's results from simpler considerations than the ones used in the original proof. Indeed, for $t \in (0,4)$, there exists a unique piecewise smooth Jordan curve γ_t around the origin, not intersecting the semi-axis $[1,\infty[$ and whose image under some function h_t lies in $\mathbb{T}$. Our construction is naturally suggested by a residue-type integral representation of $m_n(t)$ and fails when $t \geq 4$. Note that the same phenomenon happens here and in Biane's proof: γ_t is constructed upon two curves that have a non empty intersection if and only if $t < 4$, while the inverse function of τ_t is defined on the interior s of two Jordan domains whose boundaries have the same phase transition [2, p. 267]. Moreover, the function h_t appears in the integrand of our residue-type representation and coincides up to the Möbius transformation

$$z \mapsto \frac{2}{z} - 1$$

with the inverse function of τ_t used in the original proof. Once the curve γ_t is constructed, we consider a piecewise smooth parametrization z_t of γ_t and prove that the derivative of $\theta \mapsto \arg[h_t(z_t(\theta))]$ vanishes if and only if $\theta = \pm \arccos(\sqrt{t}/2)$ (note that similarly the derivative of τ_t^{-1} vanishes if and only if $z = \pm i\sqrt{(4/t) - 1}$, [2] p. 269). As a matter of fact, $\theta \mapsto \arg[h_t(z_t(\theta))]$ defines far from the critical points a local diffeomorphism so that performing local changes of variables in our integral representation yields both the absolute-continuity of μ_t with respect to the Haar measure on $\mathbb{T}$ and the description of its support.

2 A Residue-Type Representation of $m_n(t)$

A residue-type integral representation of $m_n(t), n \geq 1$ is not new in its own. Indeed, the one we derive below is an elaborated version of the one used in order to determine the decay order of $m_n(t)$ (see [3] p. 566):

$$m_n(t) = \frac{e^{-nt/2}}{2i\pi n} \int_{\gamma} e^{-ntz} \left(1 + \frac{1}{z}\right)^n dz$$

where γ is a circle around the origin. More precisely, we need to integrate by parts since then the meromorphic integrand we obtain determines in a unique way the required curve γ_t we mentioned above. To proceed, we first apply Cauchy's Residues Theorem to $z \mapsto z^k e^{n/z}$ in order to get

$$\frac{n^{k+1}}{(k+1)!} = \frac{1}{2i\pi} \int_{\gamma} z^k e^{n/z} dz,$$

then use the combinatorial identity for binomial numbers:

$$\binom{n}{k+1} = \frac{n}{k+1}\binom{n-1}{k}, \qquad n \geq 1.$$

As a result, for any $n \geq 1$

$$\begin{aligned} m_n(t) &= e^{-nt/2} \sum_{k=0}^{n-1} \frac{(-t)^k}{k!} n^{k-1} \binom{n}{k+1} \\ &= \frac{e^{-nt/2}}{2i\pi n} \int_{\gamma} \sum_{k=0}^{n-1} \binom{n-1}{k} (-tz)^k e^{n/z} dz \\ &= \frac{1}{2i\pi n} \int_{\gamma} (1 - tz)^n e^{n(1/z - t/2)} \frac{dz}{1 - tz} \end{aligned}$$

$$= \frac{1}{2i\pi n}\int_{t\gamma}\left[(1-z)e^{t(1/z-1/2)}\right]^n \frac{dz}{t(1-z)}$$
$$= \frac{1}{2i\pi n}\int_{\gamma}\left[(1-z)e^{t(1/z-1/2)}\right]^n \frac{dz}{t(1-z)}$$
$$:= \frac{1}{2i\pi n}\int_{\gamma}[h_t(z)]^n \frac{dz}{t(1-z)}.$$

Now, choose further γ such that it does not meet the semi-axis $[1,\infty[$ then $z \mapsto \log(1-z)$ is well defined for $z \in \gamma$ and is holomorphic there. Hence, setting $z = re^{i\theta}, 0 < r < 1$ and integrating by parts yield

$$m_n(t) = \frac{1}{2i\pi t}\int_{\gamma}[h_t(z)]^n \frac{h_t'(z)}{h_t(z)}\log(1-z)dz.$$

As a matter of fact, $m_n(t)$ is the residue of

$$z \mapsto \frac{1}{t}[h_t(z)]^n \frac{h_t'(z)}{h_t(z)}\log(1-z)$$

at $z = 0$ so that one may integrate along any piecewise smooth Jordan curve γ_t (possibly depending on t) around zero provided that the integrand is well defined. Assume further that we can choose γ_t such that $|h_t(\gamma_t)| \in \mathbb{T}$ and let $\theta \in [-\pi,\pi] \mapsto z_t(\theta)$ be a piecewise smooth parametrization of γ_t, then

$$m_n(t) = \frac{1}{2i\pi t}\int_{-\pi}^{\pi} e^{in\arg h_t(z_t(\theta))}\frac{h_t'(z_t(\theta))}{h_t(z_t(\theta))}\log(1-z_t(\theta))z_t'(\theta)d\theta$$
$$= \int_{-\pi}^{\pi} e^{in\theta}\nu_t(d\theta)$$

where ν_t is the push-forward of

$$\frac{h_t'(z_t(\theta))}{h_t(z_t(\theta))}z_t'(\theta)\log(1-z_t(\theta))\mathbf{1}_{[-\pi,\pi]}(\theta)\frac{d\theta}{2i\pi}$$

under the map $\theta \mapsto \arg h_t(z_t(\theta))$. Heuristically, we are attempted to conclude that $\nu_t = \mu_t$ however it is not clear at all that ν_t is a real measure and there is no guarantee even for γ_t to exist. Below, we shall prove that γ_t exists if and only if $t \in (0,4)$ and is unique. Then γ_t splits into two curves $\gamma_t^1 \cup \gamma_t^2$ where $\theta \mapsto \arg[h_t(z_t(\theta))]$ is a diffeomorphism from γ_t^i to its image, for $i = 1,2$. As a result, a change of variables shows that ν_t and μ_t coincide since their trigonometric moments do, therefore μ_t is absolutely continuous and its support is easily recovered as $h_t(\gamma_t)$.

3 Construction of the Curve γ_t

Our main result is stated as

Proposition 1. *Let $t \in (0,4)$, then there exists a unique (piecewise smooth) Jordan curve γ_t such that:*

- $h_t(\gamma_t) \in \mathbb{T}$.
- γ_t *encircles* $z = 0$ *and* $\gamma_t \cap [1, \infty[= \emptyset$.

Proof. Before coming into computations, let us point to the fact that γ_t has to be invariant under complex conjugation: this fact follows from $|h_t(\overline{z})| = |\overline{h_t(z)}| = |h_t(z)|$. It is also coherent with the fact that ρ_t shares the same invariance property since $Y^{-1} = Y^\star$ is also a free unitary Brownian motion, therefore we only consider $\theta \in [0,\pi]$. We also inform the patient reader that both polar and cartesian coordinates are used in the sequel depending on how behaves the curve defined by $|h_t(z)| = 1$ when t runs over $]0,4[$.

3.1 Polar Coordinates

Let $z = re^{i\theta}, \theta \in [0,\pi]$ then $|h_t(z)| = 1$ is equivalent to

$$g_{t,\theta}(r) := (1 + r^2 - 2r\cos\theta)e^{(2t\cos\theta)/r} = e^t.$$

We distinguish two regions:

- $\{\theta, \cos\theta < 0\}$: In this region the function $g_{t,\theta}$ is increasing and satisfies

$$\lim_{r\to 0+} g_{t,\theta}(r) = 0, \quad \lim_{r\to\infty} g_{t,\theta}(r) = \infty.$$

The monotonicity of $g_{t,\theta}$ follows obviously from

$$g'_{t,\theta}(r) = 2e^{2t\cos\theta/r}\left(r - \cos\theta - (1 + r^2 - 2r\cos\theta)\frac{t\cos\theta}{r^2}\right). \tag{1}$$

Then $g_{t,\theta}(r) = e^t$ has a unique solution $r = r_t(\theta) > 0$. Note that the implicit function Theorem together with the fact that $g'_{t,\theta}(r) > 0$ show that $\theta \mapsto r_t(\theta)$ is at least C^1 on $]\pi/2,\pi[$. Now, it is obvious that r_t does not vanish on $\{\cos\theta < 0\}$ and we shall check that it remains so on $\{\cos\theta = 0\}$. More precisely, we claim that $r_t(\theta) > \sqrt{t}$ which may be proved as follows: use

$$1 + t < e^t \quad \text{and} \quad 1 - 2\sqrt{t}\cos\theta < e^{-2\sqrt{t}\cos\theta}$$

to get

$$1 + t - 2\sqrt{t}\cos\theta \le (1+t)(1 - 2\sqrt{t}\cos\theta) < e^{t-2\sqrt{t}\cos\theta}.$$

This in turn yields

$$g_{t,\theta}(\sqrt{t}) < e^t$$

and the monotonicity of $g_{t,\theta}$ proves the claim. As a matter of fact r_t extends continuously to $\pi/2$ and one obviously has

$$r_t(\pi/2) = \sqrt{e^t - 1}.$$

• $\{\theta, \cos\theta > 0\}$: For these values of θ, observe that $g_{t,\theta}(2\cos\theta) = e^t$ for all t. However, $r_t(\theta) = 2\cos\theta$ does not fulfill our requirements since on the one hand it vanishes at $\theta = \pm\pi/2$ and on the other hand it meets $[1, \infty[$ at $\theta = 0$. Fortunately, there exists another radius r_t satisfying $g_{t,\theta}(r_t) = e^t$, $r_t(\pi/2) \neq 0$ and $r_t(0) \in]0,1[$. Indeed, letting $\theta_t := \arccos(\sqrt{t}/2) \in]0, \pi/2$, then

$$g'_{t,\theta}(2\cos\theta) = \frac{4\cos^2\theta - t}{2\cos\theta} e^t$$

is negative on $]\theta_t, \pi/2[$, positive on $[0, \theta_t[$ and

$$\lim_{r\to\infty} g_{t,\theta}(r) = +\infty$$

for any θ such that $\cos\theta \ge 0$. Besides, this radius is unique except possibly for $2 + \sqrt{3} < t < 4$ and for θ close to zero, and we keep using polar coordinates. For exceptional values of (t, θ), cartesian coordinates are more adequate and doing so we recover γ_t as the graph of some function.

▲ $\{0 < t \le 2 + \sqrt{3}\}$: When $t \in]0, 1]$, we shall prove that $g_{t,\theta}$ is a convex function. To this end, we compute the second derivative of $g_{t,\theta}$

$$\begin{aligned} g''_{t,\theta}(r) &= e^{2t\cos\theta/r}\Bigg[\left(\frac{4t\cos\theta}{r^3} + \frac{4t^2\cos^2\theta}{r^4}\right)(1 + r^2 - 2r\cos\theta) \\ &\quad + 2 - \frac{8t\cos\theta}{r^2}(r - \cos\theta)\Bigg] \\ &= \frac{e^{2t\cos\theta/r}}{r^4}\big[2r^4 - 4tr^3\cos\theta + 4t^2r^2\cos^2\theta \\ &\quad + 4rt\cos\theta(1 - 2t\cos^2\theta) + 4t^2\cos^2\theta\big]. \end{aligned}$$

The first equality shows that if $r \le \cos\theta$ then $g''_{t,\theta}(r) > 0$. For $r \ge \cos\theta$, define k_t by

$$g''_{t,\theta}(r) = \frac{e^{2t\cos\theta/r}}{r^4} k_{t,\theta}(r).$$

Then

$$k'_{t,\theta}(r) = 8r^3 - 12tr^2\cos\theta + 8t^2r\cos^2\theta + 4t\cos\theta(1 - 2t\cos^2\theta)$$

$$k''_{t,\theta}(r) = 8(3r^2 - 3tr\cos\theta + t^2\cos^2\theta) \geq 0$$

even for all $r \geq 0$. Hence

$$\begin{aligned} k'_{t,\theta}(r) \geq k'_{t,\theta}(t\cos\theta) &= 4t\cos\theta(t^2\cos^2\theta - 2t\cos^2\theta + 1) \\ &= (t\cos\theta - 1)^2 + 2t\cos\theta(1 - \cos\theta) \geq 0. \end{aligned}$$

Since $t\cos\theta \leq \cos\theta$ when $t \leq 1$, then $k'_{t,\theta}(r) \geq 0$ for all $r \geq 0$ therefore

$$k_{t,\theta}(r) \geq k_{t,\theta}(0) = 4t^2\cos^2\theta > 0$$

which yields the strict convexity of $g_{t,\theta}$ for $t \in [0,1]$. Now since

$$\lim_{r\to 0} g_{t,\theta}(r) = \lim_{r\to+\infty} g_{t,\theta}(r) = +\infty,$$

then the equation $g_{t,\theta}(r) = e^t$ admits exactly two solutions among them the trivial one $r = r(\theta) = 2\cos\theta$ which has to be discarded. The required curve γ_t is then constructed upon the non trivial solution and defines even a C^1-piecewise curve. The last claim is obvious for the regular points of $g_{t,\theta}$ by the virtue of the implicit function Theorem again. So we need to focus on the critical points of the curve: $g'_{t,\theta}(r(\theta)) = 0$. But, the strict convexity of $g_{t,\theta}$ forces then $r_t(\theta) = 2\cos\theta$ which gives after substituting in $g'_{t,\theta}$ the unique critical point $\theta = \theta_t$ for which $r_t(\theta_t) = \sqrt{t}$. Before considering the range $t \in [1, 2+\sqrt{3}]$, we point out that $\gamma_t, t \in]0,4[$ crosses the positive real semi-axis at some point $0 < x_t < 1$ that is described in Lemma 1 below. Now, let $t \in [1, 2+\sqrt{3}]$ and define

$$v_{t,\theta}(r) := r^3 - (t+1)\cos\theta r^2 + 2t\cos^2\theta r - t\cos\theta$$

so that

$$g'_{t,\theta}(r) = \frac{e^{2t\cos\theta/r}}{r^2} v_{t,\theta}(r).$$

Then the first derivative of $v_{t,\theta}$ reads

$$v'_{t,\theta}(r) = 3r^2 - 2(t+1)\cos\theta r + 2t\cos^2\theta.$$

The discriminant of this second degree polynomial is easily computed as $4(t^2 - 4t + 1)\cos^2\theta$ and is easily seen to be negative on $t \in [1, 2+\sqrt{3}]$. Consequently $v'_{t,\theta} \geq 0$ thereby $v_{t,\theta}(r) \geq v_{t,\theta}(0) = -t\cos\theta$ for any $r \geq 0$. Finally, there exists $r_0 = r_0(t,\theta)$ such that $g'_{t,\theta}(r_0) = 0$ so that the variations of $g_{t,\theta}$ are described by

r	0	r_0	$+\infty$
$v_{t,\theta}(r)$		$-\ 0\ +$	
$g_{t,\theta}(r)$	$+\infty$	$\searrow\quad\nearrow$	$+\infty$

The conclusion follows in similar way to the previous range of times $t \in]0, 1]$.

▲ $2+\sqrt{3} < t < 4$: This part of the proof needs more care since for small amplitudes of θ, $\theta \mapsto r_t(\theta)$ may be a multi-valued function (this is seen from computer-assisted pictures). This multivalence happens precisely in the interior of the region bounded by the curve $\theta \mapsto 2\cos\theta$ or $0 \le \theta < \theta_t$, and cartesian coordinates are more adequate for our purposes. Nonetheless, we shall keep use of polar coordinates outside the latter curve where the existence of a required radius is ensured by the same arguments evoked above. It then remains to prove uniqueness on $]2\cos\theta, +\infty[$, $\theta \in]\theta_t, \pi/2]$. To this end, one easily sees that the largest root of $v'_{t,\theta}(r)$ is given by

$$R_t^+(\theta) := \frac{t+1+\sqrt{t^2-4t+1}}{3}\cos\theta$$

and that $t \mapsto R_t^+(\theta)$ is increasing. Thus, $R_t^+(\theta) \le R_4(\theta) = 2\cos\theta$ yielding $v'_{t,\theta}(r) > 0$ for $r > 2\cos\theta$. This entails

$$v_{t,\theta}(r) \ge v_{t,\theta}(2\cos\theta) = 4\cos\theta(\cos^2\theta - t/4)$$

which is negative on $]\theta_t, \pi/2]$. Therefore the variations of $g_{t,\theta}$ are summarized below

r	$2\cos\theta$	r_0	$+\infty$
$v_{t,\theta}(r)$		$-\ 0\ +$	
$g_{t,\theta}(r)$	e^t	$\searrow\quad\nearrow$	$+\infty$

for some $r_0 = r_0(t,\theta) > 2\cos\theta$, whence the uniqueness follows. Hence, the obtained branch of γ_t is smooth and its endpoints are $i\sqrt{e^t-1}$ and $\sqrt{t}\,e^{i\arccos(\sqrt{t}/2)}$.

Remark 1. The equation $g_{t,\theta}(r) = e^t$ has no solution in the region $\{r > 2\cos\theta,\ 0 \le \theta < \theta_t\}$. Indeed, $v_{t,\theta}(r) \ge v_{t,\theta}(2\cos\theta) > 0$ so that $g_{t,\theta}$ is increasing.

3.2 Cartesian Coordinates

The curve $\theta \mapsto 2\cos\theta, \theta \in [0, \pi/2]$ is the graph of the function $x \mapsto y = \sqrt{x(2-x)}, x \in [0, 2]$. Hence, we shall restrict ourselves to the region

$$\left\{0 \leq y < \sqrt{x(2-x)}, 0 < x < 2\right\}.$$

Besides, the branch of γ_t, if it exists, would meet the axis $\{y = 0\}$ at a solution of

$$k_t(x) := (x-1)^2 e^{t[(2/x)-1]} = 1, 0 < x < 2,$$

whose needed properties are collected in the following Lemma.

Lemma 1. *For every $t \in]0,4[$, the above equation admits a unique solution $x_t \in]0,1[$ and the map $t \mapsto x_t$ is increasing. In particular $x_t > 3 - \sqrt{5}$.*

Proof. Let $t \in]0,4[$ then

$$k_t'(x) = \frac{2(x-1)(x^2 - tx + t)}{x^2} e^{t[(2/x)-1]}$$

and the polynomial $x^2 - tx + t$ is obviously positive since its discriminant is negative. As a result, we get the variations of k_t

x	0		1		2
$k_t'(x)$		$-$		$+$	
$k_t(x)$	$+\infty$	$\searrow$	0	$\nearrow$	1

This asserts the existence of a unique value $x_t \in]0,1[$ solving the equation $k_t(x) = 1$. For the variations of $t \mapsto x_t$, we write

$$x_t' = -\frac{\partial_t k_t(x_t)}{\partial_x k_t(x_t)} = \frac{x_t(1-x_t)(2-x_t)}{2(x_t^2 - tx_t + t)} > 0$$

so that $x_t > x_3$ and the Lemma is proved by noting that $k_3(3-\sqrt{5}) = (\sqrt{5}-1)^2 e^{3(1+\sqrt{5})/2} \geq 1$. $\square$

Now, we rewrite $g_{t,\theta}(r) = e^t$ using cartesian coordinates as

$$(1 + x^2 + y^2 - 2x)e^{2tx/(x^2+y^2)} = e^t$$

and denote $g_{t,x}(y)$ the LHS. In this way, $g_{t,x}(0) = k_t(x)e^t \leq e^t$ for $x \in [x_t, 2]$ since $k_t(x) < 1$ for $x \in [x_t, 2]$, while $g_{t,x}(0) > e^t$ for $x \notin [0, x_t[$. We shall prove that for each $x \in [x_t, t/2] \subset [x_t, 2[$, the equation $g_{t,x}(y) = e^t$ admits a unique solution $\left\{0 \leq y < \sqrt{x(2-x)}\right\}$ while it has none when $x \notin]x_t, t/2[$. This in turn finishes the proof of our main result. To proceed, we first compute

$$g'_{t,x}(y) = \frac{2y}{(x^2+y^2)^2}\Big(y^4 + (2x^2 - 2xt)y^2 + x^4 - 2x^3t + 4x^2t - 2xt\Big)e^{2tx/(x^2+y^2)}$$
$$:= \frac{2y}{(x^2+y^2)^2}e^{2tx/(x^2+y^2)}w_{t,x}(y^2).$$

The discriminant of $w_{t,x}(y)$ is given by $4tx(2+(t-4)x)$ and is positive since $t \geq 3$ and $x \leq 2$. Thus $w_{t,x}(y)$ has two roots

$$y_t^{\pm} := x(t-x) \pm \sqrt{tx(2+(t-4)x)}$$

satisfying the following properties:

Lemma 2. *Let $2+\sqrt{3} < t < 4$, then:*

1. *For $x \in [x_t, 2]$, then* $0 < y_t^-(t) \leq y_t^+$.
2. *For $x \in [0, t/2]$ then* $y_t^-(t) \leq x(2-x) \leq y_t^+$.
3. *For $x \in [t/2, 2]$, then* $x(2-x) \leq y_t^- \leq y_t^+$

Proof. Since $x \in]0,2[$, then $y_t^+ \geq 0$ and the first property (1) is equivalent to

$$y_t^+ y_t^- = x^4 - 2x(x-1)^2 t > 0.$$

But for any $x > 0$, the function $t \mapsto x^4 - 2x(x-1)^2 t$ is decreasing for positive t therefore

$$x^4 - 2x(x-1)^2 t > x^4 - 8x(x-1)^2 = x(x-2)\big(x-(3-\sqrt{5})\big)\big(x-(3+\sqrt{5})\big)$$

for any $t < 4$. Consequently $x^4 - 2x(x-1)^2 t > 0$ for $x \in [3-\sqrt{5}, 2]$ in particular for $x \in [x_t, 2]$ since $x_t > 3-\sqrt{5}$ by the virtue of Lemma 1.
Since $y_t^+ > x(2-x)$, then the second property (2) is equivalent to

$$x^2(t-2)^2 \leq tx(2+(t-4)x)$$

which is in turn equivalent to $t - 2x \geq 0$. We are done. □

It remains to discuss the variations of $g_{t,x}$ according to:
⋆ $x \in [x_t, t/2]$: Using properties (1) and (2) stated in Lemma 2 we get

y	0		$(y_t^-)^{\frac{1}{2}}$		$(y_t^+)^{\frac{1}{2}}$		$+\infty$
$g'_{t,x}(y)$		$+$		$-$		$+$	
$g_{t,x}(y)$	$g_{t,x}(0)$	↗		↘		↗	$+\infty$

Since $g_{t,x}(0) \leq e^t$, then the equation $g_{t,x}(y) = e^t$ has a unique solution $y_t(x)$ lying in the interval $[0, \sqrt{x(2-x)}[$. This allows to construct a curve $x \mapsto y_t(x)$ for $x \in [x_t, t/2]$ which is continuous since the function $g_{t,x}$ depends continuously on the parameter x. By the virtue of the implicit function Theorem, it is even at least C^1-piecewise curve since the derivative $g'_{t,x}(y)$ vanishes only in a finite set.
⋆ $x \in [t/2, 2]$: Using properties (1) and (3) of Lemma 2, we conclude that $g_{t,x}$ is increasing on $[0, \sqrt{x(2-x)}]$. Since $g_{t,x}(\sqrt{x(2-x)}) = e^t$ then the equation $g_{t,x}(y) = e^t$ has no solution in $[0, \sqrt{x(2-x)}[$.
⋆ $x \in]0, x_t[$: Since $x_t \in]0, 1[$ and $t/2 > 1 + (\sqrt{3}/2) > x_t$, then we make use of property (2) of Lemma 2. But the issue depends on whether or not y_t^- is positive. Assume y_t^- is negative then $g_{t,x}$ is decreasing on $[0, \sqrt{x(2-x)}[$ and thus the equation $g_{t,x}(y) = e^t$ has no solution in this interval. Otherwise $y_t^- > 0$ and $g_{t,x}$ keeps the same variations as in the range $x \in [x_t, t/2]$:

y	0		$(y_t^-)^{\frac{1}{2}}$		$\sqrt{x(2-x)}$
$g'_{t,x}(y)$		$+$		$-$	
$g_{t,x}(y)$	$g_{t,x}(0)$	↗		↘	e^t

We remark that $g_{t,x}(0) = e^t\, k(t,x)$ and according to the variations of $x \mapsto k(t,x)$ we get $k(t,x) > 1$ for $0 < x < x_t$. Consequently the equation $g_{t,x}(y) = e^t$ has no solution in $]0, x_t[$.
Finally, the above discussion shows the set $\{(x,y), g_{t,x}(y) = e^t, x \in]0,2[, 0 \leq y \leq \sqrt{x(2-x)}\}$ is described by a unique C^1-piecewise graph joining the points $(x_t, 0)$ and $z = \sqrt{t}e^{i \arccos(\sqrt{t}/2)}$. □

Remark 2. For $t = 4$, $g'_{t,\theta}(2\cos\theta) = 0$ if and only if $\theta = \theta_t = 0$. Thus both curves whose radii solve $g_{t,\theta}(r) = e^t$ meet at $\theta = 0$ thereby satisfy $r_t(0) = 2 > 1$. When $t > 4$, they even become disconnected.

4 Critical Points of h_t

Let z_t be a piecewise smooth parametrization of γ_t and consider $\theta \mapsto \arg[h_t(z_t(\theta))], \theta \in [-\pi, \pi]$. Using the invariance of γ_t under complex conjugation, we restrict our attention to $\theta \in [0, \pi]$. If $\theta \in \{0, \pi\}$ then $\arg(1-z) = 0, z \in \gamma_t$ since $r_t(0) \in]0, 1[$ therefore $\arg[h_t(z_t(0))] = 0$. Thus we discard these two values and consider $\theta \in (0, \pi)$. Then $\arg(1-z) \in (-\pi, 0)$ and we need to look for critical points of

$$\arg[h_t(z)] = \cot^{-1}\left[\frac{r\cos\theta - 1}{r\sin\theta}\right] - \pi - \frac{t}{r}\sin\theta$$

under the constraint $z = z_t = r_t(\theta)e^{i\theta} \in \gamma_t \cap \mathbb{C}^+$. For ease of notations, we shall omit the dependence on t of the radius of $\gamma_t, t \in]0,4[$ and write simply r_θ. Hence

$$\frac{d}{d\theta}\arg[h_t(r_\theta e^{i\theta})] = \frac{r_\theta^2 - r_\theta\cos\theta - r'_\theta\sin\theta}{r_\theta^2 - 2r_\theta\cos\theta + 1} - t\frac{r_\theta\cos\theta - r'_\theta\sin\theta}{r_\theta^2}$$

which vanishes if and only if

$$r_\theta\Big[r_\theta^3 - r_\theta^2\cos\theta - t\cos\theta(r_\theta^2 - 2r_\theta\cos\theta + 1)\Big] = r'_\theta\sin\theta\Big[(r_\theta^2 - t(r_\theta^2 - 2r_\theta\cos\theta + 1)\Big].$$

By the virtue of (1), the LHS may be written as

$$\frac{1}{2}r_\theta^3\partial_r(g_{t,\theta})(r_\theta)e^{-(2t\cos\theta)/r_\theta}$$

while

$$\partial_\theta(g_\theta)(r_\theta) = \frac{2\sin\theta}{r_\theta}\Big[r_\theta^2 - t(r_\theta^2 - 2r_\theta\cos\theta + 1)\Big]e^{(2t\cos\theta)/r_\theta}.$$

It follows that the following equality holds at any critical point

$$r_\theta^2\partial_r(g_{t,\theta})(r_\theta) = r'(\theta)\partial_\theta(g_{t,\theta})(r_\theta).$$

Now comes the constraint $g_{t,\theta}(r_\theta) = e^t$ that we shall differentiate with respect to θ to get

$$r'(\theta)\partial_r(g_{t,\theta})(r_\theta) + \partial_\theta(g_{t,\theta})(r_\theta) = 0.$$

Both identities yield

$$-r_\theta^2[\partial_r(g_{t,\theta})]^2(r_\theta) = [\partial_\theta(g_{t,\theta})]^2(r_\theta).$$

Since $r_\theta \neq 0$ then a critical value θ must satisfy

$$\partial_r(g_{t,\theta})(r_\theta) = \partial_\theta(g_{t,\theta})(r_\theta) = 0$$

which can not occur unless

$$r_\theta = 2\cos\theta.$$

As a result $\theta = \theta_t$ and one easily derives using $2\cos\theta_t = \sqrt{t}$

$$\arg[h_t(2\cos\theta_t e^{i\theta_t})] = 2\theta_t - \pi - \frac{1}{2}\sqrt{t(4-t)}$$
$$= 2\arccos\frac{\sqrt{t}}{2} - \pi - \frac{1}{2}\sqrt{t(4-t)}.$$

Finally

$$\cos\left[2\arccos\frac{\sqrt{t}}{2}-\pi\right]=-\cos\left[2\arccos\frac{\sqrt{t}}{2}\right]=1-\frac{t}{2}$$

whence we deduce that

$$\arg[h_t(2\cos\theta_t e^{i\theta_t})]=-\arccos\left(1-\frac{t}{2}\right)-\frac{1}{2}\sqrt{t(4-t)}=-\beta(t).$$

A similar analysis shows that $\theta\mapsto\arg[h_t(z_t(\theta))]$ has a unique critical point when $z_t\in\gamma_t\cap\mathbb{C}^-$. It is precisely given by $\theta=-\theta_t$ and

$$\arg[h_t(2\cos\theta_t e^{-i\theta_t})]=\beta(t).$$

We now proceed to the description of $\mu_t, t\in(0,4)$.

5 Description of $\mu_t, t\in(0,4)$

We have already seen that there are exactly two critical points of $\theta\mapsto\arg[h_t(z_t(\theta))], \theta\in[-\pi,\pi]$. This fact leads easily to:

Proposition 2. *There exists a partition $\gamma_t=\gamma_t^1\cup\gamma_t^2$ with*

$$\gamma_t^1\subset\{z,|z-1|\le 1\},\gamma_t^2\subset\{z,|z-1|\ge 1\}$$

and such that the maps $h_{t,1}\equiv h_t:\gamma_t^1\to h_t(\gamma_t)$ and $h_{t,2}\equiv h_t:\gamma_t^2\to h_t(\gamma_t)$ are diffeomorphisms. Moreover, let $e^{i\phi}\in h_t(\gamma_t)$ then the equation $h_t(z)=e^{i\phi}, z\in\gamma_t$ has exactly two solutions given by $z\in\gamma_t^1$ and

$$\frac{\overline{z}}{\overline{z}-1}\in\gamma_t^2.$$

Proof. The curves γ_t^1 and γ_t^2 are given by

$$\gamma_t^1=\{z\in\gamma_t,|z-1|\le 1\}$$
$$\gamma_t^2=\{z\in\gamma_t,|z-1|\ge 1\}.$$

It is clear that the critical points are located in the circle $\{z,|z-1|=1\}$ and therefore they are the end points of the curves γ_t^1 and γ_t^2. Therefore by the previous analysis of $\theta\mapsto\arg[h_t(z_t(\theta))]$ we deduce that $h_{t,1}, h_{t,2}$ are diffeomorphisms. Finally, for any $z\in\gamma_t$

$$h_t\left[\frac{\overline{z}}{\overline{z}-1}\right]=\frac{1}{\overline{h_t(z)}}=h_t(z).$$

We have used in the last identity the fact that $h_t(z) \in \mathbb{T}$. We point out that the möbius transform $z \mapsto \frac{z}{z-1}$ is a bijective map from $\gamma_{t,1}$ to $\gamma_{t,2}$. □

Thus we obviously have[1]

$$m_n(t) = \frac{1}{2i\pi t}\int_{\gamma_t^1}[h_t(z)]^n\frac{h_t'(z)}{h_t(z)}\log(1-z)dz + \frac{1}{2i\pi t}\int_{\gamma_t^2}[h_t(z)]^n\frac{h_t'(z)}{h_t(z)}\log(1-z)dz.$$

We perform the change of variables in the first integral: $h_{t,1}(z) = h_t(z) = e^{i\theta}$ then

$$i(d\theta) = \frac{h_t'(z)}{h_t(z)}dz.$$

Since $\arg[h_t(z_t(\theta))]$ reaches its minimum at $\theta = \theta_t$, then

$$\frac{1}{2i\pi t}\int_{\gamma_t^1}[h_t(z)]^n\frac{h_t'(z)}{h_t(z)}\log(1-z)dz = -\frac{1}{2\pi t}\int_{-\beta(t)}^{\beta(t)} e^{in\theta}\log(1-h_{t,1}^{-1}(e^{i\theta}))d\theta.$$

Similarly we get

$$\frac{1}{2i\pi t}\int_{\gamma_t^2}[h_t(z)]^n\frac{h_t'(z)}{h_t(z)}\log(1-z)dz = \frac{1}{2\pi t}\int_{-\beta(t)}^{\beta(t)} e^{in\theta}\log(1-h_{t,2}^{-1}(e^{i\theta}))d\theta,$$

consequently

$$m_n(t) = \frac{1}{2\pi t}\int_{-\beta(t)}^{\beta(t)} e^{in\theta}\log\left[\frac{1-h_{t,2}^{-1}(e^{i\theta})}{1-h_{t,1}^{-1}(e^{i\theta})}\right]d\theta.$$

The last part of Proposition 2 shows that:

$$h_{t,1}^{-1}(e^{i\theta}) = \overline{\left(\frac{h_{t,2}^{-1}(e^{i\theta})}{h_{t,2}^{-1}(e^{i\theta})-1}\right)}$$

implying that

$$\frac{1-h_{t,2}^{-1}(e^{i\theta})}{1-h_{t,1}^{-1}(e^{i\theta})} = |h_{t,2}^{-1}(e^{i\theta})-1|^2.$$

Together with $|h_{t,2}^{-1}(e^{i\theta})-1| \geq 1$ yield

$$m_n(t) = \frac{1}{\pi t}\int_{-\beta(t)}^{\beta(t)} e^{in\theta}\log|h_{t,2}^{-1}(e^{i\theta})-1|\,d\theta.$$

[1] γ_t is parametrized from θ_t to $2\pi - \theta$ counter-clockwise.

Thus

$$m_{-n}(t) = \overline{m_n(t)} = \frac{1}{\pi t}\int_{-\beta(t)}^{\beta(t)} e^{in\theta} \log|h_{t,2}^{-1}(e^{i\theta}) - 1|\, d\theta,\ n \geq 1$$

whence we deduce that spectral measure μ_t is given by

$$d\mu_t(\theta) = \frac{2}{t}\mathbf{1}_{[-\beta(t),\beta(t)]}(\theta)\log|h_{t,2}^{-1}(e^{i\theta}) - 1|\frac{d\theta}{2\pi} := \rho_t(\theta)d\theta.$$

Moreover, ρ_t is continuous since $\rho_t(-\beta(t)) = \rho_t(\beta(t)) = 0$ which follows from $h_{t,2}^{-1}(e^{\pm i\beta_t}) = 1 + e^{\mp 2i\theta_t}$.

Remark 3. Set

$$Z = \frac{2}{z} - 1$$

then

$$h_t(z) = h_t\left(\frac{2}{Z+1}\right) = \frac{Z-1}{Z+1}e^{tZ/2} = \tau_t^{-1}(Z)$$

where the last equality holds for

$$\Gamma_t := \{\Re(Z) > 0, |\tau_t^{-1}(Z)| < 1\} = \{|z-1| < 1, |h_t(z)| < 1\}$$

and extends to $\overline{\Gamma_t}$ (see [2]). It follows that

$$h_{t,1}^{-1} = \frac{2}{\tau_t + 1}$$

on the closed unit disc and one derives

$$\begin{aligned}\frac{2}{t}\log|h_{t,2}^{-1}(e^{i\theta}) - 1| &= -\frac{2}{t}\log|1 - h_{t,1}^{-1}(e^{i\theta})| \\ &= -\frac{2}{t}\log\left|\frac{\tau_t(e^{i\theta}) - 1}{\tau_t(e^{i\theta}) + 1}\right| \\ &= -\frac{2}{t}\log\left|e^{-t\tau_t(e^{i\theta})/2}\right| = \Re[\tau(e^{i\theta})]\end{aligned}$$

as stated in [2] p. 270.

Acknowledgements This work was partially supported by Agence Nationale de la recherche grant ANR-09-BLAN-0084-01.

References

1. P. Biane, Free Brownian Motion, Free Stochastic Calculus and Random Matrices. Fields Institute Communications, **12**, (American Mathematical Society Providence, RI, 1997), pp. 1–19
2. P. Biane, Segal-Bargmann transform, functional calculus on matrix spaces and the theory of semi-circular and circular systems. J. Funct. Anal. **144**(1), 232–286 (1997)
3. T. Lévy, Schur-Weyl duality and the heat kernel measure on the unitary group. Adv. Math. **218**(2), 537–575 (2008)

Another Failure in the Analogy Between Gaussian and Semicircle Laws

Nathalie Eisenbaum

Abstract We establish a new characterization of the Gaussian law that is not transferable to the non commutative probability spaces.

Keywords Gaussian law • Semicircle law • Free poisson distribution • Free probability theory • Free convolution • R-transform • Dynkin's isomorphism theorem

AMS 2000 Classification: 60J25, 60J55, 60G15, 60E07

1 Introduction and Main Results

Dynkin's isomorphism theorem and its variants relate the law of the local time process of a given transient symmetric Markov process with the law of a centered Gaussian process with covariance the Green function of this Markov process. We remind in particular the following version established in [4] for the local time process $(L_t^x, x \in E, t \geq 0)$ of a symmetric recurrent Markov process Z with state space E. Denote by o an element of E, by T_o its first hitting time by Z and set: $\tau(r) = \inf\{s \geq 0 : L_s^o > r\}$. The Green function of Z killed at T_o is a positive definite function on $E \times E$. Let $(\eta_x, x \in E)$ be a centered Gaussian process with covariance this Green function, independent of Z. We then have the following identity in law for every real r

$$(\frac{1}{2}(\eta_x + r)^2 \,, x \in E) \overset{(\text{law})}{=} (\frac{1}{2}\eta_x^2 \, + \, L^x_{\tau(\frac{r^2}{2})} \,, x \in E)$$

N. Eisenbaum (✉)
CNRS: UMR7599, Université Paris 6, 4 Place Jussieu 75252, cedex Paris 05, France
e-mail: nathalie.eisenbaum@upmc.fr

C. Donati-Martin et al. (eds.), *Séminaire de Probabilités XLIV*, Lecture Notes in Mathematics 2046, DOI 10.1007/978-3-642-27461-9_10,

which can easily be reformulated as follows: for every a, b in $\mathbb{R}$ such that $|b| \geq |a|$

$$(\frac{1}{2}(\eta_x + b)^2\,,\, x \in E) \stackrel{\text{(law)}}{=} (\frac{1}{2}(\eta_x + a)^2 \,+\, L^x_{\tau(\frac{b^2-a^2}{2})}\,,\, x \in E). \tag{1}$$

One immediate consequence of this isomorphism theorem is the existence of a real random variable X such that for every a, b in $\mathbb{R}$ such that $|b| \geq |a|$ there exists a random variable ℓ depending on (a, b) only via $(b^2 - a^2)$, independent of X such that

$$(X + b)^2 \stackrel{\text{(law)}}{=} (X + a)^2 \,+\, \ell\,. \tag{A}$$

The following theorem states that the above property actually characterizes the centered Gaussian variables.

Theorem 1. *A real centered random variable X satisfies Property A if and only if X has a centered Gaussian law.*

Seeking for a free analogue of the isomorphism theorem, one asks first about one of its consequences: the free analogue of Property A. To enunciate the answer (for a complete exposure on free probability, one can consult [7]), we consider a non commutative probability space $(\mathscr{A}, \phi)$ such that $\mathscr{A}$ is a C*-algebra and $\phi : \mathscr{A} \longrightarrow \mathbb{C}$ is a positive linear functional with $\phi(1) = 1$. An element x of $\mathscr{A}$ is called a random variable. A family of C*- subalgebras $\{\mathscr{A}_i, i \in I\}$ of $(\mathscr{A}, \phi)$ is said to be free if $\phi(x_1 x_2 ... x_n) = 0$ whenever $x_j \in \mathscr{A}_{i_j}$ with $i_1 \neq i_2 \neq ... \neq i_n$ and $\phi(x_j) = 0$ for $1 \leq j \leq n$. For $\{x_i, i \in I\}$ a family of random variables, $\{x_i, i \in I\}$ is said to be free if the family of the subalgebras generated respectively by $\{1, x_i\}$ is free.

One associates to each random variable x a linear functional μ_x on $\mathbb{C}[X]$ the algebra of polynomials, defined by $\mu_x(P(X)) = \phi(P(x)))$ for any $P \in \mathbb{C}[X]$. A random variable x is said to be centered if $\mu_x(X) = 0$. If x is self-adjoint then μ_x extends to a probability measure on $\mathbb{R}$. In particular the semicircle law centered at m and of radius $r > 0$ defined by its density $\frac{2}{\pi r^2} 1_{(m-r,m+r)} \sqrt{r^2 - (t-m)^2} dt$, corresponds to an important non commutative random variable. Indeed the central limit theorem for free random variables holds with limit distribution equal to a semicircle law. Hence, in non commutative probability the part of the Gaussian law is played by the semicircle law. A priori, one would expect from the semicircle law to satisfy the free analogue of Property A. But we have the following proposition.

Proposition 1. *There is no non null centered non commutative variable x such that for every a, b in $\mathbb{R}$ such that $|b| \geq |a|$, there exists a variable y depending of (a, b) only via $(b^2 - a^2)$ such that $\{x, y\}$ is free and*

$$(x \,+\, b)^2 \stackrel{\text{(law)}}{=} (x \,+\, a)^2 \,+\, y. \tag{2}$$

Proposition 1 shows that Theorem 1, minimal consequence of the isomorphism theorem, does not have a free analogue. Moreover Theorem 1 provides a characterization of the Gaussian law that is not working for the semi-circle law. This is not the only example of such failure. In [2], Bercovici and Voiculescu show that the characterization of the Gaussian law given by Cramér Theorem is not transferable to the semi-circle law. In [1], Benaych-Georges shows similarly that Raikov's characterization of the Poisson law is neither transferable to the free Poisson law.

One consequence of Property A for a centered Gaussian variable X, is that for every real m, $(X + m)^2$ is always infinitely divisible. In view of Proposition 1, one can then ask whether the free analogue of this weaker consequence of the isomorphism theorem holds for the semi-circle law. The answer is presented by the proposition below.

Proposition 2. *Let x be a non commutative variable with a centered semicircular law. Let m be a real number. Then $(x+m)^2$ is infinitely divisible for the free additive convolution if and only if $m = 0$.*

The original isomorphism theorem due to Dynkin holds for a symmetric transient Markov process Z with a state space E. Denote o an element of E, λ_o its last visit by Z, $\mathbb{P}_o$ the probability under which Z starts from o a.s. and $(L_t^x, x \in E, t \geq 0)$ the local times process of Z. Let $(\phi_x, x \in E)$ a centered Gaussian process with covariance the Green function of Z, independent of Z. Then under $\mathbb{P}_o$, we have:

$$(L_{\lambda_o}^x + \frac{1}{2}\psi_x^2, x \in E)$$

has the law of $(\frac{1}{2}\psi_x^2, x \in E)$ under the probability $\mathbb{E}(\frac{\psi_o^2}{\mathbb{E}(\psi_o^2)}, .)$.

As for the previous isomorphism theorem, one deduces the existence of a nonnegative integrable real random variable X such that there exists an independent random variable ℓ verifying:

$$(X + \ell) \text{ has the law of } X \text{ under } \mathbb{E}(\frac{X}{\mathbb{E}(X)}, .). \qquad \textbf{(B)}$$

But Property B does not characterize the squared centered Gaussian variables. Indeed, as an immediate consequence of Lemma 3.1 in [3], we have:

A nonnegative random variable has Property B
if and only if it is infinitely divisible.

For a given non commutative variable x with distribution μ on $\mathbb{R}^+$ such that $0 < \int_{\mathbb{R}^+} t\mu(dt) < \infty$, define the variable $T(x)$ with distribution $\frac{t\mu(dt)}{\int_{\mathbb{R}^+} s\mu(ds)}$. A natural free analogue of Property B for the Gaussian law would be the existence of a variable y such that $\{x, y\}$ is free and

$$T(x^2) \overset{\text{(law)}}{=} x^2 + y \qquad (3)$$

for x with a standard semicircular law. The distribution of the free variable x^2 is called the free Poisson distribution (or the Marchenko-Pastur distribution) with parameter 1.
We show in Sect. 2 that such a variable does not exist. To do so, we will first note that

$$T(x^2) \stackrel{(\text{law})}{=} x + 2\,.$$

2 The Proofs

We adopt the following notation. For x non commutative variable, its law is denoted by μ_x. For X classical random variable, its law is denoted by $\mathscr{L}(X)$.

Proof of Theorem 1. Any centered Gaussian variable X satisfies Property A thanks to the identity (1).
Conversely, let X be a random variable satisfying (A). One writes $\ell = \ell(b^2 - a^2)$. Since (A) gives:

$$(X + a)^2 \stackrel{(\text{law})}{=} X^2 + \ell(a^2),$$

with $\ell(a^2)$ independent of X, we obtain

$$\mathscr{L}((X + b)^2) = \mathscr{L}(X^2 + \ell(a^2)) * \mathscr{L}(\ell(b^2 - a^2)).$$

Consequently for every a, c in $\mathbb{R}$ we have:

$$\mathscr{L}(\ell(a^2 + c^2)) = \mathscr{L}(\ell(a^2)) * \mathscr{L}(\ell(c^2)), \tag{4}$$

Now denote by $X_1, X_2, ..., X_n$, n independent copies of X. For every $(a_i)_{1\leq i\leq n}$ element of $\mathbb{R}^n$, we have thanks to (A) and (4)

$$\mathscr{L}(\sum_{i=1}^{n}(X_i + a_i)^2) = \mathscr{L}(\sum_{i=1}^{n} X_i^2) * \mathscr{L}(\ell(a_1^2)) * ... * \mathscr{L}(\ell(a_n^2)).$$

Hence

$$\mathscr{L}(\sum_{i=1}^{n}(X_i + a_i)^2) = \mathscr{L}(\sum_{i=1}^{n} X_i^2) * \mathscr{L}(\ell(\sum_{i=1}^{n} a_i^2)),$$

which shows that the law of $\sum_{i=1}^{n}(X_i + a_i)^2$ depends of $(a_i)_{1\leq i\leq n}$ only via $(\sum_{i=1}^{n} a_i^2)$. This is precisely the characterization of the Gaussian law established by Kagan and Shalayaevski [6]. □

Proof of Proposition 1. Let x be a centered variable satisfying (2). We denote by $y(b^2 - a^2)$ the variable y of (2). One easily obtain for any real numbers a and b

$$\mu_{y(a^2)} \boxplus \mu_{y(b^2)} = \mu_{y(a^2+b^2)} \tag{5}$$

Hence $y(a^2)$ is infinitely divisible for the free additive convolution.
Now let $\{x_1, x_2, ..., x_n\}$ be a free family of self-adjoints elements identically distributed as x. Then, thanks to (2), we have for every $(a_k)_{1\leq k\leq n}$ element of $\mathbb{R}^n$

$$\mu_{\sum_{k=1}^n (x_k+a_k)^2} = \mu_{\sum_{k=1}^n x_k^2} \boxplus \mu_{y(a_1^2)} \boxplus ... \boxplus \mu_{y(a_n^2)} = \mu_{\sum_{k=1}^n x_k^2} \boxplus \mu_{y(\sum_{k=1}^n a_k^2)}.$$

Consequently the law of $\sum_{k=1}^n (x_k + a_k)^2$ depends of $(a_k)_{1\leq k\leq n}$ only via $\sum_{k=1}^n a_k^2$. This is the characterization of the centered semicircular law established by Hiwatashi et al. (Theorem 3.2 in [5]). Hence x must have a centered semicircular law.
Consider now x a variable with a centered semicircle law of radius $r > 0$. Assume that x satisfies (2). For every real m, one has then:

$$(x+m)^2 \overset{(\text{law})}{=} x^2 + y(m^2),$$

hence $(x + m)^2$ has to be infinitely divisible. Proposition 2 shows that this is not true. Consequently since the only possible distribution does not have the requested property (2), no distribution has this property. □

Proof of Proposition 2. It is well known that x^2 is infinitely divisible. Fix $m \neq 0$. It is clearly sufficient to consider the case where x has a semicircular law of radius 1. Hiwatashi et al. [5] have computed the R-transform of $(x + m)^2$:

$$\mathscr{R}_{(x+m)^2}(z) = \frac{1}{(4-z)} + \frac{4m^2}{(2-z)^2}. \tag{6}$$

For $z \in \mathbb{C}$ with $z = u + iv$ such that $v > 0$, $Im(\mathscr{R}_{(x+m)^2}(z))$ has the same sign as

$$P(v^2) = v^4 + 2(2-u)(4m^2 + 2 - u)v^2 + (2-u)(8m^2(4-u)^2 + (2-u)^3).$$

The discriminant of P is equal to

$$(2-u)[4m^2(2-u)(4m^2 + 2(2-u)) - 8m^2(4-u)^2]$$

which is positive for $2 < u < 2 + 2m^2$. Hence under this condition, P has two distinct roots. Their sum is equal to $S = 2(u-2)(4m^2 + 2 - u)$, which is positive. This implies that at least one root is positive and hence that one can choose $v > 0$ such that $P(v^2) < 0$. For this choice of (u, v) we have $Im(z) > 0$ and $Im(\mathscr{R}_{(x+m)^2}(z) < 0$. Hence thanks to [7] (Theorem 3.7.2 (1)), we know that $(x+m)^2$ can not be infinitely divisible. □

With the notation of the introduction, we have the following proposition under the assumption that the state ϕ is faithful.

Proposition 3. *Let x be a standard semicircular variable. There is no random variable y, such that $\{x, y\}$ is free and*

$$T(x^2) \stackrel{(\text{law})}{=} x^2 + y.$$

Proof of Proposition 3. The density of x is given by :

$$w(t) = \frac{1}{2\pi}\sqrt{4-t^2}1_{[-2,2]}(t).$$

The density of x^2 is given by:

$$u(t) = \frac{1}{2\pi t}\sqrt{4t-t^2}1_{(0,4]}(t).$$

One easily checks that for every f complex polynomial in one variable, we have

$$\mathbb{E}[f(x+2)] = \mathbb{E}[x^2 f(x^2)],$$

or equivalently: $x + 2 \stackrel{(\text{law})}{=} T(x^2)$.
Assume the existence of y such that $\{x, y\}$ is free and:

$$x^2 + y \stackrel{(\text{law})}{=} x + 2. \tag{7}$$

For any non commutative variable z, denote by $var(z)$ the quantity $\phi(z^2) - [\phi(z)]^2$. We have: $var(x+2) = var(x) = 1$. Besides, since $\{x, y\}$ is free, we have:

$$var(x^2 + y) = var(x^2) + var(y) = 1 + var(y).$$

Consequently, thanks to (7), we obtain: $var(y) = 0$, which leads to the existence of a real constant c such that: $x^2 \stackrel{(\text{law})}{=} x + c$. This last identity is obviously false. □

Acknowledgements We thank the referee for improvements on the original version.

References

1. F. Benaych-Georges, Failure of the Raikov theorem for free random variables. Séminaire de Probabilités XXXVIII, (Lecture Notes in Mathematics 1857, Springer, 2004), pp. 313–320
2. H. Bercovici, D. Voiculescu, Superconvergence to the central limit and failure of the Cramér theorem for free random variables. Probab. Theor. Relat. Field. **103**, 215–222 (1995)

3. N. Eisenbaum, A Cox process involved in the Bose-Eistein condensation. Annales Henri Poincaré **9**, 1123–1140 (2008)
4. N. Eisenbaum, H. Kaspi, M.B. Marcus, J. Rosen, Z. Shi, A Ray-Knight theorem for symmetric Markov processes. Ann. Probab. **28**(4), 1781–1796 (2000)
5. O. Hiwatashi, T. Kuroda, M. Nagisa, H. Yoshida, The free analogue of noncentral chi-square distributions and symmetric quadratic forms in free random variables. Math. Z. **230**(1), 63–77 (1999)
6. A.M. Kagan, O.V. Chalaevskii, A characterization of the normal law by a property of the non-central chi-square distribution. (Russian) Litovsk. Mat. Sb. **7**, 57–58 (1967)
7. D.V. Voiculescu, K.J. Dykema, A. Nica, *Free Random Variables. A Non Commutative Probability Approach to Free Products with Applications to Random Matrices, Operator Algebras and Harmonic Analysis on Free Groups*. CRM Monograph Series, 1. (American Mathematical Society, Providence, RI, 1992)

Global Solutions to Rough Differential Equations with Unbounded Vector Fields

Antoine Lejay

Abstract We give a sufficient condition to ensure the global existence of a solution to a rough differential equation whose vector field has a linear growth. This condition is weaker than the ones already given and may be used for geometric as well as non-geometric rough paths with values in any suitable (finite or infinite dimensional) space. For this, we study the properties the Euler scheme as done in the work of A.M. Davie.

Keywords Controlled differential equations • Rough paths • Euler scheme • Global solution to differential equation • Rough differential equation

AMS Classification: 60H10, 65C30

1 Introduction

Initiated a decade ago, the theory of rough paths imposed itself as a convenient tool to define stochastic calculus with respect to a large class of stochastic processes out of the range of semi-martingales (fractional Brownian motion, ...) and also allows one to define pathwise stochastic differential equations [6, 8, 10, 11, 13, 16, 18].

The idea to define integrals of differential forms along irregular paths or solutions of differential equations driven by irregular paths is to extend properly such paths in a suitable non-commutative space. These extensions encode in some sense the "iterated integrals" of the paths. Let us denote by x the driving path which lives in a

A. Lejay (✉)
Project-team TOSCA, Institut Elie Cartan Nancy (Nancy-Université, CNRS, INRIA), Boulevard des Aiguillettes B.P. 239 F-54506 Vandœuvre-lès-Nancy, France
e-mail: Antoine.Lejay@iecn.u-nancy.fr

C. Donati-Martin et al. (eds.), *Séminaire de Probabilités XLIV*, Lecture Notes in Mathematics 2046, DOI 10.1007/978-3-642-27461-9_11,

tensor space and Lie group $T(\mathbb{R}^d) := \mathbb{R} \oplus \mathbb{R}^d \oplus (\mathbb{R}^d \otimes \mathbb{R}^d)$ and that projects onto some continuous path of finite p-variations on $\mathbb{R}^d$ with $p \in [2,3)$. Such a path x is called a *p-rough path*.

The goal of this article is to study global existence of solutions to the rough differential equation (RDE)

$$y_t = y_0 + \int_0^t f(y_s)\,\mathrm{d}x_s \tag{1}$$

when f is not bounded. What is called a solution to (1) needs to be properly defined. Indeed, there exist two notions of solutions we will deal with (See Definitions 1 and 2 below).

Let us recall here the already existing results. The end of the introduction contains a short discussion about the differences between our result and the ones presented here.

Linear case: The special case of a linear vector field f was studied for example in the articles [1,6,9,12,15].

The original approach of T. Lyons: In the original approach, uniqueness and continuity of the Itô map $x \mapsto y$, where y is a solution to (1), is proved for a bounded function f which is twice differentiable with bounded derivatives and such that $\nabla^2 f$ is γ-Hölder continuous with $2+\gamma > p$ [6,10,11,13,16,18]. One knows from [2] that these conditions on f are essentially sharp. As pointed out soon after by A.M. Davie [2], existence of a solution to (1) is granted when f is bounded with a bounded derivative which is γ-Hölder continuous, $2+\gamma > p$ (See also [6,17] for example). However, in this case, one has to restrict to finite-dimensional spaces and several solutions may exist [2].

The approach of A.M. Davie: Providing an alternative approach to the one of T. Lyons based on a fixed point theorem, A.M. Davie studied in [2] the Euler scheme defined by

$$y_{i+1}^n = y_i^n + f(y_i^n)x_{t_i,t_{i+1}}^1 + f\cdot\nabla f(y_i^n)x_{t_i,t_{i+1}}^2$$

where $x_{s,t} = x_s^{-1} \otimes x_t$ and x^1 (resp. x^2) is the projection of x onto $\mathbb{R}^d$ (resp. $\mathbb{R}^d \otimes \mathbb{R}^d$). With conditions only on the regularity of the vector field, local existence is shown. Global existence is granted if there exist two positive increasing functions A and D on $[1,+\infty)$ with $D(R) \leq R^{1+\gamma}A(R)$, $2 \leq p < \gamma + 2 < 3$ such that

$$|f(y)| \leq D(R),\ |\nabla f(y) - \nabla f(z)| \leq A(R)|y-z|^\gamma,\ \forall y,z \text{ s.t. } |y|,|z| \leq R,$$

and

$$\int_1^{+\infty} \left(\frac{1}{A(R)^{p-1}D(R)^{1+p\gamma}}\right)^{\frac{1}{1+\gamma}} \mathrm{d}R = +\infty. \tag{2}$$

Besides, if (2) is not satisfied for some functions A and D as above, it is possible to construct a vector field f and a driver x such that the solution to (1) explodes in a finite time.

Applied to functions f with a γ-Hölder continuous derivative ∇f with $2+\gamma>p$, one may take $A(R)$ equal to a constant. Then global existence is granted for example if $D(R) = R^\delta$ with $\delta < (1+\gamma)/(1+\gamma p)$, but an explosion may occur for a function f is in class of functions if $\delta > (1+\gamma)/(1+\gamma p)$, which is the case for $\delta = 1$.

The approach by P. Friz and N. Victoir: In [5, 6], P. Friz and N. Victoir provide an alternative construction of the solutions of RDE that relies on sub-Riemannian geodesics and hence of geometric rough paths. The case of geometric rough paths shall be considered using $(p, p/2)$-rough paths [17]. In [6]*Exercise 10.61, they show that it is not necessary that the vector field is bounded, provided that f is Lipschitz continuous, ∇f is γ-Hölder continuous and $f \cdot \nabla f$ is also γ-Hölder continuous.

Using a fixed point approach: In [12], we have also studied the existence of a global solution, by using the approach on fixed point theorem.

With this approach, the RDE (1) is solved up to a finite "short time" horizon. Global existence follows from the convergence of a series related to the sum of the horizons. The complete conditions are cumbersome to write, but if $h(R) = \sup_{|y|\leq R} |f(y)|$, then global existence is granted provided that f has a bounded derivative ∇f which is γ-Hölder continuous with $2+\gamma > p$ and $h(R) \sim_{R\to\infty} R^\delta$, $0 < \delta \leq 1/p$ or $h(R) \sim_{R\to\infty} \log(R)$.

The case of a "regular enough" driver (also called Young case): If $1 \leq p < 2$, then global existence is granted if the vector field f is γ-Hölder continuous with $1+\gamma > p$ [14].

The one-dimensional approach of H. Doss and H. Sussmann: If $d = 1$, H. Doss [4] and H. Sussmann [20] have shown that a solution to an equation of type (1) may be defined by considering the solution to the ODE $\mathrm{d}z_t = f(z_t)\,\mathrm{d}t$ and setting $y_t = z_{x_t}$. Local existence to $\mathrm{d}z_t = f(z_t)\,\mathrm{d}t$ is granted provided that f is continuous. Global existence holds under a linear growth condition on f as it follows easily from an application of the Gronwall inequality [19] which provides a global bound on z: If $|f(z)| \leq A + B|z|$, then $|z_t| \leq (|z_0| + At)\exp(Bt)$ for any $t \geq 0$. This approach may be generalized for a multi-dimensional driver x when f has vanishing Lie brackets.

The situation is more intricate when x lives in a space of dimension bigger than 2. In particular, the asymptotic behavior of ∇f plays a fundamental role.

If x is a p-rough path (a rough path of finite p-variation) with $p \in [2,3)$, then for any function ϕ of finite $p/2$-variations with values in $\mathbb{R}^d \otimes \mathbb{R}^d$, $z = x + \phi$ is also a p-rough path. In addition, as shown in [17] for the general case, the solution to $y_t = y_0 + \int_0^t f(y_s)\,\mathrm{d}z_s$ is also solution to

$$y_t = y_0 + \int_0^t f(y_s)\,\mathrm{d}x_s + \int_0^t f \cdot \nabla f(y_s)\,\mathrm{d}\phi_s. \tag{3}$$

Consider any rough path z living above the path 0 on $\mathbb{R}^d$ with $\phi(t) = ct$ for a matrix c (if c is anti-symmetric, then such a rough path is the limit of a sequence of smooth paths lifted in the tensor space by their iterated integrals). Then y is solution to

$$y_t = y_0 + \int_0^t f \cdot \nabla f(y_s) c \, \mathrm{d}s.$$

Thus, explosion may occurs according to the behavior of $f \cdot \nabla f$. In particular, f may grow linearly, but $f \cdot \nabla f$ may grow faster that linearly, and an explosion may occur. Of course, if ∇f is bounded and f grows linearly, then $f \cdot \nabla f$ also grows linearly.

Example 1 (M. Gubinelli). Consider the solution y of the RDE $y_t = a + \int_0^t f(y_s) \, \mathrm{d}x_s$ living in $\mathbb{R}^2$ and driven by the rough path $x_t = (1, 0, (1 \otimes 1)t)$ with values in $1 \oplus \mathbb{R} \oplus (\mathbb{R} \otimes \mathbb{R})$. This rough path lies above the constant path at $0 \in \mathbb{R}$ and has only a pure area part which proportional to t. Note that this rough path can only be seen as a p-rough path with $p > 2$ [13]. Then y is also a solution to $y_t = a + \int_0^t (f \cdot \nabla f)(y_s) \, \mathrm{d}s$ (See [17]). The vector field $f \in \mathbb{R}^2 \to \mathrm{L}(\mathbb{R}, \mathbb{R}^2)$ given by

$$f(\xi) = (\sin(\xi_2)\xi_1, \xi_1), \ \xi = (\xi_1, \xi_2) \in \mathbb{R}^2$$

has a linear growth but

$$(f \cdot \nabla f)(\xi) = (\sin^2(\xi_2)\xi_1 + \xi_1^2 \cos(\xi_2), \sin(\xi_2)\xi_1)$$

has a quadratic growth. Take the initial point $a = (a_1, 0)$ with $a_1 > 0$. Then $(y_t)_2 = 0$ and $(y_t)_1 = a_1 + \int_0^t (y_s)_1^2 \, \mathrm{d}s$ so that $(y_t)_1 \to +\infty$ in finite time equal to $1/a_1$. This proves that explosion may occur in a finite time.

It proves that in the context of RDE, the growth of $f \nabla f$ is important. The previous example involves a pure non-geometric rough path. Yet a slight modification of this example show that this may happen also for geometric rough paths.

Example 2. Let us consider

$$y_t^n = a + \int_0^t \begin{bmatrix} (y^n)_s^1 \sin((y^n)_s^2) \\ 0 \end{bmatrix} \mathrm{d}\kappa_s^n + \int_0^t \begin{bmatrix} 0 \\ (y^n)_s^1 \end{bmatrix} \mathrm{d}\lambda_s^n \tag{4}$$

with

$$(\kappa^n, \lambda^n)_t = n^{-1/2}[\cos(2\pi n t), \sin(2\pi n t)].$$

The smooth rough path x^n living above (κ^n, λ^n) converges in p-variation with $p > 2$ to the rough path in $1 \oplus \mathbb{R}^2 \oplus (\mathbb{R}^2 \otimes \mathbb{R}^2)$ defined by

$$x_{s,t} = \left(1, 0, \begin{bmatrix} 0 & 2\pi \\ -2\pi & 0 \end{bmatrix} (t - s)\right).$$

This is the standard example which show the discontinuity of the Lévy area [18]. Let us assume that the solutions y^n to (4) is uniformly bounded on $[0, T]$ for some $T > 0$. By the continuity theorem (one may assume that f, ∇f and $\nabla^2 f$ are bounded), the solution y^n to (4) converges to the solution to

$$y_t = a + 2\pi \int_0^t \begin{bmatrix} -(y_s^1)^2 \cos(y_s^2) \\ y_s^1 \sin(y_s^2) \end{bmatrix} \mathrm{d}s.$$

Hence, we are in the same situation as Example 1 if $a^2 = 0$ and then $y^2 \equiv 0$. If $f(\xi) = (f_1(\xi), f_2(\xi))$ with $f_1(\xi) = (\xi^1 \sin(\xi^2), 0)$ and $f_2(\xi) = (0, \xi^1)$, then ∇f is not bounded although f grows linearly. Since y^n converges to y which explodes in a finite time $1/a^1$, y^n cannot be uniformly bounded on any time interval $[0, T]$ for T bigger than $1/a^1$. Note however that the p-variation norm of the rough path (x^n) remains bounded. Unlike the situations where the Doss-Sussmann approach may be used, bounds on the solution to RDE driven by a p-rough path with $p > 2$ cannot be derived from the sole information on the linear growth of f.

Although we use some general ideas already used in the context of rough paths theory (See for example [2,5,6,8,12,14], ...), we should note that:

- Our conditions on the vector field are weaker than the ones already given in the literature. In particular, we show that for existence, f and ∇f need to be continuous and ∇f needs to be bounded. Yet the Hölder regularity of ∇f plays no role here, while $f\nabla f$ shall be Hölder continuous. Even if the computations are very close to the one of [2,5], we then give a simple natural condition on the vector field as well as a simple bound on the solution.

The class of functions f with a bounded derivative ∇f which is γ-Hölder continuous is different from the class of functions f with a bounded derivative ∇f and such that $f\nabla f$ is γ-Hölder continuous.

Example 3. Let us consider $f(x) = \sin(x^2)/x$ for $|x| \geq 1$. The first and second order derivatives of f are

$$\nabla f(x) = 2\cos(x^2) - \frac{\sin(x)^2}{x^2}$$

$$\text{and } \nabla^2 f(x) = -4x\sin(x^2) - \frac{2\cos(x)\sin(x)}{x} + \frac{\sin(x^2)}{x^3}.$$

Hence, ∇f is not uniformly γ-Hölder continuous on $\{x; |x| \geq 1\}$ whatever $\gamma \in (0, 1]$. Yet

$$f(x)\nabla f(x) = \frac{2\cos(x^2)\sin(x^2)}{x} - \frac{\sin^2(x^2)}{x^3}$$

and it is easily checked that $f\nabla f$ has a bounded derivative on $\{x; |x| \geq 1\}$ and is then uniformly Lipschitz continuous (and thus γ-Hölder continuous) on $\{x; |x| \geq 1\}$.

The class of vector fields f with a bounded derivatives ∇f and such that $f\nabla f$ is γ-Hölder continuous enjoys the property to be stable under some change of variables. This is not necessarily the case for vector fields f with bounded first and second-order derivatives ∇f and $\nabla^2 f$.

For a toy example, let us consider the simple case of a function $f : \mathbb{R} \to \mathbb{R}$ such that ∇f is bounded and $L = f\nabla f$ has a bounded derivative. Assume that for a smooth path x, $y_t = a + \int_0^t f(y_s)\,\mathrm{d}x_s$ has a solution y which remains in $(1, +\infty)$. Set $z_t = \log(y_t)$. Then z is solution to $z_t = \log(a) + \int_0^t g(z_s)\,\mathrm{d}x_s$ with $g(z) = e^{-z} f(e^z)$. The vector field g has the interesting property to remain bounded on $(0, +\infty)$ and one may then hope to deduce some bounds on the solution y from some bound on the solution z.
Since $L = f\nabla f$ has a bounded derivative, g has a bounded derivative on $(0, +\infty)$ and $g\nabla g$ has also a bounded derivative, which means that $g\nabla g$ is globally Lipschitz. Yet

$$g''(z) = e^{-z} f(e^z) - f'(e^z) + e^z f''(e^z).$$

This means that g' is uniformly Lipschitz on $(0, +\infty)$—a condition required to deal with RDE with a regularity index $\gamma = 1$—only if $yf''(y)$ remains bounded, which is a stronger condition than assuming that f'' remains bounded.
This kind of computations may be carried to the multi-dimensional case using polar coordinates and an exponential change of variable of the radial component.
This explains the failure of the attempt carried by some persons, including the author of the present article, to get a global bound by such a change of variable under the sole assumption that ∇f is uniformly γ-Hölder and bounded.

- The condition on the vector field appears naturally when one uses the approach by the Euler scheme proposed by A.M. Davie in [2]. This is not the case when one uses the notion of solution proposed by T. Lyons because the term $f\nabla f$ is somewhat hidden in the cross iterated integral between the solution y and the driver x. The idea used in this article to get a global bound in closely related to the one in [12] where a fixed point approach was used. Yet in this article [12], we did not succeed article to obtain a global bound in a general. In addition, the superfluous assumption on the Hölder regularity of ∇f was used and necessary.
- The use of sub-Riemaniann geodesics and the reduction to smooth drivers was the core ideas of [5,6]. Here, there is no need to restrict to geometric rough paths and the results may be used for any finite-dimensional Banach spaces and even infinite-dimensional in some cases. Hence, the structure of the underlying spaces plays no role here.

Example 4. The most natural example of a non-geometric rough path is the one B living above a d-dimensional Brownian path W with $B^1 = W$ and $B^{2,i,j}_{s,t} = \int_s^t (W^i_r - W^i_s)\,\mathrm{d}B^j_r$. Here the integral has to be understood in the Itô sense. A geometric rough path $\widetilde{B}$ may be constructed above the Brownian path W with $\widetilde{B}^1 = W$ and $\widetilde{B}^{2,i,j}_{s,t} = \int_s^t (W^i_r - W^i_s) \circ \mathrm{d}W^j_r$, where the stochastic integral is the Stratonovich one. Hence, $\widetilde{B}$ and B are linked by $B^{2,i,j}_{s,t} = \widetilde{B}^{2,i,j}_{s,t} + \frac{1}{2}(t-s)\delta_{i,j}$. RDE driven by B correspond to Itô SDE while the ones driven by $\widetilde{B}$ corresponds to Stratonovich SDE.
When one use the Brownian rough path B, the Euler scheme presented here corresponds indeed to the Milstein scheme. In the very beginning of rough paths

theory, the rate of convergence in this case has been studied by J. Gaines and T. Lyons [7] with aim at developing simulation algorithms with adaptative stepsize.

In view of (3), the regularity condition on $f\nabla f$ is also the one which is necessary to deal with a non-geometric p-rough path seen as a $(p, p/2)$-geometric rough path using the decomposition of the space introduced in [17].

• The notion of solution to a RDE introduced by T. Lyons [18] cannot be used here (See Definition 1), so that we use the notion of solution introduced by A.M. Davie in [2] (see Definition 2) which is similar to the one proposed by M. Gubinelli [8]. We show in Sect. 10 that the two notions of solutions coincide when ∇f is γ-Hölder continuous.

• A general conclusion to draw from the cited works on global existence for non bounded vector fields f is that a variety of results could be given according to the behaviour at infinity of f and its derivatives. Hence, the growth of f is not the only factor to look at.

2 Notations and Hypotheses

Let ω be a *control*. By this, we mean a function defined from $\Delta^2 := \{0 \leq s \leq t \leq T\}$ to $\mathbb{R}_+$ which is continuous close to its diagonal and super-additive

$$\omega(s,r) + \omega(r,t) \leq \omega(s,t),\ 0 \leq s < r < t \leq T.$$

For the sake of simplicity, we assume here that ω is continuous and that $\omega(s,t) > 0$ as soon as $t > s$.

Let x be a path with values in $T(\mathbb{R}^d) = \mathbb{R} \oplus \mathbb{R}^d \oplus (\mathbb{R}^d \otimes \mathbb{R}^d)$. We set $x_{s,t} = x^{-1} \otimes x_t$. The part of x in $\mathbb{R}^d$ is denoted by x^1 and the part in $\mathbb{R}^d \otimes \mathbb{R}^d$ is denoted by x^2. Then x is a *rough path of finite p-variation controlled by ω* when the quantity

$$\|x\| := \sup_{0 \leq s < t \leq T} \max \left\{ \frac{|x^1_{s,t}|}{\omega(s,t)^{1/p}}, \frac{|x^2_{s,t}|}{\omega(s,t)^{2/p}} \right\}$$

is finite for a fixed p.

If $\omega(s,t) = t - s$, then we work indeed with paths that are $1/p$-Hölder continuous.

Throughout all this article, we consider only the case where $p \in [2,3)$. The case $p \leq 2$ is covered for example by [14].

For a path y with values in $\mathbb{R}^m$, we set

$$\|y\| := \sup_{0 \leq s < t \leq T} \frac{|x^1_{s,t}|}{\omega(s,t)^{1/p}}.$$

A vector field is an application f which is linear from $\mathbb{R}^m$ to $\mathrm{L}(\mathbb{R}^d, \mathbb{R}^m)$, the space of linear applications from $\mathbb{R}^d$ to $\mathbb{R}^m$. With indices, we set $f(x) = e_i f_j^i(x)\widehat{e}_j^*$, where $\{e_i\}_{i=1,\dots,m}$ is the canonical basis of $\mathbb{R}^m$, and $\{\widehat{e}_j^*\}_{j=1,\dots,d}$ is the dual of the canonical basis of $\mathbb{R}^d$. We set

$$f \cdot \nabla f(x) := \sum_{i=1}^m e_i \sum_{k=1}^d \frac{\partial f_j^i}{\partial x_k}(x) f_\ell^k(x) \widehat{e}_\ell^* \otimes \widehat{e}_j^*,$$

which means that $f \cdot \nabla f$ is an application from $\mathbb{R}^m$ to $\mathrm{L}(\mathbb{R}^d \otimes \mathbb{R}^d, \mathbb{R}^m)$.

Hypothesis 1. The function f is continuously differentiable from $\mathbb{R}^m$ to $\mathrm{L}(\mathbb{R}^d, \mathbb{R}^m)$ and is such that ∇f is bounded and $F(x) := f(x) \cdot \nabla f(x)$ is γ-Hölder continuous with norm $H_\gamma(F)$ from $\mathbb{R}^m$ to $L(\mathbb{R}^m \otimes \mathbb{R}^d, \mathbb{R}^m)$.

Note that this hypothesis is slightly different from the one given usually to prove existence to solutions of RDE where it is assumed that ∇f is γ-Hölder continuous.

Hypothesis 2. We assume that $p \in [2, 3)$ and that $\theta := (2+\gamma)/p$ is greater than 1.

Definition 1 (Solution in the sense of Lyons). We call by a *solution to* (1) *in the sense of Lyons* the projection onto $\mathbb{R}^m$ of a rough path z of finite p-variations controlled by ω with values in $T(\mathbb{R}^m \oplus \mathbb{R}^d)$ which solves the following equation

$$z_t = z_0 + \int_0^t g(z_s)\, \mathrm{d}z_s, \tag{5}$$

where z projects onto $T(\mathbb{R}^d)$ as x, onto $\mathbb{R}^m$ as y, and g is the differential form $g(y, x) = \mathrm{d}x + f(y)\,\mathrm{d}x$. The integral in (5) shall be understood as the "rough integral", that is as an integral in the sense of rough path.

Note that the definitions of z involves the "cross-iterated integrals" between x and y, and requires that the derivative of f is γ-Hölder continuous.

Under Hypothesis 1, it is not compulsory that ∇f is γ-Hölder continuous, so that we shall use another notion of solution, since we cannot use a fixed point theorem that relies on the definition of a rough integral.

3 Solution in the Sense of Davie

The notion of solution of (1) we use is the notion of solution in the sense of Davie, introduced in [2].

Definition 2 (Solution in the sense of Davie). A solution of (1) in the sense of Davie is a continuous path y from $[0, T]$ to $\mathbb{R}^m$ of finite p-variation controlled by ω such that for some constant L,

$$|y_t - y_s - \mathfrak{D}(s,t)| \le L\omega(s,t)^\theta,\ \forall (s,t) \in \Delta^2 \tag{6}$$

with

$$\mathfrak{D}(s,t) = f(y_s)x^1_{s,t} + F(y_s)x^2_{s,t}.$$

The next propositions, which assume the existence of a solution in the sense of Davie, will be proved below in Sect. 4

Proposition 1. *Let y be a solution of* (1) *in the sense of Davie under Hypotheses 1 and 2. Then $\mathfrak{D}(s,t)$ is an almost rough path whose associated rough path is y and for a family of partitions $\{\{t_i^n\}_{i=0,\dots,n}\}_{n\in\mathbb{N}^*}$ of $[0,t]$ whose meshes decrease to 0,*

$$y_t = y_0 + \lim_{n\to\infty} \sum_{i=0}^{n} \mathfrak{D}(t_i^n, t_{i+1}^n). \tag{7}$$

Proposition 2 (Boundedness of the solution). *Let y be a solution to* (1) *in the sense of Davie under Hypotheses 1 and 2. Then $\|y\|$ and $\|y\|_\infty$ are bounded by some constants that depend only on $\|\nabla f\|_\infty$, $H_\gamma(F)$, $|f(y_0)|$, $|y_0|$, $\|x\|$, $\omega(0,T)$, γ and p. More precisely, there exist some constants C_1 depending only on $\|\nabla f\|_\infty$, $H_\gamma(F)$, $\|x\|$, p and γ and C_2 depending only on $|f(0)|$, $\|\nabla f\|_\infty$, $H_\gamma(F)$, $\|x\|$, p, γ such that*

$$\sup_{t\in[0,T]} |y_t| \le R(T)|y_0| + C_2(R(T)-1)$$

with

$$R(T) \le \exp(C_1 \max\{\omega(0,T)^{1/p}, \omega(0,T)\}).$$

4 Proofs of Propositions 1 and 2

Let us set for $0 \le s \le r \le t \le T$,

$$\mathfrak{D}(s,r,t) := \mathfrak{D}(s,t) - \mathfrak{D}(s,r) - \mathfrak{D}(r,t).$$

Then the main idea of the proofs (as well as the proofs of the other theorems) are the following: First, we find a function $C(\|y\|, L, T)$ such that $|\mathfrak{D}(s,r,t)| \le C(\|y\|, L, T)\omega(s,t)^\theta$ for all $(s,t) \in \Delta^2$. From the sewing Lemma (See for example [13]*Theorem 5, p. 89), after having shown that y is the rough path corresponding to the almost rough path $(\mathfrak{D}(s,t))_{(s,t)\in\Delta^2}$, we get that for some universal constant M,

$$|y_t - y_s - \mathfrak{D}(s,t)| \le MC(\|y\|, L, T)\omega(s,t)^\theta.$$

Then, after having estimated $|\mathfrak{D}(s,t)\| \le C'(\|y\|, L, T)\omega(s,t)^{1/p}$, we get an inequality of type

$$\|y\| \le MC(\|y\|, L, T)\omega(0,T)^{\theta-1/p} + C'(\|y\|, L, T).$$

A careful examination of the functions $C(\|y\|, L, T)$ and $C'(\|y\|, L, T)$ shows that indeed L itself depends on y_0, $\|y\|$ and T and that

$$\|y\| \leq A(y_0, T) + B(T)\|y\|$$

with $B(T)$ decreasing to 0 as T decreases to 0. Then, choosing T small enough implies that $\|y\|$ is bounded in small time and then for any time T using the arguments presented in appendix. This idea is the central one used for example in [12, 14].

Let us set

$$B_1 := \sup_{t\in[0,T]} |F(y_t)|,\ B_2(s,t) := y_t - y_s - f(y_s)x^1_{s,t},$$
$$B_3(a,b) := f(b) - f(a),\ B_4(a,b) := F(b) - F(a),$$
$$\text{and } B_5(s,t) := f(y_t) - f(y_s) - F(y_s)x^1_{s,t}$$

as well as

$$\mu := \omega(0,T)^{1/p}.$$

Remark 1. The quantity μ will be used to denote the "short time" horizon and is the central quantity for getting our estimates.

Since F is γ-Hölder continuous,

$$B_1 \leq |F(y_0)| + H_\gamma(F)\|y\|^\gamma \mu^\gamma.$$

If y is a solution in the sense of Davie with a constant L,

$$|B_2(s,t)| \leq L\omega(s,t)^\theta + B_1\|x\|\omega(s,t)^{2/p}.$$

Since f is Lipschitz continuous,

$$|B_3(a,b)| \leq \|\nabla f\|_\infty |b-a|.$$

In addition,

$$|B_4(a,b)| \leq H_\gamma(F)|b-a|^\gamma.$$

Finally,

$$\begin{aligned} f(y_t) - f(y_s) &= \int_0^1 \nabla f(y_s + \tau y_{s,t}) y_{s,t}\,\mathrm{d}\tau \\ &= \int_0^1 \nabla f(y_s + \tau y_{s,t}) f(y_s) x^1_{s,t}\,\mathrm{d}\tau + \int_0^1 \nabla f(y_s + \tau y_{s,t}) B_2(s,t)\,\mathrm{d}\tau \\ &= -\int_0^1 \nabla f(y_s + \tau y_{s,t})(f(y_s + \tau y_{s,t}) - f(y_s)) x^1_{s,t}\,\mathrm{d}\tau \end{aligned}$$

$$+\int_0^1 F(y_s+\tau y_{s,t})x^1_{s,t}\,\mathrm{d}\tau+\int_0^1 \nabla f(y_s+\tau y_{s,t})B_2(s,t)\,\mathrm{d}\tau$$

$$=\int_0^1 F(y_s+\tau y_{s,t})x^1_{s,t}\,\mathrm{d}\tau+\int_0^1 \nabla f(y_s+\tau y_{s,t})(B_2(s,t)$$
$$-B_3(y_s,y_s+\tau y_{s,t})x^1_{s,t})\,\mathrm{d}\tau$$

$$=F(y_s)x^1_{s,t}+\int_0^1 B_4(y_s,y_s+\tau y_{s,t})x^1_{s,t}\,\mathrm{d}\tau$$
$$+\int_0^1 \nabla f(y_s+\tau y_{s,t})(B_2(s,t)-B_3(y_s,y_s+\tau y_{s,t})x^1_{s,t})\,\mathrm{d}\tau. \quad (8)$$

This proves that

$$|B_5(s,t)|\le H_\gamma(F)\|y\|^\gamma\|x\|\omega(s,t)^{(1+\gamma)/p}+\|\nabla f\|^2_\infty\|x\|\|y\|\omega(s,t)^{2/p}$$
$$+\|\nabla f\|_\infty L\omega(s,t)^\theta+\|\nabla f\|_\infty|F(y_0)|\|x\|\omega(s,t)^{2/p}$$
$$+H_\gamma(F)\|y\|^\gamma\omega(0,T)^{\gamma/p}\|x\|\omega(s,t)^{2/p}.$$

Lemma 1. *For any* $0\le s\le r\le t\le T$,

$$|\mathfrak{D}(s,r,t)|\le(C_3(\mu)\|y\|^\gamma+C_4(\mu)\|y\|+C_5(\mu)L+C_6(\mu,y_0))\omega(s,t)^\theta$$

with

$$C_3(\mu):=H_\gamma(F)\big(\|x\|^2(1+\mu^{1+\gamma})+\|x\|\big),$$
$$C_4(\mu):=\|\nabla f\|^2_\infty\|x\|^2\mu,$$
$$C_5(\mu):=\|\nabla f\|_\infty\|x\|\mu$$
$$\text{and } C_6(\mu,y_0):=\|\nabla f\|_\infty|F(y_0)|\|x\|^2\mu^{1-\gamma}\le|f(y_0)|\|\nabla f\|^2_\infty\|x\|^2\mu^{1-\gamma}.$$

Proof. With $x^2_{s,t}=x^2_{s,r}+x^2_{r,t}+x^1_{s,r}\otimes x^1_{r,t}$,

$$\mathfrak{D}(s,r,t)=(f(y_s)-f(y_r))x^1_{r,t}+F(y_s)x^1_{s,r}\otimes x^1_{r,t}+(F(y_s)-F(y_r))x^2_{r,t}.$$

Hence

$$|\mathfrak{D}(s,r,t)|\le|B_5(s,r)|\|x\|\omega(s,t)^{1/p}+|B_4(y_s,y_r)|\|x\|\omega(s,t)^{2/p}.$$

This proves the result. □

Lemma 2. *For any* $(s,t)\in\Delta^2$,

$$|\mathfrak{D}(s,t)|\le(C_7(\mu,y_0)+C_8(\mu)\|y\|^\gamma+C_9(\mu)\|y\|)\omega(s,t)^{1/p}$$

with

$$\begin{aligned}C_7(\mu, y_0) &:= (|f(y_0)| + |F(y_0)|\mu)\|x\| \leq |f(y_0)|(1 + \|\nabla f\|_\infty \mu)\|x\|,\\ C_8(\mu) &:= H_\gamma(F)\|x\|\mu^\gamma,\\ \textit{and } C_9(\mu) &:= \|\nabla f\|_\infty \|x\|\mu.\end{aligned}$$

Proof. This follows from

$$\begin{aligned}|\mathfrak{D}(s,t)| \leq |f(y_0)|\|x\|\omega(s,t)^{1/p} + \|y\|\|x\|\|\nabla f\|_\infty \omega(s,t)^{2/p}\\ +|F(y_0)|\|x\|\omega(s,t)^{2/p} + \|y\|^\gamma \|x\| H_\gamma(F)\omega(s,t)^{(1+\gamma)/p}.\end{aligned}$$

This proves the Lemma. □

We have now all the required estimates to prove Propositions 1 and 2.

Proof (Proof of Proposition 1). It follows from Lemma 1 that $\{\mathfrak{D}(s,t)\}_{(s,t)\in\Delta^2}$ is an almost rough path. From the sewing lemma (See for example [13]*Theorem 5, p. 89), there exists a path $\{z_t\}_{t\in[0,T]}$ as well as a constant M depending only on θ such that

$$|z_t - z_s - \mathfrak{D}(s,t)| \leq M\omega(s,t)^\theta,\ \forall (s,t) \in \Delta^2. \tag{9}$$

This function is unique in the class of functions satisfying (9) and with (6), z is equal to y. Equality (7) follows from the very construction of z. □

Proof (Proof of Proposition 2). We assume first that y is a solution in the sense of Davie with constant L. Let us note that

$$|y_t - y_s| \leq |y_t - y_s - \mathfrak{D}(s,t)| + |\mathfrak{D}(s,t)| \leq L\omega(s,t)^\theta + |\mathfrak{D}(s,t)|.$$

With Lemma 2, if $\mu = \omega(0,T)^{1/p}$ is small enough such that

$$C_8(\mu) \leq \frac{1}{4} \text{ and } C_9(\mu) \leq \frac{1}{4}, \tag{10}$$

then

$$\|y\| \leq L\mu^{1+\gamma} + C_7(\mu, y_0) + \frac{1}{4}\|y\| + \frac{1}{4}\|y\|^\gamma.$$

Since $\gamma \leq 1$, if $\|y\| \geq 1$, then $\|y\|^\gamma \leq \|y\|$ and then for μ small enough (note that the choice of μ does not depend on y_0),

$$\|y\| \leq 2\max\{1, L\mu^{1+\gamma} + C_7(\mu, y_0)\}. \tag{11}$$

Since $C_7(\mu, y_0)$ decreases with μ, the boundedness of $\|y\|$ and $\|y\|_\infty$ also hold, with a different constant, for any time T by applying Proposition 6 in appendix.

Now, if y is a solution in the sense of Davie with a constant L, it is also a solution in the sense of Davie with the constant

$$L' := \sup_{(s,t)\in\Delta^2,\ s\neq t} \frac{|y_t - y_s - \mathfrak{D}(s,t)|}{\omega(s,t)^\theta}.$$

From the sewing Lemma, there exists a universal constant M depending only on θ such that

$$L' \leq M \sup_{(s,t)\in\Delta^2,\ s\neq t,\ r\in(s,t)} \frac{|\mathfrak{D}(s,r,t)|}{\omega(s,t)^\theta}.$$

From Lemma 1 and the inequalities $a^\gamma \leq 1 + a$ for $a \geq 0$ and $\gamma \in [0,1]$ as well as $|f(y_0)| \leq |f(0)| + \|\nabla f\|_\infty |y_0|$,

$$\begin{aligned} L' \leq& M(C_3(\mu)\|y\|^\gamma + C_4(\mu)\|y\| + C_5(\mu)L' + C_6(\mu, y_0)) \\ \leq& C_{10}(\mu) + C_{13}(\mu)\|y\| + MC_5(\mu)L' + C_{12}(\mu)|y_0|, \end{aligned}$$

with

$$\begin{aligned} C_{10}(\mu) &:= MC_3(\mu) + |f(0)|C_{11}(\mu), \\ C_{11}(\mu) &:= M\|\nabla f\|_\infty^2 \|x\|^2 \mu^{1-\gamma}, \\ C_{12}(\mu) &:= \|\nabla f\|_\infty C_{11}(\mu), \\ \text{and } C_{13}(\mu) &:= M(C_3(\mu) + C_4(\mu)). \end{aligned}$$

If

$$MC_5(\mu) \leq 1/2, \tag{12}$$

then

$$L' \leq 2C_{10}(\mu) + 2C_{13}(\mu)\|y\| + 2C_{12}(\mu)|y_0|.$$

With (11), under conditions (10) and (12) on μ,

$$\begin{aligned} \|y\| \leq 2 + 4C_{10}(\mu)\mu^{1+\gamma} + 4C_{13}(\mu)\mu^{1+\gamma}\|y\| + 4C_{12}(\mu)|y_0|\mu^{1+\gamma} \\ + C_7(\mu, y_0), \end{aligned}$$

Under the additional condition that

$$4C_{13}(\mu)\mu^{1+\gamma} \leq \frac{1}{2}, \tag{13}$$

we get that

$$\|y\| \leq C_{14}(\mu) + C_{15}(\mu)|y_0|$$

with

$$\begin{aligned} C_{14}(\mu) &:= 4 + 8C_{10}(\mu)\mu^{1+\gamma} + 2|f(0)|(1 + \|\nabla f\|_\infty \mu)\|x\|, \\ C_{15}(\mu) &:= 8C_{12}(\mu)\mu^{1+\gamma} + 2\|\nabla f\|_\infty (1 + \|\nabla f\|_\infty \mu)\|x\|. \end{aligned}$$

Due to the dependence of the constants with respect to μ, conditions (10), (12) and (13) hold true if $\mu \leq K$ for a constant K depending only on $\|\nabla f\|_\infty$, $H_\gamma(F)$, $\|x\|$, p and γ.

Indeed, (11) also holds for $y_{|[S,S']}$ on any time interval $\omega(S,S')$ provided that $\omega(S,S')^{1/p}$ is small enough. The result follows from Proposition 7 in appendix.

□

5 Existence of a Solution

The existence of a solution is proved thanks to the Euler scheme, which allows one to study to define a family of paths that is uniformly bounded with the uniform norm.

Let us fix a partition $\{t_i\}_{i=0,\dots,n}$ of $[0,T]$ and let us set $x_{i,j} := x_{t_i}^{-1} \otimes x_{t_j}$ and $\omega_{i,j} := \omega(t_i,t_j)$.

Let us consider the Euler scheme

$$y_{i+1} = y_i + f(y_i)x^1_{i,i+1} + F(y_i)x^2_{i,i+1},\ i = 0,\dots,n-1$$

as well as the family $\{y_{i,j}\}_{0\leq i<j\leq n}$ defined by

$$y_{i,j} := f(y_i)x^1_{i,j} + F(y_i)x^2_{i,j}.$$

We set

$$\|y\|_{\star,0,n} := \sup_{0\leq i<j\leq n} \frac{|y_j - y_i|}{\omega_{i,j}^{1/p}}. \tag{14}$$

Lemma 3. *If $\mu := \omega_{0,n}^{1/p}$ is small enough so that*

$$C_5(\mu) = \|\nabla f\|_\infty \|x\|\mu \leq \frac{1-K}{4} \text{ with } K := 2^{1-\theta} < 1 \tag{15}$$

and

$$L := 4\frac{C_6(\mu,y_0) + C_3(\mu)\|y\|^\gamma_{\star,0,n} + C_4(\mu)\|y\|_{\star,0,n}}{1-K}$$

then

$$|y_{i,k} - y_k - y_i| \leq L\omega^\theta_{i,k} \tag{16}$$

for all $0 \leq i \leq k \leq n$.

Proof. The proof of this Lemma follows along the lines the one of Lemma 2.4 in [2] and relies on an induction on $k-i$.

Clearly, (16) is true for $k = i+1$. Fix $m > 1$ and let us assume that (16) holds for any $i < k$ such that $k - i < m$.

Let us choose $i < k$ such that $k - i = m$, $m \geq 2$. Let j be the index such that $\omega_{i,j} \leq \omega_{i,k}/2$ and $\omega_{i,j+1} > \omega_{i,k}/2$. Since ω is super-additive, $\omega_{j+1,k} \leq \omega_{i,k}/2$ and then

$$\omega_{i,j}^{\theta} + \omega_{j+1,k}^{\theta} \leq 2^{1-\theta}\omega_{i,k}^{\theta} = K\omega_{i,k}^{\theta}. \tag{17}$$

We set

$$y_{i,j,k} := y_{i,k} - y_{i,j} - y_{j,k}.$$

For j as above, since $y_{j,j+1} - y_j - y_{j+1} = 0$, and using (17),

$$\begin{aligned}
|y_{i,k} - y_k + y_i| &\leq |y_{i,j,k}| + |y_{i,j} - y_j + y_i| + |y_{j,k} - y_k + y_j| \\
&\leq |y_{i,j,k}| + |y_{j,j+1,k}| + |y_{j+1,k} - y_k + y_{j+1}| + |y_{i,j} - y_i + y_j| \\
&\quad + |y_{j,j+1} - y_{j+1} + y_j| \\
&\leq |y_{i,j,k}| + |y_{j,j+1,k}| + LK\omega_{i,k}^{\theta}.
\end{aligned}$$

Since $x_{i,k}^2 = x_{i,j}^2 + x_{j,k}^2 + x_{i,j}^1 \otimes x_{j,k}^1$,

$$y_{i,j,k} = (f(y_j) - f(y_i))x_{j,k}^1 + (F(y_j) - F(y_i))x_{j,k}^2 + F(y_i)x_{i,j}^1 \otimes x_{j,k}^1.$$

Using the same computations as (8),

$$\begin{aligned}
y_{i,j,k} &= (f(y_i) - f(y_j))x_{i,j}^1 - F(y_i)x_{i,j}^1 \otimes x_{j,k}^1 + (F(y_i) - F(y_j))x_{j,k}^2 \\
&= \int_0^1 (F(y_i + \tau(y_j - y_i)) - F(y_i))x_{i,j}^1 \otimes x_{j,k}^1 \,\mathrm{d}\tau \\
&\quad + \int_0^1 \nabla f(y_i + \tau(y_j - y_i))(y_j - y_i - y_{i,j} + F(y_i)x_{i,j}^2)x_{i,j}^1 \,\mathrm{d}\tau \\
&\quad - \int_0^1 \int_0^1 \tau \nabla f(y_i + \tau(y_j - y_i))\nabla f(y_i + \rho\tau(y_j - y_i))(y_j - y_i)\,\mathrm{d}\rho\,\mathrm{d}\tau x_{i,j}^1 \otimes x_{j,k}^1 \\
&\quad + (F(y_i) - F(y_j))x_{j,k}^2.
\end{aligned} \tag{18}$$

We then face the same estimates as the one in the proof of Lemma 1, where we replace the fact that y is a solution in the sense of Davie with a constant L by our induction hypothesis on $|y_{i,j} - y_j + y_i|$. Then

$$|y_{i,j,k}| \leq (C_3(\mu)\|y\|_{\star,0,n}^{\gamma} + C_4(\mu)\|y\|_{\star,0,n} + C_5(\mu)L + C_6(\mu, y_0))\omega_{i,k}^{\theta}.$$

The results follows from our choice of μ and L. □

The next lemma is the equivalent of Proposition 2 for the Euler scheme.

Lemma 4. *For n such that $t_n = T$, $\|y\|_{\star,0,n}$ is bounded by a constant that depends only on $\omega(0,T)$, $\|x\|$, $\|\nabla f\|_{\infty}$, $H_{\gamma}(F)$, γ, p and $|f(0)|$.*

Proof. Using Lemma 2,

$$|y_{i,j}| \leq (C_7(\mu, y_0) + C_8(\mu)\|y\|_{\star,0,n}^{\gamma} + C_9(\mu)\|y\|_{\star,0,n})\omega_{i,j}^{1/p}$$

and then with Lemma 3,

$$\|y\|_{\star,0,n} \leq (C_7(\mu, y_0) + C_8(\mu)\|y\|_{\star,0,n}^{\gamma} + C_9(\mu)\|y\|_{\star,0,n}) + 4\mu^{1+\gamma}\frac{C_6(\mu, y_0) + C_3(\mu)\|y\|_{\star,0,n}^{\gamma} + C_4(\mu)\|y\|_{\star,0,n}}{1-K}.$$

In addition to (15), we choose μ small enough so that

$$C_8(\mu) + 4(1-K)^{-1}\mu^{1+\gamma}C_3(\mu) \leq \frac{1}{4} \tag{19}$$

and

$$C_9(\mu) + 4(1-K)^{-1}\mu^{1+\gamma}C_3(\mu) \leq \frac{1}{4}. \tag{20}$$

Since $\gamma \leq 1$, $\|y\|_{\star,0,n}^{\gamma} \leq \|y\|_{\star,0,n}$ when $\|y\|_{\star,0,n} \leq 1$. Hence,

$$\|y\|_{\star,0,n} \leq \max\{1, 2C_7(\mu, y_0) + 8\mu^{1+\gamma}C_6(\mu, y_0))\}.$$

This proves that for a choice of μ (or equivalently T or n) small enough depending only on $\|x\|$, $\|\nabla f\|_\infty$, $H_\gamma(F)$, γ and p, then $\|y\|_{\star,0,n}$ is bounded by a constant that depends only on $\|x\|$, $\|\nabla f\|_\infty$, $H_\gamma(F)$, γ, p, $|f(y_0)|$ and $|F(y_0)|$. However, $|F(y_0)|$ and $|f(y_0)|$ are bounded by some constants that depends only on $|f(0)|$ and $\|\nabla f\|_\infty$.

The result is proved by finding a sequence $n_0 = 0 < n_1 < \cdots < n_N$ such that $\omega_{n_1,n_i+1} \leq \mu^p$ with $t_{n_N} = T$ and μ satisfying (15), (19) and (20). Since ω is continuous close to its diagonal, there exists such a finite number N of intervals, and this number depends only on the choice of μ, and then on $\|x\|$, $\|\nabla f\|_\infty$, $H_\gamma(F)$, γ and p (and not on y_0 nor $f(0)$). Finally, it is easily shown that

$$\|y\|_{\star,0,n_N} \leq N^{p-1}\sum_{i=1}^{N-1}\|y\|_{\star,n_i,n_{i+1}},$$

which proves the result by applying the result on the successive time intervals $[t_{n_i}, t_{n_{i+1}}]$ and replacing y_0 by y_{t_i}. □

Finally, is order to interpolate the Euler scheme and to get a good control, we shall add an hypothesis on ω, which is trivially satisfied in the case of $\omega(s,t) = t-s$, that is for Hölder continuous rough paths.

Hypothesis 3. We assume that there exists a continuous, increasing function ϕ such that

$$\frac{\omega(s,t)}{\phi(t)-\phi(s)} \text{ and } \frac{\phi(t)-\phi(s)}{\omega(s,t)}$$

are bounded for $0 \leq s < t \leq T$.

Proposition 3. *Under Hypotheses 1, 2 and 3, there exists a least a solution in the sense of Davie to* (1).

Proof. From the family $\{y_i\}$, we construct a path from $[0, T]$ to $\mathbb{R}^m$ by

$$y_t = y_i + \frac{\phi(t)-\phi(t_i)}{\phi(t_{i+1})-\phi(t_i)} y_{i,i+1}, \ t \in [t_i, t_{i+1}]. \tag{21}$$

From standard computations, there exists a constant C_{16} which depends only on p and the lower and upper bounds of $\omega(s,t)/(\phi(t)-\phi(s))$ such that

$$\|y\| \leq C_{16} \|y\|_{\star,0,n}.$$

With Lemma 4, we have a uniform bound in $\|y\|_{\star,0,n}$, and then on the constant L when μ (or n) is small enough. Hence, for any partition satisfying Hypothesis 3, the path y has a p-variation which is bounded by a constant that does not depend on the choice of the partition.

Now, let y^n be a family of paths constructed along an increasing family of partitions Π^n whose meshes decrease to 0.

Then there exists a subsequence $\{y^{n_k}\}_{k\geq 1}$ of $\{y^n\}_{n\in\mathbb{N}^*}$ which converges in q-variation for $q > p$ to some path y of finite p-variation.

For any (s,t) in $\cap_{n\geq 0}\Pi^n$,

$$|y_t - y_s - f(y_s)x^1_{s,t} - F(y_s)x^2_{s,t}| \leq L\omega(s,t)^\theta.$$

Since $\cap_{n\geq 0}\Pi^n$ is dense in $[0, T]$ and y is continuous, this proves that y is the solution to (16) in the sense of Davie, at least when T is small enough.

The passage from a solution on $[0, T]$ with T small enough to a global solution is done by using the arguments of Lemma 7, Lemma 8 and Proposition 6. □

6 Distance Between Two Solutions and Uniqueness

We now consider a more stringent assumption than Hypothesis 1.

Hypothesis 4. The function f is twice continuously differentiable from $\mathbb{R}^m$ to $L(\mathbb{R}^d, \mathbb{R}^m)$ and is such that ∇f, $\nabla^2 f$ are bounded and $F(x) := f(x) \cdot \nabla f(x)$ is such that ∇F is γ-Hölder continuous with constant $H_\gamma(\nabla F)$ from $\mathbb{R}^m$ to $L(\mathbb{R}^m \otimes \mathbb{R}^d, \mathbb{R}^m)$.

We consider two rough paths u and v of finite p-variation controlled by ω, $p \in [2,3)$, as well as two vector fields f and g satisfying Hypothesis 4. Let y and z be respectively some solutions to $y_t = y_0 + \int_0^t f(y_s)\,\mathrm{d}u_s$ and $z_t = z_0 + \int_0^t g(z_s)\,\mathrm{d}v_s$.

We have seen that y and z remain is a ball of radius R that depends only on $\|\nabla f\|_\infty$, $\|\nabla g\|_\infty$, $\|\nabla F\|_\infty$, $\|\nabla G\|_\infty$ (since F and $G = g \cdot \nabla g$ are Lipschitz continuous), $\|u\|$, $\|v\|$, y_0, z_0, $|f(0)|$, $|g(0)|$, $\omega(0,T)$, γ and p.

Definition 3. We say that a constant C satisfies Condition (S) if it depends only on the above quantities, as well as $H_\gamma(\nabla F)$, $H_\gamma(\nabla G)$, $\|\nabla^2 f\|_\infty$ and $\|\nabla^2 g\|_\infty$.

We then set for some functions h, h',

$$\delta_R(h,h') = \sup_{z \in B(0,R)} |h(z) - h'(z)|.$$

We also set

$$\delta(u,v) = \|u - v\| \text{ and } \delta(y,z) = \sup_{0 \le s < t \le T} \frac{|y_{s,t} - z_{s,t}|}{\omega(s,t)^{1/p}}.$$

Theorem 1. *Under Hypotheses 2 and 4, there exists some constant C_{17} satisfying Condition* (S) *such that for all $(s,t) \in \Delta^2$,*

$$\begin{aligned} |y_{s,t} - z_{s,t}| &\le C_{17}\omega(s,t)^{1/p}(|y_0 - z_0| \\ &\quad + \delta(u,v) + \delta_R(f,g) + \delta_R(F,G) + \delta_R(\nabla f, \nabla g) + \delta_R(\nabla F, \nabla G)). \end{aligned} \tag{22}$$

The proof of the following corollary is then immediate from the previous estimate.

Corollary 1. *Under Hypotheses 2 and 4, there exists a unique solution in the sense of Davie to* (1).

Theorem 1 proves that the Itô map which sends x to the unique solution to (1) is locally Lipschitz continuous in y_0, f and x.

The next lemma is the main estimate of the proof and show why extra regularity shall be assumed on F. This lemma is already well-known (See Lemma 3.5 in [2]) but we recall its proof which is straightforward.

Lemma 5. *Let h be a function of class $\mathscr{C}^1$ from $\mathbb{R}^m$ to $\mathbb{R}$ such that ∇h is γ-Hölder and bounded. Then for all a, b, c, d in $\mathbb{R}^m$,*

$$\begin{aligned} &|h(a) - h(b) + h(c) - h(d)| \\ &\qquad \le |a - b - c + d|\|\nabla h\|_\infty + H_\gamma(\nabla h)(|a - b|^\gamma + |c - d|^\gamma)|b - d|. \end{aligned}$$

Proof. The result follows from

$$\begin{aligned}
&|h(a)-h(b)+h(c)-h(d)| \\
&= \left|\int_0^1 \nabla h(c+\tau(a-c))(a-c)\,\mathrm{d}\tau - \int_0^1 \nabla h(d+\tau(b-d))(b-d)\,\mathrm{d}\tau\right| \\
&= \left|\int_0^1 \nabla h(c+\tau(a-c))(a-c-b+d)\,\mathrm{d}\tau\right| \\
&\quad + \left|\int_0^1 \big(\nabla h(c+\tau(a-c)) - \nabla h(d+\tau(b-d))\big)(b-d)\,\mathrm{d}\tau\right| \\
&\le |a-b-c+d|\|\nabla h\|_\infty + H_\gamma(\nabla h)\int_0^1 |(1-\tau)(c-d)+\tau(a-b)|^\gamma |b-d|\,\mathrm{d}\tau,
\end{aligned}$$

since $(x+y)^\gamma \le x^\gamma + y^\gamma$ for $x, y \ge 0$ and $\gamma \in [0,1]$. □

Let us denote by $\mathfrak{d}$ the operator which, applied to an expression involving y, f and u, takes the difference between this expression and the similar expression with y replaced by z, f replaced by g and u replaced by v. For example,

$$\mathfrak{d}(f(y_s)x^1_{s,t}) = f(y_s)u^1_{s,t} - g(z_s)v^1_{s,t}.$$

If $\alpha(y, f, u)$ and $\beta(y, f, u)$ are two expressions, then

$$\mathfrak{d}(\alpha(y,f,u)\beta(y,f,u)) = \mathfrak{d}(\alpha(y,f,u))\beta(y,f,u) + \alpha(z,g,v)\mathfrak{d}(\beta(y,f,u)). \quad (23)$$

Proof (Proof of Theorem 1). In the proof, we assume without loss of generalities that $\gamma < 1$.

Let us choose a constant A such that

$$|\mathfrak{d}(\mathfrak{D}(s,r,t))| \le A\omega(s,t)^\theta \quad (24)$$

and

$$|\mathfrak{d}(\mathfrak{D}(s,t))| \le A\omega(s,t)^{1/p}. \quad (25)$$

From the sewing lemma on the difference of two almost rough paths (See for example [13]*Theorem 6, p. 95), there exists some universal constant M (depending only on θ) such that

$$|\mathfrak{d}(y_{s,t} - \mathfrak{D}(s,t))| \le MA\omega(s,t)^\theta. \quad (26)$$

Our aim is to obtain some estimate on A.

Let us note first that

$$\begin{aligned}
|\mathfrak{d}F(y_s)u^2_{s,t}| \le{}& (|F(y_s) - F(z_s)| + |F(z_s) - G(z_s)|\|u\|\omega(s,t)^{2/p} \\
&+ (|G(z_0)| + \|\nabla G\|_\infty \|z\|_\mu)\delta(u,v)\omega(s,t)^{2/p}.
\end{aligned}$$

Since $|y_s - z_s| \leq \delta(y,z)\mu + |y_0 - z_0|$, with Lemma 5,

$$|F(y_s) - F(z_s)| \leq \|\nabla F\|_\infty(\delta(y,z)\mu + |y_0 - z_0|),$$

then for some constant C_{18} satisfying Condition (S),

$$|\eth F(y_s)u^2_{s,t}| \leq C_{18}(\delta_R(F,G) + |y_0 - z_0| + \mu\delta(y,z) + \delta(u,v))\omega(s,t)^{2/p}.$$

With (26),

$$\begin{aligned}|\eth B_2(s,t)| &= |\eth(y_{s,t} - f(y_s)u^1_{s,t})| = |\eth(y_{s,t} - \mathfrak{D}(s,t) + F(y_s)u^2_{s,t})|\\ &\leq MA\omega(s,t)^\theta + |\eth F(y_s)u^2_{s,t}|\\ &\leq MA\omega(s,t)^\theta + C_{18}(\delta_R(F,G) + |y_0 - z_0| + \mu\delta(y,z) + \delta(u,v))\omega(s,t)^{2/p}.\end{aligned}$$

Since f is differentiable, for $\tau \in [0,1]$,

$$f(\tau y_t + (1-\tau)y_s) - f(y_s) = \int_0^1 \nabla f(y_s + \rho(\tau y_t + (1-\tau)y_s))\tau y_{s,t}\,\mathrm{d}\rho. \quad (27)$$

Hence

$$\begin{aligned}&|f(\tau y_t + (1-\tau)y_s) - f(y_s) - g(\tau y_t + (1-\tau)y_s) + g(y_s)|\\ &\qquad\leq \delta_R(\nabla f, \nabla g)\|y\|\omega(s,t)^{1/p}. \quad (28)\end{aligned}$$

With (28) and (27) for $\tau \in [0,1]$,

$$\begin{aligned}&\eth(f(\tau y_t + (1-\tau)y_s) - f(y_s)) \leq \delta_R(\nabla f, \nabla g)\|y\|\omega(s,t)^{1/p}\\ &+ 2\|\nabla^2 f\|_\infty(\delta(y,z)\mu + |y_0 - z_0|)\|y\|\omega(s,t)^{1/p} + \delta(y,z)\|\nabla f\|_\infty\omega(s,t)^{1/p}.\end{aligned}$$

With Lemma 5 and (27)–(28) applied to F and G, for $\tau \in [0,1]$,

$$\begin{aligned}|\eth B_4(y_s, y_s + \tau y_{s,t})| &= \eth(F(\tau y_t + (1-\tau)y_s) - F(y_s))\\ &\leq \delta_R(\nabla F, \nabla G)\|y\|\omega(s,t)^{1/p} + 2\|\nabla F\|_\infty\delta(y,z)\omega(s,t)^{1/p}\\ &\quad + 2H_\gamma(\nabla F)(\|y\|^\gamma + \|z\|^\gamma)\omega(s,t)^{\gamma/p}(\mu\delta(y,z) + |y_0 - z_0|).\end{aligned}$$

Besides,

$$\eth(\nabla f(y_s + \tau y_{s,t})) \leq \delta_R(\nabla f, \nabla g) + \|\nabla^2 f\|_\infty(\delta(y,z)\mu + |y_0 - z_0|).$$

In addition,

$$|\eth(F(y_0))| \leq \delta_R(F,G) + \|\nabla F\|_\infty|y_0 - z_0|.$$

Combining all these estimates using (23) with the ones given in Sect. 4 in some lengthy computations,

$$|\eth(\mathfrak{D}(s,r,t))| \le C_{19}\omega(s,t)^{\theta}\big(MA\mu + \mu\delta(y,z) + \delta(y,z)\mu^{1-\gamma} + |z_0 - y_0| + \delta(u,v) + \delta_R(F,G) + \delta_R(\nabla F, \nabla G) + \delta_R(\nabla f, \nabla g)\big), \quad (29)$$

where C_{19} satisfies Condition (S) (Note that C_{19} decreases with T).

We have also

$$\begin{aligned}|\eth(\mathfrak{D}(s,t))| &\le \delta_R(f,g)(\|y\|\mu + |y_0|)\|u\|\omega(s,t)^{1/p} \\ &+(\|\nabla f\|_\infty + \|\nabla F\|_\infty)(\mu\delta(y,z) + |y_0 - z_0|)\|u\|(1+\mu)\omega(s,t)^{1/p} \\ &+(\|g\|_\infty + \|G\|_\infty)\delta(u,v)\omega(s,t)^{2/p} + \delta_R(F,G)\|u\|\omega(s,t)^{2/p} \\ &\le C_{20}(\delta_R(F,G) + \delta_R(f,g) + \delta(u,v) + |y_0 - z_0| + \mu\delta(y,z))\omega(s,t)^{1/p},\end{aligned}$$

where C_{20} satisfies Condition (S) and decreases when T decreases. With (26),

$$|\eth y_{s,t}| \le |\eth(y_{s,t} - \mathfrak{D}(s,t))| + |\eth\mathfrak{D}(s,t)| \le MA\omega(s,t)^{\theta} + C_{20}(\delta_R(F,G) + \delta_R(f,g) + \delta(u,v) + |y_0 - z_0| + \mu\delta(y,z))\omega(s,t)^{1/p}. \quad (30)$$

Let us choose A such that an equality holds in either (24) or (25). If $|\eth(y_{s,t})| = A$, then from (30),

$$A \le MA\mu^{1+\gamma} + C_{20}(\delta_R(F,G) + \delta_R(f,g) + \delta(u,v) + |y_0 - z_0| + \mu\delta(y,z)).$$

If $|\eth(\mathfrak{D}(s,r,t))| = A$, then from (29),

$$A \le C_{19}\big(MA\mu + \mu\delta(y,z) + \delta(y,z)\mu^{1-\gamma} + |z_0 - y_0| + \delta(u,v) + \delta_R(F,G) + \delta_R(\nabla F,\nabla G) + \delta_R(\nabla f,\nabla g)\big).$$

In any case, we may choose μ small enough in function of C_{19} or of M (which depends only on θ) such that

$$\begin{aligned}A \le\ & 2C_{19}\big(\mu\delta(y,z) + \delta(y,z)\mu^{1-\gamma} \\ &+ |z_0 - y_0| + \delta(u,v) + \delta_R(F,G) + \delta_R(\nabla F,\nabla G) + \delta_R(\nabla f,\nabla g)\big) \\ &+ 2C_{20}(\delta_R(F,G) + \delta_R(f,g) + \delta(u,v) + |y_0 - z_0| + \mu\delta(y,z)).\end{aligned}$$

Injecting this inequality in (30), we get that

$$\delta(y,z) \le C_{21}(B + (\mu + \mu^{2-\gamma} + \mu^{2(1-\gamma)})\delta(y,z) + |y_0 - z_0|)$$

with

$$B := \delta_R(F,G) + \delta_R(f,g) + \delta(u,v) + \delta_R(\nabla f, \nabla g) + \delta_R(\nabla F, \nabla G)$$

and C_{21} depends only on C_{19}, C_{20}, M and $\omega(0,T)$.

Again, choosing μ small enough in function of C_{19}, C_{20}, M and $\omega(0,T)$ gives the required bound in $\delta(y,z)$ in short time. As the choice of μ does not depend on $|y_0 - z_0|$ and μ satisfies Condition (S), Proposition 7 may be applied to $y - z$. □

7 Distance Between Two Euler Schemes

Let us give two rough paths u and v of finite p-variations, as well as a partition $\{t_i\}_{i=0}^n$ of $[0,T]$.

We use the same notations and conventions as in Sect. 5. Again, we set $\mu := \omega_{0,n}^{1/p}$.

For $0 \leq i < j \leq n$, we set

$$z_{i+1} = z_i + g(z_i)v^1_{i,i+1} + G(z_i)v^2_{i,i+1},\ z_j = z_i + g(z_i)v^1_{i,j} + G(z_i)v^2_{i,j},$$

For a family $\{\epsilon_{i,j}\}_{i,j=1,\dots,n}$ such that $\sup_{0\leq i\leq j\leq n} |\epsilon_{i,j}|/\omega^\theta_{i,j}$ is finite, we set

$$y_{i+1} = y_i + f(y_i)u^1_{i,i+1} + F(y_i)u^2_{i,i+1} + \epsilon_{i,i+1},\ i = 0,\dots,n-1 \tag{31}$$

$$y_j = y_i + f(y_i)u^1_{i,j} + F(y_i)u^2_{i,j} + \epsilon_{i,j},\ 0 \leq i < j \leq n.$$

In addition, define

$$\alpha := \sup_{i=0,\dots,n-1} \frac{|\epsilon_{i,i+1}|}{\omega^\theta_{i,i+1}}.$$

Here, there are three cases of interest: (a) Both z and y are given by some Euler schemes and then $\epsilon_{i,j} = 0$ for all $0 \leq i < j \leq n$. (b) The path y is a solution to $y_t = y_0 + \int_0^t f(y_s)\,\mathrm{d}u_s$ and then $\epsilon_{i,j} = y_j - y_i - f(y_i)u^1_{i,j} - F(y_i)u^2_{i,j}$, while $v = u$ and $g = f$. (c) While $v = u$ and $g = f$, the family $\{y_i\}_{i=0,\dots,n}$ is given by the Euler scheme with respect to a partition $\{t'_i\}_{i=0,\dots,m} \subset \{t_i\}_{i=0,\dots,n}$.

We assume that z and y belong to the ball of radius R and that $\|z\|_{\star,0,n}$ and $\|y\|_{\star,0,n}$ (defined by (14)) are bounded by R'. In any cases, R and R' depend only on $\|\nabla f\|_\infty$, $\|\nabla F\|_\infty$, $\|\nabla g\|_\infty$, $\|\nabla G\|_\infty$, $\|u\|$, $\|v\|$, $|f(0)|$, $|g(0)|$, $|y_0|$, $|z_0|$, $\omega(0,T)$, γ and p.

Definition 4. We say that *a constant C satisfies condition* ($\mathrm{S_e}$) if it depends only on the quantities listed above as well as $H_\gamma(\nabla F)$, $H_\gamma(\nabla G)$, $\|\nabla^2 f\|_\infty$, $\|\nabla^2 g\|_\infty$ and $\sup_{0\leq i\leq j\leq n} |\epsilon_{i,j}|/\omega^\theta_{i,j}$.

Theorem 2. *If f and g satisfies Hypothesis 4, then for some constant C_{22} satisfying Condition* (S_e),

$$\|z - y\|_{\star,0,n} \leq C_{22}(\alpha + |z_0 - y_0| + \delta_R(f, g) + \delta_R(\nabla f, \nabla g) + \delta_R(\nabla F, \nabla G) + \delta_R(F, G) + \delta(u, v)).$$

Proof. We set

$$A := \max_{0 \leq i < j \leq n} \frac{|y_j - y_i - y_{i,j} - z_j - z_i - z_{i,j}|}{\omega_{i,j}^{\theta}}. \tag{32}$$

We have

$$|F(y_i) - F(z_i)| \leq \|\nabla F\|_\infty(|y_0 - z_0| + \|z - y\|_{\star,0,n}\mu)$$

and

$$|\nabla f(y_i + \tau(y_j - y_i)) - \nabla f(z_i + \tau(z_j - z_i))| \leq 2\|\nabla^2 f\|_\infty(|y_0 - z_0| + \|z - y\|_{\star,0,n}\mu).$$

Besides, with Lemma 5,

$$|F(y_j) - F(y_i) - F(z_j) + F(z_i)| \leq \|\nabla F\|_\infty \|z - y\|_{\star,0,n}\omega_{i,j}^{1/p} + H_\gamma(\nabla F)(\|y\|^\gamma + \|z\|^\gamma)(|y_0 - z_0| + \mu\|z - y\|_{\star,0,n})\omega_{i,j}^{\gamma/p}.$$

With (18) and considering the difference between $y_{i,j,k}$ and $z_{i,j,k}$ for $i < j < k$,

$$|y_{i,j,k} - z_{i,j,k}| \leq C_{23}\omega_{i,k}^{\theta}(|y_0 - z_0| + A\mu + (\mu + \mu^{1-\gamma})\|z - y\|_{\star,0,n} + B_1).$$

with C_{23} satisfying Condition (S_e) and

$$B_1 := \delta(u, v) + \delta_R(F, G) + \delta_R(\nabla f, \nabla g) + \delta_R(\nabla F, \nabla G).$$

On the other hand, for $i = 0, \dots, n-1$, $z_{i+1} - z_i - z_{i,i+1} = 0$ and

$$|y_{i+1} - y_i - y_{i,i+1}| \leq \alpha\omega_{i,i+1}^{\theta}.$$

For j given as in the proof of Lemma 3, we have for $K := 2^{1-\theta}$,

$$
\begin{aligned}
&\frac{|y_k - y_i - y_{i,k} - z_k - z_i - z_{i,k}|}{\omega_{i,k}^{\theta}} \\
&\qquad \leq KA + \alpha + \frac{|y_{i,j,k} - z_{i,j,k}|}{\omega_{i,k}^{\theta}} + \frac{|y_{j,j+1,k} - z_{j,j+1,k}|}{\omega_{i,k}^{\theta}} \\
&\qquad \leq KA + \alpha + C_{24}(|y_0 - z_0| + A\mu + (\mu + \mu^{1-\gamma})\|z - y\|_{\star,0,n} + B_1)
\end{aligned}
$$

with $C_{24} = 2C_{23}$. From the definition of A (see (32)),

$$
A \leq 2\alpha + KA + C_{24}(|y_0 - z_0| + A\mu + (\mu + \mu^{1-\gamma})\|z - y\|_{\star,0,n} + B_1). \tag{33}
$$

On the other hand, since both f and F are Lipschitz continuous, for some constant C_{25} satisfying Condition ($\mathrm{S_e}$),

$$
\begin{aligned}
\|z - y\|_{\star,0,n} &\leq \sup_{0 \leq i < j \leq n} \frac{|y_j - y_i - y_{i,j} - z_j + z_i + z_{i,j}|}{\omega_{i,j}^{1/p}} + \frac{|y_{i,j} - z_{i,j}|}{\omega_{i,j}^{1/p}} \\
&\leq A\mu^{1+\gamma} + C_{25}(|y_0 - z_0| + \mu\|z - y\|_{\star,0,n}) + B_2
\end{aligned}
$$

with

$$
B_2 := C_{26}(\delta_R(F, G) + \delta_R(f, g) + \delta(u, v)),
$$

for some constant C_{26} satisfying Condition ($\mathrm{S_e}$). If μ is small enough so that $C_{25}\mu \leq 1/2$, then

$$
\|z - y\|_{\star,0,n} \leq 2C_{25}|y_0 - z_0| + 2A\mu^{1+\gamma} + 2B_2. \tag{34}
$$

Injecting this in (33),

$$
A \leq 2\alpha + KA + C_{27}(|y_0 - z_0| + AC_{28}(\mu)) + B_3
$$

with a $C_{28}(\mu)$ satisfying Condition ($\mathrm{S_e}$) for fixed μ and that decreases to 0 as μ decreases to 0, C_{27} satisfying Condition ($\mathrm{S_e}$), and $B_3 = C_{29}(B_1 + B_2)$ for some constant C_{29} satisfying Condition ($\mathrm{S_e}$). For $K + C_{27}C_{28}(\mu) < (1 + K)/2 < 1$, then

$$
A \leq \frac{2}{1 - K}(2\alpha + C_{27}|y_0 - z_0| + B_3).
$$

Using the inequality on (34), this leads to the required inequality for a value of μ small enough. Thus usual arguments proves now that this is true for any time horizon T up to changing the constants. □

8 Rate of Convergence of the Euler Scheme

Let us consider now the solution to $y_t = y_0 + \int_0^t f(y_s)\,\mathrm{d}x_s$ as well as the associated Euler scheme

$$e_{i+1} = e_i + f(e_i)x^1_{i,i+1} + F(e_i)x^2_{i,i+1},\ e_0 = y_0$$

when f satisfies Hypothesis 4.

We are willing to estimate the difference between y and e. Our key argument is provided by Theorem 2 above.

Proposition 4. *Assume Hypothesis 4 on f. For $\delta := \sup_{0\leq i<n} \omega_{i,i+1}$ and $p \in [2,3)$,*

$$\sup_{i=0,\dots,n} |e_i - y_i| \leq C_{30}\delta^{(3-p)/p-\eta}. \tag{35}$$

where C_{30} depends only on y_0, $|f(0)|$, $\|x\|$, $\|\nabla f\|_\infty$, $\|\nabla^2 f\|_\infty$, $\|\nabla F\|_\infty$, $H_\gamma(\nabla F)$, $\omega(0,T)$, p and γ.

It follows that the rate of convergence is smaller than $(3-p)/p$ and belongs to $(0, 1/2)$. In addition, when p increases to 3, the rate of convergence decreases to 0. This rate is similar to the one given by A.M. Davie [2]. (See also [3] for the convergence of the Milstein scheme for the fractional Brownian motion).

Proof. Let us note first that since y and e remains bounded, if ∇F is γ-Hölder continuous, then it is locally γ'-Hölder continuous for any $\gamma' < \gamma$. Since the constraint $2+\gamma' > p$ is in force, we set $\gamma' = p - 2 + \eta p$ for some $0 < \eta < (3-p)/p < 1/2$ and $\theta' := (2+\gamma')/p = 1+\eta$.

Since F is Lipschitz continuous, then y is a solution in the sense of Davie with $\theta = 3/p$. This way,

$$|y_t - y_s - f(y_s)x^1_{s,t} - F(y_s)x^2_{s,t}| \leq L\omega(s,t)^{3/p}.$$

It follows that $\{y_i\}_{i=0}^n$ is solution to (31) with

$$\epsilon_{i,i+1} = y_{i+1} - y_i - f(y_i)x^1_{i,i+1} - F(y_i)x^2_{i,i+1}.$$

Then

$$\alpha := \sup_{0\leq i<n} \frac{|\epsilon_{i,i+1}|}{\omega_{i,i+1}^{\theta'}} \leq L\delta^{3/p-\theta'} = L\delta^{(3-p)/p-\eta}.$$

Applying Theorem 2 with our choice of θ' leads to

$$\|e - y\| \leq C_{31}\delta^{(3-p)/p-\eta},$$

where C_{31} depends only on y_0, $|f(y_0)|$, $\|x\|$, $\|\nabla f\|_\infty$, $\|\nabla^2 f\|_\infty$, $\|\nabla F\|_\infty$, $H_\gamma(\nabla F)$, $\omega(0,T)$, p and γ. Then (35) is immediate. □

Let $\{\{t_i^n\}_{i=0,\dots,n}\}_{n\in\mathbb{N}^*}$ be an increasing family of partitions such that the mesh $\sup_{i=0,\dots,n}\omega(t_i^n,t_{i+1}^n)$ converges to 0. Let $\{\{e_i^n\}_{i=0,\dots,n}\}_{n\in\mathbb{N}^*}$ be the corresponding Euler schemes for a rough path x and a vector field f satisfies Hypothesis 4. With Lemmas 3, 4 and 8,

$$\sup_{n\in\mathbb{N}^*}\sup_{0\leq i<j\leq n}\frac{|e_i^n-e_j^n-f(e_i^n)x^1_{t_i^n,t_j}-F(e_i^n)x^2_{t_i^n,t_j^n}|}{\omega(t_i^n,t_{j+1}^n)^\theta}<+\infty$$

and with Proposition 3 under Hypothesis 3, the interpolation of the Euler scheme e^n constructed as in (21) has a p-variation norm $\|e^n\|$ which is bounded. Using the same proof as above, we get the following corollary.

Corollary 2. *Under Hypotheses 2, 3 and 4, the family of interpolated Euler schemes $\{e^n\}_{n\in\mathbb{N}^*}$ for a Cauchy sequence of the uniform norm and the q-variation norm for any $q>p$.*

Remark 2. Here the dimension d of the space plays no role so that this argument may be used for an infinite dimensional rough path. This Corollary allows one to define the solution to (1) as the limit of the sequence $\{e^n\}_{n\in\mathbb{N}^*}$.

9 Case of Geometric Rough Paths

We now consider the case where x is a geometric rough path of finite p-variation controlled by ω and the vector field f satisfies Hypothesis 4. This means that there exists a sequence of rough paths x^n converging to x in p-variation such that x^n lives above a piecewise smooth path z^n in $\mathbb{R}^d$ and $x_t^n=1+z_t^n+\int_0^t(z_s^n-z_0^n)\otimes z_s^n\,\mathrm{d}s$. Such a path x^n is called a smooth rough path.

In order to simplify the analysis, we assume that $\omega(s,t)=t-s$ and then we are dealing with α-Hölder continuous paths with $\alpha=1/p$.

The core idea of P. Friz and N. Victoir was to consider a family of smooth rough paths $(\widetilde{x}^n)_{n\geq 1}$ such that, given a family of partitions $\{\{t_i^n\}_{i=0,\dots,n}\}_{n\in\mathbb{N}^*}$, $\widetilde{x}^n$ converges to x in the β-Hölder norm for any $\beta<\alpha$ and $\widetilde{x}^n_{t_i^n}=x_{t_i^n}$ for $i=0,\dots,n$.

For this, they used sub-Riemannian geodesics. In [13], we have provided an alternative construction using some segments and some loops.

Let $\widetilde{z}^n$ be the projection of $\widetilde{x}^n$ onto $\mathbb{R}^d$, and let y^n be the solution of the ODE $y_t^n=y_0+\int_0^t f(y_s^n)\,\mathrm{d}\widetilde{z}_s^n$. As $\widetilde{z}^n$ is piecewise smooth, one knows that this solution corresponds to the one of the solution of the RDE $y_t^n=y_0+\int_0^t f(y_s^n)\,\mathrm{d}\widetilde{x}_s^n$ in the sense of Davie or in the sense of Lyons (See Sect. 10 below).

Let $\widetilde{e}^n$ be the Euler scheme associated to y^n with the partition $\{t_i^n\}_{i=0,\dots,n}$:

$$\widetilde{e}_{i+1}^n=\widetilde{e}_i^n+f(\widetilde{e}_i^n)\widetilde{x}^{1,n}_{t_i^n,t_{i+1}^n}+F(\widetilde{e}_i^n)\widetilde{x}^{2,n}_{t_i^n,t_{i+1}^n},\ i=0,\dots,n,$$

and e^n the Euler scheme

$$e^n_{i+1} = e^n_i + f(e^n_i)x^{1,n}_{t^n_i,t^n_{i+1}} + F(e^n_i)x^{2,n}_{t^n_i,t^n_{i+1}},\ i = 0,\dots,n.$$

The key observation from P. Friz and N. Victoir is that from the very construction of $\widetilde{x}^n$, $\widetilde{x}^n_{t^n_i,t^n_{i+1}} = \widetilde{x}^n_{t^n_i,t^n_{i+1}}$ and then $\widetilde{e}^n = e^n$.

With Proposition 4, for some $\eta \in (0, 3\alpha - 1)$,

$$\|e^n - y^n\|_\star := \sup_{0\leq i<j\leq n} \frac{|e^n_j - e^n_i - y^n_j + y^n_i|}{\omega^{1/p}_{i,j}} = \|\widetilde{e}^n - y^n\| \leq C_{32}\delta^{3\alpha-1-\eta}$$

where C_{32} depends only on $\|x\|$, f, T, α and η, and $\delta := \sup_{i=0,\dots,n+1} \omega(t^n_i, t^n_{i+1})$.

On the other hand with Theorem 1,

$$\|y^n - y^m\|_\star \leq \|y^n - y^m\| \leq C_{33}\|\widetilde{x}^m - \widetilde{x}^n\|,$$

where C_{33} depends only on $\|x\|$, f, T, α and η (when $\widetilde{x}^n$ is such that $\|\widetilde{x}^n\| \leq A\|x\|$ which is the case with the constructions mentioned above).

Since $(\widetilde{x}^n)_{n\in\mathbb{N}^*}$ is a Cauchy sequence in the space of β-Hölder continuous functions, $\beta < \alpha$, it follows that (y^n) is a Cauchy sequence and converges to some element y.

We then obtain the convergence of e^n to y in the sense that $\|e^n - y\|_\star$ decreases to 0 as n tends to infinity.

10 Solution in the Sense of Davie and Solution in the Sense of Lyons

The notion of solution in the sense of Davie (Definition 2) is different of the solution in the sense of Lyons (Definition 1), as the iterated integrals of y and the cross-iterated integrals between y and x are not constructed, while they are part of the solution in the sense of Lyons.

However, once a solution in the sense of Davie is constructed, it is easy to construct a rough paths with values in $T_1(\mathbb{R}^d \oplus \mathbb{R}^m)$.

Lemma 6. *A solution y in the sense of Davie — which is a path with values in $\mathbb{R}^m$ — may be lifted to a rough path with values in $T_1(\mathbb{R}^d \oplus \mathbb{R}^m)$, as the rough path $\widetilde{y}$ associated to the almost rough path*

$$h_{s,t} := 1 + x_{s,t} + y_{s,t} + f(y_s) \otimes f(y_s) \cdot x^2_{s,t} + f(y_s) \otimes 1 \cdot x^2_{s,t} + 1 \otimes f(y_s) \cdot x^2_{s,t}. \quad (36)$$

Besides, the map $y \mapsto \widetilde{y}$ is locally Lipschitz continuous.

Proof. It is easily checked that

$$|h_{s,r,t} - h_{s,r} \otimes h_{r,t}| \leq C_{34}\omega(s,t)^\theta$$

with $\theta > 1$ and then that it is an almost rough path. With Proposition 1, the associated rough path $\widetilde{y}$ projects onto (x, y) in $\mathbb{R}^d \oplus \mathbb{R}^m$. The local Lipschitz continuity of $y \mapsto \widetilde{y}$ follows from the same kind of computation as the one of the proof of Theorem 1. □

Proposition 5. *Let us assume that f is a vector field with a bounded derivative which is γ-Hölder continuous, such that a solution to (1) exists in the sense of Lyons. Let us assume also that $f \cdot \nabla f$ is γ-Hölder continuous, so that a solution to (1) exists in the sense of Davie, lifted as a rough path with values in $T_1(\mathbb{R}^d \oplus \mathbb{R}^m)$ as above. Then the two solutions coincide.*

Proof. Let y be a solution in the sense of Davie of (1), which is lifted as a rough path $\widetilde{y}$ using Lemma 6. Let us consider

$$u_{s,t} := f(y_s)x^1_{s,t} + \nabla f(y_s)\widetilde{y}^\times_{s,t},$$

where $\widetilde{y}^\times_{s,t}$ is the projection onto $\mathbb{R}^m \otimes \mathbb{R}^d$ of $\widetilde{y}_{s,t}$ (or roughly speaking, it is the cross-iterated integral between y and x).

Since $\widetilde{y}^\times_{s,t} = y_{s,r} \otimes x^1_{r,t}$,

$$\begin{aligned} |u_{s,t} - u_{s,r} - u_{r,t}| &\leq |(f(y_r) - f(y_s))x^1_{r,t} - \nabla f(y_s)y_{s,r} \otimes x^1_{r,t})| + |\nabla f(y_r) - \nabla f(y_s)||\widetilde{y}^\times_{s,r}| \\ &\leq \left| \int_0^1 (\nabla f(y_s + \tau y_{s,r}) - \nabla f(y_s))y_{s,r} \otimes x^1_{r,t}\,\mathrm{d}\tau \right| + H_\gamma(\nabla f)\|y\|^\gamma \|z\|\omega^\theta_{s,t} \\ &\leq H_\gamma(\nabla f)(\|y\|^{1+\gamma}\|x\| + \|y\|^\gamma\|z\|)\omega(s,t)^\theta. \end{aligned}$$

Then $\{u_{s,t}\}_{(s,t)\in\Delta^2}$ is an almost rough path in $\mathbb{R}^m$ whose associated rough path v satisfies

$$|v_t - v_s - u_{s,t}| \leq C_{35}\omega(s,t)^\theta.$$

From the very construction, v is the rough integral $v_t = v_0 + \int_0^t g(z_s)\,\mathrm{d}z_s$ with z is a rough path lying above $(x, y, y^\times) \in (\mathbb{R} \oplus \mathbb{R}^d \oplus \mathbb{R}^d, \mathbb{R}^m, \mathbb{R}^m \otimes \mathbb{R}^d)$ and $g(y,x) = \mathrm{d}x + f(y)\,\mathrm{d}x$. Thus $y = v$ when $v_0 = y_0$. With Lemma 6, this is also true for the iterated integrals. This proves that the a solution in the sense of Davie is also a solution in the sense of Lyons.

If z is a solution in the sense of Lyons, by construction, for a constant L and all $(s,t) \in \Delta^2$,

$$|z_t - z_s - f(z_s)x^1_{s,t} - \nabla f(z_s)z^\times_{s,t}| \leq L\omega(s,t)^\theta$$
$$\text{and } |z^\times_{s,t} \leq 1 \otimes f(z_s)x^2_{s,t}| \leq L\omega(s,t)^\theta,$$

where $z^\times$ lives in $\mathbb{R}^m \otimes \mathbb{R}^d$. Hence, it is immediate the a solution in the sense of Lyons is also a solution in the sense of Davie. □

11 From Local to Global Theorems

We present here some results which allows one to pass from local estimates to global estimates and then show the existence for any horizon T provided some uniform estimates.

Lemma 7. *Let $\mu > 0$ be fixed. Then there exists a finite number of times $0 = T_0 < T_1 < \cdots < T_{N+1}$ with $T_N \leq T < T_{N+1}$ and $\omega(T_i, T_{i+1}) = \mu$.*

Proof. Let us extend ω on $D_+ := \{(s,t) \,|\, 0 \leq s \leq t\}$ by

$$\omega(s,t) = \begin{cases} \omega(s,T) + T - s & \text{if } s \leq T \leq t, \\ t - s & \text{if } T \leq s \leq t. \end{cases}$$

Let us note that ω is continuous on D_+, and that $\omega(s,t) \xrightarrow[t\to\infty]{} +\infty$. Then for any $\mu > 0$, for any $s \geq 0$, there exists a value $\tau(s)$ such that $\omega(s,\tau(s)) = \mu$, since $\omega(s,s) = 0$.

For $T > 0$ fixed, let us set $T_0 = 0$ and $T_{i+1} = \tau(T_i)$. Then there exists a finite number N such that $\omega(0, T_{N-1}) \leq \omega(0,T) \leq \omega(T_0, T_{N+1})$ and $\omega(T_i, T_{i+1}) = \mu$. This follows from the fact that

$$N\mu = \sum_{i=0}^{N-1} \omega(T_i, T_{i+1}) \leq \omega(0, T_N)$$

and then $\omega(0, T_N) \xrightarrow[N\to\infty]{} +\infty$. This proves the lemma. □

For a path y and times $0 \leq S < S' \leq T$, let us set

$$\|y\|_{[S,S']} := \sup_{S \leq s < t \leq S'} \frac{|y_t - y_s|}{\omega(s,t)^{1/p}}.$$

Lemma 8. *Let us assume that a continuous path y on $[S, S'']$ is a solution in the sense of Davie on time interval $[S, S']$ and $[S', S'']$ respectively with constants L_1 and L_2. Then y is a solution in the sense of Davie on $[S, S'']$ with a new constant L_3 that depends only on L_1, L_2, $H_\gamma(F)$, $\|\nabla f\|_\infty$, $\|x\|$, $\|y\|_{|[S,S']}$, $\omega(0,T)$, γ and p.*

Proof. First, it is classical that

$$\|y\|_{[S,S'']} \leq 2^{1-1/p} \max\{\|y\|_{[S,S']}, \|y\|_{[S',S'']}\},$$

so that y is of finite p-variation over $[S, S'']$. For $s \in [S, S']$ and $t \in [S', S'']$,

$$|y_t - y_s - \mathfrak{D}(s,t)| \leq |y_t - y_{S'} - \mathfrak{D}(S',t)| + |y_{S'} - y_s - \mathfrak{D}(s,S')| + |\mathfrak{D}(s,S',t)|.$$

Let us note that

$$|\mathfrak{D}(s,S',t)| \leq C_{36}\omega(s,t)^{\theta}$$

where C_{36} depends only on L_1, $\|y\|_{|[S,S']}$, $H_\gamma(F)$, $\|\nabla f\|_\infty$, $|F(y_s)|$, $\|x\|$, $\omega(0,T)$, γ and p. Hence, since $\omega(s,S')^\theta + \omega(S',t) \leq \omega(s,t)$,

$$|y_t - y_s - \mathfrak{D}(s,t)| \leq (\max\{L_1,L_2\} + C_{36})\omega(s,t)^{\theta}.$$

This proves the result. □

Proposition 6. *Let us assume that a solution to* (1) *exists on any time interval* $[S,S']$ *with a condition that* $\omega(S,S') \leq C_{37}$ *where* C_{37} *does not depend on* S. *Then a solution exists for any time* T.

Proof. Using the sequence $\{T_i\}_{i=0,\dots,N}$ of times given by Lemma 7, it is sufficient to solve (1) successively on $[T_i, T_{i+1}]$ with initial condition y_{T_i} and to invoke Lemma 8 to prove the existence of a solution in the sense of Davie in $[0,T]$. □

Proposition 7. *Let* y *be a path of finite* p*-variations such that for some constants* A, B *and* K,

$$\text{when } \omega(S,S') \leq K,\ \|y\|_{[S,S']} \leq B|y_S| + A.$$

Then

$$\sup_{t\in[0,T]} |y_t| \leq R(T)|y_0| + (R(T)-1)\frac{A}{B} \tag{37}$$

with

$$R(T) = \exp(B(1+K^{-1})^{1-1/p}\max\{\omega(0,T), \omega(0,T)^{1/p}\}),$$

and the p*-variation norm of* $\|y\|_{[0,T]}$ *depends only on* A, B, $\omega(0,T)$, K *and* p.

Proof. Remark first that

$$\sup_{t\in[S,S']} |y_t| \leq |y_S|(1 + B\omega(S,S')^{1/p}) + A\omega(S,S')^{1/p}.$$

Let us choose an integer N such that $\mu := \omega(0,T)/N \leq K$ and construct recursively T_i with $T_0 = 0$ and $\omega(T_i, T_{i+1}) = \mu$. Then

$$|y_{T_{i+1}}| \leq |y_{T_i}|(1 + \mu^{1/p}B) + A\mu^{1/p}.$$

From a classical result easily proved by an induction,

$$|y_T| \leq \exp(N\mu^{1/p}B)|y_0| + \frac{A}{B}(\exp(N\mu^{1/p}B) - 1).$$

As this is true for any T, surely (37) holds.

Choosing N such that $\omega(0,T)/N \le K$ and $\omega(0,T)/(N-1) > K$,

$$N \le \frac{\omega(0,T)}{K} + 1.$$

This way

$$N\mu^{1/p} \le \left(\frac{\omega(0,T)}{K} + 1\right)^{1-\frac{1}{p}} \omega(0,T)^{\frac{1}{p}} \le \begin{cases} \widetilde{K}\omega(0,T)^{1/p} & \text{if } \omega(0,T) \le 1, \\ \widetilde{K}\omega(0,T) & \text{if } \omega(0,T) > 1. \end{cases}$$

with $\widetilde{K} := (1 + K^{-1})^{1-1/p}$.

Finally,

$$\|y\|_{[0,T]} \le N^{1-1/p} \max_{i=0,\dots,N-1} \|y\|_{[T_i,T_{i+1}]},$$

which proves the last statement □

Acknowledgements This work has been supported by the ANR ECRU founded by the French Agence Nationale pour la Recherche. The author is also indebted to Massimiliano Gubinelli for interesting discussions about the topic. The author also wish to thank the anonymous referee for having suggested some improvements in the introduction.

References

1. S. Aida, Notes on proofs of continuity theorem in rough path analysis, 2006. Preprint of Osaka University
2. A.M. Davie, Differential equations driven by rough paths: an approach via discrete approximation. Appl. Math. Res. Express. AMRX **2**(abm009), 40 (2007)
3. A. Deya, A. Neuenkirch, S. Tindel, A Milstein-type scheme without Levy area terms for sdes driven by fractional brownian motion, 2010. arxiv:1001.3344
4. H. Doss, Liens entre équations différentielles stochastiques et ordinaires. Ann. Inst. H. Poincaré Sect. B (N.S.) **13**(2), 99–125 (1977)
5. P. Friz, N. Victoir, Euler estimates of rough differential equations. J. Differ. Eqn. **244**(2), 388–412 (2008)
6. P. Friz, N. Victoir, *Multidimensional Stochastic Processes as Rough Paths. Theory and Applications* (Cambridge University Press, Cambridge, 2010)
7. J.G. Gaines, T.J. Lyons, Variable step size control in the numerical solution of stochastic differential equations. SIAM J. Appl. Math. **57**(5), 1455–1484 (1997)
8. M. Gubinelli, Controlling rough paths. J. Funct. Anal. **216**(1), 86–140 (2004)
9. Y. Inahama, H. Kawabi, Asymptotic expansions for the Laplace approximations for Itô functionals of Brownian rough paths. J. Funct. Anal. **243**(1), 270–322 (2007)
10. T. Lyons, M. Caruana, T. Lévy, Differential equations driven by rough paths. In *École d'été des probabilités de Saint-Flour XXXIV — 2004*, ed. by J. Picard, vol. 1908, *Lecture Notes in Mathematics* (Springer, New York, 2007)
11. A. Lejay, An introduction to rough paths. In *Séminaire de probabilités XXXVII*, vol. 1832 of *Lecture Notes in Mathematics* (Springer, New York, 2003), pp. 1–59
12. A. Lejay. On rough differential equations. Electron. J. Probab. **14**(12), 341–364 (2009)

13. A. Lejay, Yet another introduction to rough paths. In *Séminaire de probabilités XLII*, vol. 1979 of *Lecture Notes in Mathematics* (Springer, New York, 2009), pp. 1–101
14. A. Lejay, Controlled differential equations as Young integrals: a simple approach. J. Differ. Eqn. **249**, 1777–1798 (2010)
15. T. Lyons, Z. Qian, Flow of diffeomorphisms induced by a geometric multiplicative functional. Probab. Theor. Relat. Field **112**(1), 91–119 (1998)
16. T. Lyons, Z. Qian, *System Control and Rough Paths*, Oxford Mathematical Monographs (Oxford University Press, Oxford, 2002)
17. A. Lejay, N. Victoir, On (p,q)-rough paths. J. Differ. Eqn. **225**(1), 103–133 (2006)
18. T.J. Lyons, Differential equations driven by rough signals. Rev. Mat. Iberoamericana **14**(2), 215–310 (1998)
19. D.S. Mitrinović, J.E. Pečarić, A.M. Fink, *Inequalities involving functions and their integrals and derivatives*, vol. 53 of *Mathematics and its Applications (East European Series)*. (Kluwer, Dordrecht, 1991)
20. H.J. Sussmann, On the gap between deterministic and stochastic ordinary differential equations. Ann. Probability **6**(1), 19–41 (1978)

Asymptotic Behavior of Oscillatory Fractional Processes

Renaud Marty and Knut Sølna

Abstract In this paper we consider the antiderivative of the product of a fractional random process and a periodic function. We establish that the rescaled process constructed in this way converges to a Brownian motion whose variance depends on the frequency of the periodic function and the Hurst parameter. We also prove that for two different frequencies the limits are independent. Finally, we discuss applications to wave propagation in random media.

Keywords Fractional processes • Brownian motion • Waves in random media

AMS Classification: 60F17, 60G10, 60G15

1 Introduction

Fractional processes have attracted a lot of attention in recent years. They provide relevant models for long-range dependence or antipersistent random behavior [11].

Fractional Brownian motion is the most famous fractional process. It satisfies many relevant properties such as self-similarity, Gaussianity and stationarity of the increments.

R. Marty (✉)
Institut Elie Cartan de Nancy, Nancy-Université, CNRS, INRIA, B.P. 239,
F-54506 Vandoeuvre-lès-Nancy Cedex, France
e-mail: renaud.marty@iecn.u-nancy.fr

K. Sølna
University of California at Irvine, Irvine, CA 92697-3875, USA
e-mail: ksolna@math.uci.edu

C. Donati-Martin et al. (eds.), *Séminaire de Probabilités XLIV*, Lecture Notes in Mathematics 2046, DOI 10.1007/978-3-642-27461-9_12,

Another crucial property for applications is the invariance principle [11]. Let $H \in (0,1)$, $B_H = \{B_H(z)\}_z$ be a fractional Brownian motion with Hurst index H. Depending on the value of H, the fractional white noise

$$\delta B_H = \{B_H(z+1) - B_H(z)\}_z$$

(which is stationary) satisfies various "memory-properties". For $H \in (1/2, 1)$ the process δB_H is long-range dependent, for $H = 1/2$ it is mixing, and for $H \in (0, 1/2)$ it is antipersistent. Let consider a stationary and Gaussian process $\mu = \{\mu(z)\}_z$ that satisfies roughly the same range property as the fractional white noise δB_H, that is, it has the same decay rate of the integrated correlation. Then, the invariance principle says that when ε goes to 0, the rescaled process

$$\left\{\varepsilon^{2H} \int_0^{z/\varepsilon^2} \mu(y)dy\right\}_z = \left\{\frac{1}{\varepsilon^{2-2H}} \int_0^z \mu\left(\frac{y}{\varepsilon^2}\right) dy\right\}_z$$

converges in distribution to the fractional Brownian motion B_H. Notice that the invariance principle is of great importance. In particular it makes the increments of B_H serve as universal stationary Gaussian models with various memory properties depending on the index H [11].

Applications of fractional processes to wave propagation in random media have recently been studied in [5, 9, 10]. Here they are used to model very complex and long range media. In addition to random phenomena, due to the nature of waves, periodic modulation arises naturally in harmonic descriptions associated with wave phenomena [2]. This then leads to the study of evolution problems driven by periodic and random processes. A simple example is the asymptotic study of the process defined by

$$v_p^\varepsilon = \left\{\frac{1}{\varepsilon^\gamma} \int_0^z \mu\left(\frac{y}{\varepsilon^2}\right) p\left(\frac{y}{\varepsilon^\tau}\right) dy\right\}_z \tag{1}$$

where τ and γ are two positive constants and p is a periodic function. Such a question have been investigated in particular cases in [8]. It was proven that if $p = \cos(\omega\times\cdot)$ (or $\sin(\omega\times\cdot)$), $\tau \in (0,2)$, $\gamma = 1 + (2-\tau)(1/2 - H)$ and μ is a Gaussian and stationary process that is essentially a fractional white noise, then v_p^ε converges in distribution to a Brownian motion. The result with $H = 1/2$ had been known for many years and applied to wave propagation in mixing media, see [2], while the results with $H \in (0,1)$ allowed for the study of waves in fractional media.

In the present paper we aim to generalize the result presented above. Our framework is based on a Gaussian process μ defined in terms of a general harmonizable representation. We first consider functions $p = \cos(\omega\times\cdot)$ (or $\sin(\omega\times\cdot)$) and $\tau \in (0,2)$. The proof of the convergence of v_p^ε in [8] was based on the particular form of the covariance of μ and the cases $H < 1/2$, $H = 1/2$ or $H > 1/2$ had to be considered separately. Here we give a new proof that unifies the three cases for H and that is based on the harmonizable representation and thus more general. We also consider general period functions p. We prove that under general assumptions on p the process converges to a Brownian motion whose covariance depends on the Fourier coefficients of p.

A second point we address is the correlation of the limits for different frequencies. Let ω_1 and ω_2 be two real numbers satisfying $\omega_1 \neq \omega_2$ and $\omega_1 \neq -\omega_2$. It is well known in the mixing case ($H = 1/2$), that if we denote respectively by B^{ω_1} and B^{ω_2} the limits of v_p^ε with $p = \cos(\omega_1 \times \cdot)$ and $p = \cos(\omega_2 \times \cdot)$ respectively, then the limits B^{ω_1} and B^{ω_2} are independent [2]. This asymptotic decorrelation is a crucial property in the applications to waves. Here we generalize this result to every $H \in (0, 1)$.

The assumption $\tau \in (0, 2)$ corresponds to models for wave propagations where the phases are slowly varying with respect to the random fluctuations [2]. Another problem of interest is when phases oscillate at the same speed as the random medium fluctuations. Physically this corresponds to a strong "resonance" between the waves and the fluctuations. This motivates studying the case $\tau = 2$ and then, in the mixing case, one obtains different results from those of the case $\tau \in (0, 2)$. In particular, the effective second moment coefficient can be written as $\int_0^\infty \mathbb{E}[\mu(0)\mu(y)]dy$ if $\tau \in (0, 2)$ and $\int_0^\infty \cos(\omega y)\mathbb{E}[\mu(0)\mu(y)]dy$ if $\tau = 2$ [2]. In the present paper we analyze the specific behavior when $\tau = 2$ for the full range of $H \in (0, 1)$.

In conclusion, we discuss some possible extensions of our work for applications to wave propagation. In particular we give a conjecture that would generalize the study of wave propagation in mixing [1, 2] or long-range [9] media to a general fractional media.

Our paper is organized as follows. In Sect. 2 we establish the main results and we prove them in Sect. 3. In Sect. 4 we extend our results to general periodic functions. In Sect. 5 we discuss applications to the study of wave propagation in random media.

2 Assumptions and Main Result

2.1 Assumptions

Here we describe the process that we will be working with throughout the paper. We define μ for every $z \geq 0$ by

$$\mu(z) = \int_{\mathbb{R}} \exp(iz\zeta)\psi(\zeta)|\zeta|^{1/2-H}\widehat{W}(d\zeta) \tag{2}$$

where $H \in (0, 1)$ and ψ is a complex-valued function and $\widehat{W}(d\zeta)$ is the Fourier transform of a real Gaussian measure. We assume that ψ is continuous and even and satisfies $|\psi(\zeta)| = \mathscr{O}_{|\zeta|\to\infty}(|\zeta|^{-1})$. Thus μ is a centered and stationary Gaussian process and its covariance takes the form

$$r(z) = \mathbb{E}[\mu(0)\mu(z)] = \int_{\mathbb{R}} \exp(iz\zeta)|\psi(\zeta)|^2|\zeta|^{1-2H}\,d\zeta. \tag{3}$$

An integration with respect to z gives

$$\int_0^Z r(z)\,dz \sim \sigma(H)Z^{2H-1} \tag{4}$$

as $Z\to\infty$ with

$$\sigma(H) = 2|\psi(0)|^2 \int_0^\infty \frac{\sin(\zeta)}{\zeta^{2H}}\,d\zeta.$$

We deduce from (4), depending on the value of H, that the process μ satisfies one of the three following properties: $\int_0^\infty r(z)\,dz = \infty$ if $H \in (1/2, 1)$, $\int_0^\infty r(z)\,dz \in (0,\infty)$ if $H = 1/2$ and $\int_0^\infty r(z)\,dz = 0$ if $H \in (0, 1/2)$. In other words μ is long-range dependent when $H > 1/2$, antipersistent when $H < 1/2$ and satisfies a general mixing property when $H = 1/2$.

We conclude this section by two examples of processes satisfying the assumptions cited above. These examples are defined in terms of the (non-standard) fractional Brownian motion defined for every z by

$$B_H(z) = \frac{1}{\rho(H)} \int_{\mathbb{R}} \frac{e^{iz\zeta} - 1}{i\zeta|\zeta|^{H-1/2}} \widehat{W}(d\zeta)$$

where $\rho(H)$ is a normalizing constant defined by

$$\rho(H)^2 = \int_{\mathbb{R}} \frac{|e^{i\zeta} - 1|^2}{|\zeta|^{2H+1}}\,d\zeta.$$

The first example we present is the fractional white noise δB_H defined as

$$\delta B_H(z) = B_H(z+1) - B_H(z) = \frac{1}{\rho(H)} \int_{\mathbb{R}} \exp(iz\zeta) \frac{e^{i\zeta} - 1}{i\zeta|\zeta|^{H-1/2}} \widehat{W}(d\zeta).$$

In this example

$$\psi(\zeta) = \frac{e^{i\zeta} - 1}{\rho(H)i\zeta}.$$

The second example is the fractional Ornstein-Uhlenbeck (OU) process. Recall that the stationary fractional OU process X_H can be written in terms of a the fractional Brownian motion B_H as

$$X_H(z) = B_H(z) - e^{-z}\int_{-\infty}^{z} e^{y} B_H(y)\,dy$$
$$= \frac{1}{\rho(H)}\int_{\mathbb{R}} \frac{\exp(iz\zeta)}{1+i\zeta}\frac{\widehat{W}(d\zeta)}{|\zeta|^{H-1/2}}$$

so the corresponding function ψ is

$$\psi(\zeta) = \frac{1}{\rho(H)(1+i\zeta)}.$$

2.2 Main Result

Here we consider a general process μ as defined in the previous section and *throughout the paper we assume that* $H \in (0,1)$. We introduce the two processes $v_c^{\omega,\varepsilon}$ and $v_s^{\omega,\varepsilon}$ defined by

$$v_c^{\omega,\varepsilon}(z) = \frac{1}{\varepsilon^{\gamma}}\int_0^z \mu\left(\frac{y}{\varepsilon^2}\right)\cos\left(\omega\frac{y}{\varepsilon^{\tau}}\right)dy \tag{5}$$

and

$$v_s^{\omega,\varepsilon}(z) = \frac{1}{\varepsilon^{\gamma}}\int_0^z \mu\left(\frac{y}{\varepsilon^2}\right)\sin\left(\omega\frac{y}{\varepsilon^{\tau}}\right)dy \tag{6}$$

with μ defined as in (2), $\omega \neq 0$, while γ and τ are constants and ε is a small parameter. We assume that $\tau \in (0,2]$ and γ is given by

$$\gamma = 1 + (2-\tau)(1/2 - H). \tag{7}$$

Notice that in the particular case $\tau = 2$ we have $\gamma = 1$.

We fix an n-dimensional vector of frequencies $\Omega = (\omega_1, ..., \omega_n)$ such that $\omega_j \neq 0$ for every j and $\omega_j \neq \omega_k$ and $\omega_j \neq -\omega_k$ for every $j \neq k$. We define the $2n$-dimensional process $V^{\Omega,\varepsilon}$ by

$$V^{\Omega,\varepsilon} = \left(v_c^{\omega_1,\varepsilon}, v_s^{\omega_1,\varepsilon}, v_c^{\omega_2,\varepsilon}, v_s^{\omega_2,\varepsilon}, \cdots, v_c^{\omega_n,\varepsilon}, v_s^{\omega_n,\varepsilon}\right). \tag{8}$$

We state the first main theorem of this paper.

Theorem 1. *The finite-dimensional distributions of the process* $V^{\Omega,\varepsilon}$ *converge to those of a* $2n$*-dimensional Brownian motion* B^{Ω} *defined by*

$$B^{\Omega} = \left(B_c^{\omega_1}, B_s^{\omega_1}, B_c^{\omega_2}, B_s^{\omega_2}, \cdots, B_c^{\omega_n}, B_s^{\omega_n}\right)$$

whose coordinates are independent and satisfy

$$\mathbb{E}[(B_c^{\omega_j}(1))^2] = \mathbb{E}[(B_s^{\omega_j}(1))^2] = C(\omega_j)^2 \tag{9}$$

for every j and where the second moment coefficient is

$$C(\omega)^2 = \frac{\widetilde{\psi}(\omega)}{2} \int_{-\infty}^{\infty} \frac{|e^{i\zeta} - 1|^2}{|\zeta|^2} d\zeta = \pi \widetilde{\psi}(\omega) \tag{10}$$

with $\widetilde{\psi}(\omega) = |\psi(0)|^2 |\omega|^{1-2H}$ *if* $\tau \in (0,2)$ *and* $\widetilde{\psi}(\omega) = |\psi(\omega)|^2 |\omega|^{1-2H}$ *if* $\tau = 2$.

Notice that the cases $\tau \in (0,2)$ (the slow-phase case) and $\tau = 2$ (the fast-phase case) are dramatically different. The second moment depends on the whole function ψ if $\tau = 2$ while it does not if $\tau \in (0,2)$. When $\tau = 2$ the analogue in the wave case is that the wave resolves the full spectrum of the fluctuations. This theorem generalizes the diffusion-approximation theory set forth in [2]. Indeed, if we consider the case $H = 1/2$, we have

$$\begin{aligned}\int_0^A r(z)\cos(\omega z)\,dz &= \int_0^A \cos(\omega z) \int_{\mathbb{R}} \exp(iz\zeta)|\psi(\zeta)|^2\,d\zeta\,dz \\ &= \frac{1}{2}\int_{\mathbb{R}} |\psi(\zeta/A + \omega)|^2 \frac{e^{i\zeta} - 1}{i\zeta} d\zeta \\ &\quad + \frac{1}{2}\int_{\mathbb{R}} |\psi(\zeta/A - \omega)|^2 \frac{e^{i\zeta} - 1}{i\zeta} d\zeta.\end{aligned}$$

Then, because ψ is continuous and even we get

$$\int_0^\infty r(z)\cos(\omega z)\,dz = |\psi(\omega)|^2 \int_{\mathbb{R}} \frac{\sin(\zeta)}{\zeta} d\zeta = \pi |\psi(\omega)|^2.$$

It then follows that the constant $C(\omega)$ (depending on the value of τ) corresponds to the effective coefficients found in the diffusion-approximation presented in [2].

To conclude this section we make a remark about the super-fast case that corresponds to $\tau > 2$. We can observe, following the lines of the proof of the theorem, that the process $V^{\Omega,\varepsilon}$ converges to $0_{\mathbb{R}^{2n}}$ as already observed in the classical approximation-diffusion theory (for $H = 1/2$, see for instance [3]).

3 Proof of Theorem 1

In the first part of the proof we introduce some notation. Next we establish that, for a fixed frequency ω, the 2-dimensional process $(v_c^{\omega,\varepsilon}, v_s^{\omega,\varepsilon})$ defined by (5) and (6) converges in the sense of the finite dimensional distributions to the 2-dimensional

Brownian motion $(B_c^{\omega}, B_s^{\omega})$ defined by (9). In the third part we prove that for two different frequencies ω_1 and ω_2 the limits $(B_c^{\omega_1}, B_s^{\omega_1})$ and $(B_c^{\omega_2}, B_s^{\omega_2})$ are independent. In the proof we use technical lemmas whose derivations are postponed to the end of this section.

3.1 Preliminaries

We introduce the complex-variable function g defined for every $u \in \mathbb{C}$ by

$$g(u) = \frac{\exp(u) - 1}{u}.$$

By a direct integration we obtain

$$\int_0^z \exp\left(\frac{iy\zeta}{\varepsilon^2}\right) \cos\left(\omega \frac{y}{\varepsilon^\tau}\right) dy = \frac{z}{2}\left\{g\left(iz\left(\frac{\zeta}{\varepsilon^2} - \frac{\omega}{\varepsilon^\tau}\right)\right) + g\left(iz\left(\frac{\zeta}{\varepsilon^2} + \frac{\omega}{\varepsilon^\tau}\right)\right)\right\}$$

and

$$\int_0^z \exp\left(\frac{iy\zeta}{\varepsilon^2}\right) \sin\left(\omega \frac{y}{\varepsilon^\tau}\right) dy = \frac{iz}{2}\left\{g\left(iz\left(\frac{\zeta}{\varepsilon^2} - \frac{\omega}{\varepsilon^\tau}\right)\right) - g\left(iz\left(\frac{\zeta}{\varepsilon^2} + \frac{\omega}{\varepsilon^\tau}\right)\right)\right\}.$$

Then we can deduce that for every ω, ε and z we have

$$v_c^{\omega,\varepsilon}(z) = \int_{\mathbb{R}} g_c^{\omega,\varepsilon}(z,\zeta)\widehat{W}(d\zeta) \tag{11}$$

and

$$v_s^{\omega,\varepsilon}(z) = \int_{\mathbb{R}} g_s^{\omega,\varepsilon}(z,\zeta)\widehat{W}(d\zeta) \tag{12}$$

where

$$g_c^{\omega,\varepsilon}(z,\zeta) = \frac{z}{2\varepsilon^\gamma}\frac{\psi(\zeta)}{|\zeta|^{H-1/2}}\left\{g\left(iz\left(\frac{\zeta}{\varepsilon^2} - \frac{\omega}{\varepsilon^\tau}\right)\right) + g\left(iz\left(\frac{\zeta}{\varepsilon^2} + \frac{\omega}{\varepsilon^\tau}\right)\right)\right\}$$

and

$$g_s^{\omega,\varepsilon}(z,\zeta) = \frac{iz}{2\varepsilon^\gamma}\frac{\psi(\zeta)}{|\zeta|^{H-1/2}}\left\{g\left(iz\left(\frac{\zeta}{\varepsilon^2} - \frac{\omega}{\varepsilon^\tau}\right)\right) - g\left(iz\left(\frac{\zeta}{\varepsilon^2} + \frac{\omega}{\varepsilon^\tau}\right)\right)\right\}.$$

It is convenient to introduce the integral

$$I_1^\varepsilon(\omega_1,\omega_2,z_1,z_2) = \frac{z_1z_2}{4\varepsilon^{2\gamma}}\int_{\mathbb{R}}\frac{|\psi(\zeta)|^2}{|\zeta|^{2H-1}}g\left(iz_1\left(\frac{\zeta}{\varepsilon^2}-\frac{\omega_1}{\varepsilon^\tau}\right)\right)g\left(-iz_2\left(\frac{\zeta}{\varepsilon^2}-\frac{\omega_2}{\varepsilon^\tau}\right)\right)d\zeta,$$

and

$$\begin{aligned}
I_2^\varepsilon(\omega_1,\omega_2,z_1,z_2) &= I_1^\varepsilon(\omega_1,-\omega_2,z_1,z_2),\\
I_3^\varepsilon(\omega_1,\omega_2,z_1,z_2) &= I_1^\varepsilon(-\omega_1,\omega_2,z_1,z_2),\\
I_4^\varepsilon(\omega_1,\omega_2,z_1,z_2) &= I_1^\varepsilon(-\omega_1,-\omega_2,z_1,z_2).
\end{aligned}$$

3.2 Convergence at a Fixed Frequency

Now we can state the convergence of $(v_c^{\omega,\varepsilon}, v_s^{\omega,\varepsilon})$.

Lemma 1. *The finite dimensional distributions of* $(v_c^{\omega,\varepsilon}, v_s^{\omega,\varepsilon})$ *converge to those of* (B_c^ω, B_s^ω).

Proof. We first consider $v_c^{\omega,\varepsilon}$. For every z_1 and z_2 we have

$$\begin{aligned}
\mathbb{E}[v_c^{\omega,\varepsilon}(z_1)v_c^{\omega,\varepsilon}(z_2)] &= \int_{\mathbb{R}} g_c^{\omega,\varepsilon}(z_1,\zeta)\bar{g}_c^{\omega,\varepsilon}(z_2,\zeta)\,d\zeta\\
&= \sum_{j=1}^{4} I_j^\varepsilon(\omega,\omega,z_1,z_2).
\end{aligned}$$

By the substitution $\zeta \to \varepsilon^2\zeta + \omega\varepsilon^{2-\tau}$ we can write

$$I_1^\varepsilon(\omega,\omega,z_1,z_2) = \frac{z_1z_2}{4}\int_{\mathbb{R}}\frac{\left|\psi\left(\varepsilon^2\zeta+\omega\varepsilon^{2-\tau}\right)\right|^2}{|\varepsilon^\tau\zeta+\omega|^{2H-1}}g\left(iz_1\zeta\right)g\left(-iz_2\zeta\right)d\zeta$$

where we use $\gamma = 1+(2-\tau)(1/2-H)$. It now follows from Lemma 3 below that we have the limit

$$\lim_{\varepsilon\to 0}\int_{\mathbb{R}}\frac{\left|\psi\left(\varepsilon^2\zeta+\omega\varepsilon^{2-\tau}\right)\right|^2}{|\varepsilon^\tau\zeta+\omega|^{2H-1}}g\left(iz_1\zeta\right)g\left(-iz_2\zeta\right)d\zeta = \widetilde{\psi}(\omega)\int_{\mathbb{R}}g\left(iz_1\zeta\right)g\left(-iz_2\zeta\right)\,d\zeta$$

where $\widetilde{\psi}(\omega) = |\psi\,(0)\,|^2|\omega|^{1-2H}$ if $\tau\in(0,2)$ and $\widetilde{\psi}(\omega) = |\psi\,(\omega)\,|^2|\omega|^{1-2H}$ if $\tau = 2$. We deal with the integral $I_4^\varepsilon(\omega,\omega,z_1,z_2)$ in the same way so that we get

$$\lim_{\varepsilon\to 0}(I_1^\varepsilon(\omega,\omega,z_1,z_2)+I_4^\varepsilon(\omega,\omega,z_1,z_2)) = \frac{z_1z_2\widetilde{\psi}(\omega)}{2}\int_{\mathbb{R}}g\left(iz_1\zeta\right)g\left(-iz_2\zeta\right)\,d\zeta. \tag{13}$$

Now we deal with $I_2^\varepsilon(\omega,\omega,z_1,z_2)$. By the same substitution as for $I_1^\varepsilon(\omega,\omega,z_1,z_2)$ we have

$$I_2^\varepsilon(\omega,\omega,z_1,z_2) = \frac{z_1 z_2}{4}\int_{\mathbb{R}} \frac{|\psi\left(\varepsilon^2\zeta + \omega\varepsilon^{2-\tau}\right)|^2}{|\varepsilon^\tau\zeta+\omega|^{2H-1}} g\left(iz_1\zeta\right) g\left(-iz_2\left(\zeta+2\omega\varepsilon^{-\tau}\right)\right) d\zeta.$$

It is now enough to prove that the last integral converges to 0 and this is due to Lemma 4 below. We deal with the integral $I_3^\varepsilon(\omega,\omega,z_1,z_2)$ in the same way so that we obtain again by Lemma 4

$$\lim_{\varepsilon\to 0}(I_2^\varepsilon(\omega,\omega,z_1,z_2) + I_3^\varepsilon(\omega,\omega,z_1,z_2)) = 0 \tag{14}$$

and thus, using (13),

$$\lim_{\varepsilon\to 0}\mathbb{E}[v_c^{\omega,\varepsilon}(z_1)v_c^{\omega,\varepsilon}(z_2)] = \frac{z_1 z_2\widetilde{\psi}(\omega)}{2}\int_{\mathbb{R}} g\left(iz_1\zeta\right) g\left(-iz_2\zeta\right) d\zeta,$$

which means that the finite dimensional distributions of $v_c^{\omega,\varepsilon}$ converge to those of the Brownian motion B_c^ω with the scaling (9). Using the same argument we also prove that the finite dimensional distribution of $v_s^{\omega,\varepsilon}$ converge to those of the Brownian motion B_s^ω with scaling (9). Now it remains to establish the independence between B_c^ω and B_s^ω. We can write for every z_1 and z_2

$$\begin{aligned}\mathbb{E}[v_c^{\omega,\varepsilon}(z_1)v_s^{\omega,\varepsilon}(z_2)] &= \int_{\mathbb{R}} g_c^{\omega,\varepsilon}(z_1,\zeta)\bar{g}_s^{\omega,\varepsilon}(z_2,\zeta)\, d\zeta \\ &= i\sum_{j=1}^{4}(-1)^j I_j^\varepsilon(\omega,\omega,z_1,z_2).\end{aligned}$$

From the asymptotic study of $I_j^\varepsilon(\omega,\omega,z_1,z_2)$ for $j\in\{1,2,3,4\}$ just above, we get

$$\lim_{\varepsilon\to 0}\mathbb{E}[v_c^{\omega,\varepsilon}(z_1)v_s^{\omega,\varepsilon}(z_2)] = 0,$$

which proves that B_c^ω and B_s^ω are independent and this concludes the proof. □

3.3 Asymptotic Decorrelation for Different Frequencies

Now we establish the asymptotic decorrelation between the vectors $(v_c^{\omega_1,\varepsilon}, v_s^{\omega_1,\varepsilon})$ and $(v_c^{\omega_2,\varepsilon}, v_s^{\omega_2,\varepsilon})$ when $\omega_1\neq\omega_2$ and $\omega_1\neq-\omega_2$.

Lemma 2. *Let z_1 and $z_2\in\mathbb{R}$ and ω_1 and ω_2 such that $\omega_1\neq\omega_2$ and $\omega_1\neq-\omega_2$. We have*

$$\lim_{\varepsilon\to 0}\mathbb{E}\left[\begin{pmatrix} v_c^{\omega_1,\varepsilon}(z_1)v_c^{\omega_2,\varepsilon}(z_2) & v_c^{\omega_1,\varepsilon}(z_1)v_s^{\omega_2,\varepsilon}(z_2)\\ v_s^{\omega_1,\varepsilon}(z_1)v_c^{\omega_2,\varepsilon}(z_2) & v_s^{\omega_1,\varepsilon}(z_1)v_s^{\omega_2,\varepsilon}(z_2)\end{pmatrix}\right] = \begin{pmatrix} 0 & 0\\ 0 & 0\end{pmatrix}.$$

Proof. We consider $\omega_1 \neq \omega_2$ and $\omega_1 \neq -\omega_2$. To prove the lemma it is sufficient to establish the limit of $\mathbb{E}[v_c^{\omega_1,\varepsilon}(z_1)v_c^{\omega_2,\varepsilon}(z_2)]$ and $\mathbb{E}[v_c^{\omega_1,\varepsilon}(z_1)v_s^{\omega_2,\varepsilon}(z_2)]$. The asymptotic study of $\mathbb{E}[v_s^{\omega_1,\varepsilon}(z_1)v_s^{\omega_2,\varepsilon}(z_2)]$ and $\mathbb{E}[v_s^{\omega_1,\varepsilon}(z_1)v_c^{\omega_2,\varepsilon}(z_2)]$ respectively will be similar. Let us start by $\mathbb{E}[v_c^{\omega_1,\varepsilon}(z_1)v_c^{\omega_2,\varepsilon}(z_2)]$. We have

$$\mathbb{E}[v_c^{\omega_1,\varepsilon}(z_1)v_c^{\omega_2,\varepsilon}(z_2)] = \sum_{j=1}^{4} I_j^{\varepsilon}(\omega_1,\omega_2,z_1,z_2)$$

As in the case of $I_2^{\varepsilon}(\omega,\omega,z_1,z_2)$ in the proof of Lemma 1 we can write by the substitution $\zeta\to\varepsilon^2\zeta+\omega_1\varepsilon^{2-\tau}$

$$I_1^{\varepsilon}(\omega_1,\omega_2,z_1,z_2) = \frac{z_1z_2}{4}\int_{\mathbb{R}}\frac{|\psi\left(\varepsilon^2\zeta+\omega_1\varepsilon^{2-\tau}\right)|^2}{|\varepsilon^{\tau}\zeta+\omega_1|^{2H-1}}g\left(iz_1\zeta\right)g\left(-iz_2\left(\zeta+(\omega_1-\omega_2)\varepsilon^{-\tau}\right)\right)d\zeta.$$

Using Lemma 4 and the condition $\omega_1 \neq \omega_2$ we get $\lim_{\varepsilon\to 0} I_1^{\varepsilon}(\omega_1,\omega_2,z_1,z_2)=0$. Using the same approach and also either $\omega_1 \neq \omega_2$ or $\omega_1 \neq -\omega_2$ we obtain $\lim_{\varepsilon\to 0} I_j^{\varepsilon}(\omega_1,\omega_2,z_1,z_2) = 0$ for $j = 2$, 3 and 4 also, thus

$$\lim_{\varepsilon\to 0}\mathbb{E}[v_c^{\omega_1,\varepsilon}(z_1)v_s^{\omega_2,\varepsilon}(z_2)] = 0.$$

It remains to study $\mathbb{E}[v_c^{\omega_1,\varepsilon}(z_1)v_s^{\omega_2,\varepsilon}(z_2)]$. We have

$$\begin{aligned}\mathbb{E}[v_c^{\omega_1,\varepsilon}(z_1)v_s^{\omega_2,\varepsilon}(z_2)] &= \int_{\mathbb{R}} g_c^{\omega_1,\varepsilon}(z_1,\zeta)\bar{g}_s^{\omega_2,\varepsilon}(z_2,\zeta)\,d\zeta\\ &= i\sum_{j=1}^{4}(-1)^j I_j^{\varepsilon}(\omega_1,\omega_2,z_1,z_2).\end{aligned}$$

Using the same calculation as just above for $\mathbb{E}[v_c^{\omega_1,\varepsilon}(z_1)v_c^{\omega_2,\varepsilon}(z_2)]$ we get

$$\lim_{\varepsilon\to 0}\mathbb{E}[v_c^{\omega_1,\varepsilon}(z_1)v_s^{\omega_2,\varepsilon}(z_2)] = 0,$$

which concludes the proof. □

3.4 Technical Lemmas

Lemma 3. *We define* $\widetilde{\psi}(\omega) = |\psi(0)|^2|\omega|^{1-2H}$ *if* $\tau \in (0,2)$ *and* $\widetilde{\psi}(\omega) = |\psi(\omega)|^2|\omega|^{1-2H}$ *if* $\tau = 2$. *We have the following convergence*

$$\lim_{\varepsilon\to 0}\int_{\mathbb{R}} \frac{\left|\psi\left(\varepsilon^2\zeta+\omega\varepsilon^{2-\tau}\right)\right|^2}{|\varepsilon^\tau\zeta+\omega|^{2H-1}} g(iz_1\zeta)\, g(-iz_2\zeta)\, d\zeta = \widetilde{\psi}(\omega)\int_{\mathbb{R}} g(iz_1\zeta)\, g(-iz_2\zeta)\, d\zeta$$

for every z_1 *and* $z_2 \in \mathbb{R}$ *and* $\omega \neq 0$.

Proof. Without loss of generality and for simplicity of presentation we assume that $\omega > 0$. We define for a and b in $[-\infty,\infty]$ the ε-dependent integral

$$S^\varepsilon(a,b) = \int_a^b \frac{\left|\psi\left(\varepsilon^2\zeta+\omega\varepsilon^{2-\tau}\right)\right|^2}{|\varepsilon^\tau\zeta+\omega|^{2H-1}} g(iz_1\zeta)\, g(-iz_2\zeta)\, d\zeta$$

then, we need to study the convergence of $S^\varepsilon(-\infty,\infty)$. We fix $\eta > 0$ sufficiently small and write

$$S^\varepsilon(-\infty,\infty) = S^\varepsilon(-\infty,-\eta\varepsilon^{-\tau}) + S^\varepsilon(-\eta\varepsilon^{-\tau},\eta\varepsilon^{-\tau}) + S^\varepsilon(\eta\varepsilon^{-\tau},\infty).$$

By means of the substitution $\zeta\to\zeta/\varepsilon^\tau$ we get

$$S^\varepsilon(-\infty,-\eta\varepsilon^{-\tau}) = \varepsilon^\tau\int_{-\infty}^{-\eta} \frac{\left|\psi\left(\varepsilon^{2-\tau}(\zeta+\omega)\right)\right|^2}{z_1z_2\zeta^2\,|\zeta+\omega|^{2H-1}} \left(e^{iz_1\zeta\varepsilon^{-\tau}}-1\right)\left(e^{-iz_2\zeta\varepsilon^{-\tau}}-1\right)d\zeta$$

so that

$$|S^\varepsilon(-\infty,-\eta\varepsilon^{-\tau})| \le \varepsilon^\tau\int_{-\infty}^{-\eta} \frac{\|\psi\|_\infty^2}{z_1z_2\zeta^2\,|\zeta+\omega|^{2H-1}}\,d\zeta\to 0$$

as $\varepsilon\to 0$. We deal with $S^\varepsilon(\eta\varepsilon^{-\tau},\infty)$ in the same way and therefore it remains only to study $S^\varepsilon(-\eta\varepsilon^{-\tau},\eta\varepsilon^{-\tau})$. We have

$$S^\varepsilon(-\eta\varepsilon^{-\tau},\eta\varepsilon^{-\tau}) = \int_{-\infty}^{\infty} \Psi^\varepsilon_{\omega,\eta}(\zeta)\,d\zeta$$

where

$$\Psi^\varepsilon_{\omega,\eta}(\zeta) = 1_{[-\eta\varepsilon^{-\tau},\eta\varepsilon^{-\tau}]}(\zeta)\frac{\left|\psi\left(\varepsilon^2\zeta+\omega\varepsilon^{2-\tau}\right)\right|^2}{|\varepsilon^\tau\zeta+\omega|^{2H-1}} g(iz_1\zeta)\, g(-iz_2\zeta)\,.$$

For ω, η and ζ fixed we have

$$\lim_{\varepsilon\to 0} \Psi^{\varepsilon}_{\omega,\eta}(\zeta) = \widetilde{\psi}(\omega) g\,(i z_1 \zeta)\, g\,(-i z_2 \zeta)\,.$$

Moreover, observing that

$$|\varepsilon^{\tau}\zeta + \omega|^{1-2H} \le \max\{(\omega-\eta)^{1-2H}, (\omega+\eta)^{1-2H}\}$$

for every $\zeta \in [-\eta\varepsilon^{-\tau}, \eta\varepsilon^{-\tau}]$, we get

$$\left|\Psi^{\varepsilon}_{\omega,\eta}(\zeta)\right| \le \|\psi\|^2_{\infty} \max\{(\omega-\eta)^{1-2H}, (\omega+\eta)^{1-2H}\} |g\,(i z_1 \zeta)\, g\,(-i z_2 \zeta)\,|.$$

The function $\zeta \mapsto g\,(i z_1 \zeta)\, g\,(-i z_2 \zeta)$ is absolutely integrable, thus, by the bounded convergence theorem,

$$\lim_{\varepsilon\to 0} \int_{-\infty}^{\infty} \Psi^{\varepsilon}_{\omega,\eta}(\zeta) d\zeta = \widetilde{\psi}(\omega) \int_{-\infty}^{\infty} g\,(i z_1 \zeta)\, g\,(-i z_2 \zeta)\, d\zeta,$$

which concludes the proof. □

Lemma 4. *We have the following convergence*

$$\lim_{\varepsilon\to 0} \int_{\mathbb{R}} \frac{|\psi\,(\varepsilon^2\zeta + \omega\varepsilon^{2-\tau})\,|^2}{|\varepsilon^{\tau}\zeta + \omega|^{2H-1}} g\,(i z_1 \zeta)\, g\,(-i z_2\,(\zeta + 2\omega\varepsilon^{-\tau}))\; d\zeta = 0$$

for every z_1 and $z_2 \in \mathbb{R}$ and $\omega \neq 0$.

Proof. Without loss of generality and for simplicity of presentation we assume that $\omega = 1$. We have

$$\int_{\mathbb{R}} \frac{|\psi\,(\varepsilon^2\zeta + \varepsilon^{2-\tau})\,|^2}{|\varepsilon^{\tau}\zeta + 1|^{2H-1}} |g\,(i z_1 \zeta)\, g\,(-i z_2\,(\zeta + 2\varepsilon^{-\tau}))|\, d\zeta \tag{15}$$

$$\le \|\psi\|^2_{\infty} \int_{\mathbb{R}} \frac{|g\,(i z_1 \zeta)\, g\,(-i z_2\,(\zeta + 2\varepsilon^{-\tau}))|}{|\varepsilon^{\tau}\zeta + 1|^{2H-1}} d\zeta$$

$$= \|\psi\|^2_{\infty} \int_{\mathbb{R}} \frac{|\exp(i z\zeta) - 1|\,|\exp(-i z_1(\zeta + 2\varepsilon^{-\tau})) - 1|}{|i z\zeta|\,|-i z_2(\zeta + 2\varepsilon^{-\tau})|\,|\varepsilon^{\tau}\zeta + 1|^{2H-1}} d\zeta$$

$$= \varepsilon^{\tau} \frac{\|\psi\|^2_{\infty}}{|z_1 z_2|} \int_{\mathbb{R}} \frac{|\exp(i z_1 \zeta\varepsilon^{-\tau}) - 1|\,|\exp(-i z_2(\zeta + 2)\varepsilon^{-\tau}) - 1|}{|\zeta||\zeta + 2|\,|\zeta + 1|^{2H-1}} d\zeta.$$

We next define for a and b in $[-\infty, \infty]$ the ε-dependent integral

$$J^{\varepsilon}(a,b) = \varepsilon^{\tau} \int_a^b \frac{|\exp(i z\zeta\varepsilon^{-\tau}) - 1|\,|\exp(-i z(\zeta + 2)\varepsilon^{-\tau}) - 1|}{|\zeta||\zeta + 2|\,|\zeta + 1|^{2H-1}} d\zeta.$$

We fix a real number $0 < \eta \ll 1$ sufficiently small and consider separately the integrals $J^\varepsilon(-\infty, -2-\eta)$, $J^\varepsilon(-2-\eta, -2+\eta)$, $J^\varepsilon(-2+\eta, -\eta)$, $J^\varepsilon(-\eta, \eta)$ and $J^\varepsilon(\eta, \infty)$, which are defined such that

$$\varepsilon^\tau \int_{\mathbb{R}} \frac{|\exp(iz\zeta\varepsilon^{-\tau}) - 1|\,|\exp(-iz(\zeta+2)\varepsilon^{-\tau}) - 1|}{|\zeta||\zeta+2|\,|\zeta+1|^{2H-1}} d\zeta \tag{16}$$
$$= J^\varepsilon(-\infty, -2-\eta) + J^\varepsilon(-2-\eta, -2+\eta)$$
$$+ J^\varepsilon(-2+\eta, -\eta) + J^\varepsilon(-\eta, \eta) + J^\varepsilon(\eta, \infty).$$

Using $|\exp(iu) - 1| \leq 2$ for every $u \in \mathbb{R}$, we get

$$J^\varepsilon(-\infty, -2-\eta) + J^\varepsilon(-2+\eta, -\eta) + J^\varepsilon(\eta, \infty) \tag{17}$$
$$\leq 4\varepsilon^\tau \left(\int_{-\infty}^{-2-\eta} + \int_{-2+\eta}^{-\eta} + \int_{\eta}^{\infty} \right) \frac{d\zeta}{|\zeta||\zeta+2|\,|\zeta+1|^{2H-1}}.$$

Using $|\exp(iu) - 1| \leq \min(|u|, 2)$ for every $u \in \mathbb{R}$, we get

$$J^\varepsilon(-2-\eta, -2+\eta) + J^\varepsilon(-\eta, \eta)$$
$$\leq 2|z| \left(\int_{-2-\eta}^{-2+\eta} \frac{d\zeta}{|\zeta|\,|\zeta+1|^{2H-1}} + \int_{-\eta}^{\eta} \frac{d\zeta}{|\zeta+2|\,|\zeta+1|^{2H-1}} \right)$$
$$= \mathcal{O}(\eta) \text{ (independent of } \varepsilon). \tag{18}$$

Then using (15), (16), (17) and (18) we get

$$\lim_{\varepsilon \to 0} \int_{\mathbb{R}} \frac{|\psi(\varepsilon^2\zeta + \varepsilon^{2-\tau})|^2}{|\varepsilon^\tau \zeta + 1|^{2H-1}} |g(iz\zeta)\, g(-iz(\zeta + 2\varepsilon^{-\tau}))|\, d\zeta = 0,$$

which concludes the proof. □

4 Extension to More General Periodic Functions

4.1 Assumptions and Results

In this section we generalize the above results and study the asymptotic behavior of the quantity defined for every $z \geq 0$ by

$$v_p^\varepsilon(z) = \frac{1}{\varepsilon^\gamma} \int_0^z \mu\left(\frac{y}{\varepsilon^2}\right) p\left(\frac{y}{\varepsilon^\tau}\right) dy \tag{19}$$

where p is a zero-mean periodic function. More precisely we assume that there exists $\omega > 0$ and two sequences $\{p_c(k), k \geq 1\}$ and $\{p_s(k), k \geq 1\}$ such that for every $z \geq 0$

$$p(z) = \sum_{k=1}^{\infty} \left(p_c(k)\cos(k\omega z) + p_s(k)\sin(k\omega z)\right).$$

We assume that the Fourier series of the right-hand side of the identity above is uniformly convergent and that the coefficients $\{p_c(k), k \geq 1\}$ and $\{p_s(k), k \geq 1\}$ satisfy

$$\sum_{k=1}^{\infty} k^{1/2-H} \left(|p_c(k)| + |p_s(k)|\right) < \infty. \tag{20}$$

Notice that Assumption 20 implies

$$\sum_{k=1}^{\infty} k^{1-2H} \left(p_c(k)^2 + p_s(k)^2\right) < \infty. \tag{21}$$

Theorem 2. *Under the assumptions above, as $\varepsilon \to 0$ the finite dimensional distributions of v_p^ε converge to those of the Brownian motion B_p satisfying*

$$\mathbb{E}[B_p(z_1)B_p(z_2)] = \begin{cases} \min\{z_1, z_2\} \dfrac{|\psi(0)|^2}{2|\omega|^{2H-1}} \displaystyle\sum_{l=1}^{\infty} l^{1-2H} \left(p_c(l)^2 + p_s(l)^2\right) & \text{if } \tau < 2, \\ \min\{z_1, z_2\} \dfrac{|\omega|^{1-2H}}{2} \displaystyle\sum_{l=1}^{\infty} \dfrac{|\psi(l\omega)|^2}{l^{2H-1}} \left(p_c(l)^2 + p_s(l)^2\right) & \text{if } \tau = 2. \end{cases}$$

Next we consider another zero-mean periodic function $\tilde{p}$ defined for every z by

$$\tilde{p}(z) = \sum_{k=1}^{\infty} \left(\tilde{p}_c(k)\cos(k\tilde{\omega} z) + \tilde{p}_s(k)\sin(k\tilde{\omega} z)\right). \tag{22}$$

where the Fourier series of the right-hand side is uniformly convergent, $\tilde{\omega} > 0$ and the two sequences $\{\tilde{p}_c(k), k \geq 1\}$ and $\{\tilde{p}_s(k), k \geq 1\}$ satisfy

$$\sum_{k=1}^{\infty} k^{1/2-H} \left(|\tilde{p}_c(k)| + |\tilde{p}_s(k)|\right) < \infty. \tag{23}$$

From this function we define the process $v_{\tilde{p}}^\varepsilon$ for every $z \geq 0$ by

$$v_{\tilde{p}}^{\varepsilon}(z) = \frac{1}{\varepsilon^{\gamma}} \int_0^z \mu\left(\frac{y}{\varepsilon^2}\right) \tilde{p}\left(\frac{y}{\varepsilon^{\tau}}\right) dy. \tag{24}$$

By applying Theorem 2 we get that the finite dimensional distributions of $v_{\tilde{p}}^{\varepsilon}$ converge to those of a Brownian motion. Now we address the question of the asymptotic correlation or decorrelation of the processes v_p^{ε} and $v_{\tilde{p}}^{\varepsilon}$.

Theorem 3. *Let z_1 and z_2 be two positive real numbers. Under the assumptions above, as $\varepsilon \to 0$ the covariance $\mathbb{E}[v_p^{\varepsilon}(z_1) v_{\tilde{p}}^{\varepsilon}(z_2)]$ converges to $\mathscr{C}(z_1, z_2)$ defined as*

$$\mathscr{C}(z_1,z_2) = \begin{cases} \min\{z_1,z_2\} \dfrac{|\psi(0)|^2}{2|n\omega|^{2H-1}} \displaystyle\sum_{l=1}^{\infty} l^{1-2H} \left(p_c(nl)\tilde{p}_c(ml) + p_s(nl)\tilde{p}_s(ml)\right) \\ \quad \text{if } \tilde{\omega}/\omega = n/m \in \mathbb{Q} \text{ with } \gcd(n,m)=1 \text{ and } \tau < 2, \\ \min\{z_1,z_2\} \dfrac{|n\omega|^{1-2H}}{2} \displaystyle\sum_{l0=1}^{\infty} \dfrac{|\psi(nl\omega)|^2}{l^{2H-1}} \left(p_c(nl)\tilde{p}_c(ml) + p_s(nl)\tilde{p}_s(ml)\right) \\ \quad \text{if } \tilde{\omega}/\omega = n/m \in \mathbb{Q} \text{ with } \gcd(n,m)=1 \text{ and } \tau = 2, \\ 0 \text{ if } \tilde{\omega}/\omega \notin \mathbb{Q}. \end{cases}$$

4.2 Proof of Theorems 2 and 3

First we establish a technical lemma regarding the quantities $I_j^{\varepsilon}(\omega_1, \omega_2, z_1, z_2)$ defined in Sect. 3.

Lemma 5. *For every z_1 and z_2, there exists a constant $C_I(z_1, z_2) > 0$ such that for every $j \in \{1,2,3,4\}$, $\varepsilon \in (0,1)$ and $\omega > 0$*

$$|I_j^{\varepsilon}(\omega, \omega, z_1, z_2)| \le (\omega^{1-2H} + \omega^{-2H}) C_I(z_1, z_2).$$

As a consequence, for every ω and z, there exists a constant $\tilde{C}(\omega, z) > 0$ such that for every k and ε

$$\mathbb{E}\left[v_c^{k\omega,\varepsilon}(z)^2\right] + \mathbb{E}\left[v_s^{k\omega,\varepsilon}(z)^2\right] \le \tilde{C}(\omega, z) k^{1-2H}$$

Proof. Using the notation introduced in the proof of Lemma 3 and taking $\eta = \omega/2$ we have

$$\begin{aligned} & I_1^{\varepsilon}(\omega, \omega, z_1, z_2) \\ &= \frac{z_1 z_2}{4} \int_{\mathbb{R}} \frac{\left|\psi\left(\varepsilon^2\zeta + \omega\varepsilon^{2-\tau}\right)\right|^2}{|\varepsilon^{\tau}\zeta + \omega|^{2H-1}} g(iz_1\zeta)\, g(-iz_2\zeta)\, d\zeta \\ &= \frac{z_1 z_2}{4} \left(S^{\varepsilon}\left(-\infty, -\frac{\omega}{2\varepsilon^{\tau}}\right) + S^{\varepsilon}\left(-\frac{\omega}{2\varepsilon^{\tau}}, \frac{\omega}{2\varepsilon^{\tau}}\right) + S^{\varepsilon}\left(\frac{\omega}{2\varepsilon^{\tau}}, \infty\right)\right). \end{aligned} \tag{25}$$

By making the substitution $\zeta \to \zeta/\varepsilon^\tau$ and subsequently $\zeta \to \omega\zeta$ we obtain

$$\left|S^\varepsilon\left(-\infty, -\frac{\omega}{2\varepsilon^\tau}\right)\right| \leq 4\varepsilon^\tau \frac{\|\psi\|_\infty^2}{z_1 z_2} \int_{-\infty}^{-\omega/2} \frac{d\zeta}{\zeta^2 |\zeta + \omega|^{2H-1}}$$
$$\leq 4\omega^{-2H} \frac{\|\psi\|_\infty^2}{z_1 z_2} \int_{-\infty}^{-1/2} \frac{d\zeta}{\zeta^2 |\zeta + 1|^{2H-1}}$$

We deal with $S^\varepsilon(\omega/2\varepsilon^\tau, \infty)$ in a similar way. Regarding $S^\varepsilon(-\omega/2\varepsilon^\tau, \omega/2\varepsilon^\tau)$ we have

$$\left|S^\varepsilon\left(-\frac{\omega}{2\varepsilon^\tau}, \frac{\omega}{2\varepsilon^\tau}\right)\right| \leq \frac{\|\psi\|_\infty^2}{z_1 z_2} \int_{-\omega/2\varepsilon^\tau}^{\omega/2\varepsilon^\tau} \frac{|e^{iz_1\zeta} - 1||e^{-iz_2\zeta} - 1|}{\zeta^2 |\varepsilon^\tau \zeta + \omega|^{2H-1}} d\zeta.$$

Moreover, we have

$$|\varepsilon^\tau \zeta + \omega|^{1-2H} \leq \max\{(\omega/2)^{1-2H}, (3\omega/2)^{1-2H}\}$$
$$= \omega^{1-2H} \sup\{(1/2)^{1-2H}, (3/2)^{1-2H}\}$$

for every $\zeta \in [-\omega/2\varepsilon^\tau, \omega/2\varepsilon^\tau]$, thus

$$\left|S^\varepsilon\left(-\frac{\omega}{2\varepsilon^\tau}, \frac{\omega}{2\varepsilon^\tau}\right)\right| \leq \omega^{1-2H} \max\{(1/2)^{1-2H}, (3/2)^{1-2H}\} \frac{\|\psi\|_\infty^2}{z_1 z_2}$$
$$\times \int_{-\infty}^{\infty} \frac{|e^{iz_1\zeta} - 1||e^{-iz_2\zeta} - 1|}{\zeta^2} d\zeta.$$

Hence, the lemma is proved for $I_1^\varepsilon(\omega, \omega, z_1, z_2)$. We deal with $I_4^\varepsilon(\omega, \omega, z_1, z_2)$ in the same way. Regarding $I_2^\varepsilon(\omega, \omega, z_1, z_2)$ (and $I_3^\varepsilon(\omega, \omega, z_1, z_2)$) we observe that by Cauchy-Schwarz inequality we have

$$|I_2^\varepsilon(\omega, \omega, z_1, z_2)| \leq \sqrt{I_1^\varepsilon(\omega, \omega, z_1, z_1) I_4^\varepsilon(\omega, \omega, z_2, z_2)}.$$

Thus, using the bounds for $I_1^\varepsilon(\omega, \omega, z_1, z_2)$ (and $I_4^\varepsilon(\omega, \omega, z_1, z_2)$) completes the proof of the first part of the lemma. To prove the second part, we let

$$\tilde{C}(\omega, z) = 8(\omega^{1-2H} + \omega^{-2H}) C_I(z, z).$$

□

The following lemma deals with a uniform bound for the second moment of $v_p^\varepsilon(z)$ for every z.

Lemma 6. *For every z and ω there exists a sequence $\{c_k(\omega, z)\}_k$ of positive numbers such that for every k and ε*

$$|p_c(k)|\sqrt{\mathbb{E}\left[v_c^{k\omega,\varepsilon}(z)^2\right]} + |p_s(k)|\sqrt{\mathbb{E}\left[v_s^{k\omega,\varepsilon}(z)^2\right]} \leq c_k(\omega, z)$$

and

$$\sum_{k=1}^{\infty} c_k(\omega, z) < \infty.$$

Proof. It is a direct consequence of (20) and Lemma 5. □

Proof. (Theorems 2 and 3)
We have

$$v_p^{\varepsilon}(z) = \frac{1}{\varepsilon^{\gamma}} \int_0^z \mu\left(\frac{y}{\varepsilon^2}\right) \sum_{k=1}^{\infty} \left(p_c(k)\cos\left(k\omega\frac{y}{\varepsilon^{\tau}}\right) + p_s(k)\sin\left(k\omega\frac{y}{\varepsilon^{\tau}}\right)\right) dy.$$

Because the convergence of the Fourier series is uniform we can change the order of the integral and the infinite sum to obtain

$$v_p^{\varepsilon}(z) = \sum_{k=1}^{\infty} \left(p_c(k) v_c^{k\omega,\varepsilon}(z) + p_s(k) v_s^{k\omega,\varepsilon}(z)\right). \tag{26}$$

Moreover, using Lemma 6 we get

$$\begin{aligned} &|p_c(k)\tilde{p}_c(j)|\mathbb{E}\left[|v_c^{k\omega,\varepsilon}(z_1) v_c^{j\tilde{\omega},\varepsilon}(z_2)|\right] \\ &\qquad \leq |p_c(k)\tilde{p}_c(j)| \sqrt{\mathbb{E}\left[v_c^{k\omega,\varepsilon}(z_1)^2\right]\mathbb{E}\left[v_c^{j\tilde{\omega},\varepsilon}(z_2)^2\right]} \\ &\qquad \leq c_k(\omega, z_1) c_j(\tilde{\omega}, z_2) \end{aligned} \tag{27}$$

where $c_k(\omega, z_1)$ and $c_j(\tilde{\omega}, z_2)$ are defined as in Lemma 6. Then we can write

$$\begin{aligned} \mathbb{E}[v_p^{\varepsilon}(z_1) v_{\tilde{p}}^{\varepsilon}(z_2)] &= \sum_{k=1}^{\infty}\sum_{j=1}^{\infty} p_c(k)\tilde{p}_c(j)\mathbb{E}\left[v_c^{k\omega,\varepsilon}(z_1) v_c^{j\tilde{\omega},\varepsilon}(z_2)\right] \\ &+ \sum_{k=1}^{\infty}\sum_{j=1}^{\infty} p_c(k)\tilde{p}_s(j)\mathbb{E}\left[v_c^{k\omega,\varepsilon}(z_1) v_s^{j\tilde{\omega},\varepsilon}(z_2)\right] \\ &+ \sum_{k=1}^{\infty}\sum_{j=1}^{\infty} p_s(k)\tilde{p}_c(j)\mathbb{E}\left[v_s^{k\omega,\varepsilon}(z_1) v_c^{j\tilde{\omega},\varepsilon}(z_2)\right] \\ &+ \sum_{k=1}^{\infty}\sum_{j=1}^{\infty} p_s(k)\tilde{p}_s(j)\mathbb{E}\left[v_s^{k\omega,\varepsilon}(z_1) v_s^{j\tilde{\omega},\varepsilon}(z_2)\right]. \end{aligned} \tag{28}$$

We first deal with the convergence of the first term of the right hand side of the identity above. We know from Theorem 1 that for each (j,k),

$$\lim_{\varepsilon\to 0}\mathbb{E}\left[v_c^{k\omega,\varepsilon}(z_1)v_c^{j\tilde{\omega},\varepsilon}(z_2)\right]=\frac{C(j\omega)}{2}\inf(z_1,z_2)\mathbb{I}_{\{j\omega=k\tilde{\omega}\}}. \tag{29}$$

Combining the bounded convergence theorem, (29), (27) and Lemma 6 then implies

$$\lim_{\varepsilon\to 0}\sum_{k=1}^{\infty}\sum_{j=1}^{\infty}p_c(k)\tilde{p}_c(j)\mathbb{E}\left[v_c^{k\omega,\varepsilon}(z_1)v_c^{j\tilde{\omega},\varepsilon}(z_2)\right]$$
$$=\frac{\min(z_1,z_2)}{2}\sum_{k=1}^{\infty}\sum_{j=1}^{\infty}C(j\omega)^2\mathbb{I}_{\{j\omega=k\tilde{\omega}\}}p_c(k)\tilde{p}_c(j).$$

Proceeding in the same way for the other terms of the right hand side of (28) we get

$$\lim_{\varepsilon\to 0}\mathbb{E}[v_p^{\varepsilon}(z_1)v_{\tilde{p}}^{\varepsilon}(z_2)]=\frac{\min(z_1,z_2)}{2}\sum_{k=1}^{\infty}\sum_{j=1}^{\infty}C(j\omega)^2\mathbb{I}_{\{j\omega=k\tilde{\omega}\}}$$
$$\times\left(p_c(k)\tilde{p}_c(j)+p_s(k)\tilde{p}_s(j)\right). \tag{30}$$

If $\tilde{\omega}/\omega\in\mathbb{R}-\mathbb{Q}$ then it is obvious that

$$\lim_{\varepsilon\to 0}\mathbb{E}[v_p^{\varepsilon}(z_1)v_{\tilde{p}}^{\varepsilon}(z_2)]=0. \tag{31}$$

Let us now consider that $\tilde{\omega}/\omega\in\mathbb{Q}$. Then there exist two positive integers m and n such that $\gcd(m,n)=1$ and $\tilde{\omega}m=\omega n$. If $j\omega=k\tilde{\omega}$ then $j=kn/m$ so that there exists an integer l such that $k=ml$ and $j=nl$. We then conclude

$$\lim_{\varepsilon\to 0}\mathbb{E}[v_p^{\varepsilon}(z_1)v_{\tilde{p}}^{\varepsilon}(z_2)]=\frac{\min(z_1,z_2)}{2}\sum_{l=1}^{\infty}C(nl\omega)^2\left(p_c(ml)\tilde{p}_c(nl)+p_s(ml)\tilde{p}_s(nl)\right). \tag{32}$$

□

5 Waves in Randomly Layered Media

In this section we return to the discussion of applications to waves, the first motivation of our work. We consider wave propagation in randomly layered media. In the model that we consider the governing equations are the Euler equations giving conservation of moments and mass:

$$\rho^\varepsilon(z)\frac{\partial u^\varepsilon}{\partial t}(z,t) + \frac{\partial p^\varepsilon}{\partial z}(z,t) = 0\,,$$
$$\frac{1}{K^\varepsilon(z)}\frac{\partial p^\varepsilon}{\partial t}(z,t) + \frac{\partial u^\varepsilon}{\partial z}(z,t) = 0\,,$$

where t is the time, z is the depth into the medium, p^ε is the pressure and u^ε the particle velocity. The medium parameters are the density ρ^ε and the bulk-modulus K^ε (reciprocal of the compressibility). We assume that ρ^ε is a constant identically equal to one in our non-dimensionalized units and $1/K^\varepsilon$ is modeled as

$$\frac{1}{K^\varepsilon(z)} = \begin{cases} 1 + \varepsilon^\kappa \mu\left(\frac{z}{\varepsilon^2}\right) & \text{for } z \in [0, Z]\,, \\ 1 & \text{for } z \in \mathbb{R} - [0, Z]\,, \end{cases}$$

where $\kappa > 0$. We introduce the right- and left-going waves:

$$A^\varepsilon = p^\varepsilon + u^\varepsilon \quad \text{and} \quad B^\varepsilon = u^\varepsilon - p^\varepsilon\,,$$

The boundary conditions are of the form

$$A^\varepsilon(z=0,t) = f(t/\varepsilon^\tau) \quad \text{and} \quad B^\varepsilon(z=Z,t) = 0\,,$$

for a positive real number $\tau > 0$ and a source function f. In order to deduce a description of the transmitted pulse, we open a window of size ε^τ in the neighborhood of the travel time of the homogenized medium and define the processes

$$a^\varepsilon(z,s) = A^\varepsilon(z, z+\varepsilon^\tau s) \quad \text{and} \quad b^\varepsilon(z,s) = B^\varepsilon(z, -z+\varepsilon^\tau s)\,. \tag{33}$$

Observe that the background or homogenized medium in our scaling has a constant speed of sound equal to unity and that the medium is matched so that in the frame introduced in (33) the pulse shape is constant if $\mu \equiv 0$ or if we consider the homogenized medium [2].

An important question is the study of the asymptotics of $a^\varepsilon(Z,s)$ when ε goes to 0 [2]. In order to address this question we introduce next the Fourier transforms $\widehat{a}^\varepsilon$ and $\widehat{b}^\varepsilon$ of a^ε and b^ε respectively:

$$\widehat{a}^\varepsilon(z,\omega) = \int e^{i\omega s} a^\varepsilon(z,s)ds \quad \text{and} \quad \widehat{b}^\varepsilon(z,\omega) = \int e^{i\omega s} b^\varepsilon(z,s)ds\,,$$

that satisfy

$$\frac{d\widehat{a}^\varepsilon}{dz} = \frac{i\omega}{2\varepsilon^{\tau-\kappa}}\mu\left(\frac{z}{\varepsilon^2}\right)\left(\widehat{a}^\varepsilon - e^{-2i\omega z/\varepsilon^\tau}\widehat{b}^\varepsilon\right), \qquad \widehat{a}^\varepsilon(0,\omega) = \widehat{f}(\omega),$$

$$\frac{d\widehat{b}^\varepsilon}{dz} = \frac{i\omega}{2\varepsilon^{\tau-\kappa}}\mu\left(\frac{z}{\varepsilon^2}\right)\left(e^{2i\omega z/\varepsilon^\tau}\widehat{a}^\varepsilon - \widehat{b}^\varepsilon\right), \qquad \widehat{b}^\varepsilon(Z,\omega) = 0.$$

Following [1, 2] we express the previous system of equations in term of the propagator $P_\omega^\varepsilon(z)$ which can be written as

$$P_\omega^\varepsilon(z) = \begin{pmatrix} \alpha_\omega^\varepsilon(z) & \overline{\beta_\omega^\varepsilon}(z) \\ \beta_\omega^\varepsilon(z) & \overline{\alpha_\omega^\varepsilon}(z) \end{pmatrix},$$

and which satisfies

$$\frac{dP_\omega^\varepsilon}{dz}(z) = \frac{1}{\varepsilon^{\tau-\kappa}}\mathscr{H}_\omega\left(\frac{z}{\varepsilon^\tau}, \frac{z}{\varepsilon^2}\right)P_\omega^\varepsilon(z), \qquad P_\omega^\varepsilon(z=0) = \begin{pmatrix} 1 & 0 \\ 0 & 1 \end{pmatrix}, \tag{34}$$

with

$$\mathscr{H}_\omega(z_1, z_2) = \frac{i\omega}{2}\mu(z_2)\begin{pmatrix} 1 & -e^{-2i\omega z_1} \\ e^{2i\omega z_1} & -1 \end{pmatrix}.$$

Defining next the transmission coefficient T_ω^ε and the reflection coefficient R_ω^ε by

$$T_\omega^\varepsilon(z) = \frac{1}{\overline{\alpha_\omega^\varepsilon}(z)} \quad \text{and} \quad R_\omega^\varepsilon(z) = \frac{\beta_\omega^\varepsilon(z)}{\overline{\alpha_\omega^\varepsilon}(z)}, \tag{35}$$

we can then write

$$a^\varepsilon(Z,s) = \frac{1}{2\pi}\int e^{-is\omega}T_\omega^\varepsilon(Z)\widehat{f}(\omega)\,d\omega, \tag{36}$$

and

$$b^\varepsilon(0,s) = \frac{1}{2\pi}\int e^{-is\omega}R_\omega^\varepsilon(Z)\widehat{f}(\omega)\,d\omega. \tag{37}$$

Hence we shall study the asymptotics of the propagator P_ω^ε in order to characterize a^ε as ε goes to 0.

What we wish to point out here is the dramatic influence of the range parameter H of μ. Let us recall two cases that have already been studied in previous works. For the moment we restrict ourself to the cases of slow phases $\tau \in (0,2)$.

- *The case $H = 1/2$ or the mixing case.* This corresponds to the framework studied in [1] and further developed in [2]. With $\tau \in (0,2)$ we assume $\kappa = \tau - 1$ to get a non-trivial asymptotic behavior. We then have the convergence as ε goes to 0

$$a^\varepsilon(Z,s) \longrightarrow \widetilde{a}(Z,s) := (G * f)(s-B), \tag{38}$$

where G is a centered Gaussian density and B a Gaussian random variable.

- *The case $H > 1/2$ or the long-range case.* Results on the propagation in a long-range medium have been derived in [9]. An important property of such a model is that there exists a constant $c_H > 0$ such that

$$\mathbb{E}[\mu(0)\mu(z)] \sim c_H z^{2H-2}.$$

In this case, we can prove that

$$a^\varepsilon(Z,s) \longrightarrow \widetilde{a}(Z,s) := f(s-B)\,, \tag{39}$$

where B a Gaussian random variable.

In these two cases one of the most crucial step in the proof is the convergence of the propagator P^ε_ω. This consists of studying the asymptotics of the random differential equation (34). To do so, different tools are used depending on whether $H = 1/2$ or $H > 1/2$. If $H = 1/2$ then the diffusion-approximation and martingales techniques have been used [1,2]. If $H > 1/2$, because of the long-range property, the martingale approach cannot be used. In [9] the theory of rough paths [6, 7] that we describe below was used.

A motivation for this work is to find a general framework dealing with the full range of $H \in (0,1)$ and in particular to address the case $H \in (0,1/2)$. Based on Theorem 1 we next present our conjecture regarding $H \in (0,1/2)$. We consider μ defined as throughout this paper, but to fix the ideas we may assume that μ is defined by $\mu(z) = B_H(z+1) - B_H(z)$ where $H \in (0,1/2)$. Now we assume that

$$\tau - \kappa = \gamma = 1 + (2-\tau)(1-2H)/2$$

and $\kappa < 1$ so that $\tau \in (0,2)$. Note that the case $\gamma = 0$ corresponds to the so called strong fluctuations regime, with medium fluctuations being of order one, see [2]. Recall that the propagator P^ε_ω satisfies (34) that we can write in the form

$$dP^\varepsilon_\omega(z) = \frac{i\omega}{2}\left(F_1 P^\varepsilon_\omega \circ dv^{0,\varepsilon}_c(z) + F_2 P^\varepsilon_\omega \circ dv^{2\omega,\varepsilon}_c(z) + F_3 P^\varepsilon_\omega \circ dv^{2\omega,\varepsilon}_s(z)\right), \tag{40}$$

where $F_1 = \begin{pmatrix} 1 & 0 \\ 0 & -1 \end{pmatrix}$, $F_2 = \begin{pmatrix} 0 & -1 \\ 1 & 0 \end{pmatrix}$ and $F_3 = \begin{pmatrix} 0 & i \\ i & 0 \end{pmatrix}$ and

$$v^{0,\varepsilon}_c(z) = \frac{1}{\varepsilon^\gamma}\int_0^z \mu\left(\frac{y}{\varepsilon^2}\right) dy = \frac{\varepsilon^{\tau(1-2H)/2}}{\varepsilon^{2-2H}}\int_0^z \mu\left(\frac{y}{\varepsilon^2}\right) dy. \tag{41}$$

Because of $H < 1/2$, (41) and the invariance principle that establishes that

$$\lim_{\varepsilon \to 0} \left\{ \frac{1}{\varepsilon^{2-2H}} \int_0^z \mu\left(\frac{y}{\varepsilon^2}\right) dy \right\}_z =^{\text{distribution}} B_H,$$

we can then observe the convergence

$$\lim_{\varepsilon \to 0} v_c^{0,\varepsilon} =^{\text{distribution}} 0. \tag{42}$$

Let $\Omega = (\omega_1, \cdots, \omega_n)$ be a collection of frequencies. Because of (42) and Theorem 1, as ε goes to 0, we may expect that the propagator vector $P_\Omega^\varepsilon = (P_{\omega_1}^\varepsilon, \cdots, P_{\omega_n}^\varepsilon)$ converges to $P_\Omega = (P_{\omega_1}, \cdots, P_{\omega_n})$ where the asymptotic propagator P_ω is solution of

$$dP_\omega(z) = \frac{iC(\omega)\omega}{2}\Big(F_2 P_\omega \circ dB_c^\omega(z) + F_3 P_\omega \circ dB_s^\omega(z)\Big)$$

where $C(\omega)^2$ is defined by (10). Then we follow the same procedure as in [2] pp. 180–184 or as in [4] using complex martingales to get the limit distribution written as

$$\widetilde{a}(Z,s) =^{\text{distribution}} \frac{1}{2\pi} \int_{-\infty}^{\infty} \exp\left(-i\omega s - \frac{\omega^2 C(\omega)^2}{4} Z\right) \widehat{f}(\omega) d\omega.$$

Hence, we get the following conjecture.

Conjecture 1. Under the above assumptions, as ε goes to 0, $\{a^\varepsilon(Z,s)\}_s$ converges to the random process $\{\widetilde{a}(Z,s)\}_s$ that can be written as

$$\widetilde{a}(Z,s) = (G_Z * f)(s), \tag{43}$$

where G_Z is such that its Fourier transform is

$$\widehat{G_Z}(\omega) = \exp\left(-\frac{\pi |\psi(0)|^2}{4} |\omega|^{3-2H} Z\right).$$

Note that this means that we in the antipersistent case have a modification in the pulse shape only, while we in the long range case have a correction in the travel time only and in the mixing case a modification in both. The technical challenge in the full proof would be the rigorous convergence of P_Ω^ε to P_Ω as $\varepsilon \to 0$. This could be based on the use of the theory of rough paths initiated by T. Lyons [7] and for which a good reference is [6]. To summarize, the main theorem establishes that the solution x of the differential system $dx = F(x)dw$ driven by a continuous and multidimensional noise $w = (w_1, .., w_n)$ is a continuous function of the noise w for a topology based on the q-variations of iterated integrals of w for q large enough.

This would imply that the convergence of P^{ε}_{Ω} could be reduced to the convergence of the noises of the form $V^{\Omega,\varepsilon}$ and their iterated integrals.

References

1. J.F. Clouet, J.P. Fouque, Spreading of a pulse travelling in random media. Ann. Appl. Probab. **4**, 1083–1097 (1994)
2. J.P. Fouque, J. Garnier, G. Papanicolaou, K. Solna, *Wave Propagation and Time Reversal in Randomly Layered Media* (Springer, New York, 2007)
3. J. Garnier, A multi-scaled diffusion-approximation theorem. Applications to wave propagation in random media. ESAIM Probab. Statist. **1**, 183–206 (1997)
4. J. Garnier, G. Papanicolaou, Analysis of pulse propagation through an one-dimensional random medium using complex martingales. Stoch. Dyn. **8**, 127–138 (2008)
5. J. Garnier, K. Solna, Pulse propagation in random media with long-range correlation. SIAM Multiscale Model. Simul. **7**, 1302–1324 (2009)
6. A. Lejay, An introduction to rough paths, *Séminaire de Probabilités XXXVII*, Lecture Notes in Mathematics (Springer, New York, 2003)
7. T. Lyons, Differential equations driven by rough signals. Rev. Mat. Iberoamer. **14**(2), 215–310 (1998)
8. R. Marty, Asymptotic behavior of differential equations driven by periodic and random processes with slowly decaying correlations. ESAIM: Probab. Statist. **9**, 165–184 (2005)
9. R. Marty, K. Solna, Acoustic waves in long range random media. SIAM J. Appl. Math. **69**, 1065–1083 (2009)
10. R. Marty, K. Solna, A general framework for waves in random media with long-range correlations. Ann. Appl. Probab. **21**(1), 115–139 (2011)
11. G. Samorodnitsky, M.S. Taqqu, *Stable Non-Gaussian Random Processes* (Chapman and Hall, New York, 1994)

Time Inversion Property for Rotation Invariant Self-similar Diffusion Processes

Juha Vuolle-Apiala

Abstract We will show, using the skew product representation and the corresponding result for the radial process, that Shiga–Watanabe's time inversion property of index α, α positive, holds for all α-self-similar, rotation invariant diffusion processes on $\mathbf{R}^{\mathrm{d}}$, $\mathrm{d} \geq 2$, starting at 0.

Keywords Time inversion • Self-similar • Bessel process • Diffusion • Rotation invariant • Skew product • Radial process

AMS Classification: 60G18, 60J60, 60J25

1 Introduction

The following property is well known for Brownian motion in $\mathbf{R}^d$, $d \geq 1$, starting at 0:

$$(X_t) \text{ and } (tX_{1/t}) \text{ are equivalent diffusions under } \mathbf{P}^0. \tag{1}$$

Shiga and Watanabe [8] and Watanabe [13] showed that (1) is also true for all Bessel diffusions on non-negative real numbers.

Bessel diffusions (including Brownian motion) form exactly the class of 1/2-self-similar diffusions on the non-negative real axis; see Lamperti [5]. For α-self-similar diffusions, α strictly positive, (1) can immediately be generalized to (2):

$$(X_t) \text{ and } (t^{2\alpha} X_{1/t}) \text{ are equivalent diffusions under } \mathbf{P}^0. \tag{2}$$

It was in Vuolle-Apiala [11] showed that (2) is valid for all rotation invariant (RI) α-self-similar diffusions (X_t) on $\mathbf{R}^d$, $d \geq 2$, for which 0 is a polar set, that is,

J. Vuolle-Apiala (✉)
Mathematical Department, Åbo Akademi University, FIN-20500 Åbo, Turku, Finland
e-mail: jvuolle@abo.fi

C. Donati-Martin et al. (eds.), *Séminaire de Probabilités XLIV*, Lecture Notes in Mathematics 2046, DOI 10.1007/978-3-642-27461-9_13,

after starting at 0 X_t will never return to it. The time inversion problem on $\mathbf{R}$ was studied in Graversen, Vuolle-Apiala [3] and on finite number of rays meeting at 0 in Vuolle-Apiala [10].

We call (2) the time inversion property of index α, (1) is a special case for $\alpha = 1/2$.

In this note we will show that (2) is valid for all rotation invariant (RI) α-self-similar diffusions even if 0 is not a polar set. We have a proof which is valid, no matter if 0 is polar or non-polar. Our proof is based on:

(1) The fact that (2) is true for the radial process.
(2) The skew product representation of (X_t), which is formulated at the end of this section (formula (6)).

By α-self-similarity property, $\alpha > 0$, we mean that

$$\begin{aligned} &(X_t, \mathbf{P}^x) \text{ has the same finite dimensional distributions as} \\ &(a^{-\alpha} X_{at}, \mathbf{P}^{a^\alpha x}) \text{ for all } x \in \mathbf{R}^d, \text{ for all positive } a. \end{aligned} \tag{3}$$

By (RI) we mean

$$\begin{aligned} &(X_t, \mathbf{P}^x) \text{ has the same finite dimensional distributions as} \\ &(T^{-1}(X_t), \mathbf{P}^{T(x)}) \text{ for all rotations } T \in O(d). \end{aligned} \tag{4}$$

Obviously, (4) (and (3), of course) is fulfilled for Brownian motion on $\mathbf{R}^{\mathrm{d}}$, $\mathrm{d} \geq 2$.

Remark 1. A related problem for (X_t) under $\mathbf{P}^x$ for arbitrary x has been studied in two papers: Gallardo and Yor [1] and Lawi [6]. In both articles the following property was called a time inversion property (of index 2α according to the definition by S. Lawi):

($\star$) Assume (X_t) under $\mathbf{P}^x$ is a time-homogeneous Markov process, for all $x \in \mathbf{R}^d$. Then $(t^{2\alpha} X_{1/t})$ is a time-homogeneous Markov process under $\mathbf{P}^x$.

For $x = 0$ this implies our definition (2). Lawi also obtained a converse property: ($\star$) implies that (X_t) is α-self-similar or h-transform of an α-self-similar Markov process (see Corollary 2.5 in Lawi [6]).

In Vuolle-Apiala [11], where $\{0\}$ was assumed to be a polar set, our main tool was the skew product representation for $(X_t, \mathbf{P}^0)$:

$$\begin{aligned} &(X_t) \text{ under } \mathbf{P}^0 \text{ has the same finite dimensional distributions as} \\ &[|X_{\mathrm{t}}|, \nu_{\lambda \int_1^{\mathrm{t}} |X_h|^{-1/\alpha} dh}] \text{ under } \mathbf{P}^0 \times \mathbf{Q}, \end{aligned} \tag{5}$$

for some $\lambda > 0$, where $(\nu_s, \mathbf{Q})$ is a stationary spherical Brownian motion, independent of the radial process $(|X_t|, \mathbf{P}^0)$, defined for $-\infty < s < +\infty$ such that the law of (ν_0) is the uniform spherical distribution $m(d\theta)$. Equation (5) is a consequence of the result of Ito and McKean Jr. [4], p.275.

The radial process $(|X_t|, \mathbf{P}^{|x|})$, $x \in \mathbf{R}^d$, is an α-self-similar diffusion on $[0, \infty)$ and the time inversion property (2) for $(|X_t|, \mathbf{P}^0)$ is therefore fulfilled.

If $\{0\}$ is a polar set then the skew product (5) at 0 completely characterizes $(X_t, \mathbf{P}^0)$. If 0 is a regular point for $\{0\}$, then the skew product representation (5) at 0 is not available. However, when X_t is started outside 0, then we have the following skew product representation:

$$(X_t) \text{ under } \mathbf{P}^x \text{ has the same finite dimensional distributions as}$$
$$[|X_t|, \theta_{\lambda \int_0^t |X_h|^{-1/\alpha} dh}] \text{ under } \mathbf{P}^{|x|} \times \mathbf{Q}^{\theta},\ t < T_0, \tag{6}$$

for $x \neq 0$, for some $\lambda > 0$, where $(\theta_s, \mathbf{Q}^{\theta})$ is a spherical Brownian motion on S^{d-1}, starting at $\theta = arg(x)$, $s > 0$, independent of $(|X_t|)$ (see Graversen, Vuolle-Apiala [2] and Vuolle-Apiala [9]), $T_0 = inf\{t > 0 | X_t = 0\}$.

We apply (6) and the fact that the radial process fulfills (2) to construct a proof for (2) which is valid both in the case of $\{0\}$ as a non-polar set and in the case of $\{0\}$ as a polar set for (X).

Remark 2. All the α-self-similar (RI) diffusions on $\mathbf{R}^d$, $d \geq 2$, can be characterized as follows:

As shown in Graversen, Vuolle-Apiala [2] and Vuolle-Apiala [9], every α-self-similar (RI) diffusion on $\mathbf{R}^d \setminus \{0\}$, $d \geq 2$, killed at T_0 if T_0 is finite, can be written as a skew product (6). On the other hand, every skew product of type (6) gives an α-self-similar (RI) diffusion on $\mathbf{R}^d \setminus \{0\}$, the radial process can be chosen to be any α-self-similar diffusion on $(0, \infty)$ which is independent of the spherical process (see also Vuolle-Apiala [9], Remark on p. 965 and Lemma 2.1). As shown in Vuolle-Apiala [9], Sect. 2, 0 can be added to the state space such that the extended process (Z_t) is an α-self-similar (RI) diffusion on $\mathbf{R}^d$ which on $\mathbf{R}^d \setminus \{0\}$ behaves like (6). $\{0\}$ is polar for $(Z_t, \mathbf{P}^x)$, $x \in \mathbf{R}^d$, if and only if 0 is an entrance, non-exit boundary point of $[0, \infty)$ for $(|Z_t|)$.

2 The Main Result

Theorem 1. *Let $(X_t, \mathbf{P}^0)$ be an α-self-similar (RI) diffusion on $\mathbf{R}^d$, $d \geq 2$, $\alpha > 0$, starting at 0. Then the time inversion property (2) is valid.*

The following lemma plays a central role in the proof of Theorem 1:

Lemma 1. *Let $(X_t, \mathbf{P}^x)$, $x \in \mathbf{R}^d$, be an α-self-similar (RI) diffusion on $\mathbf{R}^d$, $d \geq 2$, $\alpha > 0$. Then*

(i) the radial process $(|X_t|, \mathbf{P}^{|x|})$ is an α-self-similar diffusion on $[0, \infty)$. Consequently, the process $(|X_t|, \mathbf{P}^0)$ fulfills the time inversion property (2).
(ii) For $x \neq 0$, $(X_t, \mathbf{P}^x)$, $t < T_0$, has the skew product representation (6).

Proof. The time inversion property for the radial process in (i) was shown in Shiga and Watanabe [8] for $\alpha = \frac{1}{2}$, the generalization for $\alpha > 0$ is obvious. For the Proof of (ii) see Graversen and Vuolle-Apiala [2], Theorem 2.2, and Vuolle-Apiala [9], Lemma 2.1. □

Proof (Proof of Theorem 1). Assume for simplicity that $\alpha = 1/2$, the general case can be shown similarly.

Since the two processes (X_t) and $(tX_{1/t})$ have continuous paths on $(0,\infty)$, standard arguments of general theory of processes, based on the monotone classes theorem (See for example [7]), allow us to only check that these two processes have same finite dimensional distributions.

We will prove that (X_t) and $(tX_{1/t})$ have the same n-dimensional distributions for any positive integer n:

Assume $n = 2$, the proof in the general case is analogous. Let I_i, $i = 1,2$, be a Borel subset of $(0,\infty)$ and $J_i, i = 1,2$, a Borel subset of S^{d-1} and let $X_t = [r_t, \phi_t]$, r_t and ϕ_t are the radial and the spherical parts of X_t, respectively. Then for $s < t$ we have

$$\mathbf{P}^0\{r_s \in I_1, r_t \in I_2, \phi_s \in J_1, \phi_t \in J_2\} = A + B,$$

where

$$\begin{aligned} A &= \mathbf{P}^0\{r_s \in I_1, r_t \in I_2, \phi_s \in J_1, \phi_t \in J_2;\ s \text{ and } t \text{ belong to different excursions of} \\ &\qquad (r_u) \text{ away from } 0\} \\ &= \mathbf{P}^0\{r_s \in I_1, r_t \in I_2, \phi_s \in J_1, \phi_t \in J_2; r_u = 0 \text{ for some } u \in (s,t)\} \end{aligned}$$

and

$$\begin{aligned} B &= \mathbf{P}^0\{r_s \in I_1, r_t \in I_2, \phi_s \in J_1, \phi_t \in J_2;\ s \text{ and } t \text{ belong to the same excursion of} \\ &\qquad (r_u) \text{ away from } 0\} \\ &= \mathbf{P}^0\{r_s \in I_1, r_s \in I_2, \phi_s \in J_1, \phi_t \in J_2; r_u \neq 0 \text{ for all } u \in (s,t)\}. \end{aligned}$$

Remark 3. (i) As shown by J. Lamperti [5], p.210, Lemma 2.5, either $T_0 < \infty$ a.s. or $T_0 = \infty$ a.s.. (Lamperti actually showed that $T_0 < \infty$ a.s. or $T_0 = \infty$ a.s. for $(X_t, \mathbf{P}^x)$, $x \neq 0$. However, if X_t starts at 0 then it immediately leaves 0 and enters $\mathbf{R}^d \setminus \{0\}$).

(ii) If $\{0\}$ is polar then obviously $A = 0$.

Now A is equal to

$$\begin{aligned} &\mathbf{P}^0\{r_s \in I_1, r_t \in I_2, \phi_s \in J_1, \phi_t \in J_2 | r_u = 0 \text{ for some } u \in (s,t)\} \\ &\qquad\qquad \times \mathbf{P}^0\{r_u = 0 \text{ for some } u \in (s,t)\}. \end{aligned}$$

Because of (RI) and independence of the excursions of (r_u, ϕ_u), away from 0 we have

$$
\begin{aligned}
\mathbf{P}^0\{r_s \in I_1, r_t \in I_2, \phi_s \in J_1, \phi_t \in J_2 | r_u = 0 \text{ for some } u \in (s,t)\}& \\
= \mathbf{P}^0\{r_s \in I_1, \phi_s \in J_1 | r_u = 0 \text{ for some } u \in (s,t)\}& \\
\times \mathbf{P}^0\{r_t \in I_2, \phi_t \in J_2 | r_u = 0 \text{ for some } u \in (s,t)\}& \\
= m(J_1)\mathbf{P}^0\{r_s \in I_1 | r_u = 0 \text{ for some } u \in (s,t)\}& \\
\times m(J_2)\mathbf{P}^0\{r_t \in I_2 | r_u = 0 \text{ for some } u \in (s,t)\}& \\
= m(J_1)m(J_2)\mathbf{P}^0\{r_s \in I_1, r_t \in I_2 | r_u = 0& \\
\text{for some } u \in (s,t)\}& \qquad (7)
\end{aligned}
$$

because of independence of the excursions of (r_u), m is the uniform spherical distribution on S^{d-1}.

Thus

$$A = m(J_1)m(J_2)\mathbf{P}^0\{r_s \in I_1, r_t \in I_2; r_u = 0 \text{ for some } u \in (s,t)\}.$$

Similarly as in Vuolle-Apiala [10], p.257, using the fact that (1) is valid for the radial process (r_t), we can now conclude that

$$
\begin{aligned}
A &= m(J_1)m(J_2)\mathbf{P}^0\{sr_{1/s} \in I_1, tr_{1/t} \in I_2; ur_{1/u} = 0 \text{ for some } u \in (s,t)\} \\
&= m(J_1)m(J_2)\mathbf{P}^0\{sr_{1/s} \in I_1, tr_{1/t} \in I_2; r_u = 0 \text{ for some } u \in (1/t, 1/s)\}.
\end{aligned}
$$

Equation (7) implies, replacing s, t by $1/s$ and $1/t$, (now $1/t < 1/s$) and I_1, I_2 by I_1/s, I_2/t, that

$$
\begin{aligned}
&\mathbf{P}^0\{r_{1/s} \in I_1/s, r_{1/t} \in I_2/t,\ \phi_{1/s} \in J_1, \phi_{1/t} \in J_2 | r_u = 0 \text{ for some } u \in (1/t, 1/s)\} \\
&\quad = m(J_1)m(J_2)\mathbf{P}^0\{sr_{1/s} \in I_1, tr_{1/t} \in I_2 | r_u = 0 \text{ for some } u \in (1/t, 1/s)\}.
\end{aligned}
$$

which further implies

$$
\begin{aligned}
A &= \mathbf{P}^0\{r_{1/s} \in I_1/s, r_{1/t} \in I_2/t,\ \phi_{1/s} \in J_1, \phi_{1/t} \in J_2 | r_u = 0 \text{ for some } u \in (1/t, 1/s))\} \\
&\quad \times \mathbf{P}^0\{r_u = 0 \text{ for some } u \in (1/t, 1/s)\} \\
&= \mathbf{P}^0\left\{sr_{1/s} \in I_1, tr_{1/t} \in I_2, \phi_{1/s} \in J_1, \phi_{1/t} \in J_2;\ 1/s \text{ and } 1/t \text{ belong to}\right. \\
&\quad \left.\text{different excursions of } (r_u) \text{ away from } 0\right\}.
\end{aligned}
$$

Now

$$
\begin{aligned}
B &= \mathbf{P}^0\{r_s \in I_1, r_t \in I_2, \phi_s \in J_1, \phi_t \in J_2; r_u \neq 0 \text{ for all } u \in (s,t)\} \\
&= \mathbf{E}^0\{\mathbf{P}^{(r_s,\phi_s}\{r_{t-s} \in I_2, \phi_{t-s} \in J_2, t-s < T_0\}; r_s \in I_1, \phi_s \in J_1\}
\end{aligned}
$$

because of the Markov property. Using the skew product representation (6) for $\alpha = \frac{1}{2}$ we obtain

$$B = \mathbf{E}^0\{\mathbf{P}^{(r_s,\phi_s)}\{r_{t-s} \in I_2, \theta_{\lambda \int_0^{t-s} r_h^{-2} dh} \in J_2\}; r_s \in I_1, \phi_s \in J_1\}.$$

This can be rewritten

$$B = \iint_{I_1 x J_1} \mathbf{P}^{(r,\phi)}\{r_{t-s} \in I_2, \theta_{\lambda \int_0^{t-s} r_h^{-2} dh} \in J_2\}\mathbf{P}^0\{r_s \in dr,\ \phi_s \in d\phi\}.$$

(RI) implies

$$B = \iint_{I_1 \mathrm{x}\, J_1} \mathbf{P}^{(r,\phi)}\{r_{t-s} \in I_2, \theta_{\lambda \int_0^{t-s} r_h^{-2} dh} \in J_2\}\mathbf{P}^0\{r_s \in dr\}m(d\phi),$$

where m is the uniform spherical distribution on S^{d-1}. This is further equal to

$$\int_0^\infty \left[\iint_{I_1 x J_1} \mathbf{P}^{(r,\phi)}\{r_{t-s} \in I_2, \theta_v \in J_2, \lambda \int_0^{t-s} r_h^{-2} dh \in dv\}\mathbf{P}^0\{r_s \in dr\}m(d\phi)\right].$$

The independence of (r_u) and (θ_v) implies

$$B = \iint_{I_1 x J_1} \mathbf{P}^0\{r_s \in dr\}m(d\phi)\left[\int_0^\infty \mathbf{Q}^\phi\{\theta_v \in J_2\}\mathbf{P}^r\{r_{t-s} \in I_2, \lambda \int_0^{t-s} r_h^{-2} dh \in dv\}\right].$$

which is equal to

$$\int_{J_1} m(d\phi)\left[\int_0^\infty \mathbf{Q}^\phi\{\theta_v \in J_2\} \int_{I_1} \mathbf{P}^0\{r_s \in dr\}\mathbf{P}^r\{r_{t-s} \in I_2, \lambda \int_0^{t-s} r_h^{-2} dh \in dv\}\right].$$

Markov property of the radial process (r_u) gives

$$B = \int_{J_1} m(d\phi)\left[\int_0^\infty \mathbf{Q}^\phi\{\theta_v \in J_2\}\mathbf{P}^0\{r_s \in I_1, \lambda \int_s^t r_h^{-2} dh \in dv,\ r_t \in I_2\}\right].$$

The time-inversion property (1) is valid for (r_u) and thus

$$\begin{aligned} B &= \int_{J_1} m(d\phi)\left[\int_0^\infty \mathbf{Q}^\phi\{\theta_v \in J_2\}\mathbf{P}^0\{s r_{1/s} \in I_1,\ \lambda \int_s^t (h r_{1/h})^{-2} dh \in dv,\ t r_{1/t} \in I_2\}\right], \\ &= \int_{J_1} m(d\phi)\left[\int_0^\infty \mathbf{Q}^\phi\{\theta_v \in J_2\}\mathbf{P}^0\{r_{1/s} \in s^{-1} I_1,\ \lambda \int_{1/t}^{1/s} r_h^{-2} dh \in dv,\ r_{1/t} \in t^{-1} I_2\}\right]. \end{aligned}$$

Because (θ_v) has a symmetric density with respect to $m(d\phi)$ (Vuolle-Apiala, Graversen [12]), we have

$$B = \int_{J_2} m(d\phi) \left[\int_0^\infty \mathbf{Q}^\phi\{\theta_v \in J_1\} \mathbf{P}^0\{r_{1/s} \in s^{-1} I_1,\ \lambda \int_{1/t}^{1/s} r_h^{-2} dh \in dv,\ r_{1/t} \in t^{-1} I_2\} \right].$$

Reversing the procedure above we get this equivalent to

$$B = \mathbf{P}^0\{r_{1/t} \in t^{-1} I_2,\ r_{1/s} \in s^{-1} I_1, \phi_{1/t} \in J_2,\ \phi_{1/s} \in J_1; r_u \neq 0 \text{ for all } u \in (1/t, 1/s)\},$$

which is further equivalent to

$$B = \mathbf{P}^0 \left\{ s r_{1/s} \in I_1,\ t r_{1/t} \in I_2,\ \phi_{1/s} \in J_1,\ \phi_{1/t} \in J_2,\ 1/s \text{ and } 1/t \text{ belong to the same excursion of } (r_u) \text{ away from } 0 \right\}. \qquad \square$$

Acknowledgements The author wants to thank an anonymous referee for useful comments and suggestions which considerably improved this paper.

References

1. L. Gallardo, M. Yor, Some new examples of Markov processes which enjoy the time-inversion property. Probab. Theor. Relat. Field. **132**(1), 150–162 (2005)
2. S.E. Graversen, J. Vuolle-Apiala, α-self-similar Markov processes. Probab. Theor. Relat. Field. **71**(1), 149–158 (1986)
3. S.E. Graversen, J. Vuolle-Apiala, On Paul Levy's arc sine law and Shiga-Watanabe's time inversion result. Probab. Math. Statist. **20**(1), 63–73 (2000)
4. K. Ito, H.P. McKean Jr., *Diffusion Processes and Their Sample Paths* (Springer, Berlin, 1974)
5. J.W. Lamperti, Semi-stable Markov processes I. Z. Wahrschein. verw. Gebiete **22**, 205–225 (1972)
6. S. Lawi, Towards a characterization of Markov processes enjoying the time-inversion property. J. Theoret. Probab. **21**(1), 144–168 (2008)
7. D. Revuz, M. Yor, *Continuous martingales and Brownian motion* (Springer, Berlin, 1991)
8. T. Shiga, S. Watanabe, Bessel diffusions as one parameter family of diffusion processes. Z. Wahrschein. verw. Gebiete **27**, 37–46 (1973)
9. J. Vuolle-Apiala, On certain extensions of a rotation invariant Markov process. J. Theoret. Probab. **16**(4), 957–969 (2003)
10. J. Vuolle-Apiala, Time inversion result for self-similar diffusions on finite number of rays, in *Contributions to Management Science, Mathematics and Modelling*, Acta Wasaensia **122**, 253–260 (University of Vaasa, 2004)
11. J. Vuolle-Apiala, Shiga-Watanabe's time inversion property for self-similar diffusion processes. Stat. Probab. Lett. **77**(9), 920–924 (2007)
12. J. Vuolle-Apiala, S.E. Graversen, Duality theory for self-similar processes. Ann. Inst. Henri Poincaré **22**, 323–332 (1986)
13. S. Watanabe, On time inversion of one-dimensional diffusion process. Z. Wahrschein. verw. Gebiete **31**, 115–124 (1975)

On Peacocks: A General Introduction to Two Articles

Antoine-Marie Bogso, Christophe Profeta, and Bernard Roynette

The aim of the following two articles, namely:

1. *Some examples of peacocks in a Markovian set-up*
2. *Peacocks obtained by renormalisation, strong and very strong peacocks*

is to provide examples of integrable processes $(X_t, t \geq 0)$ which increase in the convex order, i.e., such that:

i. For every $t \geq 0$, $\mathbb{E}[|X_t|] < \infty$,
ii. For every convex function $\psi : \mathbb{R} \to \mathbb{R}$ and every $0 \leq s < t$,

$$-\infty < \mathbb{E}[\psi(X_s)] \leq \mathbb{E}[\psi(X_t)].$$

Such processes are called *peacocks* (see [1]).

In these two papers, the notions of conditional monotonicity and that of log-concave increments play an important role.

In the first one, the emphasis is made on processes constructed from Markov processes (Sects. 2–4). We also establish a link between stochastic and convex orders (Sect. 5). The bulk of this article is taken up in Chaps. 1 and 8 of [1].

The second paper is devoted to processes obtained, from other processes, either by centering or normalisation (Sects. 1–3). The notions of strong and very strong peacock are developed in Sects. 4 and 5. This second article is a sort of continuation of the first paper.

The reader interested in questions developed in these two articles may refer to [1] for more informations and results.

A.-M. Bogso (✉) · C. Profeta · B. Roynette
Institut Elie Cartan, Université Henri Poincaré, B.P. 239, 54506 Vandœuvre-lès-Nancy Cedex, France
e-mail: antoine.bogso@iecn.u-nancy.fr; christophe.profeta@univ-evry.fr; bernard.roynette@iecn.u-nancy.fr

C. Donati-Martin et al. (eds.), *Séminaire de Probabilités XLIV*, Lecture Notes in Mathematics 2046, DOI 10.1007/978-3-642-27461-9_14,

Finally, we draw to the reader's attention that each of these two papers may be read independently.

Reference

1. F. Hirsch, C. Profeta, B. Roynette, M. Yor, *Peacocks and Associated Martingales*, vol 3 (Bocconi-Springer, New York, 2011)

Some Examples of Peacocks in a Markovian Set-Up

Antoine-Marie Bogso, Christophe Profeta, and Bernard Roynette

Abstract We give, in a Markovian set-up, some examples of processes which are increasing in the convex order (we call them peacocks). We then establish some relation between the stochastic and convex orders.

Keywords Processes increasing in the convex order • Peacocks • Conditionally monotone processes • Stochastic order • Markov process

AMS Classification: 60J25, 32F17, 60G44, 60E15

1 Introduction

1.1 Definitions

We start with a few definitions:
(a) A real-valued process $(X_t, t \geq 0)$ is said to be *increasing in the convex order* if:

$$\forall t > 0, \quad \mathbb{E}\left[|X_t|\right] < \infty$$

and, for every convex function $\psi : \mathbb{R} \longrightarrow \mathbb{R}$:

$$t \in \mathbb{R}_+ \longmapsto \mathbb{E}\left[\psi(X_t)\right] \in]-\infty, +\infty] \quad \text{is increasing.} \tag{1}$$

A.-M. Bogso (✉) · C. Profeta · B. Roynette
Institut Elie Cartan, Université Henri Poincaré, B.P. 239, 54506 Vandœuvre-lès-Nancy Cedex, France
e-mail: antoine.bogso@iecn.u-nancy.fr; christophe.profeta@univ-evry.fr; bernard.roynette@iecn.u-nancy.fr

C. Donati-Martin et al. (eds.), *Séminaire de Probabilités XLIV*, Lecture Notes in Mathematics 2046, DOI 10.1007/978-3-642-27461-9_15,

This notion plays an important role in many applied domains of probability; see, e.g., Shaked–Shanthikumar [20, 21]. We call such a process a peacock, an acronym derived from the French term: Processus Croissant pour l'Ordre Convexe. To prove (1), it suffices (see [10, Chap. 1]) to consider only the class:

$$\boldsymbol{C} := \{\psi \text{ is a convex function of } \mathscr{C}^2 \text{ class such that } \psi'' \text{ has compact support}\}.$$

Note that if $\psi \in \boldsymbol{C}$, then ψ' is a bounded function.

(b) A real-valued process $(X_t, t \geq 0)$ is called a *1-martingale* if there exists a martingale $(M_t, t \geq 0)$ (defined on a suitable filtered probability space) which has the same one-dimensional marginals as $(X_t, t \geq 0)$, that is to say, for each fixed $t \geq 0$:

$$X_t \overset{\text{(law)}}{=} M_t.$$

We say that such a martingale is associated to the process $(X_t, t \geq 0)$. From Jensen's inequality, it is clear that a 1-martingale is a peacock. Conversely, a remarkable result due to Kellerer [13] states that any peacock is a 1-martingale. However, the proofs presented in Kellerer's paper are not constructive, and in general, it is a difficult task to exhibit such a martingale.
In this paper, we shall only tackle the question of exhibiting peacocks, and mainly focus on examples derived from diffusions.

1.2 Some Examples

Let $(B_s, s \geq 0)$ be a standard Brownian motion. Carr et al. [3] proved that the process:

$$\left(A_t := \frac{1}{t}\int_0^t \exp\left(B_s - \frac{s}{2}\right) ds = \int_0^1 \exp\left(B_{ts} - \frac{st}{2}\right) ds \ , \ t \geq 0\right) \tag{2}$$

is a peacock. Baker and Yor [2] then exhibited a martingale which is associated to this peacock, and is constructed from the Wiener sheet. This example was the starting point of many recent developments which we try to synthesize; consider, for every $\lambda \geq 0$, a real-valued measurable process

$$Z_{\lambda,\cdot} := (Z_{\lambda,t}, \ t \geq 0)$$

such that

$$\forall \lambda \in \mathbb{R}_+, \ \forall t \in \mathbb{R}_+, \quad \mathbb{E}\left[e^{Z_{\lambda,t}}\right] < \infty,$$

and define, for any finite and positive measure μ on $\mathbb{R}_+$ the process:

$$\left(A_\lambda^{(\mu)} := \int_0^{+\infty} \frac{e^{Z_{\lambda,t}}}{\mathbb{E}\left[e^{Z_{\lambda,t}}\right]}\mu(dt),\ \lambda \geq 0\right). \tag{3}$$

(Taking $Z_{\lambda,t} = B_{\lambda t}$ and $\mu(ds) = 1_{[0,1]}(ds)$, we recover (2).)
Now, this raises the following natural question:

Under which conditions is the process $\left(A_\lambda^{(\mu)}, \lambda \geq 0\right)$ a peacock?

It is known that $(A_\lambda^{(\mu)}, \lambda \geq 0)$ is a peacock in the following cases:

- $Z_{\lambda,t} = \lambda t X$ with X a r.v., see [10].
- $Z_{\lambda,t} = \lambda L_t$ with $(L_t, t \geq 0)$ a Lévy process such that $\mathbb{E}\left[e^{L_1}\right] < \infty$, (see [8]).
- $Z_{\lambda,t} = G_{\lambda,t}$ with, for every $\lambda \geq 0$, $(G_{\lambda,t}, t \geq 0)$ a Gaussian process such that the function $\lambda \longrightarrow \mathbb{E}\left[G_{\lambda,t}G_{\lambda,s}\right]$ is increasing for every $s, t \geq 0$, (see [7]).

In this paper, we shall exhibit several other families of peacocks.

In Sect. 2, we introduce the notion of conditional monotonicity which will lead to a new large class of peacocks.

In Sect. 3, we give many examples, among which the processes with independent log-concave increments and the "well-reversible" diffusions at fixed times.

In Sect. 4, we present another condition, this time relying upon Laplace transforms, which implies the peacock property.

Finally, in Sect. 5, we present a result which links the stochastic and convex orders, and makes it possible to recover some of the peacocks presented above.

2 A Class of Peacocks Under the Conditional Monotonicity Hypothesis

In this section, we introduce and study the notion of conditional monotonicity, which already appear in [20, Chap. 4.B, pp. 114–126].

Definition 1 (Conditional monotonicity). A process $(X_\lambda, \lambda \geq 0)$ is said to be conditionally monotone if, for every $n \in \mathbb{N}^*$, every $i \in \{1, \dots, n\}$, every $0 \leq \lambda_1 < \dots < \lambda_n$ and every bounded Borel function $\phi : \mathbb{R}^n \longrightarrow \mathbb{R}$ which increases (resp. decreases) with respect to each of its arguments, we have:

$$\mathbb{E}[\phi(X_{\lambda_1}, X_{\lambda_2}, \dots, X_{\lambda_n})|X_{\lambda_i}] = \phi_i(X_{\lambda_i}), \tag{4}$$

where $\phi_i : \mathbb{R} \longrightarrow \mathbb{R}$ is a bounded increasing (resp. decreasing) function.

Remark 1. 1. If there is an interval I of $\mathbb{R}$ such that, for every $\lambda \geq 0$, $X_\lambda \in I$, we may assume in Definition 1 that ϕ is merely defined on I^n, and ϕ_i is defined on I.
2. Note that $(X_\lambda, \lambda \geq 0)$ is conditionally monotone if and only if $(-X_\lambda, \lambda \geq 0)$ is conditionally monotone.
3. Let $\theta : \mathbb{R} \longrightarrow \mathbb{R}$ be a strictly monotone and continuous function. It is not difficult to see that if the process $(X_\lambda, \lambda \geq 0)$ is conditionally monotone, then so is $(\theta(X_\lambda), \lambda \geq 0)$.

To prove that a process is conditionally monotone, we can restrict ourselves to bounded Borel functions ϕ increasing with respect to each of their arguments. Indeed, replacing ϕ by $-\phi$, the result then holds also for bounded Borel functions decreasing with respect to each of their arguments.

Definition 2. We denote by $\mathscr{E}_n$ the set of bounded Borel functions $\phi : \mathbb{R}^n \longrightarrow \mathbb{R}$ which are increasing with respect to each of their arguments.

Theorem 1. *Let $(X_\lambda, \lambda \geq 0)$ a real-valued process which is right-continuous, conditionally monotone and which satisfies the following integrability conditions: For every compact $K \subset \mathbb{R}_+$ and every $t \geq 0$:*

$$\Theta_{K,t} := \sup_{\lambda \in K} \exp(tX_\lambda) = \exp\left(t \sup_{\lambda \in K} X_\lambda\right) \text{ is integrable,} \tag{5}$$

and

$$k_{K,t} := \inf_{\lambda \in K} \mathbb{E}\left[\exp(tX_\lambda)\right] > 0. \tag{6}$$

We set $h_\lambda(t) = \log \mathbb{E}[\exp(tX_\lambda)]$. Then, for every finite positive measure μ on $\mathbb{R}_+$:

$$\left(A_t^{(\mu)} := \int_0^\infty e^{tX_\lambda - h_\lambda(t)} \mu(d\lambda)\ ,\ t \geq 0\right)$$

is a peacock.

Proof (of Theorem 1).

1. By (5), for every $\lambda \geq 0$ and every $t \geq 0$, $\mathbb{E}\left[\exp(tX_\lambda)\right] < \infty$. This easily implies, thanks to the dominated convergence theorem, that h_λ is continuous on $\mathbb{R}_+$, differentiable on $]0, +\infty[$, and

$$h'_\lambda(t) e^{h_\lambda(t)} = \mathbb{E}\left[X_\lambda e^{tX_\lambda}\right]. \tag{7}$$

Since $\mathbb{E}\left[e^{tX_\lambda - h_\lambda(t)}\right] = 1$, we obtain from (7):

$$\mathbb{E}\left[(X_\lambda - h'_\lambda(t)) e^{tX_\lambda - h_\lambda(t)}\right] = 0. \tag{8}$$

Moreover, we also deduce from (5) that, for every $t \geq 0$, the function $\lambda \geq 0 \longmapsto h_\lambda(t)$ is right-continuous.

2. We first consider the case

$$\mu = \sum_{i=1}^{n} a_i \delta_{\lambda_i} \tag{9}$$

where $n \in \mathbb{N}^*$, $a_1 \geq 0, \dots, a_n \geq 0$, and $0 \leq \lambda_1 < \dots < \lambda_n$. Let $\psi \in \boldsymbol{C}$. For $t > 0$, we have:

$$\frac{\partial}{\partial t}\mathbb{E}\left[\psi\left(A_t^{(\mu)}\right)\right] = \mathbb{E}\left[\psi'\left(A_t^{(\mu)}\right)\sum_{i=1}^{n} a_i\left(X_{\lambda_i} - h'_{\lambda_i}(t)\right)\exp\left(tX_{\lambda_i} - h_{\lambda_i}(t)\right)\right]$$

Setting for $i \in \{1, \dots, n\}$,

$$\Delta_i = \mathbb{E}\left[\psi'\left(A_t^{(\mu)}\right)\left(X_{\lambda_i} - h'_{\lambda_i}(t)\right)\exp\left(tX_{\lambda_i} - h_{\lambda_i}(t)\right)\right]$$

we shall show that $\Delta_i \geq 0$ for every $i \in \{1, \dots, n\}$. Note that the function

$$(x_1, \dots, x_n) \longmapsto \psi'\left(\sum_{j=1}^{n} a_j \exp\left(tx_j - h_{\lambda_j}(t)\right)\right)$$

is bounded and increases with respect to each of its arguments, i.e., belongs to $\mathscr{E}_n$. Hence, from the conditional monotonicity property of $(X_\lambda, \lambda \geq 0)$:

$$\begin{aligned}\Delta_i &= \mathbb{E}\left[\mathbb{E}\left[\psi'\left(A_t^{(\mu)}\right)(X_{\lambda_i} - h'_{\lambda_i}(t))e^{tX_{\lambda_i} - h_{\lambda_i}(t)}|X_{\lambda_i}\right]\right] \\ &= \mathbb{E}\left[(X_{\lambda_i} - h'_{\lambda_i}(t))e^{tX_{\lambda_i} - h_{\lambda_i}(t)}\phi_i(X_{\lambda_i})\right]\end{aligned}$$

where ϕ_i is a bounded increasing function. Besides, we have,

$$(X_{\lambda_i} - h'_{\lambda_i}(t))\left(\phi_i(X_{\lambda_i}) - \phi_i\left(h'_{\lambda_i}(t)\right)\right) \geq 0.$$

Therefore,

$$\begin{aligned}\Delta_i &\geq \phi_i\left(h'_{\lambda_i}(t)\right)\mathbb{E}\left[(X_{\lambda_i} - h'_{\lambda_i}(t))e^{tX_{\lambda_i} - h_{\lambda_i}(t)}\right] \\ &= 0 \quad \text{from (8).}\end{aligned}$$

3. We now assume that μ has compact support contained in a compact interval K. Since the function $\lambda \longmapsto \exp(tX_\lambda - h_\lambda(t))$ is right-continuous and bounded from above by $k_{K,t}^{-1}\Theta_{K,t}$ which is finite a.s., there exists a sequence $(\mu_n, n \geq 0)$ of measures of the form (9), with supp $(\mu_n) \subset K$, $\int \mu_n(d\lambda) = \int \mu(d\lambda)$ and for every $t \geq 0$, $\lim\limits_{n \to +\infty} A_t^{(\mu_n)} = A_t^{(\mu)}$ a.s. Moreover, from (5) and (6):

$$|A_t^{(\mu_n)}| \leq \frac{\theta_{K,t}}{k_{K,t}} \int \mu(d\lambda),$$

and from Point (2), for $0 \leq s \leq t$:

$$\mathbb{E}[\psi(A_s^{(\mu_n)})] \leq \mathbb{E}[\psi(A_t^{(\mu_n)})].$$

Therefore, since ψ is sublinear, we can apply the dominated convergence theorem and pass to the limit when $n \to +\infty$ in this last inequality to obtain that $(A_t^{(\mu)}, t \geq 0)$ is a peacock.

4. In the general case, we set $\mu_n(d\lambda) = 1_{[0,n]}(\lambda)\mu(d\lambda)$ and observe that $A^{(\mu_n)}$ is an increasing sequence of processes. Let ρ be defined by $\rho(x) = \int_0^x (x-z)\psi''(z)dz$. An integration by parts yields, for $0 \leq s \leq t$:

$$\begin{aligned}
&\mathbb{E}\left[\psi\left(A_t^{(\mu_n)}\right)\right] - \mathbb{E}\left[\psi\left(A_s^{(\mu_n)}\right)\right] \\
&\quad = \mathbb{E}\left[\psi'(0)\left(A_t^{(\mu_n)} - A_s^{(\mu_n)}\right)\right] + \mathbb{E}\left[\rho\left(A_t^{(\mu_n)}\right)\right] - \mathbb{E}\left[\rho\left(A_s^{(\mu_n)}\right)\right] \\
&\quad = \mathbb{E}\left[\rho\left(A_t^{(\mu_n)}\right)\right] - \mathbb{E}\left[\rho\left(A_s^{(\mu_n)}\right)\right] \geq 0
\end{aligned}$$

from Point (3), and since $\mathbb{E}\left[A_t^{(\mu_n)}\right] = \mathbb{E}\left[A_s^{(\mu_n)}\right]$. Now, since ρ is an increasing function on $\mathbb{R}_+$, the result follows from the monotone convergence theorem.

□

Remark 2. Let $\theta : \mathbb{R} \longrightarrow \mathbb{R}$ be a strictly monotone and continuous function, and μ denote a finite positive measure. From Remark 1, under the assumption that $(\theta(X_\lambda), \lambda \geq 0)$ still satisfies conditions (5) and (6), we obtain, denoting $h_{\lambda,\theta}(t) = \log \mathbb{E}\left[\exp\left(t\theta(X_\lambda)\right)\right]$, that the process

$$\left(A_t^{(\theta,\mu)} := \int_0^\infty e^{t\theta(X_\lambda) - h_{\lambda,\theta}(t)} \mu(d\lambda)\ ,\ t \geq 0\right)$$

is a peacock. Note that θ only needs to be continuous and strictly monotone on an interval containing the image of X_λ for every $\lambda \geq 0$.

Of course, Theorem 1 may have some practical interest only if we are able to exhibit enough examples of processes which enjoy the conditional monotonicity (4) property. Below, we shall see that there exists a large class of diffusions which enjoy this property. But to start with, let us first give a few examples which consist of processes with independent increments and Lévy processes in particular.

3 Examples of Processes Satisfying the Conditional Monotonicity Property

3.1 Processes with Independent Increments Satisfying the Conditional Monotonicity Property

We start by giving an assertion equivalent to (4) when dealing with processes with independent (not necessarily time-homogeneous) increments.

Proposition 1. *Let $(X_\lambda, \lambda \geq 0)$ be a process with independent increments. Then, the conditional monotonicity hypothesis (4) is equivalent to the following:*
For every $n \in \mathbb{N}^*$, every $0 \leq \lambda_1 < \cdots < \lambda_n$ and every function $\phi : \mathbb{R}^n \longrightarrow \mathbb{R}$ in $\mathscr{E}_n$, we have:

$$\mathbb{E}\left[\phi\left(X_{\lambda_1}, \ldots, X_{\lambda_n}\right)|X_{\lambda_n}\right] = \phi_n(X_{\lambda_n}) \tag{10}$$

where ϕ_n is an increasing bounded function.

Proof (of Proposition 1).
The proof is straightforward. Indeed, let $\phi \in \mathscr{E}_n$. For $i \in \{1, \ldots, n\}$, the hypothesis of independent increments implies:

$$\begin{aligned}
&\mathbb{E}\left[\phi\left(X_{\lambda_1}, \ldots, X_{\lambda_n}\right)|X_{\lambda_i}\right] \\
&= \mathbb{E}\left[\mathbb{E}\left[\phi\left(X_{\lambda_1}, \ldots, X_{\lambda_n}\right)|\mathscr{F}_{\lambda_i}\right]|X_{\lambda_i}\right] \\
&= \mathbb{E}\left[\mathbb{E}\left[\phi\left(X_{\lambda_1}, \ldots, , X_{\lambda_i}, X_{\lambda_{i+1}} - X_{\lambda_i} + X_{\lambda_i}, \ldots, X_{\lambda_n} - X_{\lambda_i} + X_{\lambda_i}\right)|\mathscr{F}_{\lambda_i}\right]|X_{\lambda_i}\right] \\
&= \mathbb{E}\left[\widetilde{\phi}\left(X_{\lambda_1}, \ldots, X_{\lambda_i}\right)|X_{\lambda_i}\right]
\end{aligned}$$

where

$$\widetilde{\phi}(x_1, \ldots, x_i) = \mathbb{E}\left[\phi\left(x_1, \ldots, x_i, X_{\lambda_{i+1}} - X_{\lambda_i} + x_i, \ldots, X_{\lambda_n} - X_{\lambda_i} + x_i\right)\right]$$

belongs to $\mathscr{E}_i$. □

3.1.1 The Gamma Subordinator Is Conditionally Monotone

The Gamma subordinator $(\gamma_\lambda, \lambda \geq 0)$ is characterized by:

$$\mathbb{E}\left[e^{-t\gamma_\lambda}\right] = \frac{1}{(1+t)^\lambda} = \exp\left(-\lambda \int_0^\infty (1 - e^{-tx})\frac{e^{-x}}{x}dx\right).$$

In particular, γ_λ is a gamma random variable with parameter λ. From (10), we wish to show that for every $n \in \mathbb{N}^*$, every $0 \leq \lambda_1 < \cdots < \lambda_n$ and every function $\phi : \mathbb{R}^n \longrightarrow \mathbb{R}$ in $\mathscr{E}_n$:

$$\mathbb{E}[\phi(\gamma_{\lambda_1}, \ldots, \gamma_{\lambda_n})|\gamma_{\lambda_n}] = \phi_n(\gamma_{\lambda_n}), \tag{11}$$

where ϕ_n is an increasing function. The explicit knowledge of the law of γ_λ and the fact that $(\gamma_\lambda, \lambda \geq 0)$ has time-homogeneous independent increments imply the well-known result that, given $\{\gamma_{\lambda_n} = x\}$, the vector $(\gamma_{\lambda_1}, \gamma_{\lambda_2} - \gamma_{\lambda_1}, \dots, \gamma_{\lambda_n} - \gamma_{\lambda_{n-1}})$ follows the Dirichlet law with parameters $(\lambda_1, \lambda_2 - \lambda_1, \dots, \lambda_n - \lambda_{n-1})$ on $[0, x]$. In other words, the density f_n^x of $(\gamma_{\lambda_1}, \gamma_{\lambda_2}, \dots, \gamma_{\lambda_{n-1}})$ conditionally on $\{\gamma_{\lambda_n} = x\}$ equals:

$$f_n^x(x_1, \dots, x_{n-1}) = \frac{C}{x^{\lambda_n - 1}} x_1^{\lambda_1 - 1}(x_2 - x_1)^{\lambda_2 - \lambda_1 - 1} \dots$$
$$(x_{n-1} - x_{n-2})^{\lambda_{n-1} - \lambda_{n-2} - 1}(x - x_{n-1})^{\lambda_n - \lambda_{n-1} - 1} 1_{\mathbb{S}^{n,x}},$$

where $C := C(\lambda_1, \dots, \lambda_n)$ is a positive constant and

$$\mathbb{S}^{n,x} = \{(x_1, \dots, x_{n-1}) \in \mathbb{R}^{n-1} : 0 \leq x_1 \leq \dots \leq x_{n-1} \leq x\}.$$

Hence,
$\mathbb{E}[\phi(\gamma_{\lambda_1}, \dots, \gamma_{\lambda_n}) | \gamma_{\lambda_n} = x]$

$$= \int_{\mathbb{S}^{n,x}} \phi(x_1, \dots, x_{n-1}, x) f_n^x(x_1, \dots, x_{n-1}) dx_1 \dots dx_{n-1}$$
$$= C \int_{\mathbb{S}^{n,1}} \phi(xy_1, \dots, xy_{n-1}, x) y_1^{\lambda_1 - 1}(y_2 - y_1)^{\lambda_2 - \lambda_1 - 1} \dots$$
$$(y_{n-1} - y_{n-2})^{\lambda_{n-1} - \lambda_{n-2} - 1}(1 - y_{n-1})^{\lambda_n - \lambda_{n-1} - 1} dy_1 \dots dy_{n-1}$$

after the change of variables: $x_i = xy_i$, $i = 1, \dots, n-1$. It is then clear that since ϕ increases with respect to each of its arguments, this last expression is an increasing function with respect to x.

Corollary 1. *Let $(\gamma_\lambda, \lambda \geq 0)$ be the gamma subordinator. Then, for every finite positive measure μ on $\mathbb{R}_+$, and for every $p > 0$, the process:*

$$\left(A_t^{(\mu,p)} := \int_0^\infty e^{-t(\gamma_\lambda)^p - h_{\lambda,p}(t)} \mu(d\lambda) \, , \, t \geq 0 \right) \tag{12}$$

is a peacock. Here, the function $h_{\lambda,p}$ is defined as:

$$h_{\lambda,p}(t) = \log \mathbb{E}\left[\exp\left(-t(\gamma_\lambda)^p\right)\right].$$

Proof (of Corollary 1).
By Remark 2 with $\theta(x) = -x^p$ for $x \geq 0$, the process $(X_\lambda := -\gamma_\lambda^p, \lambda \geq 0)$ is conditionally monotone. Since it is a negative process, (5) is obviously satisfied. Moreover, since $(\gamma_\lambda, \lambda \geq 0)$ is an increasing process, (6) is easily verified. Finally, Theorem 1 holds. □

Remark 3. Actually, for $p = 1$, Corollary 1 holds more generally with μ a signed measure, see [8].

3.1.2 The Simple Random Walk Is Conditionally Monotone

Let $(\varepsilon_i, i \in \mathbb{N}^*)$ be a sequence of independent and identically distributed r. v.'s such that, for every $i \in \mathbb{N}^*$:

$$\mathbb{P}(\varepsilon_i = 1) = p, \quad \mathbb{P}(\varepsilon_i = -1) = q \qquad \text{with } p, q > 0 \text{ and } p + q = 1.$$

Let $(S_n, n \in \mathbb{N})$ be the random walk defined by: $S_0 = 0$ and

$$S_n = \sum_{i=1}^{n} \varepsilon_i, \qquad \text{for every } n \in \mathbb{N}^*$$

We shall prove that $(S_n, n \in \mathbb{N})$ is conditionally monotone; i.e: for every $r \in [\![2, +\infty[\![$, every $0 \le n_1 < n_2 < \cdots < n_r < +\infty$ and every function $\phi : \mathbb{R}^{r-1} \longrightarrow \mathbb{R}$ in $\mathscr{E}_{r-1}$,

$$k \in I_{n_r} \longmapsto \mathbb{E}[\phi(S_{n_1}, S_{n_2}, \dots, S_{n_{r-1}}) | S_{n_r} = k] \text{ is an increasing function} \quad (13)$$

where $I_x \subset [\![-x, x]\!]$ denotes the set of all the values the r.v. S_x can take. It is not difficult to see that (13) holds if and only if: for every $N \in [\![2, +\infty[\![$ and every function $\phi : \mathbb{R}^{N-1} \longrightarrow \mathbb{R}$ in $\mathscr{E}_{N-1}$:

$$k \in I_N \longmapsto \mathbb{E}[\phi(S_1, \dots, S_{N-1}) | S_N = k] \text{ is an increasing function on } I_N. \quad (14)$$

We shall distinguish two cases:
1. If N and k are even, we set $N = 2n$ ($n \in [\![1, +\infty[\![$) and $k = 2x$ ($x \in [\![-n, n]\!]$). For every $n \in [\![1, +\infty[\![$ and every $x \in [\![-n, n]\!]$, let us denote by $\mathscr{J}_{2n}^{2x}$, the set of polygonal lines $\omega := (\omega_i, i \in [\![0, 2n]\!])$ such that $\omega_0 = 0$, $\omega_{p+1} = \omega_p \pm 1$, ($p \in [\![0, 2n-1]\!]$) and $\omega_{2n} = 2x$. Observe that any $\omega \in \mathscr{J}_{2n}^{2x}$ has $n + x$ positive slopes and $n - x$ negative ones. This implies that:

$$|\mathscr{J}_{2n}^{2x}| = C_{2n}^{n+x},$$

where $|\cdot|$ denotes cardinality. It is well known that, conditionally on $\{S_{2n} = 2x\}$, the law of the random vector $(S_0, S_, \dots, S_{2n})$ is the uniform law on $\mathscr{J}_{2n}^{2x}$.

Let $n \in [\![1, +\infty[\![$ and $x \in [\![-n, n]\!]$ be fixed and consider, for every $i \in [\![1, n + x + 1]\!]$ the map:

$$\Pi_i : \mathscr{J}_{2n}^{2x+2} \longrightarrow \mathscr{J}_{2n}^{2x}$$

defined by: for every $\omega \in \mathscr{J}_{2n}^{2x+2}$, $\Pi_i(\omega)$ has the same negative slopes and the same positive slopes as ω except the ith positive slope which is replaced by a negative one.

For every $\omega \in \mathscr{I}_{2n}^{2x+2}$ and every function $\phi : \mathbb{R}^{2n} \longrightarrow \mathbb{R}$ in $\mathscr{E}_{2n}$,

$$\phi(\omega) \geq \phi(\Pi_i(\omega)).$$

Summing this relation, we obtain:

$$\begin{aligned}(n+x+1) \sum_{\omega \in \mathscr{I}_{2n}^{2x+2}} \phi(\omega) &\geq \sum_{\omega \in \mathscr{I}_{2n}^{2x+2}} \sum_{i=1}^{n+x+1} \phi(\Pi_i(\omega)) \\ &= \sum_{\omega \in \mathscr{I}_{2n}^{2x}} \sum_{i=1}^{n+x+1} |\Pi_i^{-1}(\omega)| \phi(\omega) \\ &= (n-x) \sum_{\omega \in \mathscr{I}_{2n}^{2x}} \phi(\omega).\end{aligned}$$

Thus, we have proved the following:

Lemma 1. *For every $n \in \mathbb{N}^*$ and every $\phi : \mathbb{R}^{2n} \longrightarrow \mathbb{R}$ in $\mathscr{E}_{2n}$,*

$$\frac{1}{|\mathscr{I}_{2n}^{2x+2}|} \sum_{\omega \in \mathscr{I}_{2n}^{2x+2}} \phi(\omega) \geq \frac{1}{|\mathscr{I}_{2n}^{2x}|} \sum_{\omega \in \mathscr{I}_{2n}^{2x}} \phi(\omega), \tag{15}$$

which means that $(S_n, n \in \mathbb{N})$ is conditionally monotone.

2. It is not difficult to establish a similar result when k and N are odd.

Corollary 2. *For every odd and positive integer p, and for every positive finite measure $\sum_{n \in \mathbb{N}} a_n \delta_n$ on $\mathbb{N}$:*

$$\left(\sum_{n=0}^{+\infty} a_n e^{-t(S_n)^p - h_{n,p}(t)}, t \geq 0 \right) \quad \textit{is a peacock.}$$

Here, the function $h_{n,p}$ is defined by: $h_{n,p}(t) = \log \mathbb{E}\left[\exp\left(-t(S_n)^p\right)\right]$.

3.1.3 The Processes with Independent Log-Concave Increments Are Conditionally Monotone

We first introduce the notions of PF_2 and log-concave random variables (see [4]).

Definition 3 ($\mathbb{R}$-valued PF_2 r.v.'s).
An $\mathbb{R}$-valued random variable X is said to be PF_2 if:

1. X admits a probability density f,

2. For every $x_1 \geq x_2$, $y_2 \geq y_1$,

$$\det \begin{pmatrix} f(x_1 - y_1) & f(x_1 - y_2) \\ f(x_2 - y_1) & f(x_2 - y_2) \end{pmatrix} \geq 0.$$

Definition 4 ($\mathbb{Z}$-valued PF$_2$ r.v.'s).
A $\mathbb{Z}$-valued random variable X is said to be PF$_2$ if, setting $f(x) = \mathbb{P}(X = x)$ $(x \in \mathbb{Z})$, one has: for every $x_1 \geq x_2$, $y_2 \geq y_1$,

$$\det \begin{pmatrix} f(x_1 - y_1) & f(x_1 - y_2) \\ f(x_2 - y_1) & f(x_2 - y_2) \end{pmatrix} \geq 0.$$

Definition 5 ($\mathbb{R}$-valued log-**concave r.v.'s).**
An $\mathbb{R}$-valued random variable X is said to be log-concave if:

1. X admits a probability density f,
2. The set $S_f := \{f > 0\}$ is convex (i.e., is an interval) and $\log f$ is concave on S_f (i.e., the second derivative of $\log f$ (in the distribution sense) is a negative measure).

Definition 6 ($\mathbb{Z}$-valued log-**concave r.v.'s).**
A $\mathbb{Z}$-valued random variable X is said to be log-concave if, with $f(x) = \mathbb{P}(X = x)$, $(x \in \mathbb{Z})$, the set $S_f := \{f > 0\}$ is an interval of $\mathbb{Z}$, and for every $n, n-1, n+1 \in S_f$,

$$f^2(n) \geq f(n-1)f(n+1);$$

in other words, the discrete second derivative of $\log f$ is negative on S_f.

We then deduce the equivalence:

Theorem 2 (see [1] or [4]). *An $\mathbb{R}$-valued (or $\mathbb{Z}$-valued) random variable is PF$_2$ if and only if it is* log-*concave.*

Example 1. Many common density functions on $\mathbb{R}$ (or $\mathbb{Z}$) are PF$_2$. Indeed, the normal density, the uniform density, the exponential density, the negative binomial density, the Poisson density and the geometric density are PF$_2$. We refer to [1] for more examples. Note that:

1. A gamma random variable of parameter a (with density $f_a(x) = \dfrac{1}{\Gamma(a)} e^{-x} x^{a-1} 1_{[0,+\infty[}(x)$, $a > 0$) is not PF$_2$ if $a < 1$,
2. A Bernoulli random variable X such that $\mathbb{P}(X = 1) = p = 1 - \mathbb{P}(X = -1)$ is not PF$_2$.

The following result is due to Efron [5] (see also [22]).

Theorem 3. *Let $n \in [\![1, +\infty[\![$, $X_1, X_2, \ldots, X_n$ be independent $\mathbb{R}$-valued (or $\mathbb{Z}$-valued) PF$_2$ random variables, $S_n = \sum_{i=1}^n X_i$, and $\phi : \mathbb{R}^n \to \mathbb{R}$ belonging to $\mathscr{E}_n$. Then,*

$$\mathbb{E}[\phi(X_1, X_2, \ldots, X_n)|S_n = x] \quad \textit{is increasing in } x.$$

Thanks to Theorem 3, we obtain the following result:

Theorem 4. *Let* $(Z_\lambda, \lambda \in \mathbb{R}_+$ *or* $\lambda \in \mathbb{N})$ *be a* $\mathbb{R}$*-valued (or* $\mathbb{Z}$*-valued) process satisfying (5) and (6), with independent (not necessarily time-homogeneous)* PF_2 *increments. Then,* $(Z_\lambda, \lambda \geq 0)$ *is conditionally monotone, and for every positive measure* μ *on* $\mathbb{R}_+$ *(or* $\mathbb{N}$*) with finite total mass,*

$$\left(\int_0^{+\infty} e^{tZ_\lambda - h_\lambda(t)}\mu(d\lambda), t \geq 0\right) \quad \textit{is a peacock,}$$

where the function h_λ *is defined by:* $h_\lambda(t) = \log \mathbb{E}\left[e^{tZ_\lambda}\right]$.

Proof (of Theorem 4).
It suffices to show that $(Z_\lambda, \lambda \geq 0)$ satisfies (10). Let $n \in [\![1, +\infty[\![$ and $\phi : \mathbb{R}^n \to \mathbb{R}$ belonging to $\mathscr{E}_n$. For every $0 \leq \lambda_1 < \lambda_2 < \cdots < \lambda_n$ and $k \in \mathbb{R}$ (or $\mathbb{Z}$),

$$\begin{aligned}\mathbb{E}[\phi(Z_{\lambda_1}, Z_{\lambda_2}, \ldots, Z_{\lambda_n})|Z_{\lambda_n} = k] \\ = \mathbb{E}\left[\widehat{\phi}(Z_{\lambda_1}, Z_{\lambda_2} - Z_{\lambda_1}, \ldots, Z_{\lambda_n} - Z_{\lambda_{n-1}})|Z_{\lambda_n} = k\right],\end{aligned}$$

where the function $\widehat{\phi}$ is given by:

$$\widehat{\phi}(x_1, x_2, \ldots, x_n) = \phi(x_1, x_2 + x_1, \ldots, x_1 + x_2 + \cdots + x_n).$$

It is obvious that $\widehat{\phi}$ belongs to $\mathscr{E}_n$. Thus, applying Theorem 3 with: $X_1 = Z_{\lambda_1}$ and $X_{i+1} = Z_{\lambda_{i+1}} - Z_{\lambda_i}$ $i = 1, \ldots n-1$, one obtains the desired result. □

Remark 4.
1. Theorem 4 does not apply neither in the case of the Gamma subordinator, nor in the case of the random walk whose increments are Bernoulli with values in $\{-1, 1\}$. Nevertheless, its conclusion remains true in these cases, see Sects. 3.1.1 and 3.1.2.
2. We deduce from Corollary 4 that the Poisson process and the random walk with geometric increments are conditionally monotone. We shall give below a direct proof, i.e., without using Theorem 3.

3.1.4 The Poisson Process Is Conditionally Monotone

Let $(N_\lambda, \lambda \geq 0)$ be a Poisson process with parameter 1 and let $(T_n, n \geq 1)$ be its successive jumps times. Then

$$N_\lambda = \#\{i \geq 1 : T_i \leq \lambda\}.$$

In order to prove that $(N_\lambda, \lambda \geq 0)$ is conditionally monotone, we shall show that for every $0 \leq \lambda_1 < \cdots < \lambda_n$ and every function $\phi : \mathbb{R}^n \longrightarrow \mathbb{R}$ in $\mathscr{E}_n$, we have:

$$\mathbb{E}[\phi(N_{\lambda_1}, \ldots, N_{\lambda_n})|N_{\lambda_n}] = \phi_n(N_{\lambda_n}), \tag{16}$$

where $\phi_n : \mathbb{R} \longrightarrow \mathbb{R}$ increases. But, conditionally on $\{N_{\lambda_n} = k\}$, the random vector $(T_1, \ldots, T_k)$ is distributed as $(U_1, \ldots, U_k)$, $U_1, \ldots, U_k$ being the increasing rearrangement of k independent random variables, uniformly distributed on $[0, \lambda_n]$. We go from k to $k+1$ by adding one more point. Thus, with obvious notation, it is clear that: for all $\lambda \in [0, \lambda_n]$, $N_\lambda^{(k+1)} \geq N_\lambda^{(k)}$. Then, the conditional monotonicity property follows immediately.

Corollary 3. *Let $(N_\lambda, \lambda \geq 0)$ be a Poisson process and let μ be a finite positive measure on $\mathbb{R}_+$. Then, for every $p > 0$, the process:*

$$\left(A_t^{(\mu,p)} := \int_0^\infty e^{-t(N_\lambda)^p - h_{\lambda,p}(t)} \mu(d\lambda), t \geq 0\right) \tag{17}$$

is a peacock with:

$$h_{\lambda,p}(t) = \log \mathbb{E}\left[\exp\left(-t(N_\lambda)^p\right)\right].$$

3.1.5 The Random Walk with Geometric Increments Is Conditionally Monotone

Let $(\varepsilon_i, i \in [\![1, +\infty[\![)$ be a sequence of independent geometric variables with the same parameter p; i.e., such that:

$$\mathbb{P}(\varepsilon_i = k) = p^k(1-p) \qquad (k \geq 0,\ 0 < p < 1).$$

We consider the random walk $(S_n, n \in \mathbb{N})$ defined by: $S_0 = 0$ and

$$S_n = \sum_{i=1}^n \varepsilon_i, \text{ for every } n \in \mathbb{N}^*.$$

For $n \in \mathbb{N}^*$, S_n is distributed as a negative binomial random variable with parameters n and p; more precisely:

$$\mathbb{P}(S_n = k) = C_{n+k-1}^k p^n (1-p)^k, \text{ for every } k \in \mathbb{N}.$$

As in Sect. 3.1.2, we only need to prove that: for every $N \in \mathbb{N}^*$ and every function $\phi : \mathbb{R}^N \longrightarrow \mathbb{R}$ in $\mathscr{E}_N$:

$$k \longmapsto \mathbb{E}[\phi(S_1, \ldots, S_N)|S_{N+1} = k] \quad \text{is an increasing function on } \mathbb{N}. \tag{18}$$

Let $\mathbf{J}_N^k$ denote the set:

$$\mathbf{J}_N^k := \{(x_1, \ldots, x_N) \in \mathbb{N}^N : 0 \le x_1 \le \cdots \le x_N \le k\}. \qquad (19)$$

For every $k \ge 0$ and $N \ge 1$, it is well known that $|\mathbf{J}_N^k| = C_{N+k}^k$. Now, we have:

$$\begin{aligned}
&\mathbb{E}\left[\phi\left(S_1, \ldots, S_N\right)|S_{N+1} = k\right] \\
&= \sum_{(l_1,\ldots,l_N)\in\mathbf{J}_N^k} \phi\left(l_1, \ldots, l_N\right) \frac{\mathbb{P}(S_1 = l_1, \ldots, S_N = l_N, S_{N+1} = k)}{\mathbb{P}(S_{N+1} = k)} \\
&= \sum_{(l_1,\ldots,l_N)\in\mathbf{J}_N^k} \phi\left(l_1, \ldots, l_N\right) \frac{\mathbb{P}(S_1 = l_1, S_2 - S_1 = l_2 - l_1, \ldots, S_{N+1} - S_N = k - l_N)}{\mathbb{P}(S_{N+1} = k)} \\
&= \sum_{(l_1,\ldots,l_N)\in\mathbf{J}_N^k} \phi\left(l_1, \ldots, l_N\right) \frac{\mathbb{P}(S_1 = l_1)\mathbb{P}(S_2 - S_1 = l_2 - l_1)\ldots\mathbb{P}(S_{N+1} - S_N = k - l_N)}{\mathbb{P}(S_{N+1} = k)} \\
&= \sum_{(l_1,\ldots,l_N)\in\mathbf{J}_N^k} \phi\left(l_1, \ldots, l_N\right) \frac{p(1-p)^{l_1} p(1-p)^{l_2-l_1} \ldots p(1-p)^{k-l_N}}{C_{N+k}^k p^{N+1}(1-p)^k} \\
&= \frac{1}{C_{N+k}^k} \sum_{(l_1,\ldots,l_N)\in\mathbf{J}_N^k} \phi\left(l_1, \ldots, l_N\right) \\
&= \frac{1}{|\mathbf{J}_N^k|} \sum_{(l_1,\ldots,l_N)\in\mathbf{J}_N^k} \phi\left(l_1, \ldots, l_N\right).
\end{aligned}$$

Therefore, the law of the random vector $(S_1, \ldots, S_N)$ conditionally on $\{S_{N+1} = k\}$ is the uniform law on the set $\mathbf{J}_N^k$. Hence, we will obtain (18) if we prove that: for every $k \in \mathbb{N}$, every $N \in \mathbb{N}^*$ and every function $\phi : \mathbb{R}^N \longrightarrow \mathbb{R}_+$ in $\mathscr{E}_N$:

$$\frac{1}{|\mathbf{J}_N^k|} \sum_{x\in\mathbf{J}_N^k} \phi(x) \le \frac{1}{|\mathbf{J}_N^{k+1}|} \sum_{x\in\mathbf{J}_N^{k+1}} \phi(x). \qquad (20)$$

Let us notice that:

$$\mathbf{J}_N^0 = \{\underbrace{(0, \ldots, 0)}_{N \text{ times}}\}, \qquad \text{for every } N \in [\![1, +\infty[\![$$

and

$$\mathbf{J}_1^k = \{(0), (1), \ldots, (k)\}, \qquad \text{for every } k \in [\![0, +\infty[\![.$$

For $k \in [\![0, +\infty[\![$ and $N \in [\![1, +\infty[\![$, we define:

$$\mathbf{\Delta}_N^{k+1} := \mathbf{J}_N^{k+1} \setminus \mathbf{J}_N^k = \{(x_1, \ldots, x_N) \in \mathbf{J}_N^{k+1} : x_N = k + 1\}. \qquad (21)$$

and set $\mathbf{\Delta}_N^0 = \emptyset$. By Pascal's formula,

$$|\mathbf{\Delta}_N^{k+1}| = C_{k+1+N}^{k+1} - C_{k+N}^{k} = C_{N+k}^{k+1} = |\mathbf{J}_{N-1}^{k+1}|, \quad \text{(with } N \in [\![2, +\infty]\!]\text{)}.$$

On one hand, we consider, for $N \in [\![2, +\infty[\![$, the map $\Gamma : \mathbf{J}_{N-1}^{k+1} \longrightarrow \mathbf{\Delta}_N^{k+1}$ defined by:

$$\Gamma[(x_1, \dots, x_{N-1})] = (x_1, \dots, x_{N-1}, k+1). \tag{22}$$

The map Γ is bijective, and for every non empty pair of subsets G and H of $\mathbf{J}_{N-1}^{k+1}$, there is the equivalence:

$$\begin{cases} \forall f : \mathbb{R}^{N-1} \longrightarrow \mathbb{R} \in \mathscr{E}_{N-1}, \\ \\ \dfrac{1}{|G|} \sum\limits_{x \in G} f(x) \le \dfrac{1}{|H|} \sum\limits_{x \in H} f(x) \end{cases} \Longleftrightarrow \begin{cases} \forall \phi : \mathbb{R}^{N} \longrightarrow \mathbb{R} \in \mathscr{E}_{N}, \\ \\ \dfrac{1}{|\Gamma(G)|} \sum\limits_{z \in \Gamma(G)} \phi(z) \le \dfrac{1}{|\Gamma(H)|} \sum\limits_{z \in \Gamma(H)} \phi(z) \end{cases}$$

On the other hand, for $N \in [\![2, +\infty[\![$, let $\Lambda : \mathbf{\Delta}_N^{k} \longrightarrow \mathbf{\Delta}_N^{k+1}$ be the injection given by:

$$\Lambda[(x_1, \dots, x_{N-1}, k)] = (x_1, \dots, x_{N-1}, k+1). \tag{23}$$

For every $z \in \mathbf{\Delta}_N^k$ and function $\phi : \mathbb{R}^N \longrightarrow \mathbb{R}$ in $\mathscr{E}_N$,

$$\phi(z) \le \phi(\Lambda(z)).$$

Therefore, for every non empty subset K of $\mathbf{\Delta}_N^k$,

$$\frac{1}{|K|} \sum_{z \in K} \phi(z) \le \frac{1}{|\Lambda(K)|} \sum_{u \in \Lambda(K)} \phi(u). \tag{24}$$

since $|K| = |\Lambda(K)|$. Furthermore, one notices that:

$$\Gamma^{-1}[\Lambda(\mathbf{\Delta}_N^k)] = \mathbf{J}_{N-1}^k \quad \text{and} \quad \Gamma^{-1}(\mathbf{\Delta}_N^{k+1}) = \mathbf{J}_{N-1}^{k+1}$$

where Γ^{-1} denotes the inverse map of Γ. Hence, the following is easily obtained:

Lemma 2. *Let $k \in [\![1, +\infty[\![$ and $N \in [\![2, +\infty[\![$. Assume that for every function $f : \mathbb{R}^{N-1} \longrightarrow \mathbb{R}$ in $\mathscr{E}_{N-1}$:*

$$\frac{1}{|\mathbf{J}_{N-1}^k|} \sum_{x \in \mathbf{J}_{N-1}^k} f(x) \le \frac{1}{|\mathbf{J}_{N-1}^{k+1}|} \sum_{x \in \mathbf{J}_{N-1}^{k+1}} f(x). \tag{25}$$

Then, for every function $\phi : \mathbb{R}^N \to \mathbb{R}$ in $\mathscr{E}_N$,

$$\frac{1}{|\mathbf{\Delta}_N^k|}\sum_{y\in\mathbf{\Delta}_N^k}\phi(y)\leq\frac{1}{|\mathbf{\Delta}_N^{k+1}|}\sum_{y\in\mathbf{\Delta}_N^{k+1}}\phi(y). \tag{26}$$

Now, we are able to prove (20) by induction on $N \in [\![1,+\infty[\![$ and $k \in [\![0,+\infty[\![$.

Proposition 2. *Let $k \in [\![0,+\infty[\![$, $N \in [\![1,+\infty[\![$ and let $\phi : \mathbb{R}^N \to \mathbb{R}$ be any function in $\mathscr{E}_N$. Then,*

$$\frac{1}{|\mathbf{J}_N^k|}\sum_{z\in\mathbf{J}_N^k}\phi(z)\leq\frac{1}{|\mathbf{J}_N^{k+1}|}\sum_{z\in\mathbf{J}_N^{k+1}}\phi(z); \tag{27}$$

in other words, $(S_n, n \in \mathbb{N})$ is conditionally monotone.

Proof (of Proposition 2).

1. It is obvious that (27) holds for $(k,N) \in [\![0,+\infty[\![\times\{1\}$, and for $(k,N) \in \{0\}\times[\![1,+\infty[\![$.
2. Let $(k,N) \in [\![1,+\infty[\![\times[\![2,+\infty[\![$. We assume that:

$$\forall\ (l,m)\in\mathscr{D}:=[\![0,k-1]\!]\times[\![1,+\infty[\![\cup\{k\}\times[\![1,N-1]\!]$$

(see Fig. 1) and any function $f : \mathbb{R}^m \to \mathbb{R}$ in $\mathscr{E}_m$:

$$\frac{1}{|\mathbf{J}_m^l|}\sum_{x\in\mathbf{J}_m^l}f(x)\leq\frac{1}{|\mathbf{J}_m^{l+1}|}\sum_{x\in\mathbf{J}_m^{l+1}}f(x). \tag{28}$$

By taking $(l,m) = (k, N-1)$ in (28), Lemma 2 yields:

$$\frac{1}{|\mathbf{\Delta}_N^k|}\sum_{y\in\mathbf{\Delta}_N^k}\phi(y)\leq\frac{1}{|\mathbf{\Delta}_N^{k+1}|}\sum_{y\in\mathbf{\Delta}_N^{k+1}}\phi(y). \tag{29}$$

On the other hand, from the definition of $\mathbf{\Delta}_N^{k+1}$, (27) is equivalent to:

$$\frac{1}{|\mathbf{J}_N^k|}\sum_{y\in\mathbf{J}_N^k}\phi(y)\leq\frac{1}{|\mathbf{\Delta}_N^{k+1}|}\sum_{y\in\mathbf{\Delta}_N^{k+1}}\phi(y). \tag{30}$$

Using (28) with $(l,m) = (k-1, N)$, we have:

$$\frac{1}{|\mathbf{J}_N^k|}\sum_{y\in\mathbf{J}_N^k}\phi(y)\leq\frac{1}{|\mathbf{\Delta}_N^{k}|}\sum_{y\in\mathbf{\Delta}_N^{k}}\phi(y). \tag{31}$$

The comparison of (29) with (31) yields (30) which is equivalent to (27) (Fig. 1).

□

Fig. 1 $\mathcal{D} := [\![0, k-1]\!] \times [\![1, +\infty[\![\cup \{k\} \times [\![1, N-1]\!]$

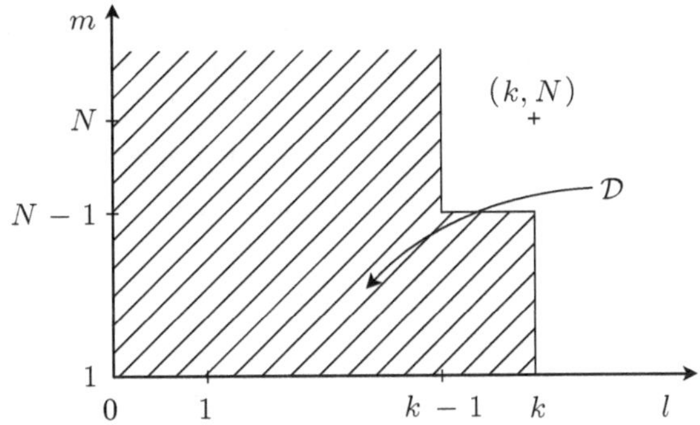

Corollary 4. *For every positive finite measure* $\sum_{n \in \mathbb{N}} a_n \delta_n$ *on* $\mathbb{N}$ *and every* $p > 0$:

$$\left(\sum_{n=0}^{+\infty} a_n e^{-t(S_n)^p - h_{n,p}(t)}, t \geq 0 \right) \quad \textit{is a peacock,}$$

where the function $h_{n,p}$ *is defined by:* $h_{n,p}(t) = \log \mathbb{E}\left[\exp\left(-t(S_n)^p\right)\right]$.

Remark 5. The result in this example has to be compared with that of Sect. 3.1.1: we replace the gamma r.v's by geometric ones.

3.2 *Diffusions Which Are "Well-Reversible" at Fixed Times Are Conditionally Monotone*

Let us now present an important class of conditionally monotone processes: that of the "well-reversible" diffusions at a fixed time.

3.2.1 The Diffusion $(X_\lambda, \lambda \geq 0; \mathbb{P}_x, x \in \mathbb{R})$

Let $\sigma : \mathbb{R}_+ \times \mathbb{R} \to \mathbb{R}$ and $b : \mathbb{R}_+ \times \mathbb{R} \to \mathbb{R}$ be two Borel measurable functions and let $(B_u, u \geq 0)$ be a standard Brownian motion starting from 0. We consider the SDE:

$$X_\lambda = x + \int_0^\lambda \sigma(s, X_s)\, dB_s + \int_0^\lambda b(s, X_s)\, ds, \ \lambda \geq 0. \tag{32}$$

We assume that:

A1. For every $x \in \mathbb{R}$, this SDE admits a unique pathwise solution $(X_\lambda^{(x)}, \lambda \geq 0)$, and furthermore the mapping $x \longmapsto (X_\lambda^{(x)}, \lambda \geq 0)$ may be chosen measurable.

As a consequence of (A1), from Yamada-Watanabe's theorem, $(X_\lambda^{(x)}, \lambda \geq 0)$ is a strong solution of (32), and it enjoys the strong Markov property; finally the transition kernel $P_\lambda(x, dy) = \mathbb{P}(X_\lambda^{(x)} \in dy)$ is measurable.
We now remark that, for $x \leq y$, the process $(X_\lambda^{(y)}, \lambda \geq 0)$ is stochastically greater than $(X_\lambda^{(x)}, \lambda \geq 0)$ in the following sense: for every $a \in \mathbb{R}$ and $\lambda \geq 0$,

$$\mathbb{P}\left(X_\lambda^{(y)} \geq a\right) \geq \mathbb{P}\left(X_\lambda^{(x)} \geq a\right). \tag{33}$$

Indeed, assuming that both $(X_\lambda^{(x)}, \lambda \geq 0)$ and $(X_\lambda^{(y)}, \lambda \geq 0)$ are defined on the same probability space, and setting

$$T = \inf\{\lambda \geq 0;\ X_\lambda^{(x)} = X_\lambda^{(y)}\}$$
$$(= +\infty \ \text{ if } \{\lambda \geq 0;\ X_\lambda^{(x)} = X_\lambda^{(y)}\} = \emptyset),$$

it is clear that, on $\{T = +\infty\}$,

$$X_\lambda^{(y)} \geq X_\lambda^{(x)} \qquad (\text{since } y \geq x)$$

while on $\{T < +\infty\}$, we have:

$$X_\lambda^{(y)} > X_\lambda^{(x)} \qquad \text{for every } \lambda \in [0, T[$$

and

$$X_\lambda^{(y)} = X_\lambda^{(x)} \qquad \text{for every } \lambda \in [T, +\infty[$$

since, as a consequence of our hypothesis (A1), (32) admits a unique strong Markovian solution.
On the other hand, (33) is equivalent to: for every bounded and increasing (resp. decreasing) function, and for every $\lambda \geq 0$:

$$x \to \mathbb{E}_x[\phi(X_\lambda)] = \int_{\mathbb{R}} P_\lambda(x, dy)\phi(y) \text{ is increasing (resp. decreasing).} \tag{34}$$

Lemma 3. *Let $((X_\lambda)_{\lambda \geq 0}, (\mathscr{F}_\lambda)_{\lambda \geq 0}, (\mathbb{P}_x)_{x \in \mathbb{R}})$ be a Markov process in $\mathbb{R}$ which satisfies (33). Then, for every $n \geq 1$, every $0 < \lambda_1 < \cdots < \lambda_n$, every $i \in \{1, \ldots, n\}$, every function $\phi : \mathbb{R}^n \to \mathbb{R}$ in $\mathscr{E}_n$, and for every $x \geq 0$,*

$$\mathbb{E}_x[\phi(X_{\lambda_1}, \ldots, X_{\lambda_n}) | \mathscr{F}_{\lambda_i}] = \widetilde{\phi}_i(X_{\lambda_1}, \ldots, X_{\lambda_i}), \tag{35}$$

where $\widetilde{\phi}_i : \mathbb{R}^i \to \mathbb{R}$ belongs to $\mathscr{E}_i$. In particular,

$$x \to \mathbb{E}_x[\phi(X_{\lambda_1}, \ldots, X_{\lambda_n})] \textit{ is increasing.} \tag{36}$$

Proof (of Lemma 3).
If $i = n$, (35) is obvious. If $i = n - 1$, then (35) is satisfied since:

$$\mathbb{E}_x[\phi(X_{\lambda_1}, \ldots, X_{\lambda_{n-1}}, X_{\lambda_n})|\mathscr{F}_{\lambda_{n-1}}]$$
$$= \int_{\mathbb{R}} \phi(X_{\lambda_1}, \ldots, X_{\lambda_{n-1}}, y) P_{\lambda_n - \lambda_{n-1}}(X_{\lambda_{n-1}}, dy)$$

and then, for $i = n-1$, (35) follows immediately from (34). Thus, Lemma 3 follows by iteration of this argument. □

Observe that as a consequence of Lemma 3, the conditional monotonicity property (4) for these diffusions is equivalent to (10).

3.2.2 Time-Reversal at a Fixed Time

Let $x \in \mathbb{R}$ fixed. We assume that:

A2. For every $\lambda > 0$, $\sigma(\lambda, \cdot)$ is a differentiable function and X_λ admits a $\mathscr{C}^{1,2}$ density function p on $]0, +\infty[\times\mathbb{R}$.

By setting

$$a(\lambda, y) := \sigma^2(\lambda, y) \qquad \text{for every } \lambda \geq 0 \text{ and } y \in \mathbb{R},$$

we define successively, for any fixed $\lambda_0 > 0$ and for $y \in \mathbb{R}$:

$$\begin{cases} \overline{a}^{\lambda_0}(\lambda, y) = a(\lambda_0 - \lambda, y), \quad (0 \leq \lambda \leq \lambda_0) \\ \overline{b}^{\lambda_0}(\lambda, y) = -b(\lambda_0 - \lambda, y) + \dfrac{1}{p(\lambda_0 - \lambda, y)} \dfrac{\partial}{\partial y}\big(a(\lambda_0 - \lambda, y)\, p(\lambda_0 - \lambda, y)\big), \\ \qquad (0 \leq \lambda < \lambda_0) \end{cases} \tag{37}$$

and the differential operator $L_\lambda^{\lambda_0}$, $(0 \leq \lambda < \lambda_0)$:

$$L_\lambda^{\lambda_0} f(x) = \frac{1}{2}\overline{a}^{\lambda_0}(\lambda, y) f''(y) + \overline{b}^{\lambda_0}(\lambda, y) f'(y) \quad \text{for } f \in \mathscr{C}_b^2.$$

Under some suitable conditions on a and b, Haussmann and Pardoux [6] (see also Meyer [17]) proved that:

A3. The process $(\overline{X}_\lambda^{\lambda_0}, 0 \leq \lambda < \lambda_0)$ obtained by time-reversing $(X_\lambda, 0 < \lambda \leq \lambda_0)$ at time λ_0:
$$(\overline{X}_\lambda^{\lambda_0}, 0 \leq \lambda < \lambda_0) := (X_{\lambda_0 - \lambda}, 0 \leq \lambda < \lambda_0)$$
is a diffusion and there exists a Brownian motion $(\overline{B}_u, 0 \leq u \leq \lambda_0)$, independent of X_{λ_0}, such that $(\overline{X}_\lambda^{\lambda_0}, 0 \leq \lambda < \lambda_0)$ solves the SDE:

$$\begin{cases} dY_\lambda &= \overline{\sigma}^{\lambda_0}(\lambda, Y_\lambda)d\overline{B}_\lambda + \overline{b}^{\lambda_0}(\lambda, Y_\lambda)d\lambda \qquad (0 \leq \lambda < \lambda_0) \\ Y_0 &= X_{\lambda_0} \qquad (\text{with } \overline{\sigma}^{\lambda_0}(\lambda, y) = \sigma(\lambda_0 - \lambda, y)). \end{cases} \tag{38}$$

Note that the coefficients $\overline{b}^{\lambda_0}$ and $\overline{\sigma}^{\lambda_0}$ depend on x.

A4. We assume furthermore that the SDE (38) admits a unique strong solution on $[0, \lambda_0[$; thus, this strong solution is strongly Markovian.

Note that, a priori, the solution of (38) is only defined on $[0, \lambda_0[$, but it can be extended on $[0, \lambda_0]$ by setting $\overline{X}^{\lambda_0}_{\lambda_0} = x$.

3.2.3 Our Hypotheses and the Main Result

Our goal here is not to give optimal hypotheses under which the assertions (A1)–(A4) are satisfied. We refer the reader to [6] or [18] for more details. Instead, we shall present two hypotheses (H1) and (H2), either of them implying the preceding assertions:

H1. We assume that:

i. The functions $(\lambda, y) \longmapsto \sigma(\lambda, y)$ and $(\lambda, y) \longmapsto b(\lambda, y)$ are of $\mathscr{C}^{1,2}$ class on $]0, +\infty[\times\mathbb{R}$, locally Lipschitz continuous in y uniformly in λ, and the solution of (32) does not explode on $[0, \lambda_0]$,

ii. There exists $\alpha > 0$ such that:

$$a(\lambda, y) \equiv \sigma^2(\lambda, y) \geq \alpha \qquad \text{for every } y \in \mathbb{R} \text{ and } 0 \leq \lambda \leq \lambda_0,$$

and

$$\frac{\partial^2 a}{\partial y^2} \in L^\infty(]0, \lambda_0] \times \mathbb{R}_+).$$

H2. We assume that:

i. The functions σ and b are of $\mathscr{C}^{1,2}$ class, locally Lipschitz continuous in y uniformly in λ, and the solution of (32) does not explode on $[0, \lambda_0]$,

ii. The functions a and b are of $\mathscr{C}^\infty$ class on $]0, +\infty[\times\mathbb{R}$ in (λ, y) and the differential operator

$$\overline{L} = \frac{\partial}{\partial \lambda} + L_\lambda$$

is hypoelliptic (see Ikeda–Watanabe [11, p. 411] for the definition and properties of hypoelliptic operators), where $(L_\lambda, \lambda \geq 0)$ is the generator of the diffusion (32):

$$L_\lambda = \frac{1}{2}a(\lambda, \cdot)\frac{d}{dy^2} + b(\lambda, \cdot)\frac{d}{dy}. \tag{39}$$

Then, under either (H1) or (H2), the assertions $(A_i)_{i=1\ldots4}$ of both Sects. 3.2.1 and 3.2.2 are satisfied, see [6]. In particular, $(\overline{X}_\lambda^{\lambda_0}, 0 \leq \lambda < \lambda_0)$ is a strong solution of (38), see Meyer [17]. Let us now give the main result of this section.

Theorem 5. *Under either (H1) or (H2), and for every* $x \in \mathbb{R}$, *the process* $(X_\lambda, \lambda > 0)$ *is conditionally monotone under* $\mathbb{P}_x$.

Proof (of Theorem 5).
Let $n \in \mathbb{N}^*$ and let $\phi : \mathbb{R}^n \to \mathbb{R}$ in $\mathscr{E}_n$. For every $0 < \lambda_1 < \cdots < \lambda_n$ and every $i \in \{1, \ldots, n\}$:

$$\begin{aligned}
&\mathbb{E}_x \left[\phi(X_{\lambda_1}, \ldots, X_{\lambda_n}) | X_{\lambda_i} = z\right] \\
&\quad = \mathbb{E}_x \left[\mathbb{E}_x \left[\phi(X_{\lambda_1}, \ldots, X_{\lambda_n}) | \mathscr{F}_{\lambda_i}\right] | X_{\lambda_i} = z\right] \\
&\quad = \mathbb{E}_x \left[\widetilde{\phi}_i(X_{\lambda_1}, \ldots, X_{\lambda_i}) | X_{\lambda_i} = z\right] \\
&\quad \quad \text{(by Lemma 3, where } \widetilde{\phi}_i : \mathbb{R}^i \to \mathbb{R} \text{ belongs to } \mathscr{E}_i \\
&\quad = \overline{\mathbb{E}}_x \left[\widetilde{\phi}_i(\overline{X}^{\lambda_i}_{\lambda_i - \lambda_1}, \ldots, \overline{X}^{\lambda_i}_0) | \overline{X}^{\lambda_i}_0 = z\right] \quad \text{(by time-reversal at } \lambda_i) \\
&\quad = \overline{\mathbb{E}}_z \left[\widetilde{\phi}_i(\overline{X}^{\lambda_i}_{\lambda_i - \lambda_1}, \ldots, \overline{X}^{\lambda_i}_{\lambda_i - \lambda_{i-1}}, z)\right]
\end{aligned}$$

and, by applying (36) to the reversed process $(\overline{X}_\lambda^{\lambda_i}, 0 \leq \lambda < \lambda_i)$, this last expression is a bounded function which increases with respect to z. □

Remark 6. Observe that we were careful to exclude the point $\lambda_1 = 0$ in Theorem 5, since "well-reversible" diffusions can be only reversed a priori on $]0, \lambda_0]$: $(\overline{X}_\lambda^{\lambda_0}, 0 \leq \lambda < \lambda_0) := (X_{\lambda_0 - \lambda}, 0 \leq \lambda < \lambda_0)$.

Corollary 5. *Let* $(X_\lambda, \lambda > 0)$ *the unique strong solution of (32), taking values in* $\mathbb{R}_+$, *where* b *and* σ *satisfy either (H1) or (H2). Then, for every finite positive measure* μ *on* $]0, +\infty[$ *and for every* $p > 0$, *the process:*

$$\left(A_t^{(\mu,p)} := \int_0^\infty e^{-t(X_\lambda)^p - h_{\lambda,p}(t)} \mu(d\lambda), t \geq 0\right) \tag{40}$$

is a peacock, with:

$$h_{\lambda,p}(t) = \log \mathbb{E}_x \left[\exp\left(-t(X_\lambda)^p\right)\right].$$

Proof (of Corollary 5).
Let $\varepsilon > 0$ and define $\mu^{(\varepsilon)}$ to be the restriction of μ to the interval $[\varepsilon, +\infty[$: $\mu^{(\varepsilon)} := \mu_{|[\varepsilon, +\infty[}$. As $(X_\lambda, \lambda \geq \varepsilon)$ is a continuous positive process, conditions (5) and (6) are satisfied, and we may apply Theorems 5 and 1 to obtain that, for every $\psi \in \boldsymbol{C}$ and every $0 \leq s \leq t$:

$$\mathbb{E}\left[\psi(A_s^{(\mu^{(\varepsilon)})})\right] \leq \mathbb{E}\left[\psi(A_t^{(\mu^{(\varepsilon)})})\right].$$

Then, proceeding as in Point (4) of the proof of Theorem 1, the result follows by letting ε tend to 0. □

3.2.4 A Few Examples of Diffusions Which are "Well-Reversible" at Fixed Times

Example 2 (Brownian motion with drift ν).
We take $\sigma \equiv 1, b(s,y) = \nu$ and $X_\lambda = x + B_\lambda + \nu\lambda$. Then,

$$p(t,x,y) = \frac{1}{\sqrt{2\pi t}} \exp\left(-\frac{(y-(x+\nu t))^2}{2t}\right),$$

and $(\overline{X}_\lambda^{\lambda_0}, 0 \leq \lambda < \lambda_0)$ is the solution of:

$$Y_\lambda = \overline{X}_0^{\lambda_0} + \overline{B}_\lambda + \int_0^\lambda \frac{x - Y_u}{\lambda_0 - u} du$$

with $(\overline{B}_\lambda, 0 \leq \lambda < \lambda_0)$ independent from $\overline{X}_0^{\lambda_0} = X_{\lambda_0}^{(x)}$.
See Jeulin–Yor [12] for similar computations.

Example 3 (Bessel processes of dimension $\delta \geq 2$).
We take $\sigma \equiv 1$ and $b(s,y) = \dfrac{\delta - 1}{2y}$, with $\delta = 2(\nu+1), \delta \geq 2$. Then,
i. For $x > 0$:

$$p(t,x,y) = \frac{1}{t}\frac{y^{\nu+1}}{x^\nu} \exp\left(-\frac{x^2+y^2}{2t}\right) I_\nu\left(\frac{xy}{t}\right),$$

where I_ν denote the modified Bessel function of index ν (see Lebedev [14, p.110] for the definition of I_ν), and $(\overline{X}_\lambda^{\lambda_0}, 0 \leq \lambda < \lambda_0)$ is the solution of:

$$Y_\lambda = \overline{X}_0^{\lambda_0} + \overline{B}_\lambda + \int_0^\lambda \left(\frac{1}{2Y_u} - \frac{Y_u}{\lambda_0 - u} + \frac{x}{\lambda_0 - u}\frac{I_\nu'}{I_\nu}\left(\frac{xY_u}{\lambda_0 - u}\right)\right) du.$$

ii. For $x = 0$:

$$p(t,0,y) = \frac{1}{2^\nu t^{\nu+1}\Gamma(\nu+1)} y^{2\nu+1} \exp\left(\frac{-y^2}{2t}\right),$$

and $(\overline{X}_\lambda^{\lambda_0}, 0 \leq \lambda < \lambda_0)$ is the solution of:

$$Y_\lambda = \overline{X}_0^{\lambda_0} + \overline{B}_\lambda + \int_0^\lambda \left(\frac{2\nu+1}{2Y_u} - \frac{Y_u}{\lambda_0 - u}\right) du$$

This examples has a strong likelihood with Bessel processes with drift, see Watanabe [24].

Example 4 (Squared Bessel processes of dimension $\delta > 0$).
We take $\sigma(s, y) = 2\sqrt{y}$ and $b \equiv \delta$. Then:
i. For $x > 0$:

$$p(t, x, y) = \frac{1}{2t}\left(\frac{y}{x}\right)^{\nu/2} \exp\left(-\frac{x+y}{2t}\right) I_\nu\left(\frac{\sqrt{xy}}{t}\right),$$

and $(\overline{X}^{\lambda_0}_\lambda, 0 \leq \lambda < \lambda_0)$ is the solution of:

$$Y_\lambda = \overline{X}^{\lambda_0}_0 + 2\int_0^\lambda \sqrt{Y_u} dB_u + 2\lambda - 2\int_0^\lambda \left(\frac{Y_u}{\lambda - u} - \frac{\sqrt{xY_u}}{\lambda_0 - u}\frac{I'_\nu}{I_\nu}\left(\frac{\sqrt{xY_u}}{\lambda_0 - u}\right)\right) du.$$

ii. For $x = 0$:

$$p(t, 0, y) = \left(\frac{1}{2t}\right)^{\delta/2} \frac{1}{\Gamma(\delta/2)} y^{\frac{\delta}{2}-1} \exp\left(-\frac{y}{2t}\right),$$

and $(\overline{X}^{\lambda_0}_\lambda, 0 \leq \lambda < \lambda_0)$ is the solution of:

$$Y_\lambda = \overline{X}^{\lambda_0}_0 + 2\int_0^\lambda \sqrt{Y_u} dB_u + \delta\lambda - \int_0^\lambda \frac{2Y_u}{\lambda - u} du.$$

Note that we could also have obtained this example by squaring the results on Bessel processes.

Remark 7. All the above examples have a strong link with initial enlargement of a filtration (by the terminal value). We refer the reader to Mansuy–Yor [16] for further examples.

4 Another Class of Markovian Peacocks

We shall introduce another set of hypotheses on the Markov process $(X_\lambda, \lambda \geq 0)$ which ensures that:

$$\left(A^{(\mu)}_t := \int_0^\infty e^{-tX_\lambda - h_\lambda(t)} \mu(d\lambda), t \geq 0\right)$$

is a peacock.

Definition 7. A right-continuous Markov process $(X_\lambda, \lambda \geq 0; \mathbb{P}_x, x \in \mathbb{R}_+)$, with values in $\mathbb{R}_+$, is said to satisfy condition (L) if both (i) and (ii) below are satisfied:

i. This process increases in the stochastic order with respect to the starting point x; in other words, for every $a \geq 0$ and $\lambda \geq 0$, and for every $0 \leq x \leq y$:

$$\mathbb{P}_y(X_\lambda \geq a) \geq \mathbb{P}_x(X_\lambda \geq a). \tag{41}$$

ii. The Laplace transform $\mathbb{E}_x[e^{-tX_\lambda}]$ is of the form:

$$\mathbb{E}_x[e^{-tX_\lambda}] = C_1(t,\lambda)\exp(-x\, C_2(t,\lambda)), \tag{42}$$

where C_1 and C_2 are two positive functions such that:

- For every $t > 0$ and $\lambda \geq 0$,

$$\frac{\partial}{\partial t}C_2(t,\lambda) > 0. \tag{43}$$

- For every $t \geq 0$ and every compact K, there exist two constants $k_K(t) > 0$ and $\widetilde{k}_K(t) < +\infty$ such that:

$$k_K(t) \leq \inf_{\lambda\in K} C_1(t,\lambda); \quad \sup_{\lambda\in K} C_2(t,\lambda) \leq \widetilde{k}_K(t). \tag{44}$$

Taking $x = 0$ in (42), we see that $C_1(\centerdot,\lambda)$ is completely monotone (and hence infinitely differentiable) on $]0,+\infty[$ and continuous at 0. Consequently, $C_2(\centerdot,\lambda)$ is also infinitely differentiable on $]0,+\infty[$ and continuous at 0. Moreover, we have for $t > 0$ and $\lambda \geq 0$:

$$\mathbb{E}_x\left[X_\lambda e^{-tX_\lambda}\right] = \left(-\frac{\partial}{\partial t}C_1(t,\lambda) + xC_1(t,\lambda)\frac{\partial}{\partial t}C_2(t,\lambda)\right)\exp\left(-xC_2(t,\lambda)\right)$$

and we introduce:

$$\begin{cases} \alpha(t,\lambda) & := -\dfrac{\partial}{\partial t}C_1(t,\lambda) \geq 0 \\ \beta(t,\lambda) & := C_1(t,\lambda)\dfrac{\partial}{\partial t}C_2(t,\lambda) > 0 \end{cases} \tag{45}$$

We can now state the main result of this section.

Theorem 6. *Let $(X_\lambda, \lambda \geq 0; \mathbb{P}_x, x \in \mathbb{R}_+)$ be a Markov process which satisfies condition (L). Then, for every $x \geq 0$ and every finite positive measure μ on $\mathbb{R}_+$,*

$$\left(A_t^{(\mu)} := \int_0^\infty e^{-tX_\lambda - h_\lambda(t)}\mu(d\lambda), t \geq 0\right)$$

is a peacock under $\mathbb{P}_x$. Here, the function h_λ is defined by:

$$h_\lambda(t) = \log\left(\mathbb{E}_x\left[e^{-tX_\lambda}\right]\right).$$

Before proving Theorem 6, let us give two examples of processes $(X_\lambda, \lambda \geq 0; \mathbb{P}_x, x \in \mathbb{R}_+)$ which satisfy condition (L).

Example 5. Let $(X_\lambda, \lambda \geq 0; \mathbb{Q}_x, x \in \mathbb{R}_+)$ be the square of a δ-dimensional Bessel process (denoted BESQ^δ, $\delta \geq 0$, see [19, Chap. XI or [23]]). This process satisfies condition (L) since:

- It is stochastically increasing with respect to x; indeed, it solves a SDE which enjoys both existence and uniqueness properties, hence the strong Markov property (see Sect. 3.2.1).
- For every $t > 0$, we have:

$$\mathbb{Q}_x\left[e^{-tX_\lambda}\right] = \frac{1}{(1+2t\lambda)^{\frac{\delta}{2}}} \exp\left(-\frac{tx}{1+2t\lambda}\right),$$

  which yields Point (ii) of Definition 7.

In particular, for $(X_t, t \geq 0)$ a squared Bessel process of dimension 0, $(A_t^{(\mu)}, t \geq 0)$ is a peacock. This case was outside the scope of Example 4.

Example 6 (A generalization of the preceding example for $\delta = 0$).
Let $(X_\lambda, \lambda \geq 0; \mathbb{P}_x, x \in \mathbb{R}_+)$ be a continuous state branching process (denoted CSBP) (see [15]).We denote by $P_\lambda(x, dy)$ the law of X_λ under $\mathbb{P}_x$, (with $x \neq 0$), and by $*$ the convolution product. Then (P_λ) satisfies:

$$P_\lambda(x, \cdot) * P_\lambda(x', \cdot) = P_\lambda(x + x', \cdot) \quad \text{for every } \lambda \geq 0,\ x \geq 0 \text{ and } x' \geq 0$$

which easily implies (41) (see [15, pp. 21–23]). On the other hand, one has:

$$\mathbb{E}_x\left[e^{-tX_\lambda}\right] = \exp(-x\, C(t, \lambda)), \tag{46}$$

where the function $C : \mathbb{R}_+ \times \mathbb{R}_+ \to \mathbb{R}_+$ satisfies:

- For every $\lambda \geq 0$, the function $C(\cdot, \lambda)$ is continuous on $\mathbb{R}_+$ and differentiable on $]0, +\infty[$, and

$$\frac{\partial C}{\partial t}(t, \lambda) > 0 \quad \text{for every } t > 0,$$

- For every $t \geq 0$ and every compact K, there exists a constant $k_K(t) < \infty$ such that:

$$\sup_{\lambda \in K} C(t, \lambda) \leq k_K(t). \tag{47}$$

Thus, $(X_\lambda, \lambda \geq 0)$ satisfies (42).

Corollary 6. *Let $(X_\lambda, \lambda \geq 0; \mathbb{P}_x, x \in \mathbb{R}_+)$ be either a BESQ^δ or a CSBP. Then, for any finite positive measure μ on $\mathbb{R}_+$, and for every $x \geq 0$:*

$$\left(A_t^{(\mu)} := \int_0^\infty e^{-tX_\lambda - h_\lambda(t)} \mu(d\lambda), t \geq 0\right)$$

is a peacock under $\mathbb{P}_x$ *with:*

$$h_\lambda(t) = \log\left(\mathbb{E}_x\left[e^{-tX_\lambda}\right]\right).$$

Remark 8. This example generalizes the previous one in the following sense. Let $(Y_t, t \geq 0)$ be a Lévy process of characteristic exponent $\psi(\lambda) = c\lambda^\alpha$, $(c > 0, \alpha \in]1,2])$:

$$\mathbb{E}\left[e^{-\lambda Y_t}\right] = \exp\left(-ct\lambda^\alpha\right).$$

We denote by $(H_t, t \geq 0)$ the height process associated to $(Y_t, t \geq 0)$. This process admits a family of local times $(L_t^a(H), t \geq 0, a \geq 0)$ and, denoting by $\tau_r(H) := \inf\{s \geq 0; L_s^0(H) > r\}$ its right-continuous inverse, it is known (see [15]) that the process $(L^a_{\tau_r(H)}, a \geq 0)$ is a stable CSBP of index α. Then, observe that for $\alpha = 2$ and $c = \frac{1}{2}$, $(Y_t := B_t, t \geq 0)$ is a standard Brownian motion started from 0, $(H_t, t \geq 0) \overset{\text{(law)}}{=} (|B_t|, t \geq 0)$ has the same law as a reflected Brownian motion, and that, from the Ray-Knight theorem, $(L^a_{\tau_r(H)}, a \geq 0)$ is a squared Bessel process of dimension 0 started from r.
We refer the interested reader to [10, Chap. 4] for a description of other peacocks constructed from CSBP, and their associated martingales.

Proof (of Theorem 6).
Let $(X_\lambda, \lambda \geq 0)$ be a process which enjoys condition (L).

1. $(-X_\lambda, \lambda \geq 0)$ being a negative process, condition (5) clearly holds. Moreover, by (44), (6) also holds. Thus, following the proof of Theorem 1, it suffices to show that $(A_t^{(\mu)}, t \geq 0)$ is a peacock when μ is a finite linear combination of Dirac measures with positive coefficients.
2. For $t \geq 0$, $a_1 \geq 0, \dots, a_n \geq 0$ and $0 \leq \lambda_1 < \cdots < \lambda_n$, we set:

$$A_t := \sum_{i=1}^n a_i e^{-tX_{\lambda_i} - h_{\lambda_i}(t)}.$$

Let $\psi \in \boldsymbol{C}$. One has:

$$\frac{\partial}{\partial t}\mathbb{E}_x[\psi(A_t)] = -\mathbb{E}_x\left[\psi'(A_t)\sum_{i=1}^n a_i e^{-tX_{\lambda_i} - h_{\lambda_i}(t)}(h'_{\lambda_i}(t) + X_{\lambda_i})\right]$$

and, we shall prove as in the proof of Theorem 1 that, for every $i \in \{1,\dots,n\}$, $\Delta_i \leq 0$, with:

$$\begin{aligned}\Delta_i &= \mathbb{E}_x\left[\psi'(A_t)e^{-tX_{\lambda_i} - h_{\lambda_i}(t)}(h'_{\lambda_i}(t) + X_{\lambda_i})\right]\\ &= \mathbb{E}_x\left[\psi'(A_t)e_{\lambda_i}(X_{\lambda_i})\right],\end{aligned}$$

and where we have set

$$e_{\lambda_i}(z) := e^{-tz-h_{\lambda_i}(t)}(h'_{\lambda_i}(t)+z).$$

We note, since $\mathbb{E}\left[e^{-tX_{\lambda_i}-h_{\lambda_i}(t)}\right]=1$, that:

$$\mathbb{E}_x[e_{\lambda_i}(X_{\lambda_i})] = 0. \tag{48}$$

Since the function

$$(x_1,\dots,x_n) \to \psi'\left(\sum_{j=0}^{n} a_j e^{-tx_j-h_{\lambda_j}(t)}\right)$$

is bounded and decreases with respect to each of its arguments, it suffices to show that: for every bounded Borel function $\phi : \mathbb{R}^n \to \mathbb{R}_+$ which decreases with respect to each of its arguments, and for every $i \in \{1,\dots,n\}$,

$$\mathbb{E}_x[\phi(X_{\lambda_1},\dots,X_{\lambda_n})e_{\lambda_i}(X_{\lambda_i})] \le 0. \tag{49}$$

3. We now show (49).
(a) We may suppose $i = n$. Indeed, thanks to (41) and to Lemma 3, we have, for $i < n$:

$$\begin{aligned}\mathbb{E}_x[\phi(X_{\lambda_1},\dots,X_{\lambda_n})e_{\lambda_i}(X_{\lambda_i})] &= \mathbb{E}_x[\mathbb{E}_x[\phi(X_{\lambda_1},\dots,X_{\lambda_n})|\mathscr{F}_{\lambda_i}]e_{\lambda_i}(X_{\lambda_i})]\\ &= \mathbb{E}_x[\widetilde{\phi}_i(X_{\lambda_1},\dots,X_{\lambda_i})e_{\lambda_i}(X_{\lambda_i})],\end{aligned}$$

where $\widetilde{\phi}_i : \mathbb{R}^i \to \mathbb{R}$ is a bounded Borel function which decreases with respect to each of its arguments.
(b) On the other hand, one has:
$\mathbb{E}_x[\widetilde{\phi}_i(X_{\lambda_1},\dots,X_{\lambda_i})e_{\lambda_i}(X_{\lambda_i})]$
$= \mathbb{E}_x[\widetilde{\phi}_i(X_{\lambda_1},\dots,X_{\lambda_i})e^{-tX_{\lambda_i}-h_{\lambda_i}(t)}(h'_{\lambda_i}(t)+X_{\lambda_i})]$
$\le \mathbb{E}_x[\widetilde{\phi}_i(X_{\lambda_1},\dots,X_{\lambda_{i-1}},-h'_{\lambda_i}(t))e_{\lambda_i}(X_{\lambda_i})]$
(since $\widetilde{\phi}_i(X_{\lambda_1},\dots,X_{\lambda_i})(h'_{\lambda_i}(t)+X_{\lambda_i}) \le \widetilde{\phi}_i(X_{\lambda_1},\dots,-h'_{\lambda_i}(t))(h'_{\lambda_i}(t)+X_{\lambda_i})$)
$= \mathbb{E}_x\left[\widetilde{\widetilde{\phi}}_i(X_{\lambda_1},\dots,X_{\lambda_{i-1}})e_{\lambda_i}(X_{\lambda_i})\right],$

where $\widetilde{\widetilde{\phi}}_i : \mathbb{R}^{i-1} \to \mathbb{R}$ is a bounded Borel function which decreases with respect to each of its arguments, and is defined by:

$$\widetilde{\widetilde{\phi}}_i(z_1,\dots,z_{i-1}) = \widetilde{\phi}_i(z_1,\dots,z_{i-1},-h'_{\lambda_i}(t)). \tag{50}$$

(c) We now end the proof of Theorem 6 by showing the following lemma.

Lemma 4. *For every $i \in \{1,\dots,n\}$ and $j \in \{0,1,\dots,i-1\}$, let $\phi : \mathbb{R}^j \to \mathbb{R}$ be a bounded Borel function which decreases with respect to each of its arguments. Then,*

$$\mathbb{E}_x[\phi(X_{\lambda_1},\dots,X_{\lambda_j})e_{\lambda_i}(X_{\lambda_i})] \le 0. \tag{51}$$

In particular,

$$\mathbb{E}_x[\phi(X_{\lambda_1},\dots,X_{\lambda_{i-1}})e_{\lambda_i}(X_{\lambda_i})] \le 0. \tag{52}$$

Proof (of Lemma 4).
We prove this lemma by induction on j.

- For $j = 0$, ϕ is constant and one has:

$$\mathbb{E}_x[\phi\, e_{\lambda_i}(X_{\lambda_i})] = \phi\, \mathbb{E}_x[e_{\lambda_i}(X_{\lambda_i})] = 0 \quad \text{(from (48))}$$

- On the other hand, if one assumes that (51) holds for $0 \le j < i-1$, then

$$\begin{aligned}
&\mathbb{E}_x[\phi(X_{\lambda_1},\dots,X_{\lambda_j},X_{\lambda_{j+1}})e_{\lambda_i}(X_{\lambda_i})] \\
&= \mathbb{E}_x[\phi(X_{\lambda_1},\dots,X_{\lambda_j},X_{\lambda_{j+1}})P_{\lambda_i-\lambda_{j+1}}e_{\lambda_i}(X_{\lambda_{j+1}})] \\
&\quad \text{(by the Markov property)} \\
&= \mathbb{E}_x[\phi(X_{\lambda_1},\dots,X_{\lambda_j},X_{\lambda_{j+1}})e^{-X_{\lambda_{j+1}}C_2(t,\lambda_i-\lambda_{j+1})-h_{\lambda_i}(t)} \\
&\qquad\qquad \cdot\big(\alpha(t,\lambda_i-\lambda_{j+1}) + X_{\lambda_{j+1}}\beta(t,\lambda_i-\lambda_{j+1})\big)] \qquad (53) \\
&\quad \big(\text{where, from (42) and (45), } \beta > 0, \text{ and } \alpha \text{ depends on } a \text{ and } h'_{\lambda_i}\big) \\
&\le \mathbb{E}_x\left[\phi\left(X_{\lambda_1},\dots,X_{\lambda_j},-\frac{\alpha(t,\lambda_i-\lambda_{j+1})}{\beta(t,\lambda_i-\lambda_{j+1})}\right)P_{\lambda_i-\lambda_{j+1}}e_{\lambda_i}(X_{\lambda_{j+1}})\right] \\
&= \mathbb{E}_x\left[\widetilde{\phi}(X_{\lambda_1},\dots,X_{\lambda_j})e_{\lambda_i}(X_{\lambda_i})\right] \le 0 \quad \text{(by the induction hypothesis)},
\end{aligned}$$

where $\widetilde{\phi} : \mathbb{R}^j \to \mathbb{R}$ is defined by:

$$\widetilde{\phi}(z_1,\dots,z_j) = \phi\left(z_1,\dots,z_j,-\frac{\alpha(t,\lambda_i-\lambda_{j+1})}{\beta(t,\lambda_i-\lambda_{j+1})}\right).$$

□

5 Stochastic and Convex Orders

The purpose of this section is different from that of the previous sections. Here, we do not look a priori for peacocks, but rather study a link between the stochastic and convex orders. As a byproduct, this will provide us with some new peacocks.

Definition 8. Let μ and ν be two probability measures on $\mathbb{R}_+$. We shall say that μ is stochastically greater than ν, and we write:

$$\mu \overset{\text{(st)}}{\geq} \nu$$

if for every $t \geq 0$,

$$F_\mu(t) := \mu([0,t]) \leq F_\nu(t) := \nu([0,t]).$$

In [10], the authors prove that if $(M_t, t \geq 0)$ is a martingale in H^1_{loc} (thus, it is a peacock), and $\alpha : \mathbb{R}_+ \longrightarrow \mathbb{R}_+$ is a continuous and strictly increasing function such that $\alpha(0) = 0$, then the process

$$\left(\frac{1}{\alpha(t)} \int_0^t M_u d\alpha(u)\ ,\ u \geq 0\right)$$

is a peacock. In other words, for every $0 \leq s \leq t$:

$$\int_0^{+\infty} M_u \left(\frac{1}{\alpha(t)} 1_{[0,t]}(u)\right) d\alpha(u) \overset{\text{(c)}}{\geq} \int_0^{+\infty} M_u \left(\frac{1}{\alpha(s)} 1_{[0,s]}(u)\right) d\alpha(u).$$

Now, it is clear that:

$$\left(\frac{1}{\alpha(t)} 1_{[0,t]}(u)\right) d\alpha(u) \overset{\text{(st)}}{\geq} \left(\frac{1}{\alpha(s)} 1_{[0,s]}(u)\right) d\alpha(u),$$

and this leads to the following question: which processes $(X_t, \geq 0)$ satisfy, for every couple of probabilities (μ, ν) such that $\mu \overset{\text{(st)}}{\geq} \nu$, the property:

$$A^{(\mu)} := \int_0^{+\infty} X_u \mu(du) \overset{\text{(c)}}{\geq} A^{(\nu)} := \int_0^{+\infty} X_u \nu(du) \quad ? \tag{54}$$

Note that such a process $(X_t, t \geq 0)$ must be a peacock. Indeed, taking for $0 \leq s \leq t$, $\mu = \delta_t$ and $\nu = \delta_s$, we deduce from (54) that $X_t \overset{\text{(c)}}{\geq} X_s$, i.e., $(X_t, t \geq 0)$ is a peacock. Here is a partial answer to this question:

Theorem 7. *Let $(X_t, t \geq 0)$ be an integrable right-continuous process satisfying both following conditions:*

i. For every bounded and increasing function $\phi : \mathbb{R} \longrightarrow \mathbb{R}_+$ and every $0 \leq s \leq t$, $\mathbb{E}[\phi(X_t)|\mathscr{F}_s]$ is an increasing function of X_s.
ii. For every $n \in \mathbb{N}^*$, every $0 \leq t_1 < \cdots < t_{n+1}$ and every $\phi : \mathbb{R}^n \to \mathbb{R}$ in $\mathscr{E}_n$, we have:

$$\mathbb{E}\left[\phi\left(X_{t_1}, \ldots, X_{t_n}\right)\left(X_{t_{n+1}} - X_{t_n}\right)\right] \geq 0.$$

Let μ and ν two probability measures on $\mathbb{R}_+$ such that $\mu \overset{(st)}{\geq} \nu$. Moreover, we assume that either:

$$\mu \text{ and } \nu \text{ have compact supports, and for every compact } K \subset \mathbb{R}_+, \sup_{t \in K} |X_t| \text{ is integrable,}$$

or:

$$\sup_{t \geq 0} |X_t| \text{ is integrable.}$$

Then:

$$A^{(\mu)} := \int_0^{+\infty} X_u \mu(du) \overset{(c)}{\geq} A^{(\nu)} := \int_0^{+\infty} X_u \nu(du).$$

Remark 9.
(a) Observe that condition (ii) implies that the process $(X_t, t \geq 0)$ is a peacock. Indeed, if ψ is a convex function of $\mathscr{C}^1$ class, then, for $0 \leq s \leq t$:

$$\mathbb{E}[\psi(X_t)] - \mathbb{E}[\psi(X_s)] \geq \mathbb{E}[\psi'(X_s)(X_t - X_s)] \geq 0.$$

In particular, $\mathbb{E}[X_t]$ does not depend on t.
(b) Note also that condition (i) implies that $(X_t, t \geq 0)$ is Markovian.

Before proving Theorem 7, we shall give some examples of processes which satisfy both conditions (i) and (ii).

Example 7. Let X be a r.v. such that for every $t \geq 0$, $\mathbb{E}[e^{tX}] < \infty$. We define $(\xi_t^X = \exp(tX - h_X(t)), t \geq 0)$ where $h_X(t) = \log \mathbb{E}[e^{tX}]$. Then, $(\xi_t^X, t \geq 0)$ satisfies the conditions of Theorem 7. Indeed, condition (i) is obvious, and condition (ii) follows from:

$$\mathbb{E}\left[\phi(\xi_{t_1}^X, \ldots, \xi_{t_n}^X)(\xi_{t_{n+1}}^X - \xi_{t_n}^X)\right] \geq \phi(e^{t_1\beta_n - h_X(t_1)}, \ldots, e^{t_n\beta_n - h_X(t_n)})\mathbb{E}\left[\xi_{t_{n+1}}^X - \xi_{t_n}^X\right] = 0,$$

where $\beta_n = \dfrac{h_X(t_{n+1}) - h_X(t_n)}{t_{n+1} - t_n}$. In particular, we recover that, if $\alpha : \mathbb{R}_+ \longrightarrow \mathbb{R}_+$ is a continuous and strictly increasing function such that $\alpha(0) = 0$, then the process

$$\left(\frac{1}{\alpha(t)} \int_0^t e^{uX - h_X(u)} d\alpha(u)\ ,\ t \geq 0\right)$$

is a peacock.

Example 8 (Martingales).
Clearly, martingales satisfy condition (ii). Here are some examples of martingales satisfying also condition (i):

(a) Let $(X_t, t \geq 0)$ be an integrable process with independent and centered increments. Then

$$\mathbb{E}[\phi(X_t)|\mathscr{F}_s] = \mathbb{E}[\phi(X_s + X_t - X_s)|\mathscr{F}_s] = \mathbb{E}[\phi(x + Z)],$$

where $X_s = x$ and $Z \overset{\text{(law)}}{=} X_t - X_s$, is an increasing function of x.

(b) Let $(L_t, t \geq 0)$ be an integrable right-continuous process with independent increments, and such that, for every $\lambda, t \geq 0$, $\mathbb{E}[e^{\lambda L_t}] < \infty$. Then, the process

$$\left(X_t := e^{\lambda L_t - h_{L_t}(\lambda)}, t \geq 0\right) \quad \text{where } h_{L_t}(\lambda) = \log \mathbb{E}\left[e^{\lambda L_t}\right]$$

is a martingale which, as in item (a), satisfies condition (i).

(c) Let $(X_t, t \geq 0)$ be a diffusion process which satisfies an equation of type

$$X_t^{(x)} = x + \int_0^t \sigma(X_s^{(x)})\, dB_s.$$

Then condition (i) follows from the stochastic comparison theorem (see Point (A1)).

Example 9 ("Well-reversible" diffusions).
Let $(Z_t, t \geq 0)$ be a "well-reversible" diffusion satisfying (32) and such that b is an increasing function. Then $(X_t := Z_t - \mathbb{E}[Z_t], t > 0)$ satisfies both conditions (i) and (ii). Indeed, condition (i) is clearly satisfied from Lemma 3. As for condition (ii), setting $h(t) = \mathbb{E}[Z_t]$, we have, with $0 < t_1 < \ldots < t_{n+1}$, by time reversal at t_{n+1}:

$$\mathbb{E}\left[\phi\left(X_{t_1}, \ldots, X_{t_n}\right)\left(X_{t_{n+1}} - X_{t_n}\right)\right]$$

$$= \overline{\mathbb{E}}\left[\phi\left(\overline{X}^{(t_{n+1})}_{t_{n+1}-t_1}, \ldots, \overline{X}^{(t_{n+1})}_{t_{n+1}-t_n}\right)\left(\overline{X}^{(t_{n+1})}_0 - \overline{X}^{(t_{n+1})}_{t_{n+1}-t_n}\right)\right]$$

$$= \overline{\mathbb{E}}\left[\mathbb{E}\left[\phi\left(\overline{X}^{(t_{n+1})}_{t_{n+1}-t_1}, \ldots, \overline{X}^{(t_{n+1})}_{t_{n+1}-t_n}\right)|\overline{\mathscr{F}}_{t_{n+1}-t_n}\right]\left(\overline{X}^{(t_{n+1})}_0 - \overline{X}^{(t_{n+1})}_{t_{n+1}-t_n}\right)\right]$$

$$= \overline{\mathbb{E}}\left[\widetilde{\phi}\left(\overline{X}^{(t_{n+1})}_{t_{n+1}-t_n}\right)\left(\overline{X}^{(t_{n+1})}_0 - \overline{X}^{(t_{n+1})}_{t_{n+1}-t_n}\right)\right]$$

where $\widetilde{\phi}$ is an increasing function,

$$= \mathbb{E}\left[\widetilde{\phi}\left(X_{t_n}\right)\left(X_{t_{n+1}} - X_{t_n}\right)\right].$$

Now from (32):

$$\mathbb{E}\left[\widetilde{\phi}\left(Z_{t_n} - h(t_n)\right)\left(\int_{t_n}^{t_{n+1}} \mu(s, Z_s) dB_s + \int_{t_n}^{t_{n+1}} b(s, Z_s) ds - h(t_{n+1}) + h(t_n)\right)\right]$$

$$= \int_{t_n}^{t_{n+1}} \mathbb{E}\left[\widetilde{\phi}\left(Z_{t_n} - h(t_n)\right)\left(b(s, Z_s) - h'(s)\right)\right] ds$$

$$= \int_{t_n}^{t_{n+1}} \mathbb{E}\left[\widetilde{\phi}\,(Z_{t_n} - h(t_n))\,(\widetilde{b}(s, Z_{t_n}) - h'(s))\right] ds$$

where $x \longmapsto \widetilde{b}(s,x) := \mathbb{E}[b(s,Z_s)|Z_{t_n} = x]$ is an increasing function such that $\mathbb{E}\left[\widetilde{b}(s,Z_{t_n})\right] = \mathbb{E}[b(s,Z_s)] = h'(s)$. Denoting by $\widetilde{b}_s^{-1}$ its right-continuous inverse, we finally obtain:

$$\mathbb{E}\left[\widetilde{\phi}\,(X_{t_n})\,(X_{t_{n+1}} - X_{t_n})\right]$$
$$\geq \int_{t_n}^{t_{n+1}} \widetilde{\phi}\left(\widetilde{b}_s^{-1}(h'(s)) - h(t_n)\right) \mathbb{E}\left[\widetilde{b}(s, Z_{t_n}) - h'(s)\right] ds = 0.$$

Example 10. Let $(B_t, t \geq 0)$ be a Brownian motion started from 0 and φ be a strictly increasing odd function of $\mathscr{C}^2$ class such that φ' is convex. It is known, see [10, Chap. 1, Sect. 5] that the process $(\varphi(B_t), t \geq 0)$ is a peacock. As a consequence of Example 9 and of Theorem 7 applied with $\mu(du) = \dfrac{1}{\alpha(t)} 1_{[0,t]}(u) d\alpha(u)$ and $\nu(du) = \dfrac{1}{\alpha(s)} 1_{[0,s]}(u) d\alpha(u)$, where $\alpha : \mathbb{R}_+ \longrightarrow \mathbb{R}_+$ is a continuous and strictly increasing function such that $\alpha(0) = 0$, we deduce that the process

$$\left(\frac{1}{\alpha(t)} \int_0^t \varphi(B_u) d\alpha(u)\ ,\ t \geq 0\right)$$

is also a peacock. Indeed, from Itô's formula, $(\varphi(B_u), u \geq 0)$ satisfies (32) with $b = \frac{1}{2}\varphi'' \circ \varphi^{-1}$ increasing.

Proof (of Theorem 7).

1. Since

$$A^{(\mu)} := \int_0^\infty X_s\, d\mu(s) = \int_0^\infty X_s\, dF_\mu(s) = \int_0^1 X_{F_\mu^{-1}(u)}\, du,$$

it suffices, by approximation of dF_μ with a linear combination of Dirac measures (as in the proof of Theorem 1), to show that for every $n \in \mathbb{N}^*$, for every $a_1, a_2, \ldots, a_n$ and for every $t_1 \geq s_1, \ldots, t_n \geq s_n$,

$$\sum_{i=1}^n a_i X_{t_i} \overset{(c)}{\geq} \sum_{i=1}^n a_i X_{s_i}. \tag{55}$$

2. Let $\psi : \mathbb{R} \to \mathbb{R}$ in $\boldsymbol{C}$. By convexity, we have:

$$\begin{aligned}\psi\left(\sum_{i=1}^{n} a_i X_{t_i}\right) &= \psi\left(\sum_{i=1}^{n} a_i X_{s_i} + \sum_{i=1}^{n} a_i (X_{t_i} - X_{s_i})\right)\\ &\geq \psi\left(\sum_{i=1}^{n} a_i X_{s_i}\right) + \psi'\left(\sum_{i=1}^{n} a_i X_{s_i}\right)\sum_{j=1}^{n} a_j (X_{t_j} - X_{s_j}).\end{aligned}$$

Then, taking the expectation leads to:

$$\begin{aligned}\mathbb{E}\left[\psi\left(\sum_{i=1}^{n} a_i X_{t_i}\right)\right] &\geq \mathbb{E}\left[\psi\left(\sum_{i=1}^{n} a_i X_{s_i}\right)\right]\\ &\quad + \mathbb{E}\left[\psi'\left(\sum_{i=1}^{n} a_i X_{s_i}\right)\sum_{j=1}^{n} a_j (X_{t_j} - X_{s_j})\right]. \qquad (56)\end{aligned}$$

We set $\phi(x_1, \dots, x_n) := \psi'\left(\sum_{i=1}^{n} a_i x_i\right)$. Thus, $\phi \in \mathscr{E}_n$. Let j be fixed and assume that:

$$0 \leq s_1 < \dots < s_j < \dots < s_{j+r} < t_j < s_{j+r+1} < \dots < s_n.$$

We write:

$$\begin{aligned}&\phi(X_{s_1}, \dots, X_{s_n})(X_{t_j} - X_{s_j})\\ &\quad= \phi(X_{s_1}, \dots, X_{s_n})(X_{t_j} - X_{s_{j+r}} + X_{s_{j+r}} - \dots + X_{s_{j+1}} - X_{s_j})\\ &\quad= \phi(X_{s_1}, \dots, X_{s_n})(X_{t_j} - X_{s_{j+r}}) + \sum_{k=0}^{r-1} \phi(X_{s_1}, \dots, X_{s_n})(X_{s_{j+k+1}} - X_{s_{j+k}})\end{aligned}$$

and we study the expectation of each term separately. From condition (i), we obtain by iteration:

$$\begin{aligned}&\mathbb{E}\left[\phi(X_{s_1}, \dots, X_{s_n})(X_{t_j} - X_{s_{j+r}})\right]\\ &\qquad= \mathbb{E}\left[\widetilde{\phi}(X_{s_1}, \dots, X_{s_{j+r}}, X_{t_j})(X_{t_j} - X_{s_{j+r}})\right]\\ &\qquad\geq \mathbb{E}\left[\widetilde{\phi}(X_{s_1}, \dots, X_{s_{j+r}}, X_{s_{j+r}})(X_{t_j} - X_{s_{j+r}})\right]\\ &\qquad\geq 0\end{aligned}$$

from condition (ii). The other terms can be dealt with in the same way.

□

Remark 10.
1. Note that, in general, the process $\left(\frac{1}{t}\int_0^t X_u du, t \geq 0\right)$ may be a peacock even if $(X_t, t \geq 0)$ is not a peacock. For example, this is the case for the process $(X_t = e^{-t}G, t \geq 0)$ where G is a centered Gaussian r.v. Similarly, $(X_u, u \geq 0)$ may be a peacock while $\left(\frac{1}{t}\int_0^t X_u du, t \geq 0\right)$ is not; for example, take the process $(X_t = (1_{[0,1]}(t) - 1_{]1,+\infty[}(t))G, t \geq 0)$ where G is a centered Gaussian r.v.
2. Theorem 7 answers partially a question raised in [9], namely, for which martingales does (54) hold?

Remark 11. In this paper, our aim has been to give several examples of peacocks. On the other hand, we did not exhibit associated martingales (see Point (b)) of the introduction). We refer the interested reader to [10] where numerous martingales associated to given peacocks are presented. However, for most of the peacocks presented in this paper, we do not know how to exhibit an associated martingale.

Acknowledgements We are grateful to F. Hirsch and M. Yor for numerous fruitful discussions during the preparation of this work.

References

1. An, M.Y., Log-concave probability distributions: Theory and statistical testing. Duke University Dept of Economics, Working Paper No. 95-03. SSRN, pp. i–29 (1997)
2. D. Baker, M. Yor, A Brownian sheet martingale with the same marginals as the arithmetic average of geometric Brownian motion. Elect. J. Prob. **14**(52), 1532–1540 (2009)
3. P. Carr, C.-O. Ewald, Y. Xiao, On the qualitative effect of volatility and duration on prices of Asian options. Finance Res. Lett. **5**(3), 162–171 (2008)
4. H. Daduna, R. Szekli, A queueing theoretical proof of increasing property of Pólya frequency functions. Stat. Probab. Lett. **26**(3), 233–242 (1996)
5. B. Efron, Increasing properties of Pólya frequency functions. Ann. Math. Stat. **36**, 272–279 (1965)
6. U.G. Haussmann, É. Pardoux, Time reversal of diffusions. Ann. Probab. **14**(4), 1188–1205 (1986)
7. F. Hirsch, B. Roynette, M. Yor, From an Itô type calculus for Gaussian processes to integrals 992 of log-normal processes increasing in the convex order. J. Math. Soc. Japan, **63**(3), 887–891 (2011)
8. F. Hirsch, B. Roynette, M. Yor. Unifying constructions of martingales associated with processes increasing in the convex order, via Lévy and Sato sheets. Expo. Math. **28**(4), 299–324 (2010)
9. F. Hirsch, B. Roynette, M. Yor, Applying Itô's motto: "Look at the infinite dimensional picture" by constructing sheets to obtain processes increasing in the convex order. Period. Math. Hungar. **61**(1), 195–211 (2010)
10. F. Hirsch, C. Profeta, B. Roynette, M. Yor. Peacocks and associated martingales, with explicit constructions. Bocconi & Springer Series, 3. Springer, Milan; Bocconi University Press, Milan, 2011. xxxii+384 pp
11. N. Ikeda, S. Watanabe, in *Stochastic Differential Equations and Diffusion Processes*. North-Holland Mathematical Library, vol. 24, 2nd edn. (North-Holland, Amsterdam, 1989)
12. T. Jeulin, M. Yor, in *Inégalité de Hardy, semimartingales, et faux-amis*. Séminaire de Probabilités, XIII (University of Strasbourg, Strasbourg, 1977/78). Lecture Notes in Math., vol. 721, pp. 332–359 (Springer, Berlin, 1979)

13. H.G. Kellerer, Markov-Komposition und eine Anwendung auf Martingale. Math. Ann. **198**, 99–122 (1972)
14. N.N. Lebedev, *Special Functions and Their Applications*. Revised English edition (trans. and ed. by R.A. Silverman) (Prentice-Hall, Englewood Cliffs, 1965)
15. J.-F. Le Gall, in *Spatial Branching Processes, Random Snakes and Partial Differential Equations*. Lectures in Mathematics ETH Zürich (Birkhäuser, Basel, 1999)
16. R. Mansuy, M. Yor, in *Random Times and Enlargements of Filtrations in a Brownian Setting*. Lecture Notes in Mathematics, vol. 1873 (Springer, Berlin, 2006)
17. P.-A. Meyer, in *Sur une transformation du mouvement brownien dûe à Jeulin et Yor*. Séminaire de Probabilités, XXVIII. Lecture Notes in Mathematics, vol. 1583, pp. 98–101 (Springer, Berlin, 1994)
18. A. Millet, D. Nualart, M. Sanz, Integration by parts and time reversal for diffusion processes. Ann. Probab. **17**(1), 208–238 (1989)
19. D. Revuz, M. Yor, in *Continuous Martingales and Brownian Motion*. Grundlehren der Mathematischen Wissenschaften [Fundamental Principles of Mathematical Sciences], vol. 293, 3rd edn. (Springer, Berlin, 1999)
20. M. Shaked, J.G. Shanthikumar, in *Stochastic Orders and Their Applications*. Probability and Mathematical Statistics (Academic, Boston, 1994)
21. M. Shaked, J.G. Shanthikumar, in *Stochastic Orders*. Springer Series in Statistics (Springer, New York, 2007)
22. J.G. Shanthikumar, On stochastic comparison of random vectors. J. Appl. Probab. **24**(1), 123–136 (1987)
23. T. Shiga, S. Watanabe, Bessel diffusions as a one-parameter family of diffusion processes. Z. Wahrscheinlichkeitstheorie und Verw. Gebiete **27**, 37–46 (1973)
24. S. Watanabe, On time inversion of one-dimensional diffusion processes. Z. Wahrscheinlichkeitstheorie und Verw. Gebiete **31**, 115–124 (1974/75)

Peacocks Obtained by Normalisation: Strong and Very Strong Peacocks

Antoine-Marie Bogso, Christophe Profeta, and Bernard Roynette

Abstract This paper contains two parts:
Part I. Let $(V_t, t \geq 0)$ be an integrable right-continuous process such that $\mathbb{E}[|V_t|] < \infty$, for every $t \geq 0$. Let us consider the three types of processes:

1. $(C_t := V_t - \mathbb{E}[V_t], t \geq 0)$,
2. $\left(N_t := \dfrac{V_t}{\mathbb{E}[V_t]}, t \geq 0\right)$, with $\mathbb{E}[V_t] > 0$ for every $t \geq 0$,
3. $\left(Q_t := \dfrac{V_t}{\alpha(t)}, t \geq 0\right)$, where $\mathbb{E}[V_t] = 0$ for every $t \geq 0$ and, $\alpha : \mathbb{R}_+ \to \mathbb{R}_+$ is a Borel function which is strictly positive.

We shall give some classes of processes $(V_t, t \geq 0)$ such that C, N or Q are peacocks, i.e.: whose one-dimensional marginals are increasing in the convex order.
Part II. We introduce the notions of strong and very strong peacocks which lead to the study of new classes of processes.

Keywords Peacocks • Conditionally monotone processes • Strong peacocks • Very strong peacocks

AMS Classification: 60J25, 32F17, 60G44, 60E15

A.-M. Bogso (✉) · C. Profeta · B. Roynette
Institut Elie Cartan, Université Henri Poincaré, B.P. 239, 54506 Vandœuvre-lès-Nancy Cedex, France
e-mail: antoine.bogso@iecn.u-nancy.fr; christophe.profeta@univ-evry.fr; bernard.roynette@iecn.u-nancy.fr

C. Donati-Martin et al. (eds.), *Séminaire de Probabilités XLIV*, Lecture Notes in Mathematics 2046, DOI 10.1007/978-3-642-27461-9_16,

1 Introduction

This article deals with processes which increase in the convex order. The investigation of this family of processes has gained renewed interest since the work of Carr, Ewald and Xiao. Indeed, they showed that in the Black-Scholes model, the price of an arithmetic average Asian call option increases with maturity, i.e.: if $(B_s, s \geq 0)$ is a Brownian motion issued from 0, the process $\left(\frac{1}{t}\int_0^t e^{B_s - \frac{s}{2}}\, ds, t \geq 0\right)$ increases in the convex order. Since, many classes of processes which increase in the convex order have been described and studied (see e.g., [11]). The aim of this paper is to complete the known results by exhibiting new families of processes which increase in the convex order. Let us start with some elementary definitions and results.

Definition 1. Let U and V be two real-valued r.v.'s. U is said to be dominated by V for the convex order if, for every convex function $\psi : \mathbb{R} \to \mathbb{R}$ such that $\mathbb{E}[|\psi(U)|] < \infty$ and $\mathbb{E}[|\psi(V)|] < \infty$, one has:

$$\mathbb{E}[\psi(U)] \leq \mathbb{E}[\psi(V)]. \tag{1}$$

We denote this order relation by:

$$U \overset{(c)}{\leq} V. \tag{2}$$

We refer to Shaked and Shanthikumar ([18] and [19]) for background on stochastics orders.

Definition 2. We denote by **C** the class of convex $\mathscr{C}^2$-functions $\psi : \mathbb{R} \to \mathbb{R}$ such that ψ'' has a compact support, and by $\mathbf{C}_+$ the class of convex functions $\psi \in \mathbf{C}$ such that ψ is positive and increasing.

We note that if $\psi \in \mathbf{C}$:

- $|\psi'|$ is a bounded function,
- There exist k_1 and $k_2 \geq 0$ such that:

$$|\psi(x)| \leq k_1 + k_2|x|. \tag{3}$$

The next result is proved in [11].

Lemma 1. *Let U and V be two integrable real-valued r.v.'s. Then, the following properties are equivalent:*

1. $U \overset{(c)}{\leq} V$
2. *For every* $\psi \in \mathbf{C}$*:* $\mathbb{E}[\psi(U)] \leq \mathbb{E}[\psi(V)]$
3. $\mathbb{E}[U] = \mathbb{E}[V]$ *and for every* $\psi \in \mathbf{C}_+$*:* $\mathbb{E}[\psi(U)] \leq \mathbb{E}[\psi(V)]$.

Definition 3.

1. A process $(Z_t, t \geq 0)$ is said to be integrable if, for every $t \geq 0$, $\mathbb{E}[|Z_t|] < \infty$.

2. A process $(Z_t, t \geq 0)$ is said to be increasing (resp. decreasing) in the convex order if, for every $s \leq t$, $Z_s \overset{(c)}{\leq} Z_t$ (resp. $Z_t \overset{(c)}{\leq} Z_s$).
3. An integrable process which is increasing (resp decreasing) in the convex order will be called a peacock (resp. a pea$\underline{\text{d}}$ock).

Note: Most of this paper is devoted to the study of peacocks. However, some statements also feature pea$\underline{\text{d}}$ock. To draw the reader's attention, we shall always underline the $\underline{\text{d}}$.
If $(Z_t, t \geq 0)$ is a peacock, then it follows from Definitions 1 and 3, applied with $\psi(x) = x$ and $\psi(x) = -x$, that $\mathbb{E}[Z_t]$ does not depend on t.
In the sequel, when two processes $(U_t, t \geq 0)$ and $(V_t, t \geq 0)$ have the same 1-dimensional marginals, we shall write

$$U_t \overset{(1.d)}{=} V_t \tag{4}$$

and say that $(U_t, t \geq 0)$ and $(V_t, t \geq 0)$ are associated.
From Jensen's inequality, every martingale $(M_t, t \geq 0)$ is a peacock; conversely, a result due to Kellerer [13] states that, for any peacock $(Z_t, t \geq 0)$, there exists (at least) one martingale $(M_t, t \geq 0)$ such that:

$$Z_t \overset{(1.d)}{=} M_t. \tag{5}$$

We say that such a martingale is associated with Z. Note that:

(a) In general, there exist several different martingales associated with a given peacock.
(b) Sometimes, one may be fortunate enough to find directly a martingale associated to the process Z, thus proving that Z is a peacock, see e.g., Example 2.

Many examples of peacocks with a description of associated martingales are given in [11]. One may also refer to [2, 9, 10] where the notions of Brownian and Lévy Sheet play an essential role in constructing associated martingales to certain peacocks.
On the contrary, we note that for most of the peacocks given in this article, the question of finding an associated martingale remains open.
A table summing-up our main results is found at the end of the paper, see Table 1.

2 Part I: Peacocks Obtained by Normalisation, Centering and Quotient

2.1 *Preliminaries*

2.1.1 The Aim of This Part

Let $(V_t, t \geq 0)$ be an integrable right-continuous process and, let us consider the three families of processes:

1. $(C_t := V_t - \mathbb{E}[V_t], t \geq 0)$,
2. $\left(N_t := \dfrac{V_t}{\mathbb{E}[V_t]}, t \geq 0\right)$, with $0 < \mathbb{E}[V_t] < +\infty$ for every $t \geq 0$,
3. $\left(Q_t := \dfrac{V_t}{\alpha(t)}, t \geq 0\right)$, where $\mathbb{E}[V_t] = 0$ for every $t \geq 0$ and $\alpha : \mathbb{R}_+ \to \mathbb{R}_+$ is a Borel function which is strictly positive.

We adopt the notation C for centering, N for normalisation and Q for quotient. We note that, for every $t \geq 0$, $\mathbb{E}[C_t] = \mathbb{E}[Q_t] = 0$ and $\mathbb{E}[N_t] = 1$. Since $\mathbb{E}[C_t]$, $\mathbb{E}[N_t]$ and $\mathbb{E}[Q_t]$ do not depend on t, it may be natural to ask under which conditions on $(V_t, t \geq 0)$ the processes C, N and Q are peacocks.
Let us first recall the following elementary lemma (see [11]).

Lemma 2. *Let U be a real-valued integrable random variable. Then, the following properties are equivalent:*

1. *For every real c, $\mathbb{E}\left[1_{\{U \geq c\}} U\right] \geq 0$,*
2. *For every bounded and increasing function $h : \mathbb{R} \to \mathbb{R}_+$:*

$$\mathbb{E}[h(U)U] \geq 0,$$

3. *$\mathbb{E}[U] \geq 0$.*

2.1.2 Some Examples

We now deal with some examples.

Example 1. i. If $V_t = \int_0^t e^{B_s - \frac{s}{2}} ds$, then Carr et al. [5] showed that:

$$\left(N_t := \frac{1}{t}\int_0^t e^{B_s - \frac{s}{2}} ds, t \geq 0\right) \text{ is a peacock,}$$

when $(B_s, s \geq 0)$ is a Brownian motion issued from 0. We note that $\mathbb{E}[V_t] = t$ for every $t \geq 0$. An associated martingale is provided in [2]. This example, called the "guiding example" in [11] has been at the origin of many generalizations, as will be clear from now on.

ii. If $V_t = \int_0^t M_s d\alpha(s)$ (resp. $V_t = \int_0^t (M_s - M_0) d\alpha(s)$), where $(M_s, s \geq 0)$ is a martingale in H^1_{loc} and $\alpha : \mathbb{R}_+ \to \mathbb{R}_+$, a continuous and increasing function such that $\alpha(0) = 0$, then it is shown in [11], Chap. 1, that:

$$\left(N_t := \frac{1}{\alpha(t)}\int_0^t M_s\, d\alpha(s), t \geq 0\right)$$

(resp.

$$\left(C_t := \int_0^t (M_s - M_0)\, d\alpha(s), t \geq 0\right))$$

is a peacock. Note that, for every $t \geq 0$, $\mathbb{E}[V_t] = \alpha(t)\mathbb{E}[N_0]$ (resp. $\mathbb{E}[V_t] = 0$). In that generality, we do not know how to associate a martingale to these peacocks. We shall generalize this result in Theorem 11 thanks to the notion of very strong peacock.

Example 2.

i. If $V_t = tX$, where X is an integrable centered real-valued r.v., then $(C_t := tX, t \geq 0)$ is a peacock (see [11], Chap. 1).
ii. If $V_t = e^{tX}$, where X is a real-valued r.v. such that, for every $t \geq 0$, $\mathbb{E}[e^{tX}] < \infty$, then, it is proved in [11] (see also Example 4) that:

$$\left(N_t := \frac{e^{tX}}{\mathbb{E}[e^{tX}]}, t \geq 0\right) \text{ is a peacock.}$$

In particular, if $(G_u, u \geq 0)$ is a measurable centered Gaussian process with covariance function $K(s,t) := \mathbb{E}[G_s G_t]$, and if ν is a positive Radon measure on $\mathbb{R}_+$, then:

$$\left(N_t^{(\nu)} := \frac{\exp\left(\int_0^t G_u \nu(du)\right)}{\mathbb{E}\left[\exp\left(\int_0^t G_u \nu(du)\right)\right]}, t \geq 0\right) \text{ is a peacock}$$

as soon as

$$t \longmapsto \gamma(t) := \int_0^t \int_0^t K(u,v)\nu(du)\nu(dv) \text{ is an increasing function.}$$

Indeed,

$$\int_0^t G_u \nu(du) \overset{\text{(law)}}{=} \sqrt{\gamma(t)}G,$$

with G a reduced normal r.v. and, if $(B_t, t \geq 0)$ is a Brownian motion issued from 0, then $\left(M_t := \exp\left(B_{\gamma(t)} - \frac{\gamma(t)}{2}\right), t \geq 0\right)$ is a martingale associated to $\left(N_t^{(\nu)}, t \geq 0\right)$.

Example 3. Let $(X_t, t \geq 0)$ be an integrable centered process and $\alpha, \beta : \mathbb{R}_+ \to \mathbb{R}_+$ be two strictly positive Borel functions. We suppose that $\left(Q_t^{(\alpha)} := \frac{X_t}{\alpha(t)}, t \geq 0\right)$ is a peacock.

(a) If $t \longmapsto \dfrac{\alpha(t)}{\beta(t)}$ is increasing, then:

$$\left(Q_t^{(\beta)} := \frac{X_t}{\beta(t)}, t \geq 0 \right) \text{ is a peacock.}$$

(b) If $\left(Q_t^{(\beta)} := \dfrac{X_t}{\beta(t)}, t \geq 0 \right)$ is a peacock and if, for every $t \geq 0$, X_t is not identically 0, then

$$t \longmapsto \frac{\alpha(t)}{\beta(t)} \text{ is decreasing.}$$

Proof (of Example 3).

(a) For every $\psi \in \mathbf{C}$ and every $0 \leq s < t$, we have:

$$\mathbb{E}\left[\psi\left(\frac{X_t}{\beta(t)}\right)\right] = \mathbb{E}\left[\psi\left(\frac{X_t}{\alpha(t)}\frac{\alpha(t)}{\beta(t)}\right)\right] \geq \mathbb{E}\left[\psi\left(\frac{X_s}{\alpha(s)}\frac{\alpha(t)}{\beta(t)}\right)\right]$$

(since $\left(Q_t^{(\alpha)} := \dfrac{X_t}{\alpha(t)}, t \geq 0 \right)$ is a peacock)

$$\geq \mathbb{E}\left[\psi\left(\frac{X_s}{\alpha(s)}\frac{\alpha(s)}{\beta(s)}\right)\right]$$

(from point (i)) of Example 2 since X_s is centered and $t \longmapsto \dfrac{\alpha(t)}{\beta(t)}$ is increasing).

$$= \mathbb{E}\left[\psi\left(\frac{X_s}{\beta(s)}\right)\right]$$

(b) Let us suppose that there exist $0 \leq s < t$ such that:

$$\frac{\alpha(s)}{\beta(s)} < \frac{\alpha(t)}{\beta(t)},$$

then,

$$\mathbb{E}\left[\psi\left(\frac{X_s}{\beta(s)}\right)\right] = \mathbb{E}\left[\psi\left(\frac{X_s}{\alpha(s)}\frac{\alpha(s)}{\beta(s)}\right)\right]$$

and we may choose ψ such that:

$$\mathbb{E}\left[\psi\left(\frac{X_s}{\alpha(s)}\frac{\alpha(s)}{\beta(s)}\right)\right] < \mathbb{E}\left[\psi\left(\frac{X_s}{\alpha(s)}\frac{\alpha(t)}{\beta(t)}\right)\right]$$

(from point (i)) of Example 2 since X_s is centered and not identically 0)

Hence,

$$\mathbb{E}\left[\psi\left(\frac{X_s}{\beta(s)}\right)\right] < \mathbb{E}\left[\psi\left(\frac{X_s}{\alpha(s)}\frac{\alpha(t)}{\beta(t)}\right)\right]$$
$$\leq \mathbb{E}\left[\psi\left(\frac{X_t}{\alpha(t)}\frac{\alpha(t)}{\beta(t)}\right)\right] \text{ (since } Q^{(\alpha)} \text{ is a peacock).}$$

This contradicts the fact that $Q^{(\beta)}$ is a pea<u>d</u>ock. □

Example 4. Let $(X_t, t \geq 0)$ be a measurable $\mathbb{R}_+$-valued process such that $0 < \mathbb{E}[X_t] < \infty$:

(a) If, for every $0 \leq s \leq t$, $x \longmapsto \frac{1}{x}\mathbb{E}[X_t|X_s = x]$ is an increasing function, then:

$$\left(\frac{X_t}{\mathbb{E}[X_t]}, t \geq 0\right) \text{ is a peacock.}$$

(b) If, for every $0 \leq s \leq t$, $x \longmapsto \frac{1}{x}\mathbb{E}[X_s|X_t = x]$ is an increasing function, then:

$$\left(\frac{X_t}{\mathbb{E}[X_t]}, t \geq 0\right) \text{ is a pea\underline{d}ock.}$$

Proof (of Example 4).

(a) For every $\psi \in \mathbf{C}_+$ and every $0 \leq s < t$, we have:

$$\mathbb{E}\left[\psi\left(\frac{X_t}{\mathbb{E}[X_t]}\right)\right] - \mathbb{E}\left[\psi\left(\frac{X_s}{\mathbb{E}[X_s]}\right)\right]$$
$$\geq \mathbb{E}\left[\psi'\left(\frac{X_s}{\mathbb{E}[X_s]}\right)\left(\frac{X_t}{\mathbb{E}[X_t]} - \frac{X_s}{\mathbb{E}[X_s]}\right)\right] \quad \text{(since } \psi \text{ is convex)}$$
$$= \mathbb{E}\left[\psi'\left(\frac{X_s}{\mathbb{E}[X_s]}\right)\frac{X_s}{\mathbb{E}[X_s]}\left(\frac{\mathbb{E}[X_s]}{\mathbb{E}[X_t]}\frac{\mathbb{E}[X_t|X_s]}{X_s} - 1\right)\right] \geq 0$$

(since $x \longmapsto \frac{1}{x}\mathbb{E}[X_t|X_s = x]$ is increasing and using a slight extension of Lemma 2)

(b) For every $\psi \in \mathbf{C}_+$ and every $0 \leq s < t$, one has:

$$\begin{aligned}
&\mathbb{E}\left[\psi\left(\frac{X_t}{\mathbb{E}[X_t]}\right)\right] - \mathbb{E}\left[\psi\left(\frac{X_s}{\mathbb{E}[X_s]}\right)\right] \\
&\leq \mathbb{E}\left[\psi'\left(\frac{X_t}{\mathbb{E}[X_t]}\right)\left(\frac{X_t}{\mathbb{E}[X_t]} - \frac{X_s}{\mathbb{E}[X_s]}\right)\right] \quad \text{(since } \psi \text{ is convex)} \\
&= \mathbb{E}\left[\psi'\left(\frac{X_t}{\mathbb{E}[X_t]}\right)\frac{X_t}{\mathbb{E}[X_t]}\left(1 - \frac{\mathbb{E}[X_t]}{\mathbb{E}[X_s]}\frac{\mathbb{E}[X_s|X_t]}{X_t}\right)\right] \leq 0 \\
&\quad \text{(since } x \longmapsto \frac{1}{x}\mathbb{E}[X_s|X_t = x] \text{ is increasing and using a slight} \\
&\quad \text{extension of Lemma 2)}
\end{aligned}$$

□

In particular,

i. Let $\phi : \mathbb{R}_+ \times \mathbb{R} \to \mathbb{R}_+$ such that:
   - For every $t \geq 0$, $x \mapsto \phi(t,x)$ is increasing on $\mathbb{R}$,
   - For every $s < t$, $x \mapsto \dfrac{\phi(t,x)}{\phi(s,x)}$ is increasing.

   Then, if X is a r.v. such that $\mathbb{E}[\phi(t,X)] < \infty$ for every $t \geq 0$,

$$\left(N_t := \frac{\phi(t,X)}{\mathbb{E}[\phi(t,X)]}, t \geq 0\right) \text{ is a peacock;}$$

ii. Let $f : \mathbb{R}_+ \to \mathbb{R}_+$ be an increasing $\mathscr{C}^1$-function such that $x \mapsto \dfrac{xf'(x)}{f(x)}$ is decreasing. If $(\gamma_t, t \geq 0)$ denotes the Gamma subordinator, then

$$\left(N_t := \frac{f(\gamma_t)}{\mathbb{E}[f(\gamma_t)]}, t \geq 0\right) \text{ is a pea\underline{d}ock.}$$

   This assertion follows from point b) above and from a well-known property of the Gamma subordinator: for every $t \geq 0$, the (Dirichlet) process $\left(\dfrac{\gamma_s}{\gamma_t}, 0 \leq s \leq t\right)$ is independent of the r.v. γ_t.

Example 5. Let $(X_t, t \geq 0)$ be a càdlàg $\mathbb{R}_+$-valued process such that

$$\mathbb{E}\left[\sup_{0 \leq u \leq t} |X_u|\right] < \infty$$

and let ν be a positive Radon measure on $\mathbb{R}_+$. Suppose that, for every $0 \leq s \leq t$, $\varphi_{s,t} : x \longmapsto \dfrac{1}{x}\mathbb{E}[X_s|X_t = x]$ is an increasing function. Then:

$$\left(N_t := \frac{\displaystyle\int_0^t X_u \nu(du)}{\mathbb{E}\left[\displaystyle\int_0^t X_u \nu(du)\right]}, t \geq 0 \right) \text{ is a peadock.} \tag{6}$$

Proof (of (6)).
By approximation, we may assume that ν is absolutely continuous with respect to the Lebesgue measure and admits a continuous density f:

$$\nu(dt) = f(t)dt. \tag{7}$$

Then, with $h(t) := \mathbb{E}\left[\displaystyle\int_0^t X_u \nu(du)\right]$ and $\psi \in \mathbf{C}$:

$$\begin{aligned}
&\frac{d}{dt}\mathbb{E}\left[\psi\left(\frac{\int_0^t X_u f(u)du}{h(t)}\right)\right] \\
&= \mathbb{E}\left[\psi'\left(\frac{\int_0^t X_u f(u)du}{h(t)}\right)\left(\frac{X_t f(t)}{h(t)} - \frac{h'(t)}{h^2(t)}\int_0^t X_u f(u)du\right)\right] \\
&\leq \frac{1}{h(t)}\mathbb{E}\left[\psi'\left(\frac{X_t f(t)}{h'(t)}\right)\left(X_t f(t) - \frac{h'(t)}{h(t)}\int_0^t X_u f(u)du\right)\right]
\end{aligned}$$

since ψ' is increasing and, if

$$\frac{X_t f(t)}{h(t)} \geq \frac{h'(t)}{h^2(t)}\int_0^t X_u f(u)du \quad \left(\text{resp. } \frac{X_t f(t)}{h(t)} \leq \frac{h'(t)}{h^2(t)}\int_0^t X_u f(u)du\right),$$

then

$$\frac{X_t f(t)}{h'(t)} \geq \frac{\int_0^t X_u f(u)du}{h(t)} \quad \left(\text{resp. } \frac{X_t f(t)}{h'(t)} \leq \frac{\int_0^t X_u f(u)du}{h(t)}\right).$$

Thus,

$$\begin{aligned}
&\frac{d}{dt}\mathbb{E}\left[\psi\left(\frac{\int_0^t X_u f(u)du}{h(t)}\right)\right] \\
&\leq \frac{1}{h(t)}\mathbb{E}\left[\psi'\left(\frac{X_t f(t)}{h'(t)}\right)\left(X_t f(t) - \frac{h'(t)}{h(t)}\int_0^t X_u f(u)du\right)\right] \\
&= \frac{1}{h(t)}\mathbb{E}\left[\psi'\left(\frac{X_t f(t)}{h'(t)}\right)\left(X_t f(t) - \frac{h'(t)}{h(t)}\int_0^t \mathbb{E}[X_u|X_t] f(u)du\right)\right]
\end{aligned}$$

$$= \frac{1}{h(t)}\mathbb{E}\left[\psi'\left(\frac{X_t f(t)}{h'(t)}\right) X_t \left(f(t) - \frac{h'(t)}{h(t)}\int_0^t \varphi_{u,t}(X_t) f(u)du\right)\right]$$

(where $\varphi_{u,t}$ is an increasing function)

≤ 0 (by Lemma 2). □

In particular, if $f : \mathbb{R}_+ \to \mathbb{R}_+$ is an increasing $\mathscr{C}^1$-function such that $x \mapsto \dfrac{xf'(x)}{f(x)}$ is increasing and if $(\gamma_t, t \geq 0)$ stands for the Gamma subordinator, then:

$$\left(N_t := \frac{\int_0^t f(\gamma_s)\,ds}{\mathbb{E}\left[\int_0^t f(\gamma_s)\,ds\right]}, t \geq 0\right) \text{ is a pea\underline{d}ock.}$$

One may compare this result with the second point of Remark 7.

Example 6. i. Let $L := (L_t, t \geq 0)$ be an integrable Lévy process and ν be a positive Radon measure on $\mathbb{R}_+^*$. Then:

(a) If L is centered, then $\left(Q_t^{(\nu)} := \dfrac{1}{\nu(]0,t])}\displaystyle\int_0^t L_u \nu(du), t \geq 0\right)$ is a peacock

(b) $\left(N_t^{(\nu)} := \dfrac{\int_0^t L_u \nu(du)}{\int_0^t u\,\nu(du)}, t \geq 0\right)$ is a pea<u>d</u>ock.

Proof (of Example 6). The assertion (a) is deduced from point (ii) of Example 1 since $(L_t, t \geq 0)$ is a centered martingale. To prove (b), we shall make some computations which are close to those in the proof of Example 5 (although here, L does not take values in $\mathbb{R}_+$). We may suppose, without loss of generality, that L is centered and, as in the proof of Example 5, that $\nu(du) = f(u)du$, where f is continuous. Let us set $h(t) := \displaystyle\int_0^t u f(u) du$. Then, for every $\psi \in \mathbf{C}$:

$$\frac{d}{dt}\mathbb{E}\left[\psi\left(\frac{\int_0^t L_u f(u)du}{h(t)}\right)\right]$$

$$= \mathbb{E}\left[\psi'\left(\frac{\int_0^t L_u f(u)du}{h(t)}\right)\left(\frac{L_t f(t)}{h(t)} - \frac{tf(t)}{h^2(t)}\int_0^t L_u f(u)du\right)\right]$$

$$\leq \frac{tf(t)}{h(t)}\mathbb{E}\left[\psi'\left(\frac{L_t}{t}\right)\left(\frac{L_t}{t} - \frac{1}{h(t)}\int_0^t L_u f(u)du\right)\right]$$

$$= \frac{tf(t)}{h(t)} \mathbb{E}\left[\psi'\left(\frac{L_t}{t}\right)\left(\frac{L_t}{t} - \frac{1}{h(t)}\int_0^t \mathbb{E}\left[L_u | \mathscr{F}_t^+\right] f(u)du\right)\right]$$

(where $\mathscr{F}_t^+ = \sigma(L_u, u \geq t)$).

Observing that $\left(\frac{L_t}{t}, t \geq 0\right)$ is an inverse martingale with respect to the filtration $(\mathscr{F}_t^+, t \geq 0)$ (i.e., for every $0 < s \leq t$, $\mathbb{E}\left[\frac{L_s}{s}\middle| \mathscr{F}_t^+\right] = \frac{L_t}{t}$, see [12]), we obtain:

$$\frac{d}{dt}\mathbb{E}\left[\psi\left(\frac{\int_0^t L_u f(u)du}{h(t)}\right)\right]$$

$$\leq \frac{tf(t)}{h(t)}\mathbb{E}\left[\psi'\left(\frac{L_t}{t}\right)\frac{L_t}{t}\left(1 - \frac{1}{h(t)}\int_0^t uf(u)du\right)\right] = 0. \qquad \square$$

ii. Therefore, the following question arises naturally: for which values of α $\left(\frac{1}{t^\alpha}\int_0^t L_u du, t \geq 0\right)$ is either a peacock (it is true for $\alpha \leq 1$) or a pea<u>d</u>ock (it is true for $\alpha \geq 2$) or neither of them?

(a) If $(L_s, s \geq 0)$ is a Brownian motion (issued from 0), then, by scaling, $\left(\frac{1}{t^\alpha}\int_0^t L_u du, t \geq 0\right)$ is a peacock (resp. a pea<u>d</u>ock) if and only if $\alpha \leq \frac{3}{2}$ (resp. $\alpha \geq \frac{3}{2}$).

(b) Let $(L_s, s \geq 0)$ be a Lévy process which is stable of index γ ($1 < \gamma \leq 2$) and symmetric. Then $\left(\frac{1}{t^\alpha}\int_0^t L_u du, t \geq 0\right)$ is a peacock (resp. a pea<u>d</u>ock) if and only if $\alpha \leq 1 + \frac{1}{\gamma}$ (resp. $\alpha \geq 1 + \frac{1}{\gamma}$). Indeed, by scaling, $\int_0^t L_u du \overset{(1.d)}{=} t^{1+\frac{1}{\gamma}} S$, where the r.v. S is symmetric and stable of index γ (see point (i) of Example 2).

(c) Let $(L_s, s \geq 0)$ be a square integrable and centered Lévy process. We have

$$\mathbb{E}\left[\left(\frac{1}{t^\alpha}\int_0^t L_u du\right)^2\right] = \frac{\mathbb{E}[L_1^2]}{3} t^{3-2\alpha}.$$

Hence:

- If $\alpha < \frac{3}{2}$, $\left(\frac{1}{t^\alpha}\int_0^t L_u du, t \geq 0\right)$ is not a peadock,
- If $\alpha > \frac{3}{2}$, $\left(\frac{1}{t^\alpha}\int_0^t L_u du, t \geq 0\right)$ is not a peacock.

In some specific situations, we may obtain simultaneously a peacock with one of its associated martingale. The results in the following example are close to those obtained in [11]. Therefore, we state them without proof and refer the reader to [11, Chap. 2].

Example 7. Let $(L_t, t \geq 0)$ be a Lévy process such that:

$$\mathbb{E}\left[\exp\left(\int_0^t L_s\, ds\right)\right] < \infty, \text{ for every } t \geq 0.$$

Then:

1.

$$\left(N_t := \frac{\exp\left(\int_0^t L_s\, ds\right)}{\mathbb{E}\left[\exp\left(\int_0^t L_s\, ds\right)\right]}, t \geq 0\right) \text{ is a peacock}$$

and

$$\left(M_t := \frac{\exp\left(\int_0^t s\, dL_s\right)}{\mathbb{E}\left[\exp\left(\int_0^t s\, dL_s\right)\right]}, t \geq 0\right)$$

is a martingale associated to $(N_t, t \geq 0)$.

2.

$$\left(\widetilde{N}_t := \frac{\exp\left(\frac{1}{t}\int_0^t L_s\, ds\right)}{\mathbb{E}\left[\exp\left(\frac{1}{t}\int_0^t L_s\, ds\right)\right]}, t \geq 0\right) \text{ is a peacock}$$

and

$$\left(\widetilde{M}_t := \frac{\exp\left(\int_0^1 W_{u,t}^{(L)}\, du\right)}{\mathbb{E}\left[\exp\left(\int_0^1 W_{u,t}^{(L)}\, du\right)\right]}, t \geq 0\right)$$

is a $(\mathcal{G}_t^{(L)}, t \geq 0)$-martingale associated to $(\widetilde{N}_t, t \geq 0)$, where $(W_{u,t}^{(L)}, u \geq 0, t \geq 0)$ is the Lévy sheet associated to $(L_t, t \geq 0)$ and

$$\mathcal{G}_t^{(L)} = \sigma\left(W_{u,s}^{(L)}, u \geq 0; 0 \leq s \leq t\right) \text{ (see [8]).}$$

2.1.3 Relation Between the Peacock Properties of C and N

The peacock properties of C and N are linked as is shown in the following:

Theorem 1. *Suppose that $(V_t, t \geq 0)$ satisfies $0 < \mathbb{E}[V_t] < +\infty$ for every $t \geq 0$ and $t \longmapsto \mathbb{E}[V_t]$ is monotone.*

1. If $t \longmapsto \mathbb{E}[V_t]$ increases, we have the implication:

$$(N_t, t \geq 0) \text{ is a peacock } \Rightarrow (C_t, t \geq 0) \text{ is a peacock.}$$

2. If $t \longmapsto \mathbb{E}[V_t]$ decreases, we have the reverse implication:

$$(C_t, t \geq 0) \text{ is a peacock } \Rightarrow (N_t, t \geq 0) \text{ is a peacock.}$$

We shall give two proofs of Theorem 1. In the first proof, we use Kellerer's theorem which is not necessary in the second one.

First Proof of Theorem 1

1. We first assume that $t \longmapsto \mathbb{E}[V_t]$ is an increasing function and that $\left(N_t := \dfrac{V_t}{\mathbb{E}[V_t]}, t \geq 0\right)$ is a peacock. Then, from Kellerer's theorem, there exists a martingale $(M_t, t \geq 0)$ such that:

$$\frac{V_t}{\mathbb{E}[V_t]} \overset{(1.d)}{=} M_t, \quad \text{or, equivalently,} \quad V_t \overset{(1.d)}{=} M_t\, \mathbb{E}[V_t].$$

We note that $\mathbb{E}[M_t] = 1$ for every $t \geq 0$.
For every $\psi \in \mathbf{C}_+$ and every $0 < s \leq t$, one has:

$$\begin{aligned}
&\mathbb{E}[\psi(C_t)] - \mathbb{E}[\psi(C_s)] \\
&= \mathbb{E}[\psi\,(V_t - \mathbb{E}[V_t])] - \mathbb{E}[\psi\,(V_s - \mathbb{E}[V_s])] \\
&= \mathbb{E}[\psi\,((M_t - 1)\mathbb{E}[V_t])] - \mathbb{E}[\psi\,((M_s - 1)\mathbb{E}[V_s])] \\
&\geq \mathbb{E}\left[\psi'\,((M_s - 1)\mathbb{E}[V_s])\,((M_t - 1)\mathbb{E}[V_t] - (M_s - 1)\mathbb{E}[V_s])\right] \text{ (by convexity)} \\
&= \mathbb{E}\left[\psi'\,((M_s - 1)\mathbb{E}[V_s])\,((M_s - 1)\mathbb{E}[V_t] - (M_s - 1)\mathbb{E}[V_s])\right]
\end{aligned}$$

(taking the conditional expectation)

$$\begin{aligned}
&= \mathbb{E}\left[\psi'\,((M_s - 1)\mathbb{E}[V_s])\,(M_s - 1)\right] \underbrace{(\mathbb{E}[V_t] - \mathbb{E}[V_s])}_{\geq 0} \\
&\geq \psi'(0)(\mathbb{E}[V_t] - \mathbb{E}[V_s])\mathbb{E}[M_s - 1] = 0 \text{ (since } \psi' \text{ is increasing).}
\end{aligned}$$

2. We now assume that $t \longmapsto \mathbb{E}[V_t]$ is a decreasing function and that $(C_t := V_t - \mathbb{E}[V_t], t \geq 0)$ is a peacock. From Kellerer's theorem, there exists a martingale $(M_t, t \geq 0)$ such that:

$$V_t - \mathbb{E}[V_t] \overset{(1.d)}{=} M_t \quad \text{or, equivalently,} \quad V_t \overset{(1.d)}{=} M_t + \mathbb{E}[V_t].$$

We note that $\mathbb{E}[M_t] = 0$ for every $t \geq 0$. Let $\psi \in \mathbf{C}_+$ and $0 < s \leq t$:

$$\begin{aligned}
\mathbb{E}[\psi(N_t)] - \mathbb{E}[\psi(N_s)] &= \mathbb{E}\left[\psi\left(\frac{V_t}{\mathbb{E}[V_t]}\right)\right] - \mathbb{E}\left[\psi\left(\frac{V_s}{\mathbb{E}[V_s]}\right)\right] \\
&= \mathbb{E}\left[\psi\left(\frac{M_t}{\mathbb{E}[V_t]} + 1\right)\right] - \mathbb{E}\left[\psi\left(\frac{M_s}{\mathbb{E}[V_s]} + 1\right)\right] \\
&\geq \mathbb{E}\left[\psi'\left(\frac{M_s}{\mathbb{E}[V_s]} + 1\right)\left(\frac{M_t}{\mathbb{E}[V_t]} - \frac{M_s}{\mathbb{E}[V_s]}\right)\right] \quad \text{(by convexity)} \\
&= \mathbb{E}\left[\psi'\left(\frac{M_s}{\mathbb{E}[V_s]} + 1\right) M_s\right] \underbrace{\left(\frac{1}{\mathbb{E}[V_t]} - \frac{1}{\mathbb{E}[V_s]}\right)}_{\geq 0} \\
&\qquad \text{(taking the conditional expectation)} \\
&\geq \psi'(1)\left(\frac{1}{\mathbb{E}[V_t]} - \frac{1}{\mathbb{E}[V_s]}\right)\mathbb{E}[M_s] = 0 \\
&\qquad \text{(since } \psi' \text{ is increasing).}
\end{aligned}$$

□

Second Proof of Theorem 1

For every $t \geq 0$, we set $\alpha(t) = \mathbb{E}[V_t]$. Then, for every convex function $\psi \in \mathbf{C}$, we have:

$$\mathbb{E}[\psi\,(N_t)] = \mathbb{E}\left[\psi\left(\frac{1}{\alpha(t)}(V_t - \alpha(t)) + 1\right)\right] = \mathbb{E}\left[\widetilde{\psi}\left(\frac{C_t}{\alpha(t)}\right)\right], \tag{8}$$

where $\widetilde{\psi}(x) := \psi(x+1)$.

1. To prove the first point, we assume without loss of generality that

$$\widetilde{\psi}(0) = \widetilde{\psi}'(0) = 0 \tag{9}$$

Let us assume that $(N_t, t \geq 0)$ is a peacock. Then, for every $0 < s \leq t$, (8) implies that:

$$\mathbb{E}\left[\widetilde{\psi}\left(\frac{C_t}{\alpha(t)}\right)\right] = \mathbb{E}\left[\psi\left(N_t\right)\right] \geq \mathbb{E}\left[\psi\left(N_s\right)\right] = \mathbb{E}\left[\widetilde{\psi}\left(\frac{C_s}{\alpha(s)}\right)\right] \geq \mathbb{E}\left[\widetilde{\psi}\left(\frac{C_s}{\alpha(t)}\right)\right] \tag{10}$$

since $\dfrac{1}{\alpha(t)} \leq \dfrac{1}{\alpha(s)}$ and, from (9), $\widetilde{\psi}$ increases on $[0, +\infty[$ and decreases on $]-\infty, 0]$. Hence,

$$\mathbb{E}\left[\widetilde{\psi}\left(\frac{C_t}{\alpha(t)}\right)\right] \geq \mathbb{E}\left[\widetilde{\psi}\left(\frac{C_s}{\alpha(t)}\right)\right],$$

where $x \longmapsto \widetilde{\psi}\left(\dfrac{x}{\alpha(t)}\right)$ stands for any convex function.

2. The second point follows from (8). Indeed, if $(C_t, t \geq 0)$ is a peacock, then for every convex function $\widetilde{\psi} \in \mathbf{C}$ such that $\widetilde{\psi}(0) = \widetilde{\psi}'(0) = 0$ and every $0 < s \leq t$, we have:

$$\begin{aligned}
\mathbb{E}\left[\psi\left(N_t\right)\right] &= \mathbb{E}\left[\widetilde{\psi}\left(\frac{C_t}{\alpha(t)}\right)\right] \\
&\geq \mathbb{E}\left[\widetilde{\psi}\left(\frac{C_s}{\alpha(t)}\right)\right] \quad \text{(since } (C_t, t \geq 0) \text{ is a peacock)} \\
&\geq \mathbb{E}\left[\widetilde{\psi}\left(\frac{C_s}{\alpha(s)}\right)\right] \quad \left(\text{since } \frac{1}{\alpha(t)} \geq \frac{1}{\alpha(s)}, C_s\right. \\
&\qquad \left.\text{is centered and } \widetilde{\psi}(0) = \widetilde{\psi}'(0) = 0\right) \\
&= \mathbb{E}\left[\psi\left(N_s\right)\right].
\end{aligned}$$

□

Illustration of Theorem 1

The next example shows that N may be a peacock while C is not.

Example 8. Let X be a real-valued random variable and let μ be the law of X. Suppose that $\mathbb{E}\left[|X|e^{tX}\right] < \infty$ for every $t \geq 0$ and supp $\mu = \mathbb{R}$. Let $\alpha(t) := \mathbb{E}\left[e^{tX}\right]$. Then:

1. $\left(N_t := \dfrac{e^{tX}}{\alpha(t)}, t \geq 0\right)$ is a peacock.
2. $\left(C_t := e^{tX} - \alpha(t), t \geq 0\right)$ is a peacock if and only if α is increasing.

We note that, if $\mathbb{E}[X] > 0$, then, from Lemma 2,

$$\alpha'(t) = \mathbb{E}[Xe^{tX}] \geq 0. \tag{11}$$

Proof (of Example 8).
The first point is a particular case of Example 4. To prove the second point, we note that

$$0 = \frac{\partial}{\partial t}\mathbb{E}[C_t] = \mathbb{E}\left[Xe^{tX} - \alpha'(t)\right], \text{ for every } t \geq 0 \tag{12}$$

and, for every convex function $\psi \in \mathbf{C}_+$:

$$\frac{\partial}{\partial t}\mathbb{E}[\psi(C_t)] = \mathbb{E}[\psi'(e^{tX} - \alpha(t))(Xe^{tX} - \alpha'(t))]. \tag{13}$$

i. Let us suppose, on one hand, that α is increasing. The function $f_t : x \longmapsto xe^{tx} - \alpha'(t)$ has exactly one zero $a \geq 0$ and

$$f_t(x) > 0, \text{ for every } x > a. \text{ see Fig. 1.}$$

Indeed, the derivative function f_t' of f_t is strictly positive on $[0, \infty[$; hence, f_t is a continuous and strictly increasing function, i.e., a bijection map from $[0, \infty[$ to $[-\alpha'(t), \infty[$, $0 \in [-\alpha'(t), \infty[$ since $\alpha'(t) \geq 0$ for every $t \geq 0$ and, $f_t^{-1}(0) = a$; moreover, $f_t(x) < 0$ for every $x < 0$; therefore, distinguishing the cases $X \leq a$ and $X \geq a$, we have:

$$\frac{\partial}{\partial t}\mathbb{E}[\psi(C_t)] \geq \psi'\left(e^{ta} - \alpha(t)\right)\mathbb{E}\left[Xe^{tX} - \alpha'(t)\right] = 0.$$

Then, $\left(C_t := e^{tX} - \alpha(t), t \geq 0\right)$ is a peacock if α increases.
ii. On the other hand, if α is not increasing, then there exists $t_0 > 0$ such that $\alpha'(t_0) < 0$. The function $f_{t_0} : x \longmapsto xe^{t_0 x} - \alpha'(t_0)$ has exactly two zeros $a_1 < a_2 < 0$ and

$$f_{t_0}(x) \text{ is } \begin{cases} \text{strictly positive if } x < a_1, \\ \text{strictly negative if } a_1 < x < a_2, \\ \text{strictly positive if } x > a_2. \end{cases} \quad \text{see Fig. 2}$$

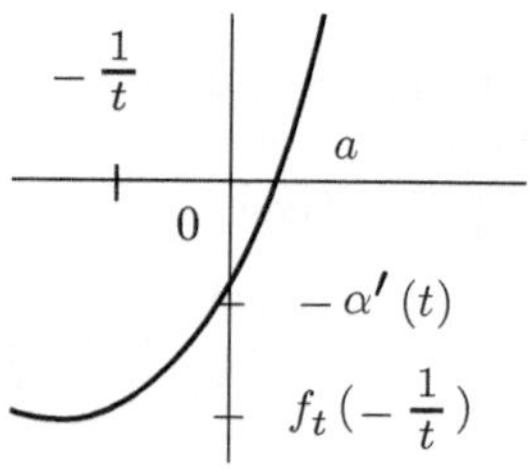

Fig. 1 Graph of f_t when α is strictly increasing and $t > 0$

Fig. 2 Graph of f_{t_0}

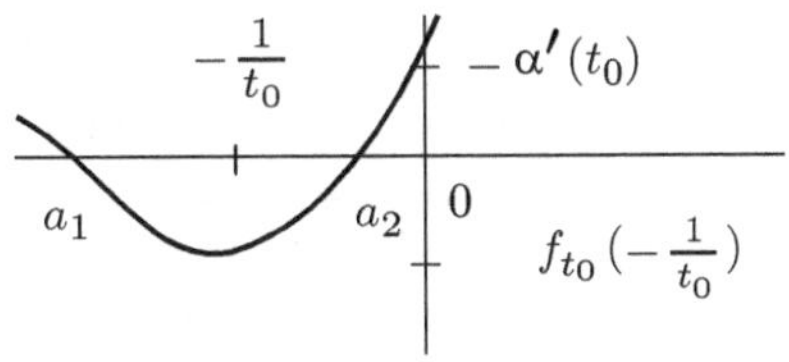

Denoting by μ the law of X, we then observe that:

$$\int_{-\infty}^{a_1} f_{t_0}(x)\mu(dx) > 0 \tag{14}$$

and, (12) implies:

$$\int_{-\infty}^{\infty} f_{t_0}(x)\mu(dx) = 0. \tag{15}$$

From (14) and (15), we deduce that:

$$\mathbb{E}\left[1_{\{X>a_1\}}\left(Xe^{t_0X} - \alpha'(t_0)\right)\right] = \int_{a_1}^{\infty} f_{t_0}(x)\mu(dx) < 0 \quad (\text{since supp } \mu = \mathbb{R}).$$

Then, the result follows by taking $\psi'\left(e^{tx} - \alpha(t)\right) = 1_{[a_1,\infty[}(x)$ in (13). □

Let us note that, if α is increasing and $\left(\dfrac{e^{tX}}{\alpha(t)}, t \geq 0\right)$ is a peacock, then, by Theorem 1, $\left(C_t := e^{tX} - \alpha(t), t \geq 0\right)$ is a peacock. This provides another proof of point (i) of the preceding proof.

3 Peacocks Obtained from Conditionally Monotone Processes

3.1 Definition of Conditionally Monotone Processes and Examples

Let us first introduce the notion of conditional monotonicity, which appears in [17, Chap. 4.B, pp. 114–126] and is studied in [3].

Definition 4 (Conditional monotonicity). A process $(X_\lambda, \lambda \geq 0)$ is said to be conditionally monotone if, for every $n \in \mathbb{N}^*$, every $i \in \{1, \dots, n\}$, every $0 < \lambda_1 < \cdots < \lambda_n$ and every bounded Borel function $\phi : \mathbb{R}^n \longrightarrow \mathbb{R}$ which increases (resp. decreases) with respect to each of its arguments, we have:

$$\mathbb{E}[\phi(X_{\lambda_1}, X_{\lambda_2}, \dots, X_{\lambda_n})|X_{\lambda_i}] = \phi_i(X_{\lambda_i}), \tag{16}$$

where $\phi_i : \mathbb{R} \longrightarrow \mathbb{R}$ is a bounded increasing (resp. decreasing) function.

To prove that a process is conditionally monotone, we may restrict ourselves to bounded Borel functions ϕ which are increasing with respect to each of their arguments. Indeed, replacing ϕ by $-\phi$, the result then also holds for bounded Borel functions which are decreasing with respect to each of their arguments.

Definition 5. We denote by $\mathscr{E}_n$ the set of bounded Borel functions $\phi : \mathbb{R}^n \longrightarrow \mathbb{R}$ which are increasing with respect to each of their arguments.

Remark 1.
1. Note that $(X_\lambda, \lambda \geq 0)$ is conditionally monotone if and only if $(-X_\lambda, \lambda \geq 0)$ is conditionally monotone.
2. Let $\theta : \mathbb{R} \longrightarrow \mathbb{R}$ be a strictly monotone and continuous function. It is not difficult to see that if the process $(X_\lambda, \lambda \geq 0)$ is conditionally monotone, then so is $(\theta(X_\lambda), \lambda \geq 0)$.

In [3], the authors exhibited a number of examples of processes enjoying the conditional monotonicity (16) property. Among them are:

i. The processes with independent and log-concave increments,
ii. The Gamma subordinator,
iii. The well-reversible diffusions at a fixed time, such as, for example:
   - Brownian motion with drift ν,
   - Bessel processes of dimension $\delta \geq 2$,
   - Squared Bessel processes of dimension $\delta > 0$.

We refer the reader to [3] and ([11], Chap. 1, Sect. 4) for more details. The next lemma follows immediately from Definition 4.

Lemma 3. *Let $(X_t, t \geq 0)$ be a real-valued right-continuous process which is conditionally monotone and, let $q : \mathbb{R}_+ \times \mathbb{R} \to \mathbb{R}$ be a continuous function such that, for every $s \geq 0$, $q_s : x \longmapsto q(s,x)$ is increasing. Then, for every positive function $\phi \in \mathscr{E}_1$, every positive Radon measure ν on $\mathbb{R}_+$ and every $t > 0$:*

$$\mathbb{E}\left[\phi\left(\int_0^t q(s, X_s)\,\nu(ds)\right)\bigg| X_t\right] = \phi_t(X_t), \tag{17}$$

where ϕ_t is an increasing function.

3.2 Peacocks Obtained by Centering Under a Conditional Monotonicity Hypothesis

Theorem 2. *Let $(X_t, t \geq 0)$ be a real-valued right-continuous process which is conditionally monotone. Let $q : \mathbb{R}_+ \times \mathbb{R} \to \mathbb{R}_+$ be a positive and continuous function such that, for every $s \geq 0$, $q_s : x \longmapsto q(s,x)$ is increasing and*

$\mathbb{E}[q(s, X_s)] > 0$. *Let* $\theta : \mathbb{R}_+ \to \mathbb{R}_+$ *a positive, increasing and convex* $\mathscr{C}^1$*-function satisfying:*

$$\forall\, t \geq 0, \quad \mathbb{E}\left[\theta\left(\int_0^t q(s, X_s)ds\right)\right] < \infty \tag{18}$$

and

$$\forall\, a > 0, \quad \mathbb{E}\left[\sup_{0<t\leq a} q(t, X_t)\theta'\left(\int_0^t q(s, X_s)ds\right)\right] < \infty. \tag{19}$$

Then:

$$\left(C_t := \theta\left(\int_0^t q(s, X_s)ds\right) - h(t), t \geq 0\right) \text{ is a peacock,}$$

where $h(t) := \mathbb{E}\left[\theta\left(\int_0^t q(s, X_s)ds\right)\right]$.

Proof (of Theorem 2).
For every convex function $\psi \in \mathbf{C}_+$, we have:

$$\begin{aligned}
\frac{d}{dt}\mathbb{E}[\psi(C_t)] &= \mathbb{E}\left[\psi'(C_t)\left(q(t, X_t)\theta'\left(\int_0^t q(s, X_s)ds\right) - h'(t)\right)\right] \\
&= \mathbb{E}\left[\psi'(C_t)q(t, X_t)\left(\theta'\left(\int_0^t q(s, X_s)ds\right) - \frac{h'(t)}{\mathbb{E}[q(t, X_t)]}\right)\right] \\
&\quad + \mathbb{E}\left[\psi'(C_t)h'(t)\left(\frac{q(t, X_t)}{\mathbb{E}[q(t, X_t)]} - 1\right)\right] \\
&:= K_1(t) + K_2(t).
\end{aligned}$$

Let us prove that $K_1(t) \geq 0$.
We note that, for every $t \geq 0$:

$$\mathbb{E}\left[q(t, X_t)\left(\theta'\left(\int_0^t q(s, X_s)ds\right) - \frac{h'(t)}{\mathbb{E}[q(t, X_t)]}\right)\right] = 0 \tag{20}$$

since

$$\mathbb{E}\left[q(t, X_t)\theta'\left(\int_0^t q(s, X_s)ds\right)\right] = h'(t).$$

Then, since θ' is increasing, one has:

$$\begin{aligned}
K_1(t) &= \mathbb{E}\left[\psi'(C_t)q(t, X_t)\left(\theta'\left(\int_0^t q(s, X_s)ds\right) - \frac{h'(t)}{\mathbb{E}[q(t, X_t)]}\right)\right] \\
&\geq \psi'\left(\theta \circ \theta'^{-1}\left(\frac{h'(t)}{\mathbb{E}[q(t, X_t)]}\right) - h(t)\right) \\
&\quad \times \mathbb{E}\left[q(t, X_t)\left(\theta'\left(\int_0^t q(s, X_s)ds\right) - \frac{h'(t)}{\mathbb{E}[q(t, X_t)]}\right)\right] = 0.
\end{aligned}$$

Let us prove that $K_2(t) \geq 0$.
We have:

$$K_2(t) = h'(t)\mathbb{E}\left[\psi'(C_t)\left(\frac{q(t,X_t)}{\mathbb{E}[q(t,X_t)]} - 1\right)\right]$$
$$= h'(t)\mathbb{E}\left[\mathbb{E}[\psi'(C_t)|X_t]\left(\frac{q(t,X_t)}{\mathbb{E}[q(t,X_t)]} - 1\right)\right].$$

But, by Lemma 3,

$$\mathbb{E}[\psi'(C_t)|X_t] = \mathbb{E}\left[\psi'\left(\theta\left(\int_0^t q(s,X_s)ds\right) - h(t)\right)\middle| X_t\right] = \varphi_t(X_t),$$

where φ_t is an increasing function. Hence, from Lemma 2,

$$K_2(t) = h'(t)\mathbb{E}\left[\varphi_t(X_t)\left(\frac{q(t,X_t)}{\mathbb{E}[q(t,X_t)]} - 1\right)\right] \geq 0$$

since $q_t : x \longmapsto q(t,x)$ is increasing for every $t \geq 0$. □

Example 9. Suppose $(X_t, t \geq 0)$ and $q : \mathbb{R}\times\mathbb{R}_+ \to \mathbb{R}_+$ be chosen as in Theorem 2. If we take successively $\theta(x) = x$ and $\theta(x) = e^x$, we obtain:

$$\left(\int_0^t q(s,X_s)ds - \mathbb{E}\left[\int_0^t q(s,X_s)ds\right], t \geq 0\right) \text{ is a peacock} \tag{21}$$

and

$$\left(\exp\left(\int_0^t q(s,X_s)ds\right) - \mathbb{E}\left[\exp\left(\int_0^t q(s,X_s)ds\right)\right], t \geq 0\right) \text{ is a peacock.} \tag{22}$$

3.3 Peacocks Obtained by Normalisation from a Particular Class of Conditionally Monotone Processes

We now consider the class of processes with independent and log-concave increments. (Note that these processes are conditionally monotone (see [3])). Let us recall some definitions and properties, see [1] or [6].

Definition 6 ($\mathbb{R}$-valued log**-concave r.v.'s).**
An $\mathbb{R}$-valued random variable X is said to be log-concave if:

1. X admits a probability density g,
2. The function $\log g$ is concave; i.e., the second derivative of $\log g$ (in the distribution sense) is a negative measure.

Definition 7 ($\mathbb{Z}$-valued log-concave r.v.'s).
A $\mathbb{Z}$-valued random variable X is said to be log-concave if, with $g(n) = \mathbb{P}(X = n)$ ($n \in \mathbb{Z}$), one has: for every $n \in \mathbb{Z}$,

$$g^2(n) \geq g(n-1)g(n+1);$$

in other words, the discrete second derivative of $\log g$ is negative.

Example 10. Many common density functions on $\mathbb{R}$ (or $\mathbb{Z}$) are log-concave. Indeed, the normal density, the uniform density, the exponential density, the Poisson density and the geometric density are log-concave.

The following properties of log-concave random variables are well-known (see [16] or [7]).

Lemma 4. *An $\mathbb{R}$-valued (resp. $\mathbb{Z}$-valued) random variable X is log-concave if and only if its probability density g satisfies:*

1. *The support of g is an (finite or infinite) interval $I \subset \mathbb{R}$ (resp. $I \subset \mathbb{Z}$),*
2. *For every $x_2 \geq x_1$, $y_2 \geq y_1$,*

$$\det \begin{pmatrix} g(x_1 - y_1) & g(x_1 - y_2) \\ g(x_2 - y_1) & g(x_2 - y_2) \end{pmatrix} \geq 0. \tag{23}$$

Lemma 5. *([16])*

i. *Every log-concave density is bounded.*
ii. *If g and h are two log-concave densities, then their convolution $g * h$ given by:*

$$g * h(x) = \int_{-\infty}^{\infty} g(y)h(x-y)dy$$

is also log-concave, i.e: the sum of two independent log-concave random variables is log-concave.

The main result of this section is the following:

Theorem 3. *Let $(X_t, t \geq 0)$ be a right-continuous $\mathbb{R}$-valued process with independent and log-concave increments issued from 0, and $\alpha : \mathbb{R}_+ \to \mathbb{R}_+$ a right-continuous and increasing function satisfying $\alpha(0) = 0$. Let $q : \mathbb{R}_+ \times \mathbb{R} \to \mathbb{R}$ be a continuous function such that, for every $t \geq 0$:*

i. *The variable*

$$\Theta_t := \exp\left(\alpha(t) \sup_{0 \leq s \leq t} q(s, X_s)\right) \text{ is integrable} \tag{24}$$

and,

$$\Delta_t := \mathbb{E}\left[\exp\left(\alpha(t) \inf_{0 \leq s \leq t} q(s, X_s)\right)\right] > 0, \tag{25}$$

ii. The function $x \longmapsto q(t,x)$ *is increasing (resp. decreasing).*

Then,

$$\left(N_t := \frac{\exp\left(\int_0^t q(s,X_s)\,d\alpha(s)\right)}{\mathbb{E}\left[\exp\left(\int_0^t q(s,X_s)\,d\alpha(s)\right)\right]}, t \geq 0 \right) \text{ is a peacock.}$$

In particular, if $(B_t, t \geq 0)$ *denotes a Brownian motion starting from* 0*, then*

$$\left(N_t^{(B)} := \frac{\exp\left(\int_0^t q(s,B_s)\,d\alpha(s)\right)}{\mathbb{E}\left[\exp\left(\int_0^t q(s,B_s)\,d\alpha(s)\right)\right]}, t \geq 0 \right) \text{ is a peacock.}$$

To prove Theorem 3, we need the following lemma.

Lemma 6. *Let* $(X_t, t \geq 0)$ *be a* $\mathbb{R}$*-valued process with independent and log-concave increments and* $(f_k : \mathbb{R} \to \mathbb{R}_+, k \in \mathbb{N}^*)$ *a family of strictly positive Borel functions such that, for every* $p \in \mathbb{N}^*$ *and every* $0 \leq t_1 < \cdots < t_p$*:*

$$\mathbb{E}\left[\prod_{k=1}^{p} f_k(X_{t_k})\right] < \infty.$$

Then, for every $n \geq 2$*, every* $0 \leq \lambda_1 < \lambda_2 < \cdots < \lambda_n$ *and every bounded Borel function* $\phi : \mathbb{R}^{n-1} \to \mathbb{R}_+$ *which increases (resp. decreases) with respect to each of its arguments:*

$$K(n,z) = \frac{\mathbb{E}\left[\phi(X_{\lambda_1},\ldots,X_{\lambda_{n-1}})\prod_{k=1}^{n-1} f_k(X_{\lambda_k})\middle| X_{\lambda_n} = z\right]}{\mathbb{E}\left[\prod_{k=1}^{n-1} f_k(X_{\lambda_k})\middle| X_{\lambda_n} = z\right]}$$

is an increasing (resp. decreasing) function of z.

Proof (of Lemma 6).
i. We give the proof of Lemma 6 only in the continuous case (the proof in the discrete one is similar).
ii. By truncation and regularisation, it suffices to prove Lemma 6 when the functions f_k $(k \in \mathbb{N}^*)$ are bounded and when the support of all increments' densities is $\mathbb{R}$.
iii. In this proof we deal only, without loss of generality, with bounded Borel functions which increase with respect to each of its arguments.
We now prove this lemma by induction on $n \geq 2$.

- **For n = 2:** we denote by g_1 (resp. $\widetilde{g}_2$) the density function of X_{λ_1} (resp. $X_{\lambda_2} - X_{\lambda_1}$). For every bounded and increasing Borel function $\phi : \mathbb{R} \to \mathbb{R}_+$:

$$K(2,z) = \frac{\mathbb{E}\left[\phi(X_{\lambda_1}) f_1(X_{\lambda_1}) | X_{\lambda_2} = z\right]}{\mathbb{E}\left[f_1(X_{\lambda_1}) | X_{\lambda_2} = z\right]}$$

$$= \frac{\displaystyle\int_{-\infty}^{\infty} \phi(u) f_1(u) g_1(u) \widetilde{g}_2(z-u)\, du}{\displaystyle\int_{-\infty}^{\infty} f_1(u) g_1(u) \widetilde{g}_2(z-u)\, du}.$$

Since ϕ is bounded and increasing, then, by approximation, we can restrict ourselves to the case:

$$\phi = \sum_{i=1}^{l} c_i 1_{[x_i, \infty[}$$

where $l \in \mathbb{N}^*$ and, for every $i \in [\![1, l]\!]$, c_i is a positive constant and x_i a real number. Hence, the function $z \longmapsto K(2,z)$ increases if and only if, for every $x \in \mathbb{R}$,

$$z \longmapsto \frac{\displaystyle\int_{x}^{\infty} f_1(u) g_1(u) \widetilde{g}_2(z-u)\, du}{\displaystyle\int_{-\infty}^{\infty} f_1(u) g_1(u) \widetilde{g}_2(z-u)\, du} \quad \text{is increasing.}$$

This is also equivalent to say that: for every $x \in \mathbb{R}$,

$$L(x,z) = \frac{\displaystyle\int_{x}^{\infty} f_1(u) g_1(u) \widetilde{g}_2(z-u)\, du}{\displaystyle\int_{-\infty}^{x} f_1(u) g_1(u) \widetilde{g}_2(z-u)\, du}$$

is an increasing function of z.
Since $\widetilde{g}_2$ is log-concave, then, for every $x \in \mathbb{R}$, $z \in \mathbb{R}$ and $\eta > 0$, we have:

$$\frac{\widetilde{g}_2(z+\eta-u)}{\widetilde{g}_2(z+\eta-x)} \geq \frac{\widetilde{g}_2(z-u)}{\widetilde{g}_2(z-x)}, \quad \text{for every } u \geq x$$

and

$$\frac{\widetilde{g}_2(z+\eta-u)}{\widetilde{g}_2(z+\eta-x)} \leq \frac{\widetilde{g}_2(z-u)}{\widetilde{g}_2(z-x)}, \quad \text{for every } u \leq x.$$

Therefore, for every $x \in \mathbb{R}$, $z \in \mathbb{R}$ and $\eta > 0$, one has:

$$
\begin{aligned}
L(x,z+\eta) &= \frac{\displaystyle\int_x^{\infty} f_1(u)g_1(u)\widetilde{g}_2(z+\eta-u)\,du}{\displaystyle\int_{-\infty}^{x} f_1(u)g_1(u)\widetilde{g}_2(z+\eta-u)\,du} \\
&= \frac{\displaystyle\int_x^{\infty} f_1(u)g_1(u)\frac{\widetilde{g}_2(z+\eta-u)}{\widetilde{g}_2(z+\eta-x)}\,du}{\displaystyle\int_{-\infty}^{x} f_1(u)g_1(u)\frac{\widetilde{g}_2(z+\eta-u)}{\widetilde{g}_2(z+\eta-x)}\,du} \\
&\geq \frac{\displaystyle\int_x^{\infty} f_1(u)g_1(u)\frac{\widetilde{g}_2(z-u)}{\widetilde{g}_2(z-x)}\,du}{\displaystyle\int_{-\infty}^{x} f_1(u)g_1(u)\frac{\widetilde{g}_2(z-u)}{\widetilde{g}_2(z-x)}\,du} \\
&= L(x,z);
\end{aligned}
$$

which means that $L(x,z)$ increases with z. Then, $z \longmapsto K(2,z)$ increases for every bounded and increasing Borel function $\phi : \mathbb{R} \to \mathbb{R}_+$.

- **For n $\geq$ 3:** we assume that, for every bounded Borel function $\varphi : \mathbb{R}^{n-2} \to \mathbb{R}_+$ which increases with respect to each of its arguments, $z \longmapsto K(n-1,z)$ increases and, we denote by g_{n-1} (resp. $\widetilde{g}_n$) the density function of $X_{\lambda_{n-1}}$ (resp. $X_{\lambda_n} - X_{\lambda_{n-1}}$). Since the variables $X_{\lambda_n} - X_{\lambda_{n-1}}$ and $X_{\lambda_{n-1}}$ are independent, then, for every bounded Borel function $\phi : \mathbb{R}^{n-1} \to \mathbb{R}$ which increases with respect to each of its arguments, we have:

$$
K(n,z) = \frac{\mathbb{E}\left[\phi(X_{\lambda_1},\ldots,X_{\lambda_{n-1}})\prod_{k=1}^{n-1} f_k(X_{\lambda_k})\,\middle|\, X_{\lambda_{n-1}} + X_{\lambda_n} - X_{\lambda_{n-1}} = z\right]}{\mathbb{E}\left[\prod_{k=1}^{n-1} f_k(X_{\lambda_k})\,\middle|\, X_{\lambda_{n-1}} + X_{\lambda_n} - X_{\lambda_{n-1}} = z\right]}
$$

$$
= \frac{\displaystyle\int_{-\infty}^{\infty} \mathbb{E}\left[\phi(X_{\lambda_1},\ldots,X_{\lambda_{n-2}},z-y)\prod_{k=1}^{n-2} f_k(X_{\lambda_k})\,\middle|\, X_{\lambda_{n-1}} = z-y\right] f_{n-1}(z-y)g_{n-1}(z-y)\widetilde{g}_n(y)\,dy}{\displaystyle\int_{-\infty}^{\infty} \mathbb{E}\left[\prod_{k=1}^{n-2} f_k(X_{\lambda_k})\,\middle|\, X_{\lambda_{n-1}} = z-y\right] f_{n-1}(z-y)g_{n-1}(z-y)\widetilde{g}_n(y)\,dy}.
$$

After the change of variable: $x = z - y$, we obtain:

$$K(n,z) = \frac{\displaystyle\int_{-\infty}^{\infty} \mathbb{E}\left[\phi(X_{\lambda_1},\dots,X_{\lambda_{n-2}},x)\prod_{k=1}^{n-2} f_k(X_{\lambda_k}) \,\middle|\, X_{\lambda_{n-1}} = x\right] f_{n-1}(x)g_{n-1}(x)\widetilde{g}_n(z-x)\,dx}{\displaystyle\int_{-\infty}^{\infty} \mathbb{E}\left[\prod_{k=1}^{n-2} f_k(X_{\lambda_k}) \,\middle|\, X_{\lambda_{n-1}} = x\right] f_{n-1}(x)g_{n-1}(x)\widetilde{g}_n(z-x)\,dx}.$$

For $x \in \mathbb{R}$, we define:

$$m(x) = \frac{\mathbb{E}\left[\phi(X_{\lambda_1},\dots,X_{\lambda_{n-2}},x)\prod_{k=1}^{n-2} f_k(X_{\lambda_k}) \,\middle|\, X_{\lambda_{n-1}} = x\right]}{\mathbb{E}\left[\prod_{k=1}^{n-2} f_k(X_{\lambda_k}) \,\middle|\, X_{\lambda_{n-1}} = x\right]}$$

and

$$f_*(x) = \mathbb{E}\left[\prod_{k=1}^{n-2} f_k(X_{\lambda_k}) \,\middle|\, X_{\lambda_{n-1}} = x\right].$$

Hence,

(a) Since $\phi : \mathbb{R}^{n-1} \to \mathbb{R}_+$ is bounded and increasing with respect to each of its arguments, the induction hypothesis implies that $m : \mathbb{R} \to \mathbb{R}_+$ is a bounded and increasing Borel function,
(b) We have:

$$K(n,z) = \frac{\displaystyle\int_{-\infty}^{\infty} m(x) f_*(x) f_{n-1}(x)g_{n-1}(x)\widetilde{g}_n(z-x)\,dx}{\displaystyle\int_{-\infty}^{\infty} f_*(x) f_{n-1}(x)g_{n-1}(x)\widetilde{g}_n(z-x)\,dx}.$$

Using the log-concavity of g_n and the case $n = 2$ computed above, we have: for every $y \in \mathbb{R}$,

$$z \longmapsto \frac{\displaystyle\int_{y}^{\infty} f_*(x) f_{n-1}(x)g_{n-1}(x)\widetilde{g}_n(z-x)\,dx}{\displaystyle\int_{-\infty}^{y} f_*(x) f_{n-1}(x)g_{n-1}(x)\widetilde{g}_n(z-x)\,dx}$$

is an increasing function of z. Then, the function $z \longmapsto K(n,z)$ increases for every bounded Borel function $\phi : \mathbb{R}^{n-1} \to \mathbb{R}_+$ which increases with respect to each of its arguments. □

Proof (of Theorem 3).
We prove this theorem only in the case where $x \longmapsto q(\lambda, x)$ is increasing.
Let $T > 0$ be fixed.
1. We first consider the case

$$1_{[0,T]} d\alpha = \sum_{i=1}^{r} a_i \delta_{\lambda_i}$$

where $r \in [\![2, \infty[\![$, $a_1 \geq 0, a_2 \geq 0, \ldots, a_r \geq 0$, $\sum_{i=1}^{r} a_i = \alpha(T)$ and $0 \leq \lambda_1 < \lambda_2 < \cdots < \lambda_r \leq T$.
Let us prove that:

$$\left(N_n := \exp\left(\sum_{i=1}^{n} a_i q(\lambda_i, X_{\lambda_i}) - h(n) \right), n \in [\![1, r]\!] \right) \text{ is a peacock,}$$

where

$$h(n) := \log \mathbb{E}\left[\exp\left(\sum_{i=1}^{n} a_i q(\lambda_i, X_{\lambda_i}) \right) \right], \text{ for every } n \in [\![1, r]\!].$$

We have:

$$\mathbb{E}[N_n - N_{n-1}] = 0, \text{ for every } n \in [\![2, r]\!]$$

with

$$N_n - N_{n-1} = N_{n-1} \left(e^{a_n q(\lambda_n, X_{\lambda_n}) - h(n) + h(n-1)} - 1 \right),$$

and, for every convex function $\psi \in \mathbf{C}$,

$$\begin{aligned}
&\mathbb{E}[\psi(N_n)] - \mathbb{E}[\psi(N_{n-1})] \\
&\geq \mathbb{E}\left[\psi'(N_{n-1}) N_{n-1} \left(e^{a_n q(\lambda_n, X_{\lambda_n}) - h(n) + h(n-1)} - 1 \right) \right] \\
&= \mathbb{E}\left[\mathbb{E}[N_{n-1} | X_{\lambda_n}] K(n, X_{\lambda_n}) \left(e^{a_n q(\lambda_n, X_{\lambda_n}) - h(n) + h(n-1)} - 1 \right) \right],
\end{aligned}$$

where

$$K(n, z) = \frac{\mathbb{E}[\psi'(N_{n-1}) N_{n-1} | X_{\lambda_n} = z]}{\mathbb{E}[N_{n-1} | X_{\lambda_n} = z]}.$$

The positive and bounded $\mathscr{C}^0$-function $\phi : \mathbb{R}^{n-1} \to \mathbb{R}_+$ given by:

$$\phi(x_1, \ldots, x_{n-1}) = \psi'\left[\exp\left(\sum_{i=1}^{n-1} a_i q(\lambda_i, x_i) - h(n-1) \right) \right]$$

increases with respect to each of its arguments. For $i \in \mathbb{N}^*$, let us define:

$$f_i(x) = e^{a_i q(\lambda_i, x)}, \text{ for every } x \in \mathbb{R};$$

then, for every $n \in [\![2, r]\!]$, we have:

$$N_{n-1} = e^{-h(n-1)} \prod_{k=1}^{n-1} f_k(X_{\lambda_k}).$$

Hence,

$$K(n,z) = \frac{\mathbb{E}\left[\phi(X_{\lambda_1}, \dots, X_{\lambda_{n-1}}) \prod_{k=1}^{n-1} f_k(X_{\lambda_k}) \middle| X_{\lambda_n} = z \right]}{\mathbb{E}\left[\prod_{k=1}^{n-1} f_k(X_{\lambda_k}) \middle| X_{\lambda_n} = z \right]}.$$

Moreover, for every $n \in [\![1, r]\!]$,

$$\begin{aligned}
\mathbb{E}\left[\prod_{k=1}^{n} f_k(X_{\lambda_k})\right] &= \mathbb{E}\left[\exp\left(\sum_{k=1}^{n} a_i q(\lambda_i, X_{\lambda_i})\right)\right] \\
&\leq \mathbb{E}\left[\exp\left(\sup_{0 \leq \lambda \leq T} q(\lambda, X_\lambda) \sum_{k=1}^{n} a_i\right)\right] \\
&\leq \mathbb{E}\left[\exp\left(\sup_{0 \leq \lambda \leq T} q(\lambda, X_\lambda) \sum_{k=1}^{r} a_i\right) \vee 1\right] \\
&= \mathbb{E}\left[\exp\left(\alpha(T) \sup_{0 \leq \lambda \leq T} q(\lambda, X_\lambda)\right) \vee 1\right] \\
&= \mathbb{E}[\Theta_T \vee 1] < \infty.
\end{aligned}$$

Therefore, thanks to Lemma 6, $K(n,z)$ is an increasing function of z. For $\lambda \geq 0$, we denote by q_λ^{-1}, the right-continuous inverse of $x \longmapsto q(\lambda, x)$. Let us also consider the variable:

$$V(n, X_{\lambda_n}) := K(n, X_{\lambda_n}) \mathbb{E}[N_{n-1} | X_{\lambda_n}] \left(e^{a_n q(\lambda_n, X_{\lambda_n}) - h(n) + h(n-1)} - 1\right).$$

Then:

i. If

$$X_{\lambda_n} \leq q_{\lambda_n}^{-1}\left(\frac{h(n) - h(n-1)}{a_n}\right),$$

then

$$e^{a_n q(\lambda_n, X_{\lambda_n}) - h(n) + h(n-1)} - 1 \leq 0$$

and

$$V(n, X_{\lambda_n}) \geq K\left(n, q_{\lambda_n}^{-1}\left(\frac{h(n) - h(n-1)}{a_n}\right)\right)$$
$$\mathbb{E}[N_{n-1}|X_{\lambda_n}]\left(e^{a_n q(\lambda_n, X_{\lambda_n}) - h(n) + h(n-1)} - 1\right),$$

ii. If

$$X_{\lambda_n} \geq q_{\lambda_n}^{-1}\left(\frac{h(n) - h(n-1)}{a_n}\right),$$

then

$$e^{a_n q(\lambda_n, X_{\lambda_n}) - h(n) + h(n-1)} - 1 \geq 0$$

and

$$V(n, X_{\lambda_n}) \geq K\left(n, q_{\lambda_n}^{-1}\left(\frac{h(n) - h(n-1)}{a_n}\right)\right)$$
$$\mathbb{E}[N_{n-1}|X_{\lambda_n}]\left(e^{a_n q(\lambda_n, X_{\lambda_n}) - h(n) + h(n-1)} - 1\right).$$

Thus,

$$\mathbb{E}[\psi(N_n)] - \mathbb{E}[\psi(N_{n-1})] \geq \mathbb{E}V(n, X_{\lambda_n})$$
$$\geq K\left(n, q_{\lambda_n}^{-1}\left(\frac{h(n) - h(n-1)}{a_n}\right)\right)\mathbb{E}\left[\mathbb{E}[N_{n-1}|X_{\lambda_n}]\left(e^{a_n q(\lambda_n, X_{\lambda_n}) - h(n) + h(n-1)} - 1\right)\right]$$
$$= K\left(n, q_{\lambda_n}^{-1}\left(\frac{h(n) - h(n-1)}{a_n}\right)\right)\mathbb{E}\left[N_{n-1}\left(e^{a_n q(\lambda_n, X_{\lambda_n}) - h(n) + h(n-1)} - 1\right)\right]$$
$$= K\left(n, q_{\lambda_n}^{-1}\left(\frac{h(n) - h(n-1)}{a_n}\right)\right)\mathbb{E}\left[N_n - N_{n-1}\right] = 0.$$

Hence, for every $r \in [\![2, \infty[\![$:

$$\left(N_n := \exp\left(\sum_{i=1}^{n} a_i q(\lambda_i, X_{\lambda_i}) - h(n)\right), n \in [\![1, r]\!]\right) \text{ is a peacock.}$$

2. We now set $\mu = 1_{[0,T]} d\alpha$ and, for every $0 \leq t \leq T$,

$$N_t^{(\mu)} = \frac{\exp\left(\int_0^t q(u, X_u)\mu(du)\right)}{\mathbb{E}\left[\exp\left(\int_0^t q(u, X_u)\mu(du)\right)\right]}.$$

Since the function $\lambda \in [0,T] \longmapsto q(\lambda, X_\lambda)$ is right-continuous and bounded from above by $\sup_{0\le\lambda\le T} |q(\lambda, X_\lambda)|$ which is finite a.s., there exists a sequence $(\mu_n, n \geq 0)$ of measures of type considered in (1), with $\operatorname{supp}\mu_n \subset [0,T]$, $\int \mu_n(du) = \int \mu(du)$ and, for every $0 \leq t \leq T$,

$$\lim_{n\to\infty} \exp\left(\int_0^t q(u, X_u)\mu_n(du)\right) = \exp\left(\int_0^t q(u, X_u)\mu(du)\right) \text{ a.s.} \tag{26}$$

Moreover, for every $0 \le t \le T$ and every $n \geq 0$,

$$\begin{aligned}
&\sup_{n\ge 0} \exp\left(\int_0^t q(u, X_u)\mu_n(du)\right) \\
&\le \exp\left(\sup_{0\le\lambda\le T} q(\lambda, X_\lambda) \int_0^t \mu_n(du)\right) \\
&= \exp\left(\sup_{0\le\lambda\le T} q(\lambda, X_\lambda) \int_0^T \mu_n(du)\right) \vee 1 \\
&= \exp\left(\sup_{0\le\lambda\le T} q(\lambda, X_\lambda) \int_0^T \mu(du)\right) \vee 1 = \Theta_T \vee 1
\end{aligned}$$

which is integrable from (24). Thus, the dominated convergence theorem yields

$$\lim_{n\to\infty} \mathbb{E}\left[\exp\left(\int_0^t q(u, X_u)\mu_n(du)\right)\right] = \mathbb{E}\left[\exp\left(\int_0^t q(u, X_u)\mu(du)\right)\right]. \tag{27}$$

Using (26) and (27), we obtain:

$$\lim_{n\to\infty} N_t^{(\mu_n)} = N_t^{(\mu)} \text{ a.s., for every } 0 \le t \le T. \tag{28}$$

Now, from (1),

$$\left(N_t^{(\mu_n)}, 0 \le t \le T\right) \text{ is a peacock for every } n \ge 0, \tag{29}$$

i.e., for every $0 \le s < t \le T$ and for every $\psi \in \mathbf{C}$:

$$\mathbb{E}\left[\psi(N_s^{(\mu_n)})\right] \le \mathbb{E}\left[\psi(N_t^{(\mu_n)})\right]. \tag{30}$$

Then, since

$$\sup_{0\le t\le T} \sup_{n\ge 0} \left|N_t^{(\mu_n)}\right| \le \frac{\Theta_T \vee 1}{\Delta_T \wedge 1}, \tag{31}$$

which is integrable from (24) and (25), it remains to apply the dominated convergence theorem in (30) to obtain that $(N_t^{(\mu)}, 0 \leq t \leq T)$ is a peacock for every $T > 0$. □

4 Peacocks Obtained from a Diffusion by Centering and Normalisation

Let us consider two Borel functions $\sigma : \mathbb{R}_+ \times \mathbb{R} \to \mathbb{R}$ and $b : \mathbb{R}_+ \times \mathbb{R} \to \mathbb{R}$ such that $\sigma_s(x) := \sigma(s, x)$ and $b_s(x) := b(s, x)$ are Lipschitz continuous with respect to x, locally uniformly with respect to s and $(X_t, t \geq 0)$ a process with values in an interval $I \subset \mathbb{R}$ and which solves the SDE:

$$Y_t = x_0 + \int_0^t \sigma(s, Y_s)dB_s + \int_0^t b(s, Y_s)ds \tag{32}$$

where $x_0 \in I$ and $(B_s, s \geq 0)$ denotes a standard Brownian motion started at 0. For $s \geq 0$, let $\mathscr{L}_s$ denotes the second-order differential operator:

$$\mathscr{L}_s := \frac{1}{2}\sigma^2(s, x)\frac{\partial^2}{\partial x^2} + b(s, x)\frac{\partial}{\partial x}. \tag{33}$$

The following results concern peacocks of C and N types.

4.1 Peacocks Obtained by Normalisation

Theorem 4. *Let $(X_t, t \geq 0)$ be a solution of (32) taking values in I. Let $\theta : I \to \mathbb{R}_+^*$ be an increasing $\mathscr{C}^2$-function such that:*

1. *For every $s \geq 0$:*

$$v_s : x \in I \longmapsto \frac{\mathscr{L}_s\theta(x)}{\theta(x)} \text{ is an increasing function} \tag{34}$$

2. *The process*

$$\left(M_t := \theta(X_t) - \theta(x_0) - \int_0^t \mathscr{L}_s\theta(X_s)ds,\ t \geq 0\right)$$

is a martingale.

Then:

$$\left(N_t := \frac{\theta(X_t)}{\mathbb{E}[\theta(X_t)]}, t \geq 0\right) \text{ is a peacock.} \tag{35}$$

Proof (of Theorem 4).
For every $t \geq 0$, let $h(t) = \mathbb{E}[\theta(X_t)]$. We note that h is strictly positive. Let $\psi \in \mathbf{C}$ and $0 \leq s < t$. By Itô's formula

$$\begin{aligned}\psi\left(\frac{\theta(X_t)}{h(t)}\right) = \psi\left(\frac{\theta(X_s)}{h(s)}\right) &+ \int_s^t \psi'\left(\frac{\theta(X_u)}{h(u)}\right)\left[\frac{dM_u}{h(u)} + \frac{\mathscr{L}_u\theta(X_u)du}{h(u)}\right] \\ &- \int_s^t \psi'\left(\frac{\theta(X_u)}{h(u)}\right)\frac{h'(u)\theta(X_u)}{h^2(u)}du \\ &+ \frac{1}{2}\int_s^t \psi''\left(\frac{\theta(X_u)}{h(u)}\right)\frac{1}{h^2(u)}d\langle M\rangle_u.\end{aligned}$$

Hence, it suffices to see that, for every $s \leq u < t$:

$$K(u) := \mathbb{E}\left[\psi'\left(\frac{\theta(X_u)}{h(u)}\right)\left[\frac{\mathscr{L}_u\theta(X_u)}{h(u)} - \frac{h'(u)\theta(X_u)}{h^2(u)}\right]\right] \geq 0. \tag{36}$$

We note that:

$$\mathbb{E}\left[\frac{\mathscr{L}_u\theta(X_u)}{h(u)} - \frac{h'(u)\theta(X_u)}{h^2(u)}\right] = 0 \tag{37}$$

since $u \longmapsto \frac{1}{h(u)}\mathbb{E}\left[\theta(X_u)\right]$ is constant and

$$\frac{d}{du}\mathbb{E}\left[\theta(X_u)\right] = \mathbb{E}[\mathscr{L}_u\theta(X_u)]. \tag{38}$$

Hence, for every $s \leq u < t$, since, by hypothesis (34), $x \longmapsto v_u(x)$ is increasing, we have:

$$\begin{aligned}K(u) &= \mathbb{E}\left[\psi'\left(\frac{\theta(X_u)}{h(u)}\right)\frac{\theta(X_u)}{h(u)}\left(v_u(X_u) - \frac{h'(u)}{h(u)}\right)\right] \\ &\geq \psi'\left(\frac{\theta\left(v_u^{-1}\left(\frac{h'(u)}{h(u)}\right)\right)}{h(u)}\right)\mathbb{E}\left[\frac{\theta(X_u)}{h(u)}\left(v_u(X_u) - \frac{h'(u)}{h(u)}\right)\right] = 0.\end{aligned}$$

□

4.2 Peacocks Obtained by Centering

Theorem 5. *Let $(X_t, t \geq 0)$ be a solution of (32) taking values in I. Let $\theta : I \to \mathbb{R}_+$ be an increasing $\mathscr{C}^2$-function such that:*

1. *For every $s \geq 0$, $x \longmapsto \mathscr{L}_s\theta(x)$ is increasing.*
2. *The process*

$$\left(M_t := \theta(X_t) - \theta(x_0) - \int_0^t \mathscr{L}_s\theta(X_s)ds,\ t \geq 0\right)$$

is a martingale.

Then:

$$(C_t := \theta(X_t) - \mathbb{E}[\theta(X_t)], t \geq 0) \ \textit{is a peacock.}$$

Proof (of Theorem 5).
Let $\psi \in \mathbf{C}$, $h(t) = \mathbb{E}[\theta(X_t)]$ and $0 \leq s < t$. From Itô's formula, we have:

$$\begin{aligned}
&\psi(\theta(X_t) - h(t)) - \psi(\theta(X_s) - h(s)) \\
&= \int_s^t \psi'(\theta(X_u) - h(u))[dM_u + \mathscr{L}_u\theta(X_u)du - h'(u)du] \\
&+ \frac{1}{2}\int_s^t \psi''(\theta(X_u) - h(u))d\langle M\rangle_u.
\end{aligned}$$

Hence, it is sufficient to show that, for every $s \leq u < t$:

$$\mathbb{E}\left[\psi'(\theta(X_u) - h(u))(\mathscr{L}_u\theta(X_u) - h'(u))\right] \geq 0. \tag{39}$$

But, (39) follows from:

$$\mathbb{E}[\mathscr{L}_u\theta(X_u) - h'(u)] = 0, \ \text{for every } s \leq u < t \tag{40}$$

and

$$\begin{aligned}
&\mathbb{E}\left[\psi'(\theta(X_u) - h(u))(\mathscr{L}_u\theta(X_u) - h'(u))\right] \\
&\geq \psi'\left(\theta\left[(\mathscr{L}_u\theta)^{-1}(h'(u))\right] - h(u)\right)\mathbb{E}\left[\mathscr{L}_u\theta(X_u) - h'(u)\right] = 0.
\end{aligned}$$

□

Remark 2. In Theorem 4, if we suppose furthermore that $\mathscr{L}_s\theta(x) \geq 0$ for every $s \geq 0$ and $x \in I$, then

$$(C_t := \theta(X_t) - \mathbb{E}[\theta(X_t)], t \geq 0) \ \text{is a peacock.} \tag{41}$$

Indeed, for every $t \geq 0$:

$$\theta(X_t) = \theta(x_0) + M_t + \int_0^t \mathscr{L}_u \theta(X_u) du.$$

Thus, $h : t \longmapsto \mathbb{E}[\theta(X_t)]$ is increasing and the result follows from both Theorems 1 and 4.

4.3 Peacocks Obtained from an Additive Functional by Normalisation

Let $\mathscr{A}_s$ be the space-time differential operator given by:

$$\mathscr{A}_s := \frac{\partial}{\partial s} + \frac{1}{2}\sigma^2(s,x)\frac{\partial^2}{\partial x^2} + b(s,x)\frac{\partial}{\partial x}. \tag{42}$$

We shall prove the following result:

Theorem 6. *Let $(X_t, t \geq 0)$ be a conditionally monotone process with values in I and which solves (32) and, let $q : \mathbb{R}_+ \times I \to \mathbb{R}_+$ be a strictly positive $\mathscr{C}^2$-function such that,*

1. *For every $s \geq 0$, $\mathbb{E}[q(s, X_s)] > 0$, $q_s : x \in I \longmapsto q(s,x)$ is increasing and*

$$f_s : x \in I \longmapsto \frac{\mathscr{A}_s q(s,x)}{q(s,x)} \text{ is an increasing function} \tag{43}$$

2. *The process*

$$\left(Z_t := q(t, X_t) - q(0, x_0) - \int_0^t \mathscr{A}_s q(s, X_s) ds,\ t \geq 0 \right)$$

is a martingale.

Then, for every positive Radon measure ν on $\mathbb{R}_+$:

$$\left(N_t := \frac{\displaystyle\int_0^t q(s, X_s)\nu(ds)}{\mathbb{E}\left[\displaystyle\int_0^t q(s, X_s)\nu(ds)\right]}, t \geq 0 \right) \text{ is a peacock.}$$

One may find in [11], Chap. 1, many examples of SDEs solutions which are conditionally monotone. This fact is related to the "well-reversible" property of these diffusions.

Proof (of Theorem 6).
We set:

$$\Gamma_u := \frac{1}{\mathbb{E}[q(u, X_u)]}, \text{ for every } u \geq 0. \tag{44}$$

For every $u \geq 0$, Itô's formula yields:

$$\begin{aligned}\Gamma_u q(u, X_u) &= 1 + \int_0^u \Gamma_v dZ_v + \int_0^u \left(\Gamma_v' q(v, X_v) + \Gamma_v \mathscr{A}_v q(v, X_v)\right) dv \\ &:= M_u + H_u\end{aligned}$$

where

$$\left(M_u := 1 + \int_0^u \Gamma_v dZ_v, u \geq 0\right) \text{ is a martingale} \tag{45}$$

and

$$\left(H_u := \int_0^u \left(\Gamma_v' q(v, X_v) + \Gamma_v \mathscr{A}_v q(v, X_v)\right) dv, u \geq 0\right) \text{ is a centered process} \tag{46}$$

since $\mathbb{E}[\Gamma_u q(u, X_u)] = 1$ for every $u \geq 0$ and

$$\frac{d}{du}\mathbb{E}[q(u, X_u)] = \mathbb{E}[\mathscr{A}_u q(u, X_u)]. \tag{47}$$

Hence, by setting

$$h(t) := \int_0^t \frac{1}{\Gamma_u} \nu(du) = \mathbb{E}\left[\int_0^t q(u, X_u)\nu(du)\right], \tag{48}$$

one has:

$$\begin{aligned}N_t &= \frac{1}{h(t)} \int_0^t q(u, X_u)\nu(du) = \frac{1}{h(t)} \int_0^t \Gamma_u q(u, X_u) \frac{1}{\Gamma_u} \nu(du) \\ &= \frac{1}{h(t)} \int_0^t (M_u + H_u) dh(u).\end{aligned}$$

Thus, integrating by parts, we obtain:

$$dN_t = \frac{dh(t)}{h^2(t)} \left(M_t^{(h)} + H_t^{(h)}\right) \tag{49}$$

with

$$M_t^{(h)} = \int_0^t h(u) dM_u \text{ and } H_t^{(h)} = \int_0^t h(s) dH_s.$$

Then, for every $\psi \in \mathbf{C}_+$ and every $0 \le s < t$, we have:

$$\begin{aligned}
\mathbb{E}[\psi(N_t)] - \mathbb{E}[\psi(N_s)] &= \mathbb{E}\left[\int_s^t \psi'(N_u)dN_u\right] \\
&= \mathbb{E}\left[\int_s^t \psi'(N_u)\left(M_u^{(h)} + H_u^{(h)}\right)\frac{dh(u)}{h^2(u)}\right] \\
&= \int_s^t \frac{dh(u)}{h^2(u)}\mathbb{E}\left[\int_0^u \psi''(N_v)\left(M_v^{(h)} + H_v^{(h)}\right)^2\frac{dh(v)}{h^2(v)}\right] \\
&\quad + \int_s^t \frac{dh(u)}{h^2(u)}\mathbb{E}\left[\int_0^u \psi'(N_v)\left(dM_v^{(h)} + dH_v^{(h)}\right)\right].
\end{aligned}$$

Hence, it remains to see that, for every $u > 0$:

$$\begin{aligned}
&\mathbb{E}\left[\int_0^u \psi'(N_v)dH_v^{(h)}\right] \\
&= \mathbb{E}\left[\int_0^u \psi'(N_v)h(v)\left(\Gamma_v\mathscr{A}_v q(v,X_v) + \Gamma_v' q(v,X_v)\right)dv\right] \ge 0 \qquad (50)
\end{aligned}$$

But, for every $0 \le v \le u$,

$$\begin{aligned}
K(v) &:= \mathbb{E}\left[\psi'(N_v)h(v)\left(\Gamma_v\mathscr{A}_v q(v,X_v) + \Gamma_v' q(v,X_v)\right)\right] \\
&= h(v)\Gamma_v\,\mathbb{E}\left[\psi'(N_v)q(v,X_v)\left(f_v(X_v) + \frac{\Gamma_v'}{\Gamma_v}\right)\right] \quad \text{(with } f_v \text{ defined by (43))} \\
&= h(v)\Gamma_v\,\mathbb{E}\left[\mathbb{E}[\psi'(N_v)|X_v]q(v,X_v)\left(f_v(X_v) + \frac{\Gamma_v'}{\Gamma_v}\right)\right]
\end{aligned}$$

But, by Lemma 3,

$$\mathbb{E}[\psi'(N_v)|X_v] = \mathbb{E}\left[\psi'\left(\frac{1}{h(v)}\int_0^v q(u,X_u)\nu(du)\right)\bigg| X_v\right] = \varphi_v(X_v), \qquad (51)$$

where φ_v is an increasing function and,

$$\mathbb{E}\left[q(v,X_v)\left(f_v(X_v) + \frac{\Gamma_v'}{\Gamma_v}\right)\right] = 0. \qquad (52)$$

Then,

$$K(v) \ge h(v)\Gamma_v\varphi_v\left(f_v^{-1}\left[-\frac{\Gamma_v'}{\Gamma_v}\right]\right)\mathbb{E}\left[q(v,X_v)\left(f_v(X_v) + \frac{\Gamma_v'}{\Gamma_v}\right)\right] = 0. \qquad \square$$

4.4 Peacocks Obtained by Normalisation from Markov Processes with Independent and Log-Concave Increments

Let us give two extensions of Theorem 3. The first one concern random walks and the second result deals with continuous Markov processes.

Theorem 7. *Let $(X_n, n \in \mathbb{N})$ be a random walk on $\mathbb{R}$ with independent and log-concave increments, issued from* 0 *and let P be its transition kernel. Let $\theta : \mathbb{R} \to \mathbb{R}_+^*$ be an increasing (resp. decreasing) Borel function such that:*

$$\left(M_n := \theta(X_n) \prod_{k=0}^{n-1} \frac{\theta(X_k)}{P\theta(X_k)}, n \in \mathbb{N}\right) \quad \textit{is a martingale.}$$

Let $q : \mathbb{R} \to \mathbb{R}$ be an increasing (resp. decreasing) Borel function such that

$$x \mapsto q(x) + \log \frac{P\theta(x)}{\theta(x)} \quad \textit{is increasing (resp. decreasing),} \tag{53}$$

and for every $n \geq 1$

$$\mathbb{E}\left[\left(\theta(X_n) + P\theta(X_n)e^{q(X_n)}\right) \exp\left(\sum_{k=0}^{n-1} q(X_k)\right)\right] < \infty. \tag{54}$$

Then,

$$\left(N_n := \theta(X_n) \exp\left(\sum_{k=0}^{n-1} q(X_k) - h(n)\right), n \in \mathbb{N}\right) \quad \textit{is a peacock,}$$

where $h(n) := \log \mathbb{E}\left[\theta(X_n) \exp\left(\sum_{k=0}^{n-1} q(X_k)\right)\right]$.

Proof (of Theorem 7).
1. We only consider the case where the functions θ, q and $x \mapsto q(x) + \log \dfrac{P\theta(x)}{\theta(x)}$ are increasing. We set $A_0 = 0$,

$$A_n := \exp\left(\sum_{k=0}^{n-1} q(X_k) - h(n)\right) \prod_{k=0}^{n-1} \frac{P\theta(X_k)}{\theta(X_k)}, \quad \text{for every } n \geq 1,$$

and

$$\mathscr{F}_n := \sigma(X_p, p \leq n), \quad \text{for every } n \geq 0.$$

Then, for every $n \geq 1$, A_n is $\mathscr{F}_{n-1}$-measurable and $N_n = M_n A_n$.
It follows from (54) that, for every $n \geq 1$, the variables $N_n := M_n A_n$ and $M_{n-1} A_n$ are integrable. Moreover, since $(N_n, n \geq 0)$ has a constant mean and $(M_n, n \geq 0)$ is a $(\mathscr{F}_n)$-martingale, we obtain, for $n \geq 1$:

$$\begin{aligned} 0 = \mathbb{E}[N_n - N_{n-1}] &= \mathbb{E}[M_n A_n - M_{n-1} A_{n-1}] \\ &= \mathbb{E}[A_n(M_n - M_{n-1})] + \mathbb{E}[M_{n-1}(A_n - A_{n-1})] \\ &= \mathbb{E}[M_{n-1}(A_n - A_{n-1})]. \end{aligned}$$

But,

$$\begin{aligned} M_{n-1}(A_n - A_{n-1}) &= N_{n-1}\left(\frac{A_n}{A_{n-1}} - 1\right) \\ &= N_{n-1}\left(\frac{P\theta(X_{n-1})}{\theta(X_{n-1})} e^{q(X_{n-1}) - h(n) + h(n-1)} - 1\right) \\ &= N_{n-1}\left(e^{\widetilde{q}(X_{n-1})} - 1\right), \end{aligned}$$

where, from (53), the function $x \mapsto \widetilde{q}(x) := q(x) + \log \dfrac{P\theta(x)}{\theta(x)} - h(n) + h(n-1)$ is increasing. Hence:

$$\mathbb{E}\left[N_{n-1}\left(e^{\widetilde{q}(X_{n-1})} - 1\right)\right] = 0 \tag{55}$$

2. For every $\psi \in \mathbf{C}$, one has:

$$\begin{aligned} &\mathbb{E}[\psi(N_n)] - \mathbb{E}[\psi(N_{n-1})] \geq \mathbb{E}\left[\psi'(N_{n-1})(N_n - N_{n-1})\right] \text{ (by convexity)} \\ &= \mathbb{E}\left[\psi'(N_{n-1})A_n(M_n - M_{n-1})\right] + \mathbb{E}\left[\psi'(N_{n-1})M_{n-1}(A_n - A_{n-1})\right] \\ &\text{(since } A_n \text{ is } \mathscr{F}_{n-1}\text{-integrable and } (M_n, n \geq 0) \text{ is a martingale)} \\ &= \mathbb{E}\left[\psi'(N_{n-1})M_{n-1}(A_n - A_{n-1})\right] \\ &= \mathbb{E}\left[\psi'(N_{n-1})N_{n-1}\left(e^{\widetilde{q}(X_{n-1})} - 1\right)\right] \\ &= \mathbb{E}\left[\frac{\mathbb{E}[\psi'(N_{n-1})N_{n-1}|X_{n-1}]}{\mathbb{E}[N_{n-1}|X_{n-1}]}\mathbb{E}[N_{n-1}|X_{n-1}]\left(e^{\widetilde{q}(X_{n-1})} - 1\right)\right]. \end{aligned}$$

Now, for every $n \geq 1$ and $x \in \mathbb{R}$, we define:

$$K(n, x) := \frac{\mathbb{E}\left[\psi'(N_{n-1})N_{n-1}|X_{n-1} = x\right]}{\mathbb{E}[N_{n-1}|X_{n-1} = x]}$$

and

$$V(n,x) := \mathbb{E}[N_{n-1}|X_{n-1}=x]\left(e^{\widetilde{q}(x)}-1\right).$$

We note that:

$$\forall\, n \geq 1,\ \mathbb{E}[V(n,X_{n-1})] = 0, \quad \text{(from (55))}$$

and from Lemma 6, the function $x \mapsto K(n,x)$ is increasing.
Let $(\widetilde{q})^{-1}$ denotes the right-continuous inverse of $\widetilde{q}$. Then, distinguishing the cases $X_{n-1} \geq (\widetilde{q})^{-1}(0)$ and $X_{n-1} \leq (\widetilde{q})^{-1}(0)$, we obtain

$$\forall\, n \geq 1,\ K(n,X_{n-1})V(n,X_{n-1}) \geq K\left(n,(\widetilde{q})^{-1}(0)\right)V(n,X_{n-1}),$$

and finally,

$$\begin{aligned}\mathbb{E}[\psi(N_n)] - \mathbb{E}[\psi(N_{n-1})] &\geq \mathbb{E}[K(n,X_{n-1})V(n,X_{n-1})]\\ &\geq K\left(n,(\widetilde{q})^{-1}(0)\right)\mathbb{E}[V(n,X_{n-1})] = 0.\end{aligned}$$ □

We now deal with an extension of Theorem 3 for continuous Markov processes.

Theorem 8. *Let $(X_t, t \geq 0)$ be a continuous (non necessary homogeneous) Markov process with independent and log-concave increments, issued from 0 and let $\mathscr{A}$ be the infinitesimal generator of the space-time process associated with X. Let $\theta : \mathbb{R}_+ \times \mathbb{R} \to \mathbb{R}_+^*$ be a continuous function such that $x \mapsto \theta(t,x)$ is increasing (resp. decreasing) for every $t \geq 0$ and*

$$\left(M_t := \theta(t,X_t)\exp\left(-\int_0^t \frac{\mathscr{A}\theta}{\theta}(u,X_u)\,du\right), t \geq 0\right)$$

is a continuous local martingale.

Let $q : \mathbb{R}_+ \times \mathbb{R} \to \mathbb{R}$ be a continuous function such that $x \mapsto q(t,x)$ and $x \mapsto \left(q + \frac{\mathscr{A}\theta}{\theta}\right)(t,x)$ are increasing (resp. decreasing) for every $t \geq 0$. We assume furthermore that, for every $t \geq 0$:

$$\mathbb{E}\left[\left(\sup_{0\leq s\leq t}(\theta + |q|\theta + |\mathscr{A}\theta|)(s,X_s)\right)\exp\left(\int_0^t |q|(s,X_s)\,ds\right)\right] < \infty \tag{56}$$

and

$$\mathbb{E}\left[\theta(t,X_t)\exp\left(t\sup_{0\leq s\leq t} q(s,X_s)\right)\right] < \infty. \tag{57}$$

Then,

$$\left(N_t := \theta(t,X_t)\exp\left(\int_0^t q(s,X_s)\,ds - h(t)\right), t \geq 0\right)$$

is a peacock, where $h(t) := \log \mathbb{E}\left[\theta(t,X_t)\exp\left(\int_0^t q(s,X_s)\,ds\right)\right]$.

Proof (of Theorem 8).
1. We start with

Lemma 7. *Let $(X_t, t \geq 0)$ be a right-continuous process with independent and log-concave increments issued from 0. Let $q : \mathbb{R}_+ \times \mathbb{R} \to \mathbb{R}$ and $\theta : \mathbb{R}_+ \times \mathbb{R} \to \mathbb{R}_+^*$ be two continuous functions such that, for every $t \geq 0$, $x \mapsto q(t,x)$ and $x \mapsto \theta(t,x)$ are increasing, and*

$$\mathbb{E}\left[\theta(t, X_t) \exp\left(t \sup_{0 \leq s \leq t} q(s, X_s)\right)\right] < \infty. \tag{58}$$

If we set

$$\forall\, t \geq 0,\; N_t := \theta(t, X_t) \exp\left(\int_0^t q(s, X_s)\, ds\right),$$

then, for every $t \geq 0$ and every increasing and bounded continuous function $\phi : \mathbb{R} \to \mathbb{R}$,

$$x \mapsto K(t,x) := \frac{\mathbb{E}[\phi(N_t) N_t | X_t = x]}{\mathbb{E}[N_t | X_t = x]} \text{ is increasing.}$$

The proof of this result is quite similar to that of Lemma 6.
2. To prove Theorem 8, it suffices to consider only the case where the functions $x \mapsto \theta(t,x)$, $x \mapsto q(t,x)$ and $x \mapsto \left(q + \frac{\mathscr{A}\theta}{\theta}\right)(t,x)$ are increasing for every $t \geq 0$.
The condition (56) ensures that the function h is differentiable and

$$\forall\, t \geq 0,\; h'(t) = \mathbb{E}\left[\widetilde{q}(t, X_t) N_t\right], \tag{59}$$

where

$$\forall\, (t,x) \in \mathbb{R}_+ \times \mathbb{R},\; \widetilde{q}(t,x) := \left(q + \frac{\mathscr{A}\theta}{\theta}\right)(t,x). \tag{60}$$

Indeed, if for every $t \geq 0$, we set:

$$C_t := \exp\left(\int_0^t \left(q + \frac{\mathscr{A}\theta}{\theta}\right)(u, X_u)\, du\right)$$

and

$$L_t := \theta(t, X_t) \exp\left(\int_0^t q(u, X_u)\, du\right),$$

then $L_t = M_t C_t$ and, for every $t \geq 0$, Itô's formula yields:

$$\begin{aligned} L_t - L_0 &= \int_0^t C_u\, dM_u + \int_0^t M_u\, dC_u \\ &= \widetilde{M}_t + \int_0^t \left(q + \frac{\mathscr{A}\theta}{\theta}\right)(u, X_u) M_u C_u\, du \\ &= \widetilde{M}_t + \int_0^t \widetilde{q}(u, X_u) L_u\, du, \end{aligned}$$

where

$$\left(\widetilde{M}_t := \int_0^t C_u\, dM_u, t \geq 0\right) \text{ is a continuous local martingale.}$$

Moreover, for every $0 \leq s \leq t$:

$$\begin{aligned}
|\widetilde{M}_s| &\leq |L_0| + |L_s| + \int_0^s |\widetilde{q}(u, X_u)|\, L_u\, du \\
&\leq |L_0| + (1+s)\left(\sup_{0\leq u\leq s} (\theta + |q|\theta + |\mathscr{A}\theta|)(u, X_u)\right) \exp\left(\int_0^s |q|(u, X_u)\, du\right) \\
&\leq |L_0| + (1+t)\left(\sup_{0\leq u\leq t} (\theta + |q|\theta + |\mathscr{A}\theta|)(u, X_u)\right) \exp\left(\int_0^t |q|(u, X_u)\, du\right)
\end{aligned}$$

which is integrable from (56). Thus, $\mathbb{E}\left[\sup_{0\leq s\leq t} \widetilde{M}_s\right] < \infty$ and then $(\widetilde{M}_t, t \geq 0)$ belongs to the class (DL) (see [14], Chap. IV, Definition 1.6 and Proposition 1.7). Therefore, $\left(\widetilde{M}_t, t \geq 0\right)$ is a true martingale. Hence:

$$\forall\, t \geq 0,\ \mathbb{E}[L_t] - \mathbb{E}[L_0] = \int_0^t \mathbb{E}\left[\widetilde{q}(u, X_u)L_u\right] du$$

and

$$h'(t) = \frac{d}{dt} \log \mathbb{E}[L_t] = \frac{\frac{d}{dt}\mathbb{E}[L_t]}{\mathbb{E}[L_t]} = \frac{\mathbb{E}\left[\widetilde{q}(t, X_t)L_t\right]}{\mathbb{E}[L_t]} = \mathbb{E}\left[\widetilde{q}(t, X_t)N_t\right]$$

which is equivalent to:

$$\forall\, t \geq 0,\ \mathbb{E}\left[\left(\widetilde{q}(t, X_t) - h'(t)\right) N_t\right] = 0 \text{ (since } \mathbb{E}[N_t] = 1). \tag{61}$$

Likewise, for every $t \geq 0$, we set

$$D_t := \exp\left(\int_0^t \left(q + \frac{\mathscr{A}\theta}{\theta}\right)(u, X_u)\, du - h(t)\right).$$

Then,

$$\forall\, t \geq 0,\ N_t = M_t D_t$$

and, for every $0 \leq s < t$,

$$\begin{aligned}
N_t - N_s = M_t D_t - M_s D_s &= \int_s^t D_u\, dM_u + \int_s^t M_u\, dD_u \\
&= \widetilde{M}_t - \widetilde{M}_s + \int_s^t \left[\left(q + \frac{\mathscr{A}\theta}{\theta}\right)(u, X_u) - h'(u)\right] M_u D_u\, du \\
&= \widetilde{M}_t - \widetilde{M}_s + \int_s^t \left[\widetilde{q}(u, X_u) - h'(u)\right] N_u\, du,
\end{aligned}$$

where $x \mapsto \widetilde{q}(u, x)$ is increasing for every $u \geq 0$.
3. Now, let $\psi \in \mathbf{C}$. Then, for every $0 \leq s < t$,

$$\begin{aligned}
&\mathbb{E}[\psi(N_t)] - \mathbb{E}[\psi(N_s)] \geq \mathbb{E}\left[\int_s^t \psi'(N_u)\, dN_u\right] \\
&= \mathbb{E}\left[\int_s^t \psi'(N_u)\, d\widetilde{M}_u\right] + \mathbb{E}\left[\psi'(N_u) N_u \left(\widetilde{q}(u, X_u) - h'(u)\right)\, du\right] \\
&= \int_s^t \mathbb{E}\left[\psi'(N_u) N_u \left(\widetilde{q}(u, X_u) - h'(u)\right)\right] du.
\end{aligned}$$

Moreover, for every $u \geq 0$, we have:

$$\begin{aligned}
&\mathbb{E}\left[\psi'(N_u) N_u \left(\widetilde{q}(u, X_u) - h'(u)\right)\right] \\
&= \mathbb{E}\left[\frac{\mathbb{E}[\psi'(N_u) N_u | X_u]}{\mathbb{E}[N_u | X_u]}\, \mathbb{E}[N_u | X_u] \left(\widetilde{q}(u, X_u) - h'(u)\right)\right].
\end{aligned}$$

For every $u \geq 0$ and $x \in \mathbb{R}$, let us define:

$$K(u, x) := \frac{\mathbb{E}\left[\psi'(N_u) N_u | X_u = x\right]}{\mathbb{E}[N_u | X_u = x]}$$

and

$$V(u, x) := \mathbb{E}[N_u | X_u = x] \left(\widetilde{q}(u, x) - h'(t)\right).$$

We note that:

$$\forall\, u \geq 0,\ \mathbb{E}[V(u, X_u)] = 0.$$

On the other hand, it follows from (57) and Lemma 7 that the function $x \mapsto K(u, x)$ is increasing.
For $u \geq 0$, let $(\widetilde{q}_u)^{-1}$ denotes the right-continuous inverse of the function $x \mapsto \widetilde{q}(u, x)$. Thus, distinguishing the cases $X_u \geq (\widetilde{q}_u)^{-1}(0)$ and $X_u \leq (\widetilde{q}_u)^{-1}(0)$, we obtain

$$\forall\, u \geq 0,\ K(u, X_u) V(u, X_u) \geq K\left(u, (\widetilde{q}_u)^{-1}(0)\right) V(u, X_u),$$

and finally, for every $0 \leq s < t$:

$$\begin{aligned}\mathbb{E}[\psi(N_t)] - \mathbb{E}[\psi(N_s)] &\geq \int_s^t \mathbb{E}[K(u, X_u)V(u, X_u)]\, du \\ &\geq \int_s^t K\left(u, (\overline{q}_u)^{-1}(0)\right) \mathbb{E}[V(u, X_u)]\, du = 0.\end{aligned}$$ □

Remark 3. Suppose that $(X_t, t \geq 0)$ is a Brownian motion issued from 0. Let $\bar{q} : \mathbb{R} \to \mathbb{R}_+$ satisfies

$$\int_0^\infty (1 + |x|)\bar{q}(x)\, dx < \infty \text{ and } \liminf_{x \to -\infty} |x|^{2\alpha} \bar{q}(x) > 0, \text{ for some } \alpha < 1,$$

and let θ be the unique solution of the Sturm-Liouville equation:

$$\begin{cases} \theta''(x) = \theta(x)\bar{q}(x) \\ \lim\limits_{x \to \infty} \theta'(x) = \sqrt{\dfrac{2}{\pi}}, \quad \lim\limits_{x \to -\infty} \theta(x) = 0. \end{cases}$$

Then,

$$\left(N_t := \theta(X_t) \exp\left(-\frac{1}{2} \int_0^t \bar{q}(X_u)\, du \right), t \geq 0 \right) \text{ is a local martingale,}$$

and it is shown in [15] that $(N_t, t \geq 0)$ is a true martingale. We are here in the situation of Theorem 8 with

$$q + \frac{\mathscr{A}\theta}{\theta} = -\frac{1}{2}\bar{q} + \frac{1}{2}\frac{\theta''}{\theta} = 0.$$

In other words, in this specific case, $(M_t, t \geq 0)$ is better than a peacock: it is a martingale.

5 Part II: Strong and Very Strong Peacocks

5.1 *Strong Peacocks*

5.1.1 Definition and Examples

Definition 8. An integrable real-valued process $(X_t, t \geq 0)$ is said to be a strong peacock (resp. a strong pea<u>d</u>ock) if, for every $0 \leq s < t$ and every increasing and bounded Borel function $\phi : \mathbb{R} \to \mathbb{R}$:

$$\mathbb{E}[(X_t - X_s)\phi(X_s)] \geq 0 \tag{62}$$

(resp.

$$\mathbb{E}[(X_t - X_s)\phi(X_t)] \leq 0.)$$

Remark 4.
1. The definition of a peacock involves only its 1-dimensional marginals. On the other hand, the definition of a strong peacock involves its 2-dimensional marginals.
2. If $(X_t, t \geq 0)$ is a strong peacock, then $\mathbb{E}[X_t]$ does not depend on t (it suffices to apply (62) with $\phi = 1$ and $\phi = -1$). Every strong peacock is a peacock; indeed, if $\psi \in \mathbf{C}_+$, then:

$$\mathbb{E}[\psi(X_t)] - \mathbb{E}[\psi(X_s)] \geq \mathbb{E}[\psi'(X_s)(X_t - X_s)] \geq 0.$$

3. If $(X_t, t \geq 0)$ is a strong peacock such that $\mathbb{E}\left[X_t^2\right] < \infty$ for every $t \geq 0$, then:

$$\mathbb{E}[X_s(X_t - X_s)] \geq 0, \text{ for every } 0 \leq s < t. \tag{63}$$

4. If X and Y are two processes which have the same 1-dimensional marginals, it may be possible that X is a strong peacock while Y is not. For example, let us consider $(X_t := t^{\frac{1}{4}} B_1, t \geq 0)$ and $\left(Y_t := \frac{B_t}{t^{\frac{1}{4}}}, t \geq 0\right)$, where $(B_t, t \geq 0)$ is a Brownian motion started at 0. By Lemma 2, $(X_t, t \geq 0)$ is a strong peacock while $(Y_t, t \geq 0)$ is not. Indeed, for every $a \in \mathbb{R}$ and $0 < s \leq t$:

$$\mathbb{E}\left[1_{\left\{\frac{B_s}{s^{\frac{1}{4}}} > a\right\}}\left(\frac{B_t}{t^{\frac{1}{4}}} - \frac{B_s}{s^{\frac{1}{4}}}\right)\right] = \left(\frac{1}{t^{\frac{1}{4}}} - \frac{1}{s^{\frac{1}{4}}}\right)\mathbb{E}\left[1_{\left\{B_s > as^{\frac{1}{4}}\right\}} B_s\right] < 0.$$

More generally, for every non null martingale $(M_t, t \geq 0)$ and every increasing Borel function $\alpha : \mathbb{R}_+ \to \mathbb{R}_+$, $\left(\frac{M_t}{\alpha(t)}, t \geq 0\right)$ is not a strong peacock.
5. Theorem 1 remains true if one replaces peacock by strong peacock.

Example 11. Some examples of strong peacocks:

- *Martingales:* Indeed, if $(M_t, t \geq 0)$ is a martingale with respect to some filtration $(\mathscr{F}_t, t \geq 0)$, then, for every bounded Borel function $\phi : \mathbb{R} \to \mathbb{R}$:

$$\mathbb{E}[\phi(M_s)(M_t - M_s)] = \mathbb{E}[\phi(M_s)(\mathbb{E}[M_t|\mathscr{F}_s] - M_s)] = 0.$$

- If $(M_u, u \geq 0)$ is a martingale belonging to H^1_{loc} and $\alpha : \mathbb{R}_+ \to \mathbb{R}_+$ is a strictly increasing Borel function such that $\alpha(0) = 0$, then

$$\left(\frac{1}{\alpha(t)} \int_0^t M_u d\alpha(u)\right) \text{ is a strong peacock}$$

(see [11, Chap. 1])

- The process $(tX, t \geq 0)$ where X is a centered and integrable r.v.
- The process $\left(\frac{e^{tX}}{\mathbb{E}[e^{tX}]}, t \geq 0\right)$ (see [11, Chap. 1]).

In the case of Gaussian processes, we obtain a characterization of strong peacocks using the covariance function. Indeed, one has:

Proposition 1. *A centered Gaussian peacock $(X_t, t \geq 0)$ is strong if and only if, for every $0 < s \leq t$:*

$$\mathbb{E}[X_s X_t] \geq \mathbb{E}[X_s^2]. \tag{64}$$

We note that a centered gaussian process $(X_t, t \geq 0)$ is a peacock if and only if

$$t \longmapsto \mathbb{E}\left[X_t^2\right] \text{ is increasing} \tag{65}$$

and of course, (64) implies (65); indeed, for every $0 < s \leq t$:

$$\mathbb{E}\left[X_s^2\right] \leq \mathbb{E}\left[X_s X_t\right] \leq \mathbb{E}\left[X_s^2\right]^{\frac{1}{2}} \mathbb{E}\left[X_t^2\right]^{\frac{1}{2}}, \text{ (from Schwartz's Inequality)}$$

which implies (65).

Proof (of Proposition 1).
Let $(X_t, t \geq 0)$ be a centered Gaussian strong peacock.
1. By taking $\phi(x) = x$ in (62), we have:

$$\mathbb{E}[X_s(X_t - X_s)] \geq 0, \text{ for every } 0 < s \leq t,$$

i.e.,

$$K(s,t) \geq K(s,s), \text{ for every } 0 < s \leq t.$$

2. Conversely, if (64) holds, then, for every $0 < s \leq t$ and every increasing Borel function $\phi : \mathbb{R} \to \mathbb{R}$:

$$\begin{aligned}\mathbb{E}[\phi(X_s)(X_t - X_s)] &= \mathbb{E}[\phi(X_s)(\mathbb{E}[X_t|X_s] - X_s)] \\ &= \left(\frac{K(s,t)}{K(s,s)} - 1\right)\mathbb{E}[\phi(X_s)X_s] \geq 0 \quad \text{(from Lemma 2).}\end{aligned}$$ □

Example 12. We give two examples:

- An Ornstein-Uhlenbeck process with parameter $c \in \mathbb{R}$:

$$X_t = B_t + c\int_0^t X_u du,$$

where $(B_t, t \geq 0)$ is a Brownian motion started at 0, is a peacock for every c and a strong peacock if and only if $c \geq 0$. Indeed, for every $t \geq 0$,

$$X_t = e^{ct} \int_0^t e^{-cs} dB_s$$

and, for every $0 < s \le t$, since

$$\mathbb{E}[X_s X_t] = \frac{\sinh(cs)}{c} e^{ct},$$

we have:

$$\mathbb{E}[X_s X_t] - \mathbb{E}\left[X_s^2\right] = \frac{\sinh(cs)}{c}[e^{ct} - e^{cs}].$$

Thus,

$$\mathbb{E}[X_s X_t] - \mathbb{E}\left[X_s^2\right] \ge 0 \text{ if and only if } c \ge 0.$$

- A fractional Brownian motion $(X_t, t \ge 0)$ with index $H \in [0, 1]$ is a peacock for every H and a strong peacock if and only if $H \ge \frac{1}{2}$. This follows from the fact that,

$$K(s,t) - K(s,s) = \frac{1}{2}(t^{2H} - s^{2H} - (t-s)^{2H})$$

  is positive for every $0 < s \le t$ if and only if $H \ge \frac{1}{2}$, where K denotes the covariance function of $(X_t, t \ge 0)$.

5.1.2 Upper and Lower Orthant Orders

Let $X = (X_1, X_2, \dots, X_p)$ and $Y = (Y_1, Y_2, \dots, Y_p)$ be two $\mathbb{R}^p$-valued random vectors. The following definitions are taken from Shaked and Shanthikumar [17], p. 140.

Definition 9. (Upper orthant order).
X is said to be smaller than Y in the upper orthant order (notation: $X \underset{\text{u.o}}{\le} Y$) if one of the two following equivalent conditions is satisfied:

1. For every p-tuple $\lambda_1, \lambda_2, \dots, \lambda_p$ of reals:

$$\mathbb{P}(X_1 > \lambda_1, X_2 > \lambda_2, \dots, X_p > \lambda_p) \le \mathbb{P}(Y_1 > \lambda_1, Y_2 > \lambda_2, \dots, Y_p > \lambda_p) \tag{66}$$

2. For every p-tuple $l_1, l_2, \dots, l_p$ of nonnegative increasing functions:

$$\mathbb{E}\left[\prod_{i=1}^p l_i(X_i)\right] \le \mathbb{E}\left[\prod_{i=1}^p l_i(Y_i)\right] \tag{67}$$

Definition 10. (Upper orthant order for processes).
A process $(X_t, t \geq 0)$ is smaller than a process $(Y_t, t \geq 0)$ for the upper orthant order (notation: $(X_t, t \geq 0) \underset{\text{u.o}}{\leq} (Y_t, t \geq 0)$) if, for every integer p and every $0 \leq t_1 < t_2 < \cdots < t_p$:

$$(X_{t_1}, X_{t_2}, \ldots, X_{t_p}) \underset{\text{u.o}}{\leq} (Y_{t_1}, Y_{t_2}, \ldots, Y_{t_p}). \tag{68}$$

If X and Y are two càdlàg processes, (68) is equivalent to:

$$\begin{aligned} &\text{for every } h : \mathbb{R} \longrightarrow \mathbb{R} \text{ càdlàg:} \\ &\mathbb{P}(\text{for every } t \geq 0, X_t \geq h(t)) \leq \mathbb{P}(\text{for every } t \geq 0, Y_t \geq h(t)). \end{aligned} \tag{69}$$

Definition 11. (Lower orthant order).
X is said to be smaller than Y in the lower orthant order (notation: $X \underset{\text{l.o}}{\leq} Y$) if one of the two following equivalent conditions is satisfied:

1. For every p-tuple $\lambda_1, \lambda_2, \ldots, \lambda_p$ of reals:

$$\mathbb{P}(X_1 \leq \lambda_1, X_2 \leq \lambda_2, \ldots, X_p \leq \lambda_p) \geq \mathbb{P}(Y_1 \leq \lambda_1, Y_2 \leq \lambda_2, \ldots, Y_p \leq \lambda_p) \tag{70}$$

2. For every p-tuple $l_1, l_2, \ldots, l_p$ of nonnegative decreasing functions:

$$\mathbb{E}\left[\prod_{i=1}^{p} l_i(X_i)\right] \geq \mathbb{E}\left[\prod_{i=1}^{p} l_i(Y_i)\right] \tag{71}$$

Definition 12. (Lower orthant order for processes).
A process $(X_t, t \geq 0)$ is smaller than a process $(Y_t, t \geq 0)$ for the lower orthant order (notation: $(X_t, t \geq 0) \underset{\text{l.o}}{\leq} (Y_t, t \geq 0)$) if, for every integer p and every $0 \leq t_1 < t_2 < \cdots < t_p$:

$$(X_{t_1}, X_{t_2}, \ldots, X_{t_p}) \underset{\text{l.o}}{\leq} (Y_{t_1}, Y_{t_2}, \ldots, Y_{t_p}). \tag{72}$$

If X and Y are two càdlàg processes, (72) is equivalent to:

$$\begin{aligned} &\text{for every } h : \mathbb{R} \longrightarrow \mathbb{R} \text{ càdlàg:} \\ &\mathbb{P}(\text{for every } t \geq 0, X_t \leq h(t)) \geq \mathbb{P}(\text{for every } t \geq 0, Y_t \leq h(t)). \end{aligned} \tag{73}$$

Remark 5. Observe that, if $X = (X_t, \geq 0)$ and $Y = (Y_t, \geq 0)$ are two processes such that: $X \underset{\text{l.o}}{\leq} Y \underset{\text{u.o}}{\leq} X$, then

$$(X_t, t \geq 0) \overset{(1.d)}{=} (Y_t, t \geq 0)$$

Let $(X_t, t \geq 0)$ be a real-valued measurable process and, for $t \geq 0$, let F_t denotes the distribution function of X_t. If U is uniformly distributed on $[0, 1]$, then

$$(X_t, t \geq 0) \overset{(1.d)}{=} \left(F_t^{-1}(U), t \geq 0\right).$$

Moreover, we state the following:

Proposition 2. *Let $(X_t, t \geq 0)$ a real-valued process, and for $t \geq 0$, let F_t be the distribution function of X_t.Then, if U is uniformly distributed on $[0, 1]$, one has:*

$$\left(F_t^{-1}(U), t \geq 0\right) \underset{l.o.}{\leq} (X_t, t \geq 0) \underset{u.o.}{\leq} (F_t^{-1}(U)).$$

Proof (of Proposition 2).
For every integer p, every p-tuple $\lambda_1, \lambda_2, \ldots, \lambda_p$ of reals and every $0 \leq t_1 < t_2 < \cdots < t_p$:

$$\begin{aligned}\mathbb{P}(X_{t_1} > \lambda_1, X_{t_2} > \lambda_2, \ldots, X_{t_p} > \lambda_p) &\leq \min_{i=1,2,\ldots,p} \mathbb{P}(X_{t_i} > \lambda_i)\\ &= 1 - \max_{i=1,2,\ldots,p} F_{t_i}(\lambda_i)\\ &= \mathbb{P}\left(U > \max_{i=1,2,\ldots,p} F_{t_i}(\lambda_i)\right)\\ &= \mathbb{P}\left(U > F_{t_1}(\lambda_1), U > F_{t_2}(\lambda_2), \ldots, U > F_{t_p}(\lambda_p)\right)\\ &= \mathbb{P}\left(F_{t_1}^{-1}(U) > \lambda_1, F_{t_2}^{-1}(U) > \lambda_2, \ldots, F_{t_p}^{-1}(U) > \lambda_p\right).\end{aligned}$$

On the other hand, one has:

$$\begin{aligned}\mathbb{P}(X_{t_1} \leq \lambda_1, X_{t_2} \leq \lambda_2, \ldots, X_{t_p} \leq \lambda_p) &\leq \min_{i=1,2,\ldots,p} \mathbb{P}(X_{t_i} \leq \lambda_i)\\ &= \mathbb{P}\left(F_{t_1}^{-1}(U) \leq \lambda_1, F_{t_2}^{-1}(U) \leq \lambda_2, \ldots, F_{t_p}^{-1}(U) \leq \lambda_p\right).\end{aligned}$$

□

Let us introduce some definitions.

Definition 13. For a given family of probability measures $\mu = (\mu_t, t \geq 0)$, we denote by $\mathscr{D}_\mu$ the set of real-valued processes which admit the family μ as one-dimensional marginals:

$$\mathscr{D}_\mu := \{(X_t, t \geq 0);\ \text{for every } t \geq 0,\ X_t \sim \mu_t\}.$$

In particular, if the family μ increases in the convex order, then $\mathscr{D}_\mu$ is the set of peacocks associated to μ.

The next corollary follows immediately from Proposition 2.

Corollary 1. *Let μ be a family of probability measures. Then, the process $(F_t^{-1}(U), t \geq 0)$ is an absolute maximum of $\mathscr{D}_\mu$ for the upper orthant order and an absolute minimum of $\mathscr{D}_\mu$ for the lower orthant order.*

The following result is due to Cambanis et al. [4].

Theorem 9. *Let (X_1, X_2) and (Y_1, Y_2) be two $\mathbb{R}^2$-valued random vectors such that:*

$$X_1 \overset{(law)}{=} Y_1,\ X_2 \overset{(law)}{=} Y_2 \text{ and } (X_1, X_2) \underset{l.o}{\leq} (Y_1, Y_2) \tag{74}$$

Let $k : \mathbb{R} \times \mathbb{R} \to \mathbb{R}$ be right-continuous and quasi-monotone, i.e:

$$k(x, y) + k(x', y') - k(x, y') - k(x', y) \geq 0, \text{ for every } x \leq x',\ y \leq y'. \tag{75}$$

Suppose that the expectations $\mathbb{E}[k(X_1, X_2)]$ and $\mathbb{E}[k(Y_1, Y_2)]$ exist (even if infinite valued) and either of the following conditions is satisfied:

i. k is symmetric and the expectations $\mathbb{E}[k(X_1, X_1)]$ and $\mathbb{E}[k(X_2, X_2)]$ are finite
ii. The expectations $\mathbb{E}[k(X_1, x_1)]$ and $\mathbb{E}[k(x_2, X_2)]$ are finite for some x_1 and x_2.

Then:

$$\mathbb{E}[k(X_1, X_2)] \geq \mathbb{E}[k(Y_1, Y_2)].$$

The next result is deduced from Proposition 2 and Theorem 9.

Corollary 2. *Let $X := (X_t, t \geq 0)$ be a peacock and, for every $t \geq 0$, let F_t be the distribution function of X_t. Let U be uniformly distributed on $[0, 1]$. Then:*

1. *For every real-valued process $Y := (Y_t, t \geq 0)$ such that $Y_t \overset{(1.d)}{=} X_t$ and every quasi-monotone function $k : \mathbb{R} \times \mathbb{R} \to \mathbb{R}$ satisfying the same conditions as in Theorem 9, one has:*

$$\forall (s, t) \in \mathbb{R}_+ \times \mathbb{R}_+,\ \mathbb{E}\left[k\left(F_s^{-1}(U), F_t^{-1}(U)\right)\right] \geq \mathbb{E}[k(Y_s, Y_t)]. \tag{76}$$

   In particular, for every $p \geq 1$ such that $\mathbb{E}[|X_u|^p] < \infty$, for every $u \geq 0$ and every $(s, t) \in \mathbb{R}_+ \times \mathbb{R}_+$,

$$\mathbb{E}\left[\left|F_t^{-1}(U) - F_s^{-1}(U)\right|^p\right] \leq \mathbb{E}\left[|Y_t - Y_s|^p\right], \tag{77}$$

2. *$(F_t^{-1}(U), t \geq 0)$ is a strong peacock.*

To prove Corollary 2 we may observe, for the first point, that for every $p \geq 1$, the function $k : (x, y) \longmapsto -|x - y|^p$ is quasi-monotone and, for the second point, that if $\phi : \mathbb{R} \to \mathbb{R}$ is increasing, then $k : (x, y) \mapsto \phi(x)(y - x)$ is a quasi-monotone function.

5.1.3 A Comparison Theorem for Peacocks

Let $(X_t, t \geq 0)$ be a real-valued process which is square integrable and which satisfies:

$$t \longmapsto X_t \text{ is a.s. measurable.} \tag{78}$$

For a probability measure ν on $\{(s,t);\ 0 \leq s \leq t\}$, let us define the 2-variability of $(X_t, t \geq 0)$ with respect to ν by the quantity:

$$\Pi_\nu(X) := \iint_{\{0 \leq s \leq t\}} \mathbb{E}[(X_t - X_s)^2]\nu(ds, dt).$$

Definition 14. For a family of probability measures $\mu := (\mu_t, t \geq 0)$, let $\mathscr{D}_\mu^+$ denotes the set of strong peacocks which admit μ as their one-dimensional marginals family:

$$\mathscr{D}_\mu^+ := \{(X_t, t \geq 0);\ X \text{ is a strong peacock such that, } \forall t \geq 0,\ X_t \sim \mu_t\}.$$

Given a family of probability measures $\mu := (\mu_t, t \geq 0)$ which increases in the convex order, we wish to determinate for which processes in $\mathscr{D}_\mu^+$, Π_ν attains its maximum (resp. its minimum).

Theorem 10. *Let ν be a probability measure on $\{(s,t);\ 0 \leq s \leq t\}$.*

1. *The maximum of $\Pi_\nu(X)$ in $\mathscr{D}_\mu^+$ is equal to:*

$$\max_{X \in \mathscr{D}_\mu^+} \Pi_\nu(X) = \iint_{\{0 \leq s \leq t\}} \left(\mathbb{E}\left[X_t^2\right] - \mathbb{E}\left[X_s^2\right]\right) \nu(ds, dt) \tag{79}$$

   and is attained when $(X_t, t \geq 0)$ is a martingale.
2. *The minimum of $\Pi_\nu(X)$ in $\mathscr{D}_\mu^+$ is equal to:*

$$\min_{X \in \mathscr{D}_\mu^+} \Pi_\nu(X) = \iint_{\{0 \leq s \leq t\}} \mathbb{E}\left[\left(F_t^{-1}(U) - F_s^{-1}(U)\right)^2\right] \nu(ds, dt) \tag{80}$$

   and is attained by $\left(X_t = F_t^{-1}(U), t \geq 0\right)$.

Proof (of Theorem 10).
1. Let $(X_t, t \geq 0)$ be a strong peacock. For every $0 \leq s \leq t$, one has:

$$\begin{aligned}\mathbb{E}\left[(X_t - X_s)^2\right] &= \mathbb{E}\left[X_t^2\right] + \mathbb{E}\left[X_s^2\right] - 2\mathbb{E}[X_t X_s] \\ &= \mathbb{E}\left[X_t^2\right] - \mathbb{E}\left[X_s^2\right] - 2\mathbb{E}[(X_t - X_s)X_s] \\ &\leq \mathbb{E}\left[X_t^2\right] - \mathbb{E}\left[X_s^2\right] \quad \text{(from (62))}.\end{aligned}$$

Hence, integrating against ν, we obtain:

$$\max_{X \in \mathscr{D}_\mu^+} \Pi_\nu(X) \leq \iint_{\{0 \leq s \leq t\}} \left(\mathbb{E}\left[X_t^2\right] - \mathbb{E}\left[X_s^2\right]\right) \nu(ds, dt) := M(X)$$

and $M(X)$ is clearly attained when $(X_t, t \geq 0)$ is a martingale.
2. This point is a consequence of Theorem 9 and Corollary 2. □

6 Very Strong Peacocks

6.1 Definition, Examples and Counterexamples

Definition 15. An integrable real-valued process $(X_t, t \geq 0)$ is said to be a very strong peacock (VSP) if, for every $n \in \mathbb{N}^*$, every $0 < t_1 < \cdots < t_n < t_{n+1}$ and every $\phi \in \mathscr{E}_n$, we have:

$$\mathbb{E}\left[\phi\left(X_{t_1}, \ldots, X_{t_n}\right)\left(X_{t_{n+1}} - X_{t_n}\right)\right] \geq 0. \tag{81}$$

Remark 6.
1. The definition of a strong peacock involves its 2-dimensional marginals while the definition of a very strong peacock involves all its finite-dimensional marginals.
2. Every very strong peacock is a strong peacock. But, the converse is not true. Let us give two examples:

(a) Let G_1 and G_2 be two independent, centered Gaussian r.v.'s such that $\mathbb{E}[G_1^2] = \mathbb{E}[G_2^2] = 1$, α, β be two constants satisfying $1 + 2\alpha^2 \leq \beta$ and (X_1, X_2, X_3) be the random Gaussian vector defined by:

$$X_1 = G_1 - \alpha G_2, \; X_2 = \beta G_1, \; X_3 = \beta G_1 + \alpha G_2. \tag{82}$$

Then, (X_1, X_2, X_3) is a strong peacock (from Proposition 1) which is not a very strong peacock since

$$\mathbb{E}[X_1(X_3 - X_2)] = -\alpha^2 \mathbb{E}\left[G_1^2\right] < 0.$$

(b) Likewise, let G_1 and G_2 be two symmetric, independent and identically distributed r.v.'s such that $\mathbb{E}\left[G_i^2\right] = 1$ $(i = 1, 2)$. Then, for every $\beta \geq 3$, the random vector (X_1, X_2, X_3) given by:

$$X_1 = G_1 - G_2, \; X_2 = \beta G_1, \; X_3 = \beta G_1 + G_2. \tag{83}$$

is a strong peacock for which (81) does not hold.

Proof. Since G_1 and G_2 are independent and centered, we first observe that:

$$\mathbb{E}\left[1_{\{X_2 \geq a\}}(X_3 - X_2)\right] = \mathbb{E}\left[1_{\{\beta G_1 \geq a\}} G_2\right] = 0.$$

Moreover,

$$\begin{aligned}
\mathbb{E}\left[1_{\{X_1 \geq a\}}(X_2 - X_1)\right] &= \mathbb{E}\left[1_{\{G_1 - G_2 \geq a\}}((\beta - 1)G_1 + G_2)\right] \\
&= (\beta - 1)\mathbb{E}\left[1_{\{G_1 - G_2 \geq a\}} G_1\right] + \mathbb{E}\left[1_{\{G_1 - G_2 \geq a\}} G_2\right] \\
&= \underbrace{(\beta - 1)\mathbb{E}\left[1_{\{G_2 - G_1 \geq a\}} G_2\right]}_{\text{(by interchanging } G_1 \text{ and } G_2)} + \mathbb{E}\left[1_{\{G_1 - G_2 \geq a\}} G_2\right] \\
&= \underbrace{(\beta - 2)\mathbb{E}\left[1_{\{G_2 - G_1 \geq a\}} G_2\right]}_{\geq 0 \text{ by Lemma 2, since } \beta > 2} + \mathbb{E}\left[1_{\{|G_1 - G_2| \geq a\}} G_2\right] \\
&\geq \mathbb{E}\left[1_{\{|G_1 - G_2| \geq a\}} G_2\right] = 0, \\
&\quad \text{(since } G_1 \text{ and } G_2 \text{ are symmetric)}
\end{aligned}$$

and similarly,

$$\begin{aligned}
\mathbb{E}\left[1_{\{X_1 \geq a\}}(X_3 - X_1)\right] &= \mathbb{E}\left[1_{\{G_1 - G_2 \geq a\}}((\beta - 1)G_1 + 2G_2)\right] \\
&= (\beta - 1)\mathbb{E}\left[1_{\{G_1 - G_2 \geq a\}} G_1\right] + 2\mathbb{E}\left[1_{\{G_1 - G_2 \geq a\}} G_2\right] \\
&= (\beta - 1)\mathbb{E}\left[1_{\{G_2 - G_1 \geq a\}} G_2\right] + 2\mathbb{E}\left[1_{\{G_1 - G_2 \geq a\}} G_2\right] \\
&= \underbrace{(\beta - 3)\mathbb{E}\left[1_{\{G_2 - G_1 \geq a\}} G_2\right]}_{\geq 0 \text{ by Lemma 2, since } \beta \geq 3} + 2\mathbb{E}\left[1_{\{|G_1 - G_2| \geq a\}} G_2\right] \\
&\geq 2\mathbb{E}\left[1_{\{|G_1 - G_2| \geq a\}} G_2\right] = 0.
\end{aligned}$$

Thus, (X_1, X_2, X_3) is a strong peacock. But, (X_1, X_2, X_3) is not a very strong peacock since

$$\mathbb{E}[X_1(X_3 - X_2)] = -\mathbb{E}\left[G_1^2\right] < 0.$$

□

Let us give some examples of very strong peacocks.

Example 13.
1. Each of the processes cited in Example 11 is a very strong peacock. We refer the reader to ([11], Chap. 8) for further examples.
2. Let $(\tau_t, t \geq 0)$ be an increasing process with independent increments (for example a subordinator) and $f : \mathbb{R} \to \mathbb{R}$ be a convex and increasing (or concave and decreasing) function such that $\mathbb{E}[|f(\tau_t)|] < \infty$, for every $t \geq 0$. Then, $(X_t := f(\tau_t) - \mathbb{E}[f(\tau_t)], t \geq 0)$ is a very strong peacock.

Proof. Let f be a convex and increasing function and let $n \geq 1$, $0 < t_1 < t_2 < \cdots < t_n < t_{n+1}$ and $\phi \in \mathscr{E}_n$. We first note that:

$$\widetilde{\phi} : (x_1, \dots, x_n) \longmapsto \phi\,(f(x_1) - \mathbb{E}[f(\tau_{t_1})], \dots, f(x_n) - \mathbb{E}[f(\tau_{t_n})]) \text{ belongs to } \mathscr{E}_n \tag{84}$$

and, by setting $c_n := \mathbb{E}[f(\tau_{t_{n+1}})] - \mathbb{E}[f(\tau_{t_n})]$,

$$\mathbb{E}\left[\phi(X_{t_1}, \dots, X_{t_n})(X_{t_{n+1}} - X_{t_n})\right] = \mathbb{E}\left[\widetilde{\phi}(\tau_{t_1}, \dots, \tau_{t_n})(f(\tau_{t_{n+1}}) - f(\tau_{t_n}) - c_n)\right]. \tag{85}$$

Let us prove by induction that, for every $i \in [\![1, n]\!]$, there exists a function $\varphi_i \in \mathscr{E}_i$ such that:

$$\mathbb{E}\left[\phi(X_{t_1}, \dots, X_{t_n})(X_{t_{n+1}} - X_{t_n})\right] \geq \mathbb{E}\left[\varphi_i(\tau_{t_1}, \dots, \tau_{t_i})(f(\tau_{t_{n+1}}) - f(\tau_{t_n}) - c_n)\right]. \tag{86}$$

We note that, for $i = n$, we may choose $\varphi_n = \widetilde{\phi}$. On the other hand, let us suppose that (86) holds for some $i \in [\![1, n]\!]$. Then, since τ_{t_i} is independent of $\tau_{t_{n+1}} - \tau_{t_i}$ and $\tau_{t_n} - \tau_{t_i}$, one has:

$$\begin{aligned}
&\mathbb{E}\left[\phi(X_{t_1}, \dots, X_{t_n})(X_{t_{n+1}} - X_{t_n})\right] \\
&\geq \mathbb{E}\left[\varphi_i(\tau_{t_1}, \dots, \tau_{t_i})(f(\tau_{t_{n+1}}) - f(\tau_{t_n}) - c_n)\right] \quad \text{(by induction)} \\
&= \mathbb{E}\left[\varphi_i(\tau_{t_1}, \dots, \tau_{t_i})\left(f(\tau_{t_i} + \tau_{t_{n+1}} - \tau_{t_i}) - f(\tau_{t_i} + \tau_{t_n} - \tau_{t_i}) - c_n\right)\right] \\
&= \mathbb{E}\left[\varphi_i(\tau_{t_1}, \dots, \tau_{t_i})\left(\mathbb{E}[f(\tau_{t_i} + \tau_{t_{n+1}} - \tau_{t_i})|\mathscr{F}_{t_i}] - \mathbb{E}[f(\tau_{t_i} + \tau_{t_n} - \tau_{t_i})|\mathscr{F}_{t_i}] - c_n\right)\right] \\
&\quad \text{(where } \mathscr{F}_{t_i} := \sigma(\tau_s, 0 \leq s \leq t_i)) \\
&= \mathbb{E}\left[\varphi_i(\tau_{t_1}, \dots, \tau_{t_i})\widehat{f}_i\,(\tau_{t_i})\right], \\
&\quad \text{(where } \widehat{f}_i(x) = \mathbb{E}[f(x + \tau_{t_{n+1}} - \tau_{t_i})] - \mathbb{E}[f(x + \tau_{t_n} - \tau_{t_i})] - c_n).
\end{aligned}$$

But, the function $\widehat{f}_i$ is increasing since f is convex and $\tau_{t_{n+1}} \geq \tau_{t_n}$. Hence,

$$\begin{aligned}
&\mathbb{E}\left[\phi(X_{t_1},\dots,X_{t_n})(X_{t_{n+1}} - X_{t_n})\right] \geq \mathbb{E}\left[\varphi_i\left(\tau_{t_1},\dots,\tau_{t_{i-1}},\tau_{t_i}\right)\widehat{f}_i\left(\tau_{t_i}\right)\right] \\
&\geq \mathbb{E}\left[\varphi_i\left(\tau_{t_1},\dots,\tau_{t_{i-1}},\widehat{f}_i^{-1}(0)\right)\widehat{f}_i\left(\tau_{t_i}\right)\right] \\
&= \mathbb{E}\left[\varphi_i\left(\tau_{t_1},\dots,\tau_{t_{i-1}},\widehat{f}_i^{-1}(0)\right)\left(f(\tau_{t_{n+1}}) - f(\tau_{t_n}) - c_n\right)\right],
\end{aligned}$$

i.e., (86) also holds for $i-1$ with

$$\varphi_{i-1} : (x_1,\dots,x_{i-1}) \longmapsto \varphi_i\left(x_1,\dots,x_{i-1},\widehat{f}_i^{-1}(0)\right).$$

Thus, (86) holds for every $i \in [\![1,n]\!]$. In particular, for $i = 1$, there exists $\varphi_1 \in \mathscr{E}_1$ such that:

$$\begin{aligned}
\mathbb{E}\left[\phi(X_{t_1},\dots,X_{t_n})(X_{t_{n+1}} - X_{t_n})\right] &\geq \mathbb{E}\left[\varphi_1\left(\tau_{t_1}\right)\widehat{f}_1\left(\tau_{t_1}\right)\right] \\
&\geq \varphi_1\left(\widehat{f}_1^{-1}(0)\right)\mathbb{E}\left[\widehat{f}_1\left(\tau_{t_1}\right)\right] \\
&= \varphi_1\left(\widehat{f}_1^{-1}(0)\right)\mathbb{E}\left[f(\tau_{t_{n+1}}) - f(\tau_{t_n}) - c_n\right] = 0.
\end{aligned}$$

□

6.2 *Peacocks Obtained by Quotient Under the Very Strong Peacock Hypothesis*

Lemma 8. *An integrable real-valued process is a very strong peacock if and only if, for every $n \geq 1$, every $0 < t_1 < \cdots < t_n < t_{n+1}$, every $i \leq n$ and every $\phi \in \mathscr{E}_n$:*

$$\mathbb{E}\left[\phi\left(X_{t_1},\dots,X_{t_n}\right)\left(X_{t_{n+1}} - X_{t_i}\right)\right] \geq 0. \tag{87}$$

Proof (of Lemma 8).
For every $n \geq 1$ and $i \leq n$, we shall prove by induction the following condition:

$$\mathbb{E}\left[\phi\left(X_{t_1},\dots,X_{t_n}\right)\left(X_{t_{n+1}} - X_{t_{n+1-i}}\right)\right] \geq 0 \tag{88}$$

which, of course, is equivalent to (87). If $i = 1$, we recover (81).
Now, let $1 \leq i \leq n-1$ be fixed and suppose that (81) is satisfied and that (88) holds for i. Let us prove that (88) is also true for $i+1$. One has:

$$\begin{aligned}
&\mathbb{E}\left[\phi\left(X_{t_1},\ldots,X_{t_{n-1}},X_{t_n}\right)\left(X_{t_{n+1}}-X_{t_{n+1-(i+1)}}\right)\right]\\
&=\underbrace{\mathbb{E}\left[\phi\left(X_{t_1},\ldots,X_{t_{n-1}},X_{t_n}\right)\left(X_{t_{n+1}}-X_{t_n}\right)\right]}_{\geq 0 \text{ (from (81))}}\\
&\qquad+\mathbb{E}\left[\phi\left(X_{t_1},\ldots,X_{t_{n-1}},X_{t_n}\right)\left(X_{t_n}-X_{t_{n-i}}\right)\right]\\
&\geq \mathbb{E}\left[\phi\left(X_{t_1},\ldots,X_{t_{n-1}},X_{t_n}\right)\left(X_{t_n}-X_{t_{n-i}}\right)\right]\\
&\geq \mathbb{E}\left[\phi\left(X_{t_1},\ldots,X_{t_{n-1}},X_{t_{n-i}}\right)\left(X_{t_n}-X_{t_{n-i}}\right)\right]\geq 0
\end{aligned}$$

(since ϕ belongs to $\mathscr{E}_n$ and (88) holds for $1 \leq i \leq n-1$). □

The importance of very strong peacocks lies in the following result.

Theorem 11. *Let $(X_t, t \geq 0)$ be a right-continuous and centered very strong peacock such that for every $t \geq 0$:*

$$\mathbb{E}\left[\sup_{s\in[0,t]} |X_s|\right] < \infty. \tag{89}$$

Then, for every right-continuous and strictly increasing function $\alpha : \mathbb{R}_+ \to \mathbb{R}_+$ such that $\alpha(0) = 0$:

$$\left(Q_t := \frac{1}{\alpha(t)} \int_0^t X_s \, d\alpha(s), t \geq 0\right) \text{ is a peacock.}$$

Remark 7.
1. Theorem 11 is a generalization of the case where $(X_s, s \geq 0)$ is a martingale (see Example 1).
2. Let $(\tau_s, s \geq 0)$ be a subordinator and $f : \mathbb{R}_+ \to \mathbb{R}$ be increasing, convex and such that $\mathbb{E}[|f(\tau_t)|] < \infty$, for every $t \geq 0$. Then, it follows from Theorem 11 and from the second point of Example 13 that, for every right-continuous and strictly increasing function $\alpha : \mathbb{R}_+ \to \mathbb{R}_+$ satisfying $\alpha(0) = 0$:

$$\left(Q_t := \frac{1}{\alpha(t)} \int_0^t (f(\tau_s) - \mathbb{E}[f(\tau_s)]) \, d\alpha(s), t \geq 0\right) \text{ is a peacock.}$$

Proof (of Theorem 11).
Let $T > 0$ be fixed.
1. Let us first suppose that $1_{[0,T]} d\alpha$ is a linear combination of Dirac measures and show that, for every $r \in [\![2, \infty]\!]$, every $a_1 > 0, a_2 > 0, \ldots, a_r > 0$ such that

$$\alpha(r) := \sum_{i=1}^{r} a_i = \alpha(T)$$

and every $0 < \lambda_1 < \lambda_2 < \cdots < \lambda_n \leq T$:

$$\left(Q_n := \frac{1}{\alpha(n)} \sum_{i=1}^{n} a_i X_{\lambda_i}, n \in [\![1, r]\!] \right) \text{ is a peacock.} \tag{90}$$

Let $\psi \in \mathbf{C}_+$ and $n \geq 2$. For every $n \in [\![2, r]\!]$, one has:

$$\begin{aligned}
&\mathbb{E}[\psi(Q_n)] - \mathbb{E}[\psi(Q_{n-1})] \geq \mathbb{E}[\psi'(Q_{n-1})(Q_n - Q_{n-1})] \\
&= \mathbb{E}\left[\psi'\left(\frac{1}{\alpha(n-1)} \sum_{i=1}^{n-1} a_i X_{\lambda_i}\right)\left(\frac{1}{\alpha(n)} \sum_{i=1}^{n} a_i X_{\lambda_i} - \frac{1}{\alpha(n-1)} \sum_{i=1}^{n-1} a_i X_{\lambda_i}\right)\right] \\
&= \frac{a_n}{\alpha(n)\alpha(n-1)} \sum_{i=1}^{n-1} a_i \mathbb{E}\left[\phi\left(X_{\lambda_1}, \dots, X_{\lambda_{n-1}}\right)\left(X_{\lambda_n} - X_{\lambda_i}\right)\right],
\end{aligned}$$

where

$$\phi : (x_1, \dots, x_{n-1}) \longmapsto \psi'\left(\frac{1}{\alpha(n-1)} \sum_{i=1}^{n-1} a_i x_i\right) \text{ belongs to } \mathscr{E}_{n-1}.$$

Then, the result follows from Lemma 8.
2. Let us set $\mu = 1_{[0,T]} d\alpha$ and, for every $0 \leq t \leq T$,

$$Q_t^{\mu} := \frac{1}{\mu([0,t])} \int_0^t X_u \mu(du).$$

Since the function $\lambda \in [0, T] \longmapsto X_\lambda$ is right-continuous and bounded from above by $\sup_{0 \leq \lambda \leq T} |X_\lambda|$ which is finite a.s., then there exists a sequence $(\mu_n, n \geq 0)$ of measures of type used in (1), with supp $\mu_n \subset [0, T]$, $\int \mu_n(du) = \int \mu(du)$ and, for every $0 \leq t \leq T$:

$$\lim_{n\to\infty} \int_0^t X_u \mu_n(du) = \int_0^t X_u \mu(du) \text{ a.s.} \tag{91}$$

$$\lim_{n\to\infty} \mu_n([0,t]) = \mu([0,t]). \tag{92}$$

Then, from (91) and (92), it follows that:

$$\lim_{n\to\infty} Q_t^{(\mu_n)} = Q_t^{(\mu)} \text{ a.s., for every } 0 \leq t \leq T. \tag{93}$$

But, using (1),

Table 1 Table of the main peacocks studied in this paper

Main hypothesis	Peacocks	References
$(X_t, t \geq 0)$ is conditionally monotone and θ is positive, convex and increasing	$\left(C_t := \theta\left(\int_0^t q(s, X_s)ds\right) - \gamma(t), t \geq 0\right)$ with $\gamma(t) = \mathbb{E}\left[\theta\left(\int_0^t q(s, X_s)ds\right)\right]$	Theorem 2
$(X_t, t \geq 0)$ is a process with independent and log-concave increments; θ is continuous and $x \mapsto \theta_t(x) := \theta(t, x)$ is increasing for every $t \geq 0$	$\left(N_t := \frac{\exp\left(\int_0^t q(s, X_s)\, d\alpha(s)\right)}{\mathbb{E}\left[\exp\left(\int_0^t q(s, X_s)\, d\alpha(s)\right)\right]}, t \geq 0\right)$	Theorem 3
	$\left(N_t := \frac{\theta_t(X_t) \exp\left(\int_0^t q_s(X_s)\, ds\right)}{\mathbb{E}\left[\theta_t(X_t) \exp\left(\int_0^t q_s(X_s)\, ds\right)\right]}, t \geq 0\right)$	Theorem 8
$(X_t, t \geq 0)$ solves an SDE, θ is positive and increasing	$\left(N_t := \frac{\theta(X_t)}{\mathbb{E}[\theta(X_t)]}, t \geq 0\right)$	Theorem 4
	$(C_t := \theta(X_t) - \mathbb{E}[\theta(X_t)], t \geq 0)$	Theorem 5
$(X_t, t \geq 0)$ is conditionally monotone and solves an SDE; ν is a positive Radon measure on $\mathbb{R}_+$	$\left(N_t := \frac{\int_0^t q(s, X_s)\nu(ds)}{\mathbb{E}\left[\int_0^t q(s, X_s)\nu(ds)\right]}, t \geq 0\right)$	Theorem 6
$(X_t, t \geq 0)$ is a centered very strong peacock	$\left(Q_t := \frac{1}{\alpha(t)} \int_0^t X_s\, d\alpha(s), t \geq 0\right)$	Theorem 11
$(L_t, t \geq 0)$ is a Lévy process such that the variable $\exp\left(\int_0^t L_u\, du\right)$ is integrable	$\left(N_t := \frac{\exp\left(\int_0^t L_s\, ds\right)}{\mathbb{E}\left[\exp\left(\int_0^t L_s\, ds\right)\right]}, t \geq 0\right)$	Example 7
	$\left(\widetilde{N}_t := \frac{\exp\left(\frac{1}{t}\int_0^t L_s\, ds\right)}{\mathbb{E}\left[\exp\left(\frac{1}{t}\int_0^t L_s\, ds\right)\right]}, t \geq 0\right)$	

In this table:

- $\alpha : \mathbb{R}_+ \to \mathbb{R}_+$ is a right-continuous and increasing function such that $\alpha(0) = 0$
- $q : \mathbb{R}_+ \times \mathbb{R} \to \mathbb{R}_+$ is a continuous and positive function such that, for every $s \geq 0$,

$x \mapsto q_s(x) := q(s, x)$ is increasing

$$\left(Q_t^{(\mu_n)}, 0 \leq t \leq T\right) \text{ is a peacock for every } n \geq 0, \tag{94}$$

i.e., for every $0 \leq s < t \leq T$, $\mathbb{E}\left[Q_s^{(\mu_n)}\right] = \mathbb{E}\left[Q_t^{(\mu_n)}\right]$ and, for every $\psi \in \mathbf{C}_+$:

$$\mathbb{E}\left[\psi(Q_s^{(\mu_n)})\right] = \mathbb{E}\left[\psi(Q_t^{(\mu_n)})\right]. \tag{95}$$

Moreover,

$$\sup_{0 \leq t \leq T} \sup_{n \geq 0} \left|Q_t^{(\mu_n)}\right| \leq \sup_{0 \leq \lambda \leq T} |X_\lambda| \tag{96}$$

which is integrable from (89).

Therefore, using (3), (93)–(95) and the dominated convergence Theorem, $\left(Q_t^{(\mu)},\right.$ $\left.0 \leq t \leq T\right)$ is a peacock for every $T > 0$. □

References

1. M.Y. An, *Log-concave probability distributions: Theory and statistical testing*. Papers 96-01. Centre for Labour Market and Social Research, Danmark (1996)
2. D. Baker, M. Yor, A Brownian sheet martingale with the same marginals as the arithmetic average of geometric Brownian motion. Electron. J. Probab. **14**(52), 1532–1540 (2009)
3. A.-M. Bogso, C. Profeta, B. Roynette, in *Some Examples of Peacocks in a Markovian Set-Up*, ed. by Donati-Martin. Séminaire de Probabiliés XLIV (Springer, Berlin, 2012)
4. S. Cambanis, G. Simons, W. Stout, Inequalities for $\mathbb{E}k(X, Y)$ when the marginals are fixed. Z. Wahrscheinlichkeitstheorie und Verw. Gebiete **36**, 285–294 (1976)
5. P. Carr, C.-O. Ewald, Y. Xiao, On the qualitative effect of volatility and duration on prices of Asian options. Finance Res. Lett. **5**(3), 162–171 (2008)
6. H. Daduna, R. Szekli, A queueing theoretical proof of increasing property of Pólya frequency functions. Stat. Probab. Lett. **26**(3), 233–242 (1996)
7. B. Efron, Increasing properties of Pólya frequency functions. Ann. Math. Stat. **36**, 272–279 (1965)
8. F. Hirsch, B. Roynette, M. Yor, Applying Itô's motto: "Look at the infinite dimensional picture" by constructing sheets to obtain processes increasing in the convex order. Period. Math. Hungar. **61**(1–2), 195–211 (2010)
9. F. Hirsch, B. Roynette, M. Yor, Unifying constructions of martingales associated with processes increasing in the convex order, via Lévy and Sato sheets. Expo. Math. **4**, 299–324 (2010)
10. F. Hirsch, B. Roynette, M. Yor, From an Itô type calculus for Gaussian processes to integrals 992 of log-normal processes increasing in the convex order. J. Mat. Soc. Jpn. **63**(3), 887–917 (2011)
11. F. Hirsch, C. Profeta, B. Roynette, M. Yor, Peacocks and Associated Martingales, with explicit constructions. Bocconi & Springer Series, vol. 3. Springer, Milan; Bocconie University Press, Milan, 2011. xxxii+384 pp
12. J. Jacod, P. Protter, Time reversal on Lévy processes. Ann. Probab. **16**(2), 620–641 (1988)
13. H.G. Kellerer, Markov-Komposition und eine Anwendung auf Martingale. Math. Ann. **198**, 99–122 (1972)

14. D. Revuz, M. Yor, in *Continuous Martingales and Brownian Motion*. Grundlehren der Mathematischen Wissenschaften, vol. 293, 3rd edn. (Springer, Berlin, 1999)
15. B. Roynette, P. Vallois, M. Yor, Limiting laws associated with Brownian motion perturbed by normalized exponential weights I. Studia Sci. Math. Hungar. **43**(2), 171–246 (2006)
16. I.J. Schoenberg, On Pólya frequency functions I. The totally positive functions and their Laplace transforms. J. Analyse Math. **1**, 331–374 (1951)
17. M. Shaked, J.G. Shanthikumar, in *Stochastic Orders and Their Applications*. Probability and Mathematical Statistics (Academic, Boston, 1994)
18. M. Shaked, J.G. Shanthikumar, in *Stochastic Orders*. Springer Series in Statistics (Springer, New York, 2007)
19. J.G. Shanthikumar, On stochastic comparison of random vectors. J. Appl. Probab. **24**(1), 123–136 (1987)

Branching Brownian Motion: Almost Sure Growth Along Scaled Paths

Simon C. Harris and Matthew I. Roberts

Abstract We give a proof of a result on the growth of the number of particles along chosen paths in a branching Brownian motion. The work follows the approach of classical large deviations results, in which paths of particles in $C[0,T]$, for large T, are rescaled onto $C[0,1]$. The methods used are probabilistic and take advantage of modern spine techniques.

1 Introduction and Statement of Result

1.1 Introduction

Fix a positive real number $r > 0$ and a random variable A taking values in $\{2, 3, \ldots\}$ such that $m := E[A] - 1 > 1$ and $E[A \log A] < \infty$. We consider a branching Brownian motion (BBM) under a probability measure $\mathbb{P}$, which is described as follows. We begin with one particle at the origin. Each particle u, once born, performs a Brownian motion independent of all other particles, until it dies, an event which occurs at an independent exponential time after its birth with mean $1/r$. At the time of a particle's death it is replaced (at its current position) by a random number A_u of offspring where A_u has the same distribution as A. Each of

S.C. Harris (✉)
Department of Mathematical Sciences, University of Bath, Claverton Down, Bath, Avon BA2 7AY, UK
e-mail: S.C.Harris@bath.ac.uk

M.I. Roberts
Laboratoire de Probabilités et Modèles Aléatoires, Université Paris VI, 4, Place Jussieu, 75005 Paris, France

C. Donati-Martin et al. (eds.), *Séminaire de Probabilités XLIV*, Lecture Notes in Mathematics 2046, DOI 10.1007/978-3-642-27461-9_17,

these particles, relative to its initial position, repeats (independently) the stochastic behaviour of its parent.

We let $N(t)$ be the set of particles alive at time t, and for $u \in N(t)$ and $s \leq t$ let $X_u(s)$ be the position of particle u (or its ancestor) at time s. Fix a set $D \subseteq C[0,1]$ and $\theta \in [0,1]$; then we are interested in the size of the sets

$$N_T(D,\theta) := \{u \in N(\theta T) : \exists f \in D \text{ with } X_u(t) = Tf(t/T) \;\; \forall t \in [0,\theta T]\}$$

for large T.

1.2 The Main Result

We define the class H_1 of functions by

$$H_1 := \left\{ f \in C[0,1] : \exists g \in L^2[0,1] \text{ with } f(s) = \int_0^s g(s)ds \;\; \forall s \in [0,1] \right\},$$

and to save on notation we set $f'(t) := \infty$ if $f \in C[0,1]$ is not differentiable at the point t. We then take integrals in the Lebesgue sense so that we may integrate functions that equal ∞ on sets of zero measure. We let

$$\theta_0(f) := \inf\left\{ \theta \in [0,1] : rm\theta - \frac{1}{2}\int_0^\theta f'(s)^2 ds < 0 \right\} \in [0,1] \cup \{\infty\}$$

(we think of θ_0 as the extinction time along f, the time at which the number of particles near f hits zero) and define our rate function K, for $f \in C[0,1]$ and $\theta \in [0,1]$, as

$$K(f,\theta) := \begin{cases} rm\theta - \frac{1}{2}\int_0^\theta f'(s)^2 ds & \text{if } f \in H_1 \text{ and } \theta \leq \theta_0(f) \\ -\infty & \text{otherwise.} \end{cases}$$

We expect approximately $\exp(K(f,\theta)T)$ particles whose paths up to time θT (when suitably rescaled) look like f. This is made precise in Theorem 1.

Theorem 1. *For any closed set $D \subseteq C[0,1]$ and $\theta \in [0,1]$,*

$$\limsup_{T\to\infty} \frac{1}{T} \log |N_T(D,\theta)| \leq \sup_{f\in D} K(f,\theta)$$

almost surely, and for any open set $U \subseteq C[0,1]$ and $\theta \in [0,1]$,

$$\liminf_{T\to\infty} \frac{1}{T} \log |N_T(U,\theta)| \geq \sup_{f\in U} K(f,\theta)$$

almost surely.

Sections 3 and 4 will be concerned with giving a proof of this theorem.

An almost identical result was stated by Git in [2]. We would like to give an alternative proof for two reasons.

Firstly, we believe that our proof of the lower bound is perhaps more intuitive, and certainly more robust, than that given in [2]. There are many more general setups for which our proofs will go through without too much extra work. One possibility is to allow particles to die without giving birth to any offspring (that is, to allow A to take the value 0): in this case the statement of the theorem would be conditional on the survival of the process, and we will draw attention to any areas where our proof must be adapted significantly to take account of this. There is work in progress on some further interesting cases and their applications, in particular the case where breeding occurs at the inhomogeneous rate rx^p, $p \in [0,2)$, for a particle at position x.

Secondly, there seems to be a slight oversight in the proof of Lemma 1 in [2], and that lemma is then used in obtaining both the upper and lower bounds. Although the gap seems minor at first, the complete lack of simple continuity properties of the processes involved means that almost all of the work involved in proving the upper bound is concerned with this matter. We give details of the oversight as an appendix.

Our tactic for the proof is to first work along lattice times, and then upgrade to the full result using Borel-Cantelli arguments. We begin, in Sect. 2, by introducing a family of martingales and changes of measure which will provide us with intuitive tools for our proofs. We then apply these tools to give an entirely new proof of the lower bound for Theorem 1 in Sect. 3. Finally, in Sect. 4, we take the same approach as in [2] to gain the upper bound along lattice times, and then rule out some technicalities in order to move to continuous time.

This work complements the article by Harris and Roberts [5]. Large deviation probabilities for the same model were given by Lee [6] and Hardy and Harris [3].

2 A Family of Spine Martingales

2.1 The Spine Setup

We will need to use some modern "spine" techniques as part of our proof. We only need some of the most basic spine tools, and we do not attempt to explain the details of these rigorously, but rather refer the interested reader to the article [4].

We first embellish our probability space by keeping track of some extra information about one particular infinite line of descent or *spine*. This line of descent is defined as follows: our original particle is part of the spine; when this particle dies, we choose one of its offspring uniformly at random to become part of the spine. We continue in this manner: when a spine particle dies, we choose uniformly at random between its offspring to decide which becomes part of the spine. In this way at any time $t \geq 0$ we have exactly one particle in $N(t)$ that is part of the spine. We refer to both this particle and its position with the label ξ_t; this is an abuse of notation, but it should always be clear from the context which meaning is intended. It is not hard to see that the spatial motion of the spine, $(\xi_t)_{t\geq 0}$, is a standard Brownian motion.

The resulting probability measure (on the set of *marked Galton-Watson trees with spines*) we denote by $\tilde{\mathbb{P}}$, and we find need for four different filtrations to encode differing amounts of this new information:

- $\mathscr{F}_t$ contains the all the information about the marked tree up to time t. However, it does not know which particle is the spine at any point. Thus it is simply the natural filtration of the original branching Brownian motion.
- $\tilde{\mathscr{F}}_t$ contains all the information about both the marked tree and the spine up to time t.
- $\tilde{\mathscr{G}}_t$ contains all the information about the spine up to time t, including the birth times of other particles along its path, and how many particles were born at each of these times; it does not know anything about the rest of the tree.
- $\mathscr{G}_t$ contains just the spatial information about the spine up to time t; it does not know anything about the rest of the tree.

We note that $\mathscr{F}_t \subseteq \tilde{\mathscr{F}}_t$ and $\mathscr{G}_t \subseteq \tilde{\mathscr{G}}_t \subseteq \tilde{\mathscr{F}}_t$, and also that $\tilde{\mathbb{P}}$ is an extension of $\mathbb{P}$ in that $\tilde{\mathbb{P}}|_{\mathscr{F}_\infty} = \mathbb{P}$. All of the above is covered more rigorously in [4].

Lemma 1 (Many-to-one lemma). *If $g(t)$ is $\mathscr{G}_t$-measurable and can be written*

$$g(t) = \sum_{u\in N(t)} g_u(t)\mathbb{1}_{\{\xi_t=u\}}$$

where each $g_u(t)$ is $\mathscr{F}_t$-measurable, then

$$\mathbb{E}\left[\sum_{u\in N(t)} g_u(t)\right] = e^{rmt}\tilde{\mathbb{E}}[g(t)].$$

This lemma is extremely useful as it allows us to reduce questions about the entire population down to calculations involving just one standard Brownian motion—the spine. A proof may be found in [4].

2.2 Martingales and Changes of Measure

For $f \in C[0,1]$ and $\theta \in [0,1]$ define

$$N_T(f,\varepsilon,\theta) := \{u \in N(\theta T) : |X_u(t) - Tf(t/T)| < \varepsilon T \quad \forall t \in [0,\theta T]\}$$

so that $N_T(f, \varepsilon, \theta) = N_T(B(f, \varepsilon), \theta)$. We look for martingales associated with these sets. For convenience, in this section we use the shorthand

$$N_T(t) := N_T(f, \varepsilon, t/T).$$

Since the motion of the spine is simply a standard Brownian motion under $\tilde{\mathbb{P}}$, if $f \in C^2[0, 1]$ then Itô's formula shows that for $t \in [0, T]$, the process

$$V_T(t) := e^{\pi^2 t/8\varepsilon^2 T^2} \cos\left(\frac{\pi}{2\varepsilon T}(\xi_t - Tf(t/T))\right) e^{\int_0^t f'(s/T)d\xi_s - \frac{1}{2}\int_0^t f'(s/T)^2 ds}$$

is a $\mathscr{G}_t$-martingale under $\tilde{\mathbb{P}}$. By stopping this process at the first exit time of the Brownian motion from the tube $\{(x, t) : |Tf(t/T) - x| < \varepsilon T\}$, we obtain also that

$$\zeta_T(t) := V_T(t)\mathbb{1}_{\{|Tf(s/T) - \xi_s| < \varepsilon T \ \forall s \leq t\}}$$

is a $\mathscr{G}_t$-martingale on $[0, T]$. As in [4], we may build from ζ_T a collection of $\tilde{\mathscr{F}}_t$-martingales $\tilde{\zeta}_T$ on $[0, T]$ given by

$$\tilde{\zeta}_T(t) := \prod_{v < \xi_t} A_v e^{-rmt} \zeta_T(t),$$

but these martingales will not be examined in this article—they are important only in changing measure below, and in that when we project $\tilde{\zeta}_T(t)$ back onto $\mathscr{F}_t$ we get a new set of mean-one $\mathscr{F}_t$-martingales Z_T. These processes Z_T are the main objects of interest in this section, and can be expressed for $t \in [0, T]$ as the sum

$$Z_T(t) = \sum_{u \in N_T(t)} V_T^{(u)}(t) e^{-rmt}$$

where

$$V_T^{(u)}(t) := e^{\pi^2 t/8\varepsilon^2 T^2} \cos\left(\frac{\pi}{2\varepsilon T}(X_u(t) - Tf(t/T))\right) e^{\int_0^t f'(s/T)dX_u(s) - \frac{1}{2}\int_0^t f'(s/T)^2 ds}.$$

We now define new measures, $\tilde{\mathbb{Q}}_T$, via

$$\tilde{\mathbb{Q}}_T|_{\tilde{\mathscr{F}}_t} = \tilde{\zeta}_T(t)\tilde{\mathbb{P}}|_{\tilde{\mathscr{F}}_t}$$

for $t \leq T$—and note that

$$\tilde{\mathbb{Q}}_T|_{\mathscr{F}_t} = Z_T(t)\tilde{\mathbb{P}}|_{\mathscr{F}_t} \quad \text{and} \quad \tilde{\mathbb{Q}}_T|_{\mathscr{G}_t} = \zeta_T(t)\tilde{\mathbb{P}}|_{\mathscr{G}_t}.$$

Lemma 2. *Under $\tilde{\mathbb{Q}}_T$, the spine ξ moves as a Brownian motion with drift*

$$f'(t/T) - \frac{\pi}{2\varepsilon T}\tan\left(\frac{\pi}{2\varepsilon T}(x - Tf(t/T))\right)$$

when at position x at time t; in particular,

$$|\xi_t - Tf(t/T)| \leq \varepsilon T \quad \forall t \leq T \quad \tilde{\mathbb{Q}}_T\text{-almost surely.}$$

Each particle u in the spine dies at an accelerated rate $(m+1)r$, *to be replaced by a random number* A_u *of offspring where* A_u *is taken from the* size-biased *distribution relative to A, given by* $\tilde{\mathbb{Q}}_T(A_u = k) = kP(A = k)(m+1)^{-1}$, $k = 0, 1, \ldots$ *(note that this distribution does not depend on T). All other particles, once born, behave* exactly as they would under $\mathbb{P}$*: they move like independent standard Brownian motions, die at the usual rate r, and give birth to a number of particles that is distributed like A.*

Proof. Most of this is standard in the spine literature; for example proof can be found in [4]. We will not use the precise drift of the spine except for the fact that the spine remains within the tube: to see this note that since the event is $\tilde{\mathscr{G}}_T$-measurable,

$$\tilde{\mathbb{Q}}_T(\exists t \leq T : |\xi_t - Tf(t/T)| > \varepsilon T) = \tilde{\mathbb{E}}[\zeta_T(T)\mathbb{1}_{\{\exists t \leq T : |\xi_t - Tf(t/T)| > \varepsilon T\}}] = 0$$

by the definition of $\zeta_T(T)$. □

Another important tool in this section is the *spine decomposition.*

Lemma 3 (Spine decomposition). $\tilde{\mathbb{Q}}_T$*-almost surely,*

$$\tilde{\mathbb{Q}}_T[Z_T(t)|\tilde{\mathscr{G}}_T] = \sum_{u<\xi_t} (A_u - 1)V_T(S_u)e^{-rmS_u} + V_T(t)e^{-rmt}$$

where we recall that $\{u < \xi_t\}$ *is the set of ancestors of the spine particle at time t, and* S_u *denotes the time at which particle u split into two new particles.*

A proof of the spine decomposition may be found in [4].

Lemma 4. *If* $f \in C^2[0,1]$ *then for any* $u \in N_T(t)$*, almost surely under both* $\tilde{\mathbb{P}}$ *and* $\tilde{\mathbb{Q}}_T$ *we have*

$$\left|\int_0^t f'(s/T)dX_u(s) - \int_0^t f'(s/T)^2 ds\right| \leq 2\varepsilon T \int_0^{t/T} |f''(s)|ds + \varepsilon T|f'(0)|.$$

Proof. From the integration by parts formula for Itô calculus (since for any particle $u \in N(t)$, $(X_u(s), 0 \leq s \leq t)$ is a Brownian motion under $\tilde{\mathbb{P}}$) we know that for any $g \in C^2[0,1]$ with $g(0) = 0$, under $\tilde{\mathbb{P}}$,

$$g'(t)X_u(t) = \int_0^t g''(s)X_u(s)ds + \int_0^t g'(s)dX_u(s).$$

From ordinary integration by parts,

$$\int_0^t g'(s)^2 ds = g'(t)g(t) - \int_0^t g(s)g''(s)ds.$$

Now set $g(t) = Tf(t/T)$ for $t \in [0, T]$. We note that, if $u \in N_T(t)$ then $|X_u(s) - g(s)| < \varepsilon T$ for all $s \leq t$. Thus

$$\begin{aligned}
&\left| \int_0^t f'(s/T) dX_u(s) - \int_0^t f'(s/T)^2 ds \right| \\
&= \left| \int_0^t g'(s) dX_u(s) - \int_0^t g'(s)^2 ds \right| \\
&\leq \left| g'(t)(X_u(t) - g(t)) - \int_0^t g''(s)(X_u(s) - g(s)) ds \right| \\
&\leq |g'(t) - g'(0)|\varepsilon T + |g'(0)|\varepsilon T - \int_0^t |g''(s)|\varepsilon T ds \\
&\leq 2\varepsilon T \int_0^t |g''(s)| ds + \varepsilon T |g'(0)| \\
&= 2\varepsilon T \int_0^{t/T} |f''(s)| ds + \varepsilon T |f'(0)|
\end{aligned}$$

almost surely under $\tilde{\mathbb{P}}$ and, since $\tilde{\mathbb{Q}}_T \ll \tilde{\mathbb{P}}$, almost surely under $\tilde{\mathbb{Q}}_T$. □

We now use this result to give approximations on $Z_T(t)$ under certain conditions. One of these conditions involves the seemingly unnatural assumption $f'(0) = 0$. This is caused by the fact that in this section we make no approximations to the path of the spine under $\tilde{\mathbb{Q}}_T$ except for using that it always remains within εT of our T-rescaled path—hence we are left with a rather bad estimate on its path at small times, where it will not get anywhere near εT. This does not matter to us, however, precisely because of this freedom to move within the ε-tube about f: if $f'(0) \neq 0$ then we may choose g near to f (in an appropriate way; certainly within the ε-tube) such that $g'(0) = 0$. This issue arises in Lemma 8 and rigorous details are given there.

Lemma 5. *If $f \in C^2[0,1]$, $f'(0) = 0$ and $rm\phi > \frac{1}{2}\int_0^\phi f'(s)^2 ds$ for all $\phi \in (0, \theta]$, then for small enough $\varepsilon > 0$ and any $T > 0$ and $t \leq \theta T$, there exists $\eta > 0$ such that*

$$\tilde{\mathbb{Q}}_T[Z_T(t)|\tilde{\mathscr{G}}_T] \leq \sum_{u<\xi_T} (A_u - 1)e^{\pi^2/8\varepsilon^2 T - \eta S_u} + e^{\pi^2/8\varepsilon^2 T - \eta t}$$

$\tilde{\mathbb{Q}}_T$-almost surely.

Proof. Since $rm\phi > \frac{1}{2}\int_0^\phi f'(s)^2 ds$ for all $\phi \in (0, \theta]$ and $f'(0) = 0$, we may choose $\eta > 0$ such that

$$2\eta\phi \le rm\phi - \frac{1}{2}\int_0^\phi f'(s)^2 ds \quad \forall \phi \in [0,\theta].$$

Then for any $\varepsilon > 0$ satisfying

$$2\varepsilon \int_0^\phi |f''(s)| ds \le \eta\phi \quad \forall \phi \in [0,\theta]$$

we have, by Lemma 4 (since $f'(0) = 0$ and using the fact that under $\tilde{\mathbb{Q}}_T$ the spine is always in $N_T(t)$),

$$V_T(t)e^{-rmt} \le e^{\pi^2/8\varepsilon^2 T - rmt + \frac{T}{2}\int_0^{t/T} f'(s)^2 ds + 2\varepsilon T \int_0^{t/T} |f''(s)| ds} \le e^{\pi^2/8\varepsilon^2 T - \eta t}$$

for all $t \in [0, \theta T]$. Plugging this into the spine decomposition, we get

$$\tilde{\mathbb{Q}}_T[Z_T(t)|\tilde{\mathscr{G}}_T] \le \sum_{u<\xi_T} (A_u - 1)e^{\pi^2/8\varepsilon^2 T - \eta S_u} + e^{\pi^2/8\varepsilon^2 T - \eta t}. \qquad \square$$

Proposition 1. *If $f \in C^2[0,1]$, $f'(0) = 0$ and $rm\phi > \frac{1}{2}\int_0^\phi f'(s)^2 ds$ for all $\phi \in (0,\theta]$, then for small enough $\varepsilon > 0$ the set $\{Z_T(t) : T \ge 1, t \le \theta T\}$ is uniformly integrable under $\mathbb{P}$.*

Proof. Fix $\delta > 0$. We first claim that there exists K such that

$$\sup_{T\ge 1,\, t\le\theta T} \tilde{\mathbb{Q}}_T(\tilde{\mathbb{Q}}_T[Z_T(t)|\tilde{\mathscr{G}}_T] > K) < \delta/2.$$

To see this, take an auxiliary probability space with probability measure Q, and on this space consider a sequence $A_1, A_2, \ldots$ of independent and identically distributed random variables satisfying

$$Q(A_i = k) = \frac{k\mathbb{P}(A = k)}{m+1}$$

so that the A_i have the same distribution as births A_u along the spine under $\tilde{\mathbb{Q}}_T$ (recall that there is no dependence on T). Take also a sequence $e_1, e_2, \ldots$ of independent random variables that are exponentially distributed with parameter $r(m+1)$; then set $S_n = e_1 + \ldots + e_n$ (so that the random variables S_n have the same distribution as the birth times along the spine under $\tilde{\mathbb{Q}}_T$). By Lemma 5 we have

$$\sup_{\substack{T\ge 1\\ t\le\theta T}} \tilde{\mathbb{Q}}_T(\tilde{\mathbb{Q}}_T[Z_T(t)|\tilde{\mathscr{G}}_T] > K) \le Q\left(\sum_{j=1}^\infty (A_j - 1)e^{\pi^2/8\varepsilon^2 - \eta S_j} + e^{\pi^2/8\varepsilon^2} > K\right).$$

Hence our claim holds if the random variable $\sum_{j=1}^{\infty}(A_j - 1)e^{-\eta S_j}$ can be shown to be Q-almost surely finite. Now for any $\gamma \in (0,1)$,

$$Q(\sum_n (A_n - 1)e^{-\eta S_n} = \infty) \le Q(A_n e^{-\eta S_n} > \gamma^n \text{ infinitely often})$$
$$\le Q\left(\frac{\log A_n}{n} > \log\gamma + \frac{\eta S_n}{n} \text{ infinitely often}\right).$$

By the strong law of large numbers, $S_n/n \to 1/r(m+1)$ almost surely under Q; so if $\gamma \in (\exp(-\eta/r(m+1)), 1)$ then the quantity above is no larger than

$$Q\left(\limsup_{n\to\infty}\frac{\log A_n}{n} > 0\right).$$

But this quantity is zero by Borel-Cantelli: indeed, for any T,

$$\sum_n Q\left(\frac{\log A_n}{n} > \varepsilon\right) = \sum_n Q(\log A_1 > \varepsilon n)$$
$$\le \int_0^\infty Q(\log A_1 \ge \varepsilon x)dx = Q\left[\frac{\log A_1}{\varepsilon}\right]$$

which is finite for any $\varepsilon > 0$ since (by direct calculation from the distribution of A_1 under Q) $Q[\log A_1] = \tilde{\mathbb{P}}[A\log A] < \infty$ (this was one of our assumptions at the beginning of the article). Thus our claim holds.

Now choose $M > 0$ such that $1/M < \delta/2$; then for K chosen as above, and any $T \ge 1, t \le \theta T$,

$$\begin{aligned}\tilde{\mathbb{Q}}_T(Z_T(t) > MK) &\le \tilde{\mathbb{Q}}_T(Z_T(t) > MK,\ \tilde{\mathbb{Q}}_T[Z_T(t)|\tilde{\mathscr{G}}_T] \le K) \\ &\qquad + \tilde{\mathbb{Q}}_T(\tilde{\mathbb{Q}}_T[Z_T(t)|\tilde{\mathscr{G}}_T] > K) \\ &\le \tilde{\mathbb{Q}}_T\left[\frac{Z_T(t)}{MK}\mathbb{1}_{\{\tilde{\mathbb{Q}}_T[Z_T(t)|\tilde{\mathscr{G}}_T]\le K\}}\right] + \delta/2 \\ &= \tilde{\mathbb{Q}}_T\left[\frac{\tilde{\mathbb{Q}}_T[Z_T(t)|\tilde{\mathscr{G}}_T]}{MK}\mathbb{1}_{\{\tilde{\mathbb{Q}}_T[Z_T(t)|\tilde{\mathscr{G}}_T]\le K\}}\right] + \delta/2 \\ &\le 1/M + \delta/2 \le \delta.\end{aligned}$$

Thus, setting $K' = MK$, for any $T \ge 1, t \le \theta T$,

$$\mathbb{P}[Z_T(t)\mathbb{1}_{\{Z_T(t)>K'\}}] = \tilde{\mathbb{Q}}_T(Z_T(t) > K') \le \delta.$$

Since $\delta > 0$ was arbitrary, the proof is complete. □

As our final result in this section we link explicitly the martingales Z_T with the number of particles N_T.

Lemma 6. *For any $\delta > 0$, if $f \in C^2[0,1]$, $f(0) = 0$ and ε is small enough then*

$$Z_T(\theta T) \leq |N_T(f,\varepsilon,\theta)| \exp\left(\frac{\pi^2\theta}{8\varepsilon^2 T} - rm\theta T + \frac{T}{2}\int_0^\theta f'(s)^2 ds + \delta T\right).$$

Proof. Simply plugging the result of Lemma 4 into the definition of $Z_T(\theta T)$ gives the desired inequality. □

We note here that, in fact, a similar bound can be given in the opposite direction, so that $N_T(f,\varepsilon/2,\theta)$ is dominated by $Z_T(\theta T)$ multiplied by some deterministic function of T. We will not need this bound, but it is interesting to note that the study of the martingales Z_T is in a sense equivalent to the study of the number of particles N_T.

3 The Lower Bound

3.1 The Heuristic for the Lower Bound

We want to show that $N_T(f,\varepsilon,\theta)$ cannot be too small for large T. For $f \in C[0,1]$ and $\theta \in [0,1]$, define

$$J(f,\theta) := \begin{cases} rm\theta - \frac{1}{2}\int_0^\theta f'(s)^2 ds & \text{if } f \in H_1 \\ -\infty & \text{otherwise.} \end{cases}$$

We note that J resembles our rate function K, but without the truncation at the extinction time θ_0. We shall work mostly with the simpler object J, before deducing our result involving K at the very last step. We now give a short heuristic to describe our route through the proof of the lower bound.

Step 1. Consider a small (relative to T) time ηT. How many particles are in $N_T(f,\varepsilon,\eta)$? If η is much smaller than ε, then (with high probability) no particle has had enough time to reach anywhere near the edge of the tube (approximately distance εT from the origin) before time ηT. Thus, with high probability,

$$|N_T(f,\varepsilon,\eta)| = |N(\eta T)| \approx \exp(rm\eta T).$$

Step 2. Given their positions at time ηT, the particles in $N_T(f,\varepsilon,\eta)$ act independently. Each particle u in this set thus draws out an independent branching Brownian motion. Let $N_T(u,f,\varepsilon,\theta)$ be the set of descendants of u that are in $N_T(f,\varepsilon,\theta)$. How big is this set? Since η is very small, each particle u is close to the origin. Thus we may hope to find some $q < 1$ such that

$$\mathbb{P}\left(|N_T(u, f, \varepsilon, \theta)| < \exp(J(f, \theta)T - \delta T)\right) \leq q.$$

(Of course, in reality we believe that this quantity will be exponentially small—but to begin with, the constant bound can be shown more readily.)

Step 3. If $N_T(f, \varepsilon, \theta)$ is to be small, then each of the sets $N_T(u, f, \varepsilon, \theta)$ for $u \in N_T(f, \varepsilon, \eta)$ must be small. Thus

$$\mathbb{P}\left(|N_T(f, \varepsilon, \theta)| < \exp(J(f, \theta)T - \delta T)\right) \lesssim q^{\exp(rm\eta T)},$$

and we may apply Borel-Cantelli to deduce our result along lattice times (that is, times T_j, $j \geq 0$ such that there exists $\tau > 0$ with $T_j - T_{j-1} = \tau$ for all $j \geq 1$).

Step 4. We carry out a simple tube-reduction argument to move to continuous time. The idea here is that if the result were true on lattice times but not in continuous time, the number of particles in $N_T(f, \varepsilon, \theta)$ must fall dramatically at infinitely many non-lattice times. We simply rule out this possibility using standard properties of Brownian motion.

The most difficult part of the proof is Step 2. However, the spine results of Sect. 2 will simplify our task significantly.

3.2 The Proof of the Lower Bound

We begin with Step 1 of our heuristic, considering the size of $N_T(f, \varepsilon, \eta)$ for small η.

Lemma 7. *For any continuous f with $f(0) = 0$ and any $\varepsilon > 0$, there exist $\eta > 0$, $k > 0$ and T_1 such that*

$$\mathbb{P}(\exists u \in N(\eta T) : u \notin N_T(f, \varepsilon/2, \eta)) \leq e^{-kT} \quad \forall T \geq T_1.$$

Proof. Choose η small enough that $\sup_{s\in[0,\eta]} |f(s)| < \varepsilon/4$. Then, using the many-to-one lemma and standard properties of Brownian motion,

$$\begin{aligned}
&\mathbb{P}(\exists u \in N(\eta T) : u \notin N_T(f, \varepsilon/2, \eta)) \\
&= \mathbb{P}\left(\exists u \in N(\eta T) : \sup_{s\leq\eta} |X_u(sT) - Tf(s)| \geq \varepsilon T/2\right) \\
&\leq \mathbb{P}\left[\sum_{u\in N(\eta T)} \mathbb{1}_{\{\sup_{s\leq\eta} |X_u(sT) - Tf(s)| \geq \varepsilon T/2\}}\right] \\
&\leq e^{rm\eta T} \tilde{\mathbb{P}}\left(\sup_{s\leq\eta} |\xi_{sT} - Tf(s)| \geq \varepsilon T/2\right)
\end{aligned}$$

$$\leq e^{rm\eta T}\tilde{\mathbb{P}}\left(\sup_{s\leq\eta}|\xi_{sT}|\geq \varepsilon T/4\right)$$

$$\leq \frac{16\sqrt{\eta}e^{rm\eta T-\varepsilon^2 T/32\eta}}{\varepsilon\sqrt{2\pi T}}.$$

A suitably small choice of η gives the exponential decay required. □

We now move on to Step 2, using the results of Sect. 2 to bound the probability of having a small number of particles strictly below 1. The bound given is extremely crude, and there is much room for manoeuvre in the proof, but any improvement would only add unnecessary detail.

Lemma 8. *If* $f \in C^2[0,1]$ *and* $J(f,s) > 0\ \forall s \in (0,\theta]$, *then for any* $\varepsilon > 0$ *and* $\delta > 0$ *there exists* $T_0 \geq 0$ *and* $q < 1$ *such that*

$$\mathbb{P}\left(|N_T(f,\varepsilon,\theta)| < e^{J(f,\theta)T-\delta T}\right) \leq q \quad \forall T \geq T_0.$$

Proof. Note that by Lemma 6 for small enough $\varepsilon > 0$ and large enough T,

$$|N_T(f,\varepsilon,\theta)|e^{-J(f,\theta)T+\delta T/2} \geq Z_T(\theta T)$$

and hence

$$\mathbb{P}\left(|N_T(f,\varepsilon,\theta)| < e^{J(f,\theta)T-\delta T}\right) \leq \mathbb{P}\left(Z_T(\theta T) < e^{-\delta T/2}\right).$$

Suppose first that $f'(0) = 0$. Then, again for small enough ε, by Proposition 1 the set $\{Z_T(\theta T), T \geq 1, t \in [1,\theta T]\}$ is uniformly integrable. Thus we may choose K such that

$$\sup_{T\geq 1}\mathbb{E}[Z_T(\theta T)\mathbb{1}_{\{Z_T(\theta T)>K\}}] \leq 1/4,$$

and then

$$\begin{aligned}1 = \mathbb{E}[Z_T(\theta T)] &= \mathbb{E}[Z_T(\theta T)\mathbb{1}_{\{Z_T(\theta T)\leq 1/2\}}] + \mathbb{E}[Z_T(\theta T)\mathbb{1}_{\{1/2<Z_T(\theta T)\leq K\}}]\\ &\quad + \mathbb{E}[Z_T(\theta T)\mathbb{1}_{\{Z_T(\theta T)>K\}}]\\ &\leq 1/2 + K\mathbb{P}(Z_T(\theta T) > 1/2) + 1/4\end{aligned}$$

so that

$$\mathbb{P}(Z_T(\theta T) > 1/2) \geq 1/4K.$$

Hence for large enough T,

$$\mathbb{P}\left(|N_T(f,\varepsilon,\theta)| < e^{J(f,\theta)T-\delta T}\right) \leq 1 - 1/4K.$$

This is true for all small $\varepsilon > 0$; but increasing ε only increases $|N_T(f, \varepsilon, \theta)|$ so the statement holds for all $\varepsilon > 0$. Finally, if $f'(0) \neq 0$ then choose $g \in C^2[0,1]$ such that $g(0) = g'(0) = 0$, $\sup_{s\leq\theta} |f - g| \leq \varepsilon/2$, $J(g, \phi) > 0$ for all $\phi \leq \theta$ and $J(g, \theta) > J(f, \theta) - \delta/2$ (for small η, the function

$$g(t) := \begin{cases} f(t) + at + bt^2 + ct^3 + dt^4 & \text{if } t \in [0, \eta) \\ f(t) & \text{if } t \in [\eta, 1] \end{cases},$$

with $a = -f'(0)$, $b = 3f'(0)/\eta$, $c = -3f'(0)/\eta^2$ and $d = f'(0)/\eta^3$, will work). Then as above we may choose K such that

$$\mathbb{P}(|N_T(f, \varepsilon, \theta)| < e^{J(f,\theta)T - \delta T}) \leq \mathbb{P}(|N_T(g, \varepsilon/2, \theta)| < e^{J(g,\theta)T - \delta T/2}) \leq 1 - 1/4K$$

as required. □

Our next result runs along integer times—these times are sufficient for our needs, although the following proof would in fact work for any lattice times.

Proposition 2. *Suppose that $f \in C^2[0,1]$ and $J(f, s) > 0\ \forall s \in (0, \theta]$. Then*

$$\liminf_{\substack{j\to\infty \\ j\in\mathbb{N}}} \frac{1}{j} \log |N_j(f, \varepsilon, \theta)| \geq J(f, \theta)$$

almost surely.

Proof. For any particle u, define

$$\begin{aligned} N_T(u, f, \varepsilon, \theta) &:= \{v \in N(\theta T) : u \leq v,\ \ |X_v(t) - Tf(t/T)| < \varepsilon T\ \ \forall t \in [0, \theta T]\} \\ &= \{v : u \leq v\} \cap N_T(f, \varepsilon, \theta), \end{aligned}$$

the set of descendants of u that are in $N_T(f, \varepsilon, \theta)$. Then for $\delta > 0$ and $\eta \in [0, \theta]$,

$$\begin{aligned} &\mathbb{P}\left(|N_T(f, \varepsilon, \theta)| < e^{J(f,\theta)T - \delta T} \,\middle|\, \mathscr{F}_{\eta T}\right) \\ &\leq \prod_{u \in N_T(f, \varepsilon/2, \eta)} \mathbb{P}\left(|N_T(u, f, \varepsilon, \theta)| < e^{J(f,\theta)T - \delta T} \,\middle|\, \mathscr{F}_{\eta T}\right) \\ &\leq \prod_{u \in N_T(f, \varepsilon/2, \eta)} \mathbb{P}\left(|N_T(g, \varepsilon/2, \theta - \eta)| < e^{J(f,\theta)T - \delta T}\right) \end{aligned}$$

since $\{|N_T(u, f, \varepsilon, \theta)| : u \in N_T(f, \varepsilon/2, \eta)\}$ are independent random variables, and where $g : [0,1] \to \mathbb{R}$ is any twice continuously differentiable extension of the function

$$\begin{aligned} \bar{g} : [0, \theta - \eta] &\to \mathbb{R} \\ t &\to f(t + \eta) - f(\eta). \end{aligned}$$

If η is small enough, then

$$|J(f,\theta) - J(g,\theta-\eta)| < \delta/2$$

and

$$J(g,s) > 0 \quad \forall s \in (0,\theta-\eta].$$

Hence, applying Lemma 8, there exists $q < 1$ such that for all large T,

$$\mathbb{P}\left(|N_T(g,\varepsilon/2,\theta-\eta)| < e^{J(f,\theta)T-\delta T}\right) \\ \leq \mathbb{P}\left(|N_T(g,\varepsilon/2,\theta-\eta)| < e^{J(g,\theta-\eta)T-\delta T/2}\right) \leq q.$$

Thus for large T,

$$\mathbb{P}\left(|N_T(f,\varepsilon,\theta)| < e^{J(f,\theta)T-\delta T} \,\middle|\, \mathscr{F}_{\eta T}\right) \leq q^{|N_T(f,\varepsilon/2,\eta)|}. \tag{1}$$

Now, recalling that $N(t)$ is the *total* number of particles alive at time t, it is well-known (and easy to calculate) that for $\alpha \in (0,1)$,

$$\mathbb{E}\left[\alpha^{|N(t)|}\right] \leq \frac{\alpha}{\alpha + (1-\alpha)e^{rt}}$$

(in fact this is exactly $\mathbb{E}[\alpha^{|N(t)|}]$ in the case of strictly dyadic branching). Taking expectations in (1), and then applying Lemma 7, for small η we can get

$$\begin{aligned}
&\mathbb{P}\left(|N_T(f,\varepsilon,\theta)| < e^{J(f,\theta)T-\delta T}\right) \\
&\leq \mathbb{P}\left(\exists u \in N(\eta T) : u \notin N_T(f,\varepsilon/2,\eta)\right) + \mathbb{E}\left[q^{|N(\eta T)|}\right] \\
&\leq e^{-kT} + \frac{q}{q + (1-q)e^{r\eta T}}
\end{aligned}$$

for some $k > 0$ and all large enough T. The Borel-Cantelli lemma now tells us that

$$\mathbb{P}\left(\liminf_{j\to\infty} \frac{1}{j} \log |N_j(f,\varepsilon,\theta)| < J(f,\theta) - \delta\right) = 0,$$

and taking a union over $\delta > 0$ gives the result. □

We note that our estimate on $\mathbb{E}[\alpha^{|N(t)|}]$ may not hold if we allowed the possibility of death with no offspring. In this case a more sophisticated estimate is required, taking into account the probability that the process becomes extinct.

We look now at moving to continuous time using Step 4 of our heuristic. For simplicity of notation, we break with convention by defining

$$\|f\|_\theta := \sup_{s\in[0,\theta]} |f(s)|$$

for $f \in C[0,\theta]$ or $f \in C[0,1]$ (on this latter space, $\|\cdot\|_\theta$ is not a norm, but this will not matter to us).

Proposition 3. *Suppose that $f \in C^2[0,1]$ and $J(f,s) > 0\ \forall s \in (0,\theta]$. Then*

$$\liminf_{T\to\infty} \frac{1}{T} \log |N_T(f,\varepsilon,\theta)| \geq J(f,\theta)$$

almost surely.

Proof. We claim first that for large enough $j \in \mathbb{N}$,

$$\left\{|N_j(f,\varepsilon,\theta)| > \inf_{t\in[j,j+1]} |N_t(f,2\varepsilon,\theta)|\right\}$$
$$\subseteq \left\{\exists u \in N(\theta(j+1)) : \sup_{t\in[j,j+1]} |X_u(t) - X_u(j)| > \frac{\varepsilon j}{2}\right\}.$$

Indeed, if $v \in N_j(f,\varepsilon,\theta)$, $t \in [j, j+1]$ and $s \in [0,\theta t]$ then for any descendant u of v at time θt,

$$\begin{aligned}
|X_u(s) - tf(s/t)| &\leq |X_u(s) - X_u(s\wedge\theta j)| + |X_u(s\wedge\theta j) - jf((s\wedge\theta j)/j)| \\
&\quad + |jf((s\wedge\theta j)/j) - jf(s/t)| + |jf(s/t) - tf(s/t)| \\
&\leq |X_u(s) - X_u(s\wedge\theta j)| + \varepsilon j \\
&\quad + j \sup_{\substack{x,y\in[0,\theta] \\ |x-y|\leq 1/j}} |f(x) - f(y)| + \|f\|_\theta \\
&\leq |X_u(s) - X_u(s\wedge\theta j)| + \frac{3\varepsilon}{2} j \qquad \text{for large j;}
\end{aligned}$$

so that if any particle is in $N_j(f,\varepsilon,\theta)$ but not in $N_t(f,2\varepsilon,\theta)$ then it must satisfy

$$\sup_{j\leq s\leq t} |X_u(s) - X_u(j)| \geq \varepsilon j/2.$$

This is enough to establish the claim, and we deduce via the many-to-one lemma and standard properties of Brownian motion that

$$\begin{aligned}
&\mathbb{P}(|N_j(f,\varepsilon,\theta)| > \inf_{t\in[j,j+1]} |N_t(f,2\varepsilon,\theta)|) \\
&\leq \mathbb{P}\left(\exists u \in N(\theta(j+1)) : \sup_{t\in[j,j+1]} |X_u(t) - X_u(j)| \geq \varepsilon j/2\right) \\
&= e^{rm\theta(j+1)} \tilde{\mathbb{P}}(\sup_{t\in[j,j+1]} |\xi_t - \xi_j| \geq \varepsilon j/2)
\end{aligned}$$

$$\leq \frac{8}{\varepsilon j \sqrt{2\pi}} \exp(rm\theta(j+1) - \varepsilon^2 j^2/8).$$

Since these probabilities are summable we may apply Borel-Cantelli to see that

$$\mathbb{P}(|N_j(f,\varepsilon,\theta)| > \inf_{t\in[j,j+1]} |N_t(f,2\varepsilon,\theta)| \text{ infinitely often}) = 0.$$

Now,

$$\mathbb{P}\left(\liminf_{T\to\infty} \frac{1}{T} \log |N_T(f,\varepsilon,\theta)| < J(f,\theta)\right)$$
$$\leq \mathbb{P}\left(\liminf_{j\to\infty} \frac{1}{j} \log |N_j(f,2\varepsilon,\theta)| < J(f,\theta)\right)$$
$$+ \mathbb{P}\left(\liminf_{j\to\infty} \frac{\inf_{t\in[j,j+1]} |N_t(f,\varepsilon,\theta)|}{|N_j(f,2\varepsilon,\theta)|} < 1\right)$$

which is zero by Proposition 2 and Borel-Cantelli. □

If we were including the possibility of death with no offspring then we would have to check that no particles in $N_j(f,\varepsilon,\theta)$ managed to reach the outside of the slightly altered 2ε-tube and then die before time $j+1$. The only added difficulty would be in keeping track of notation.

We are now in a position to give our lower bound in full.

Corollary 1. *For any open set $U \subseteq C[0,1]$ and $\theta \in [0,1]$, we have*

$$\liminf_{T\to\infty} \frac{1}{T} \log |N_T(U,\theta)| \geq \sup_{f\in U} K(f,\theta)$$

almost surely.

Proof. If $\sup_{f\in U} K(f,\theta) = -\infty$ then there is nothing to prove. Thus it suffices to consider the case when there exists $f \in U$ such that $\theta \leq \theta_0(f)$. Since U is open, in this case we can in fact find $f \in U$ such that $J(f,s) > 0$ for all $s \in (0,\theta]$ (if $J(f,\phi) = 0$ for some $\phi \leq \theta$, just choose η small enough that $(1-\eta)f \in U$) and such that f is twice continuously differentiable on $[0,1]$ (twice continuously differentiable functions are dense in $C[0,1]$). Thus necessarily $\sup_{g\in U} K(g,\theta) > 0$, and for any $\delta > 0$ we may further assume (by a simple argument, for example by approximating with piecewise linear functions and then smoothing) that $J(f,\theta) > \sup_{g\in U} K(g,\theta) - \delta$. Again since U is open, we may take ε such that $B(f,\varepsilon) \subseteq U$; then clearly for any T

$$N_T(f,\varepsilon,\theta) \subseteq N_T(U,\theta)$$

so by Proposition 2 we have

$$\liminf_{T\to\infty} \frac{1}{T} \log N_T(U,\theta) \geq \sup_{g\in U} K(g,\theta) - \delta$$

almost surely, and by taking a union over $\delta > 0$ we may deduce the result. □

4 The Upper Bound

Our plan is as follows: we first carry out the simple task of obtaining a bound along lattice times (Proposition 4). We then move to continuous time in Lemma 9, at the cost of restricting to open balls about fixed paths, by a tube-expansion argument similar to the tube-reduction argument used in Proposition 3 of the lower bound. In Lemma 10 we then rule out the possibility of any particles following unusual paths, which allows us to restrict our attention to a compact set, and hence a finite number of small open balls about sensible paths. Finally we draw this work together in Proposition 5 to give the bound in continuous time for any closed set D.

Our first task, then, is to establish an upper bound along integer times. As with the lower bound, these times are sufficient for our needs, although the following proof would work for any lattice times. In a slight abuse of notation, for $D \subseteq C[0,1]$ and $\theta \in [0,1]$ we define

$$J(D,\theta) := \sup_{f\in D} J(f,\theta).$$

Proposition 4. *For any closed set $D \subseteq C[0,1]$ and $\theta \in [0,1]$ we have*

$$\limsup_{\substack{j\to\infty \\ j\in\mathbb{N}}} \frac{1}{j} \log |N_j(D,\theta)| \leq J(D,\theta)$$

almost surely.

Proof. From the upper bound for Schilder's theorem (Theorem 5.1 of [7]) we have

$$\limsup_{T\to\infty} \frac{1}{T} \log \tilde{\mathbb{P}}(\xi_T \in N_T(D,\theta)) \leq -\inf_{f\in D} \frac{1}{2}\int_0^\theta f'(s)^2 ds.$$

Thus, by the many-to-one lemma,

$$\begin{aligned}\limsup_{T\to\infty} \frac{1}{T} \log \mathbb{E}\big[|N_T(D,\theta)|\big] &\leq \limsup_{T\to\infty} \frac{1}{T} \log\left(e^{rm\theta T}\tilde{\mathbb{P}}(\xi_T \in N_T(D,\theta))\right) \\ &\leq rm\theta - \inf_{f\in D} \frac{1}{2}\int_0^\theta f'(s)^2 ds \\ &= J(D,\theta).\end{aligned}$$

Applying Markov's inequality, for any $\delta > 0$ we get

$$\limsup_{T\to\infty} \frac{1}{T} \log \mathbb{P}\big(|N_T(D,\theta)| \geq e^{J(D,\theta)T+\delta T}\big) \leq \limsup_{T\to\infty} \frac{1}{T} \log \frac{\mathbb{E}\big[|N_T(D,\theta)|\big]}{e^{J(D,\theta)T+\delta T}} \leq -\delta$$

so that

$$\sum_{j=1}^{\infty} \mathbb{P}\big(|N_j(D,\theta)| \geq e^{J(D,\theta)j+\delta j}\big) < \infty$$

and hence by the Borel-Cantelli lemma

$$\mathbb{P}\left(\limsup_{j\to\infty} \frac{1}{j} \log |N_j(D,\theta)| \geq J(D,\theta) + \delta\right) = 0.$$

Taking a union over $\delta > 0$ now gives the result. □

We note that the proof by Git [2] works up to this point; the rest of the proof of the upper bound will be concerned with plugging the gap in [2].

For $D \subset C[0,1]$ and $\varepsilon > 0$, let

$$D^{\varepsilon} := \{f \in C[0,1] : \inf_{g\in D} \|f - g\| \leq \varepsilon\}.$$

Recall that we defined $N_T(f,\varepsilon,\theta) := N_T(B(f,\varepsilon),\theta)$.

Lemma 9. *If $D \subseteq C[0,1]$ and $f \in D$, then*

$$\limsup_{T\to\infty} \frac{1}{T} \log |N_T(f,\varepsilon,\theta)| \leq J(D^{2\varepsilon},\theta)$$

almost surely.

Proof. First note that

$$\begin{aligned}
&\mathbb{P}\left(\limsup_{T\to\infty} \frac{1}{T} \log |N_T(f,\varepsilon,\theta)| > J(D^{2\varepsilon},\theta) + \delta\right)\\
&\qquad \leq \mathbb{P}\left(\limsup_{j\to\infty} \frac{1}{j} \log |N_j(f,2\varepsilon,\theta)| > J(D^{2\varepsilon},\theta)\right)\\
&\qquad\qquad + \mathbb{P}\left(\limsup_{j\to\infty} \frac{1}{j} \log \sup_{t\in[j,j+1]} \frac{|N_t(f,\varepsilon,\theta)|}{|N_j(f,2\varepsilon,\theta)|} > \delta\right).
\end{aligned}$$

Since $f \in D$, the uniform closed ball of radius 2ε about f is a subset of $D^{2\varepsilon}$, so by Proposition 4,

$$\mathbb{P}\left(\limsup_{j\to\infty}\frac{1}{j}\log|N_j(f,2\varepsilon,\theta)| > J(D^{2\varepsilon},\theta)\right) = 0$$

and we may concentrate on the last term. We claim that for j large enough, for any $t \in [j, j+1]$ we have

$$N_t(f,\varepsilon,\theta j/t) \subseteq N_j(f,2\varepsilon,\theta).$$

Indeed, if $u \in N_t(f,\varepsilon,\theta j/t)$ then for any $s \leq \theta j$,

$$\begin{aligned}
&|X_u(s) - jf(s/j)|\\
&\leq |X_u(s) - tf(s/t)| + |jf(s/j) - tf(s/j)| + t\,|f(s/j) - f(s/t)|\\
&\leq t\varepsilon + \|f\|_\theta + t \sup_{\substack{x,y\in[0,\theta]\\ |x-y|\leq 1/j}} |f(x) - f(y)|
\end{aligned}$$

which is smaller than $2\varepsilon j$ for large j since f is absolutely continuous.

We deduce that for large j every particle in $N_t(f,\varepsilon,\theta)$ for any $t \in [j, j+1]$ has an ancestor in $N_j(f,2\varepsilon,\theta)$; thus, letting $N(u,s,t)$ be the set of all descendants (including, possibly, u itself) of particle $u \in N(s)$ at time t,

$$\begin{aligned}
&\mathbb{E}\left[\sup_{t\in[j,j+1]}\frac{|N_t(f,\varepsilon,\theta)|}{|N_j(f,2\varepsilon,\theta)|}\right]\\
&\leq \mathbb{E}\left[\frac{\mathbb{E}\left[\sup_{t\in[j,j+1]}|N_t(f,\varepsilon,\theta)|\,\middle|\,\mathscr{F}_{\theta j}\right]}{|N_j(f,2\varepsilon,\theta)|}\right]\\
&\leq \mathbb{E}\left[\frac{\mathbb{E}\left[\sup_{t\in[j,j+1]}\sum_{u\in N_j(f,2\varepsilon,\theta)}|N(u,\theta j,\theta t)|\,\middle|\,\mathscr{F}_{\theta j}\right]}{|N_j(f,2\varepsilon,\theta)|}\right].
\end{aligned}$$

Since $|N(u,\theta j,\theta t)|$ is non-decreasing in t, using the Markov property we get

$$\begin{aligned}
\mathbb{E}\left[\sup_{t\in[j,j+1]}\frac{|N_t(f,\varepsilon,\theta)|}{|N_j(f,2\varepsilon,\theta)|}\right] &\leq \mathbb{E}\left[\frac{\sum_{u\in N_j(f,2\varepsilon,\theta)}\mathbb{E}[|N(u,\theta j,\theta(j+1))||\mathscr{F}_{\theta j}]}{|N_j(f,2\varepsilon,\theta)|}\right]\\
&= \mathbb{E}\left[\frac{|N_j(f,2\varepsilon,\theta)|\mathbb{E}[|N(\theta)|]}{|N_j(f,2\varepsilon,\theta)|}\right]\\
&= \exp(rm\theta).
\end{aligned}$$

Hence by Markov's inequality

$$\mathbb{P}\left(\sup_{t\in[j,j+1]}\frac{|N_t(f,\varepsilon,\theta)|}{|N_j(f,2\varepsilon,\theta)|} > \exp(\delta j)\right) \le \exp(rm\theta - \delta j)$$

and applying Borel-Cantelli

$$\mathbb{P}\left(\limsup_{j\to\infty}\frac{1}{j}\log\sup_{t\in[j,j+1]}\frac{|N_t(f,\varepsilon,\theta)|}{|N_j(f,2\varepsilon,\theta)|} > \delta\right) = 0.$$

Again taking a union over $\delta > 0$ gives the result. □

If we were considering the possibility of particles dying with no offspring then $N(u,\theta j,\theta t)$ would not be non-decreasing in t, but considering instead the set of all descendants of u ever alive between times θj and θt would give us a slightly worse—but still good enough—estimate.

We move now onto ruling out extreme paths, by choosing a "bad set" F_N and showing that no particles follow paths in this set. There is a balance to be found between including enough paths in F_N that $C_0[0,1]\setminus F_N$ is compact, but not so many that we might find some (rescaled) Brownian paths within F_N at large times.

For simplicity of notation we extend the definition of $N_T(D,\theta)$ to sets $D\subseteq C[0,\theta]$ in the obvious way, setting

$$N_T(D,\theta) := \{u\in N(\theta T) : \exists f\in D \text{ with } X_u(t) = Tf(t/T)\ \ \forall t\in[0,\theta T]\}.$$

Lemma 10. *Fix $\theta\in[0,1]$. For $N\in\mathbb{N}$, let*

$$F_N := \left\{f\in C[0,\theta] : \exists n\ge N,\ u,s\in[0,\theta] \text{ with } |u-s|\le\frac{1}{n^2},\ |f(u)-f(s)| > \frac{1}{\sqrt{n}}\right\}.$$

Then for large N

$$\limsup_{T\to\infty}\frac{1}{T}\log|N_T(F_N,\theta)| = -\infty$$

almost surely.

Proof. Fix $T\ge S\ge 0$; then for any $t\in[S,T]$,

$$\begin{aligned}\{\xi_t\in N_t(F_N,\theta)\} &= \left\{\exists n\ge N,\ u,s\in[0,\theta] : |u-s|\le\frac{1}{n^2},\ \left|\frac{\xi_{ut}-\xi_{st}}{t}\right| > \frac{1}{\sqrt{n}}\right\}\\ &\subseteq \left\{\exists n\ge N,\ u,s\in[0,\theta] : |u-s|\le\frac{1}{n^2},\ \left|\frac{\xi_{uT}-\xi_{sT}}{S}\right| > \frac{1}{\sqrt{n}}\right\}.\end{aligned}$$

Since the right-hand side does not depend on t, we deduce that

$$\{\exists t \in [S,T] : \xi_t \in N_t(F_N,\theta)\}$$
$$\subseteq \left\{\exists n \geq N,\ u,s \in [0,\theta] : |u-s| \leq \frac{1}{n^2},\ \left|\frac{\xi_{uT}-\xi_{sT}}{S}\right| > \frac{1}{\sqrt{n}}\right\}.$$

Now, for $s \in [0,\theta]$, define $\pi(n,s) := \lfloor 2n^2 s\rfloor/2n^2$. Suppose we have a continuous function f such that $\sup_{s\in[0,\theta]} |f(s) - f(\pi(n,s))| \leq 1/4\sqrt{n}$. If $u, s \in [0,\theta]$ satisfy $|u-s| \leq 1/n^2$, then

$$|f(u)-f(s)|$$
$$\leq |f(u)-f(\pi(n,u))| + |f(s)-f(\pi(n,s))| + |f(\pi(n,s))-f(\pi(n,u))|$$
$$\leq \frac{1}{4\sqrt{n}} + \frac{1}{4\sqrt{n}} + \frac{2}{4\sqrt{n}} = \frac{1}{\sqrt{n}}.$$

Thus

$$\{\exists t \in [S,T] : \xi_t \in N_t(F_N,\theta)\} \subseteq \left\{\exists n \geq N,\ s \leq \theta : \left|\frac{\xi_{sT}-\xi_{\pi(n,s)T}}{S}\right| > \frac{1}{4\sqrt{n}}\right\}.$$

Standard properties of Brownian motion now give us that

$$\tilde{\mathbb{P}}(\exists t \in [S,T] : \xi_t \in N_t(F_N,\theta)) \leq \tilde{\mathbb{P}}\left(\exists n \geq N,\ s \leq \theta : |\xi_{sT} - \xi_{\pi(n,s)T}| > S/4\sqrt{n}\right)$$
$$\leq \sum_{n\geq N} 2n^2 \tilde{\mathbb{P}}\left(\sup_{s\in[0,1/2n^2]} |\xi_{sT}| > S/4\sqrt{n}\right)$$
$$\leq \sum_{n\geq N} \frac{8\sqrt{n^3 T}}{S\sqrt{\pi}} \exp\left(-\frac{S^2 n}{16T}\right).$$

Taking $S = j$ and $T = j+1$, we note that for large N,

$$\sum_{n\geq N} \frac{8\sqrt{n^3 T}}{S\sqrt{\pi}} \exp\left(-\frac{S^2 n}{16T}\right) \leq \sum_{n\geq N} \exp\left(-\frac{jn}{32}\right) \leq \exp\left(-\frac{jN}{64}\right)$$

so that (again for large N),

$$\tilde{\mathbb{P}}(\exists t \in [j, j+1] : \xi_t \in N_t(F_N,\theta)) \leq \exp(-2rmj).$$

Applying Markov's inequality and the many-to-one lemma,

$$\begin{aligned}\mathbb{P}(\sup_{t\in[j,j+1]}|N_t(F_N,\theta)|\geq 1) &\leq \mathbb{E}\left[\sup_{t\in[j,j+1]}|N_t(F_N,\theta)|\right]\\ &\leq \mathbb{E}\left[\sum_{u\in N(j+1)}\mathbb{1}_{\{\exists t\in[j,j+1],\ v\leq u\ :\ v\in N_t(F_N,\theta)\}}\right]\\ &\leq e^{rm\theta(j+1)}\tilde{\mathbb{P}}(\exists t\in[j,j+1]:\xi_t\in N_t(F_N,\theta))\\ &\leq \exp(rm\theta(j+1)-2rmj).\end{aligned}$$

Thus, by Borel-Cantelli, we have that for large enough N

$$\mathbb{P}(\limsup_{j\to\infty}\sup_{t\in[j,j+1]}|N_t(F_N,\theta)|\geq 1)=0$$

and since $|N_T(F_N,\theta)|$ is integer-valued,

$$\limsup_{T\to\infty}\frac{1}{T}\log|N_T(F_N,\theta)|=-\infty$$

almost surely. □

Now that we have ruled out any extreme paths, we check that we can cover the remainder of our sets in a suitable way.

Lemma 11. *For $\theta\in[0,1]$, let*

$$C_0[0,\theta]:=\{f\in C[0,\theta]:f(0)=0\}.$$

For each $N\in\mathbb{N}$, the set $C_0[0,\theta]\setminus F_N$ is totally bounded under $\|\cdot\|_\theta$ (that is, it may be covered by open balls of arbitrarily small radius).

Proof. Given $\varepsilon>0$ and $N\in\mathbb{N}$, choose n such that $n\geq N\vee(1/\varepsilon^2)$. For any function $f\in C_0[0,\theta]\setminus F_N$, if $|u-s|<1/n^2$ then $|f(u)-f(s)|\leq 1/\sqrt{n}\leq\varepsilon$. Thus the set $C_0[0,\theta]\setminus F_N$ is equicontinuous (and, since each function must start from 0, uniformly bounded) and we may apply the Arzelà-Ascoli theorem to say that $C_0[0,\theta]\setminus F_N$ is relatively compact, which is equivalent to totally bounded since $(C[0,\theta],\|\cdot\|_\theta)$ is a complete metric space. □

We are now in a position to give an upper bound for any closed set D in continuous time. This upper bound is not quite what we asked for in Theorem 1, but this issue—replacing J with K—will be corrected in Corollary 2.

Proposition 5. *If $D\subset C[0,1]$ is closed, then for any $\theta\in[0,1]$*

$$\limsup_{T\to\infty}\frac{1}{T}\log|N_T(D,\theta)|\leq J(D,\theta)$$

almost surely.

Proof. Clearly (since our first particle starts from 0) $N_T(D \setminus C_0[0,1], \theta) = \emptyset$ for all T, so we may assume without loss of generality that $D \subseteq C_0[0,1]$. Now, for each θ,

$$f \mapsto \begin{cases} \frac{1}{2}\int_0^\theta f'(s)^2 ds & \text{if } f \in H_1 \\ \infty & \text{otherwise} \end{cases}$$

is a good rate function on $C_0[0,\theta]$ (that is, lower-semicontinuous with compact level sets): we refer to Sect. 5.2 of [1] but it is possible to give a proof by showing directly that the function is lower-semicontinuous, then applying Jensen's inequality and the Arzelà-Ascoli theorem to prove that its level sets in $C_0[0,1]$ are compact. Hence we know that for any $\delta > 0$,

$$\{f \in C_0[0,\theta] : J(f,\theta) \geq J(D,\theta) + \delta\}$$

is compact, and since it is disjoint from

$$\{f \in C_0[0,\theta] : \exists g \in D \text{ with } f(s) = g(s)\ \forall s \in [0,\theta]\},$$

which is closed, there is a positive distance between the two sets. Thus we may fix $\delta > 0$ and choose $\varepsilon > 0$ such that $J(D^{2\varepsilon},\theta) < J(D,\theta) + \delta$. Then, by Lemma 11, for any N we may choose a finite α (depending on N) and some $f_k, k = 1,2,\ldots,\alpha$ such that balls of radius ε about the f_k cover $C_0[0,\theta] \setminus F_N$. Thus

$$\begin{aligned} &\mathbb{P}\left(\limsup_{T\to\infty} \frac{1}{T} \log |N_T(D,\theta)| > J(D,\theta) + \delta\right) \\ &\quad\leq \mathbb{P}\left(\limsup_{T\to\infty} \frac{1}{T} \log |N_T(F_N,\theta)| > J(D,\theta) + \delta\right) \\ &\qquad + \sum_{k=1}^{\alpha} \mathbb{P}\left(\limsup_{T\to\infty} \frac{1}{T} \log |N_T(f_k,\varepsilon,\theta)| > J(D^{2\varepsilon},\theta)\right). \end{aligned}$$

By Lemmas 9 and 10, for large enough N the terms on the right-hand side are all zero. As usual we take a union over $\delta > 0$ to complete the proof. □

Corollary 2. *For any closed set $D \subseteq C[0,1]$ and $\theta \in [0,1]$, we have*

$$\limsup_{T\to\infty} \frac{1}{T} \log |N_T(D,\theta)| \leq \sup_{f\in D} K(f,\theta)$$

almost surely.

Proof. Since $|N_T(D,\theta)|$ is integer valued,

$$\frac{1}{T} \log |N_T(D,\theta)| < 0 \;\Rightarrow\; \frac{1}{T} \log |N_T(D,\theta)| = -\infty.$$

Thus, by Proposition 4, if $J(D,\theta) < 0$ then

$$\mathbb{P}\left(\limsup_{T\to\infty} \frac{1}{T} \log |N_T(D,\theta)| > -\infty\right) = 0.$$

Further, clearly for $\phi \leq \theta$ and any $T \geq 0$, if $N_T(D,\phi) = \emptyset$ then necessarily we have $N_T(D,\theta) = \emptyset$. Thus if there exists $\phi \leq \theta$ with $J(D,\phi) < 0$, then

$$\mathbb{P}\left(\limsup_{T\to\infty} \frac{1}{T} \log |N_T(D,\theta)| > -\infty\right) = 0$$

which completes the proof. □

Combining Corollary 1 with Corollary 2 completes the proof of Theorem 1.

Appendix: The Oversight in [2]

In [2] it is written that under a certain assumption, setting

$$W_n = \left\{\omega \in \Omega : \limsup_{T\to\infty} \frac{1}{T} \log |N_T(D,\theta)| > J(D,\theta) + \frac{1}{n}\right\}$$

(it is not important what $J(D,\theta)$ is here) we have $\mathbb{P}(W_n) > 0$ for some n. This is correct, but the article then goes on to say "It is now clear that

$$\limsup_{T\to\infty} \frac{1}{T} \log \mathbb{E}\big[|N_T(D,\theta)|\big] \geq J(D,\theta) + \frac{1}{n} \text{ "}$$

which does not appear to be obviously true. To see this explicitly, work on the probability space [0, 1] with Lebesgue probability measure $\mathbb{P}$. Let X_T, $T \geq 0$ be the càdlàg random process defined (for $\omega \in [0,1]$ and $T \geq 0$) by

$$X_T(\omega) = \begin{cases} e^{2T} & \text{if } T - n \in [\omega - e^{-4T},\ \omega + e^{-4T}) \text{ for some } n \in \mathbb{N} \\ e^T & \text{otherwise.} \end{cases}$$

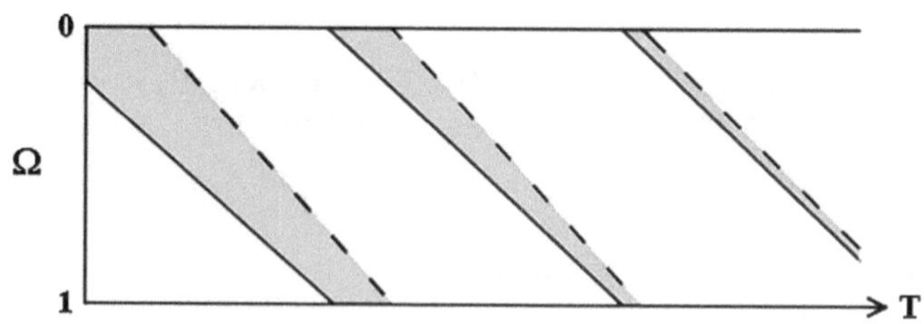

Then for every ω,

$$\limsup \frac{1}{T} \log X_T(\omega) = 2$$

but

$$\frac{1}{T} \log \mathbb{E}[X_T] \to 1.$$

Acknowledgements MIR was supported by an EPSRC studentship and by ANR MADCOF grant ANR-08-BLAN-0220-01.

References

1. A. Dembo, O. Zeitouni, *Large Deviations Techniques and Applications*. Applications of Mathematics (New York), vol. 38, 2nd edn. (Springer, New York, 1998)
2. Y. Git, *Almost Sure Path Properties of Branching Diffusion Processes*. In Séminaire de Probabilités, XXXII. Lecture Notes in Math., vol. 1686 (Springer, Berlin, 1998), pp. 108–127
3. R. Hardy, S.C. Harris, A conceptual approach to a path result for branching Brownian motion. Stoch. Process. Appl. **116**(12), 1992–2013 (2006)
4. R. Hardy, S.C. Harris, *A Spine Approach to Branching Diffusions with Applications to L^p-Convergence of Martingales*. In Séminaire de Probabilités, XLII. Lecture Notes in Math., vol. 1979 (Springer, Berlin, 2009)
5. S.C. Harris, M.I. Roberts. The unscaled paths of branching Brownian motion. Ann. Inst. Henri Poincaré Probab. Stat. (to appear)
6. T.-Y. Lee, Some large-deviation theorems for branching diffusions. Ann. Probab. **20**(3), 1288–1309 (1992)
7. S.R.S. Varadhan, *Large Deviations and Applications*. CBMS-NSF Regional Conference Series in Applied Mathematics, vol. 46 (Society for Industrial and Applied Mathematics (SIAM), Philadelphia, 1984)

On the Delocalized Phase of the Random Pinning Model

Jean-Christophe Mourrat

Abstract We consider the model of a directed polymer pinned to a line of i.i.d. random charges, and focus on the interior of the delocalized phase. We first show that in this region, the partition function remains bounded. We then prove that for almost every environment of charges, the probability that the number of contact points in $[0, n]$ exceeds $c \log n$ tends to 0 as n tends to infinity. The proofs rely on recent results of Birkner, Greven, den Hollander (2010) and Cheliotis, den Hollander (2010).

1 Introduction

Let $\tau = (\tau_i)_{i \in \mathbb{N}}$ be a sequence such that $\tau_0 = 0$ and $(\tau_{i+1} - \tau_i)_{i \geqslant 0}$ are independent and identically distributed random variables with values in $\mathbb{N}^* = \{1, 2, \ldots\}$. Let $\mathbf{P}$ be the distribution of τ, $\mathbf{E}$ the associated expectation, and $K(n) = \mathbf{P}[\tau_1 = n]$. We assume that there exists $\alpha \geqslant 0$ such that

$$\frac{\log K(n)}{\log n} \xrightarrow[n \to \infty]{} -(1 + \alpha). \tag{1}$$

As an example, one can think about the sequence τ as the sequence of arrival times at 0 of a one-dimensional simple random walk (and in this case, $\alpha = 1/2$). In a slight abuse of notation, we will look also at the sequence τ as a set, and write for instance $n \in \tau$ instead of $\exists i : n = \tau_i$.

Let $\omega = (\omega_k)_{k \in \mathbb{N}}$ be independent and identically distributed random variables. We write $\mathbb{P}$ for the law of ω, and $\mathbb{E}$ for the associated expectation. We will refer to

J.-C. Mourrat (✉)
Ecole polytechnique fédérale de Lausanne, Institut de Mathématiques, Station 8, 1015 Lausanne, Switzerland
e-mail: jean-christophe.mourrat@epfl.ch

C. Donati-Martin et al. (eds.), *Séminaire de Probabilités XLIV*, Lecture Notes in Mathematics 2046, DOI 10.1007/978-3-642-27461-9_18,

ω as the *environment*. We assume that the ω_k are centred random variables, and that they have exponential moments of all order. Let $\beta \geqslant 0, h \geqslant 0$, and $n \in \mathbb{N}^*$. We consider the probability measure $\mathbf{P}_n^{\beta,h,\omega}$ (expectation $\mathbf{E}_n^{\beta,h,\omega}$) which is defined as the following Gibbs transformation of the measure $\mathbf{P}$:

$$\frac{\mathrm{d}\mathbf{P}_n^{\beta,h,\omega}}{\mathrm{d}\mathbf{P}}(\tau) = \frac{1}{Z_n^{\beta,h,\omega}} \exp\left(\sum_{k=0}^{n-1}(\beta\omega_k - h)\mathbf{1}_{\{k\in\tau\}}\right)\mathbf{1}_{\{n\in\tau\}}.$$

In the above definition, β can be thought of as the inverse temperature, h as the disorder bias, and $Z_n^{\beta,h,\omega}$ is a normalization constant called the *partition function*,

$$Z_n^{\beta,h,\omega} = \mathbf{E}\left[\exp\left(\sum_{k=0}^{n-1}(\beta\omega_k - h)\mathbf{1}_{\{k\in\tau\}}\right)\mathbf{1}_{\{n\in\tau\}}\right].$$

At the exponential scale, the asymptotic behaviour of the partition function is captured by the *free energy* $\mathrm{F}(\beta,h)$ defined as

$$\mathrm{F}(\beta,h) = \lim_{n\to+\infty} \frac{1}{n} \log Z_n^{\beta,h,\omega}.$$

Superadditivity of the partition function implies that this limit is well defined almost surely, and that it is deterministic (see for instance [5, Theorem 4.1]). Assumption 1 implies that $\mathrm{F}(\beta,h) \geqslant 0$. It is intuitively clear that the free energy can become strictly positive only if the set $\tau \cap [0,n]$ is likely to contain many points under the measure $\mathbf{P}_n^{\beta,h,\omega}$. We thus say that we are in the *localized phase* if $\mathrm{F}(\beta,h) > 0$, and in the *delocalized phase* otherwise. One can show [7, Theorem 11.3] that for every $\beta \geqslant 0$, there exists $h_c(\beta) \geqslant 0$ such that

$$\begin{aligned} h < h_c(\beta) &\Rightarrow \text{localized phase, i.e. } \mathrm{F}(\beta,h) > 0, \\ h \geqslant h_c(\beta) &\Rightarrow \text{delocalized phase, i.e. } \mathrm{F}(\beta,h) = 0, \end{aligned}$$

and moreover, the function $\beta \mapsto h_c(\beta)$ is strictly increasing.

2 Statement of the Main Results

We focus here on the interior of the delocalized phase, that is to say when $h > h_c(\beta)$. Note that, due to the strict monotonicity of the function $h_c(\cdot)$, one sits indeed in the interior of the delocalized phase if one fixes $h = h_c(\beta_0)$ and considers any inverse temperature $\beta < \beta_0$.

By definition, the partition function is known to grow subexponentially in this region. In [2, Remark p. 417], the authors ask whether the partition function remains bounded there. We answer positively to this question, and can in fact be slightly more precise.

Theorem 1. *Let $\beta \geqslant 0$ and $h > h_c(\beta)$. For almost every environment, one has*

$$\sum_{n=1}^{+\infty} Z_n^{\beta,h,\omega} < +\infty.$$

Remark 1. This result implies that, in the interior of the delocalized phase, the *unconstrained* (or *free*) partition function $Z_{n,f}^{\beta,h,\omega}$ is also almost surely bounded (in fact, tends to 0) as n tends to infinity. Indeed, $Z_{n,f}^{\beta,h,\omega}$ is defined by

$$Z_{n,f}^{\beta,h,\omega} = \mathbf{E}\left[\exp\left(\sum_{k=0}^{n-1}(\beta\omega_k - h)\mathbf{1}_{\{k\in\tau\}}\right)\right],$$

which is equal to

$$\sum_{n'=n}^{+\infty} \mathbf{E}\left[\exp\left(\sum_{k=0}^{n-1}(\beta\omega_k - h)\mathbf{1}_{\{k\in\tau\}}\right); \tau\cap[n,n'] = \{n'\}\right] \leqslant \sum_{n'=n}^{+\infty} Z_{n'}^{\beta,h,\omega} \xrightarrow[n\to\infty]{\text{a.s.}} 0.$$

Our second result concerns the size of the set $\tau \cap [0,n]$, that we may call the set of *contact points*, under the measure $\mathbf{P}_n^{\beta,h,\omega}$. Let us write $E_{n,N}$ for the event that $|\tau\cap[0,n]| > N$ (where we write $|A|$ for the cardinal of a set A).

Theorem 2. *Let $\beta \geqslant 0$ and $h > h_c(\beta)$. For every $\varepsilon > 0$ and for almost every environment, there exists $N_\varepsilon, C_\varepsilon > 0$ such that for any $N \geqslant N_\varepsilon$ and any n:*

$$\mathbf{P}_n^{\beta,h,\omega}(E_{n,N}) \leqslant \frac{C_\varepsilon}{K(n)} e^{-N(h-h_c(\beta)-\varepsilon)}.$$

In particular, for every constant c such that

$$c > \frac{1+\alpha}{h-h_c(\beta)}$$

and for almost every environment, one has

$$\mathbf{P}_n^{\beta,h,\omega}(E_{n,c\log n}) \xrightarrow[n\to\infty]{} 0.$$

To my knowledge, results of this kind were known only under the averaged measure $\mathbb{P}\mathbf{P}_n^{\beta,h,\omega}$, and with some restrictions on the distribution of ω due to the use of concentration arguments (see [6] or [5, Sect. 8.2]). In particular, in the interior of the delocalized phase and for almost every environment, the polymer intersects the pinning line less that the simple random walk does.

It is worth comparing this result with the case when randomness of the medium is absent, that is, when $\beta = 0$. In this context, the distribution of the number of contact points of the polymer forms a tight sequence as n varies (see for instance [7, Theorem 7.3]). It is only natural to expect that a similar result holds true in the disordered case as well.

Interestingly, boundedness of the number of contact points in the delocalized phase was recently obtained for a specific model of pinning on a random interface with long-range correlations in [1], even at criticality ($h = h_c(\beta)$). In this work, the specific structure of the environment enables the authors to identify the critical point explicitly, a feature which makes the subsequent analysis more tractable.

3 Proofs

In this section, we present the proofs of Theorems 1 and 2. Although one might think at first that such an approach cannot be of much help as far as the delocalized phase is concerned, we will rely on recent results obtained in [3, 4], where the authors develop a large deviations point of view of the problem. Let us define

$$F_N^{\beta,h,\omega} = \sum_{0=l_0<l_1<\cdots<l_N} \prod_{i=0}^{N-1} K(l_{i+1} - l_i) e^{(\beta\omega_{l_i} - h)}. \tag{2}$$

Our results are based on the following fact, due to [4], that holds both in the delocalized and in the localized phases.

Lemma 1. *For almost every environment, one has*

$$\limsup_{N\to+\infty} \frac{1}{N} \log F_N^{\beta,h,\omega} = h_c(\beta) - h.$$

Proof. Although this result is not stated as a proposition in [4], the authors give all the necessary elements to prove it. Indeed, we can start from [4, (3.11)], which reads

$$\limsup_{N\to+\infty} \frac{1}{N} \log F_N^{\beta,h,\omega} = -h + S^{\mathrm{que}}(\beta; 1),$$

where $S^{\mathrm{que}}(\beta; z)$ is defined in [4, (3.10)]. We then learn from [4, (3.13)] that

$$h_c(\beta) = S^{\mathrm{que}}(\beta; 1^-),$$

so what remains to see is that

$$S^{\text{que}}(\beta; 1^-) = S^{\text{que}}(\beta; 1).$$

The proof of this fact is first obtained assuming further that the support of the distribution of ω_0 is finite, see [4, Lemma 3.2]. The general case, where one assumes only finiteness of all exponential moments, is considered in [4, Sect. 3.3]. One can start by observing that, since $z \mapsto S^{\text{que}}(\beta, z)$ is increasing (in the wide sense), one has $S^{\text{que}}(\beta; 1^-) \leqslant S^{\text{que}}(\beta; 1)$. On the other hand, it is shown in step 1 of the proof of [4, Lemma 3.3] that $S^{\text{que}}(\beta; 1) \leqslant A(\beta)$, where $A(\beta)$ is defined in [4, (3.21)]. Moreover, steps 2–4 of the proof of [4, Lemma 3.3] are devoted to the justification of the fact that $A(\beta) \leqslant S^{\text{que}}(\beta; 1^-)$. It thus follows that $S^{\text{que}}(\beta; 1) \leqslant S^{\text{que}}(\beta; 1^-)$, which finishes the proof. □

Proof of Theorem 1. The proof is close to [4, Sect. 3.2]. We can decompose Z_n the following way:

$$Z_n^{\beta,h,\omega} = \sum_{N=1}^{+\infty} \sum_{0=l_0<l_1<\cdots<l_N=n} \prod_{i=0}^{N-1} K(l_{i+1} - l_i) e^{(\beta\omega_{l_i} - h)}.$$

An interversion of sums then leads to

$$\sum_{n=1}^{+\infty} Z_n^{\beta,h,\omega} = \sum_{N=1}^{+\infty} F_N^{\beta,h,\omega},$$

and Lemma 1 ensures the almost sure convergence of the second series when $h > h_c(\beta)$. □

For an event A, let us write $Z_n^{\beta,h,\omega}(A)$ for the quantity

$$\mathbf{E}\left[\exp\left(\sum_{k=0}^{n-1}(\beta\omega_k - h)\mathbf{1}_{\{k\in\tau\}}\right)\mathbf{1}_{\{n\in\tau\}} \; ; \; A\right].$$

In words, $Z_n^{\beta,h,\omega}(A)$ is a partition function in which one integrates with respect to $\mathbf{P}$ only on the event A. In order to prove Theorem 2, we first give a refined version of Theorem 1, which goes as follows.

Proposition 1. *Let $\beta \geqslant 0$ and $h > h_c(\beta)$. For every $\varepsilon > 0$ and for almost every environment, there exist $N_\varepsilon, C_\varepsilon$ such that for any $N \geqslant N_\varepsilon$:*

$$\sum_{n=1}^{+\infty} Z_n^{\beta,h,\omega}(E_{n,N}) \leqslant C_\varepsilon e^{-N(h-h_c(\beta)-\varepsilon)}.$$

Proof. We can assume that $\varepsilon < h - h_c(\beta)$. Note that, for any n and N_0,

$$Z_n^{\beta,h,\omega}(E_{n,N_0}) = \sum_{N=N_0}^{+\infty} \sum_{0=l_0<l_1<\cdots<l_N=n} \prod_{i=0}^{N-1} K(l_{i+1}-l_i)e^{(\beta\omega_{l_i}-h)}.$$

By an interversion of sums, we obtain that

$$\sum_{n=1}^{+\infty} Z_n^{\beta,h,\omega}(E_{n,N_0}) = \sum_{N=N_0}^{+\infty} F_N^{\beta,h,\omega}.$$

By Lemma 1, there exists N_ε such that for every $N \geqslant N_\varepsilon$,

$$F_N^{\beta,h,\omega} \leqslant e^{-N(h-h_c(\beta)-\varepsilon/2)},$$

and as a consequence, for every $N_0 \geqslant N_\varepsilon$, one has

$$\sum_{n=1}^{+\infty} Z_n^{\beta,h,\omega}(E_{n,N_0}) \leqslant \sum_{N=N_0}^{+\infty} e^{-N(h-h_c(\beta)-\varepsilon/2)},$$

which implies the announced claim. □

Proof of Theorem 2. Note that

$$\mathbf{P}^{\beta,h,\omega}(E_{n,N}) = \frac{Z_n^{\beta,h,\omega}(E_{n,N})}{Z_n^{\beta,h,\omega}}.$$

The numerator can be bounded from above using Proposition 1. For the denominator, one can use the bound

$$Z_n^{\beta,h,\omega} \geqslant K(n)e^{\beta\omega_0-h},$$

which proves the desired result. □

References

1. Q. Berger, H. Lacoin, Sharp critical behavior for pinning model in random correlated environment. Preprint, To appear in Stochastic Processes and their Applications http://dx.doi.org/10.1016/j.spa.2011.12.007, arXiv:1104.4969v1 (2011)
2. M. Birkner, R. Sun, Annealed vs quenched critical points for a random walk pinning model. Ann. Inst. Henri Poincaré Probab. Stat. **46**(2), 414–441 (2010)
3. M. Birkner, A. Greven, F. den Hollander, Quenched large deviation principle for words in a letter sequence. Probab. Theory Related Fields **148** (3–4), 403–456 (2010)
4. D. Cheliotis, F. den Hollander, Variational characterization of the critical curve for pinning of random polymers. Preprint, To appear at Annals of Probability, arXiv:1005.3661v1 (2010)
5. G. Giacomin, *Random Polymer Models* (Imperial College Press, London, 2007)

6. G. Giacomin, F.L. Toninelli, Estimates on path delocalization for copolymers at selective interfaces. Probab. Theory Related Fields **133**(4), 464–482 (2005)
7. F. den Hollander, in *Random Polymers*. Ecole d'été de probabilités de Saint Flour XXXVII. Lecture Notes in Mathematics, vol. 1974 (Springer, Berlin, 2009)

Large Deviations for Gaussian Stationary Processes and Semi-Classical Analysis

Bernard Bercu, Jean-François Bony, and Vincent Bruneau

Abstract In this paper, we obtain a large deviation principle for quadratic forms of Gaussian stationary processes. It is established by the conjunction of a result of Roch and Silbermann on the spectrum of products of Toeplitz matrices together with the analysis of large deviations carried out by Gamboa, Rouault and the first author. An alternative proof of the needed result on Toeplitz matrices, based on semi-classical analysis, is also provided.

Keywords Large deviations • Gaussian processes • Distribution of eigenvalues • Toeplitz matrices

1 Introduction

For any bounded measurable real function f on the torus $\mathbb{T} = [-\pi, \pi[$, the $\ell^2(\mathbb{N})$ Toeplitz and Hankel operators are respectively defined as

$$T(f) = \left(\widehat{f}_{i-j}\right)_{i,j\geq 0} \qquad \text{and} \qquad H(f) = \left(\widehat{f}_{i+j+1}\right)_{i,j\geq 0} \tag{1}$$

where $(\widehat{f}_n)$ stands for the sequence of Fourier coefficients of f. We refer the reader to the books of Böttcher and Silbermann [2,3] for a general presentation of Toeplitz operators. A well-known identity between the product $T(f)T(g)$ and $T(fg)$ is

$$T(fg) - T(f)T(g) = H(f)H(\widetilde{g}) \tag{2}$$

B. Bercu (✉) · J.-F. Bony · V. Bruneau
Institut de Mathématiques de Bordeaux, Université Bordeaux 1, UMR CNRS 5251,
351 cours de la libération, 33405 Talence cedex, France
e-mail: Bernard.Bercu@math.u-bordeaux1.fr; Jean-Francois.Bony@math.u-bordeaux1.fr;
Vincent.Bruneau@math.u-bordeaux1.fr

C. Donati-Martin et al. (eds.), *Séminaire de Probabilités XLIV*, Lecture Notes in Mathematics 2046, DOI 10.1007/978-3-642-27461-9_19,

where $\widetilde{g}(x) = g(-x)$. The analogue of identity (2) for finite section Toeplitz matrices is given by the formula of Widom [14]

$$T_n(fg) - T_n(f)T_n(g) = P_n H(f) H(\widetilde{g}) P_n + Q_n H(\widetilde{f}) H(g) Q_n \tag{3}$$

where the projection P_n and the operator Q_n are given by

$$P_n(x_0, x_1, x_2, \ldots) = (x_0, x_1, \ldots, x_n, 0, \ldots),$$
$$Q_n(x_0, x_1, x_2, \ldots) = (x_n, x_{n-1}, \ldots, x_0, 0, \ldots),$$

and $T_n(f)$ is the finite section of order $n \geq 1$ of $T(f)$ which means that $T_n(f)$ is identified with $P_n T(f) P_n$. In other words, our operators will be considered as operators on Im P and Im P_n where P stands for the projection operator on $\ell^2(\mathbb{N})$. We clearly have $Q_n^2 = P_n$, $P_n Q_n = Q_n P_n = Q_n$, and $Q_n T_n(f) Q_n = T_n(\widetilde{f})$.

The classical Szegö theorem deals with the asymptotic behavior of the spectrum of a single Toeplitz matrix. It states that if f is a bounded measurable real function on $\mathbb{T}$, the limiting set of eigenvalues of the sequence $(T_n(f))$ is exactly

$$\sigma(T(f)) = [\operatorname{essinf} f, \operatorname{esssup} f],$$

where $\sigma(T(f))$ denotes the spectrum of the operator $T(f)$. Moreover, the empirical spectral measure of $(T_n(f))$ converges to P_f which is the image probability of the uniform measure on $\mathbb{T}$ by the application f. In other words, if $\lambda_0^n, \ldots, \lambda_n^n$ are the eigenvalues of $T_n(f)$, then for any bounded continuous real function φ

$$\lim_{n\to\infty} \frac{1}{n} \sum_{k=0}^{n} \varphi(\lambda_k^n) = \frac{1}{2\pi} \int_{\mathbb{T}} \varphi(f(x))\, dx. \tag{4}$$

In particular, the maximum eigenvalue of $T_n(f)$ converges to esssup f while the minimum eigenvalue of $T_n(f)$ converges to essinf f. One can find more details in Sect. 5.2 of [8] or in Sect. 5.4 of [2]. Our purpose is to make use of similar results for the spectrum of the product of two Toeplitz matrices $T_n(f)T_n(g)$. Several authors have investigated the asymptotic behavior of the spectrum of $T_n(f)T_n(g)$. More precisely, it was shown in Lemma 5 of [1] or Lemma 2.6 of [13] that if f and g are two bounded measurable real functions on $\mathbb{T}$, then the empirical spectral measure associated with the sequence $(T_n(f)T_n(g))$ converges to the limiting measure P_{fg}. However, the limiting set of eigenvalues of $(T_n(f)T_n(g))$ is much more difficult to understand. Via a theorem of Roch and Silbermann, we shall see that, as soon as f and $g \geq 0$ are bounded piecewise continuous real functions, the limiting set of eigenvalues of $(T_n(f)T_n(g))$ coincides with the spectrum of the limiting operator $T(f)T(g)$. In particular, the maximum and the minimum eigenvalues of $T_n(f)T_n(g)$ both converge to the maximum and minimum of the spectrum of $T(f)T(g)$.

In this paper, we make use of the previous results on Toeplitz operators to obtain a large deviation principle (LDP) for quadratic forms of Gaussian stationary processes. More precisely, consider a centered stationary real Gaussian process (X_n) with bounded piecewise continuous spectral density g. It was shown in [1] an LDP for subsequences of the empirical periodogram $(\mathscr{W}_n(f))$ integrated over a bounded piecewise continuous real function f. We can now deduce a full LDP for the sequence $(\mathscr{W}_n(f))$.

We also give an alternative proof of the theorem of Roch and Silbermann in the particular case of Toeplitz operators with continuous symbols. Our approach is based on semi-classical analysis and scattering theory by construction of quasimodes which are approximative eigenvectors. We hope that this microlocal approach can be used in other situations.

The paper is organized as follows. In Sect. 2, we recall a theorem of Roch and Silbermann. Sect. 3 is devoted to the application in probability. An enlightening example is treated in Sect. 4. Then, we give our alternative proof of the result of Roch and Silbermann in the case of Toeplitz operators with continuous symbols. This result and our functional point of view on Toeplitz operators are given in Sect. 5. The convergence of the spectrum is proved in Sect. 6. Finally, in Sect. 7, we propose an alternative proof of Coburn's theorem dealing with the essential spectrum of products of Toeplitz operators.

2 Results on Toeplitz Operators

Denote by $\mathscr{A}$ the Banach algebra of all sequences (A_n) of uniformly bounded linear operators on Im P_n endowed with the sum and the composition term by term, and the supremum of the operator norm of the elements. Let $\mathscr{B}$ be the collection of all sequences (A_n) of $\mathscr{A}$ for which one can find two bounded linear operators A and $\widetilde{A}$ in Im P such that

$$A_n \to A, \quad A_n^* \to A^*, \quad Q_n A_n Q_n \to \widetilde{A}, \quad Q_n A_n^* Q_n \to \widetilde{A}^*,$$

where $*$ stands the adjoint operator and $\to$ stands for the strong convergence. Finally, denote by $\mathscr{C}$ the smallest closed subalgebra of $\mathscr{A}$ containing the collection of all sequences $(T_n(f))$ where f are bounded piecewise continuous real functions. In fact, $\mathscr{C}$ is a subalgebra of $\mathscr{B}$ and

$$T_n(f) \to T(f), \quad Q_n T_n(f) Q_n \to T(\widetilde{f}).$$

We refer to Sect. 2.5 of [2] for more details on $\mathscr{B}$. We are now in position to state a theorem of Roch and Silbermann.

Theorem 1 (Roch–Silbermann). *Let (T_n) be a sequence of selfadjoint operators of $\mathscr{C}$. Moreover, denote the strong limits of T_n and $Q_n T_n Q_n$ by T and $\widetilde{T}$, respectively. For $\lambda \in \mathbb{R}$, the following properties are equivalent:*

(i) $\lambda \in \sigma(T) \cup \sigma(\widetilde{T})$,

(ii) λ is the limit of a sequence (λ_n) where $\lambda_n \in \sigma(T_n)$,

(iii) λ is the limit of a subsequence (λ_{n_k}) where $\lambda_{n_k} \in \sigma(T_{n_k})$.

Theorem 1 was established in [12] together with several examples of application. It is given, in its present form, in Theorem 4.16 of [2].

A direct application of this result is as follows. First of all, let us introduce some notations. Let f and g be two bounded piecewise continuous real functions with $g \geq 0$. From Lemma 3 below, the sequence $(T_n(g)^{1/2})$ as well as $(T_n(g)^{1/2}T_n(f)T_n(g)^{1/2})$ belong to $\mathscr{C}$,

$$T_n(g)^{1/2}T_n(f)T_n(g)^{1/2} \to T(g)^{1/2}T(f)T(g)^{1/2},$$
$$Q_nT_n(g)^{1/2}T_n(f)T_n(g)^{1/2}Q_n \to T(\widetilde{g})^{1/2}T(\widetilde{f})T(\widetilde{g})^{1/2}.$$

On Im P_n, we clearly have

$$\sigma\big(T_n(f)T_n(g)\big) = \sigma\big(T_n(g)^{1/2}T_n(f)T_n(g)^{1/2}\big),$$

with the same multiplicity. Moreover, by Lemma 7, we also have on Im P

$$\begin{aligned}\sigma\big(T(g)^{1/2}T(f)T(g)^{1/2}\big) &= \sigma\big(T(f)T(g)\big) = \sigma\big(T(\widetilde{f})T(\widetilde{g})\big)\\ &= \sigma\big(T(\widetilde{g})^{1/2}T(\widetilde{f})T(\widetilde{g})^{1/2}\big).\end{aligned}$$

Denote the maximum and minimum eigenvalues of $T_n(f)T_n(g)$ by

$$\lambda^n_{\max}(f,g) = \max\sigma\big(T_n(f)T_n(g)\big),$$
$$\lambda^n_{\min}(f,g) = \min\sigma\big(T_n(f)T_n(g)\big).$$

In addition, denote the extrema of the spectrum of $T(f)T(g)$ by

$$\lambda_{\max}(f,g) = \max\sigma\big(T(f)T(g)\big),$$
$$\lambda_{\min}(f,g) = \min\sigma\big(T(f)T(g)\big).$$

One can observe that, in general, we do not know if $\lambda_{\max}(f,g)$ and $\lambda_{\min}(f,g)$ are eigenvalues.

Corollary 1. *Assume that f and g are two bounded piecewise continuous real functions on $\mathbb{T}$ with $g \geq 0$. Then, the limiting sets of eigenvalues of the sequence $(T_n(f)T_n(g))$ are given by $\sigma(T(f)T(g))$. In particular,*

$$\lim_{n\to\infty}\lambda^n_{\max}(f,g) = \lambda_{\max}(f,g), \tag{5}$$

$$\lim_{n\to\infty}\lambda^n_{\min}(f,g) = \lambda_{\min}(f,g). \tag{6}$$

In Sect. 4, we shall show via an example related to Gaussian autoregressive process that it is not true in general that for two bounded continuous real functions f and g, $\lambda_{\max}(f,g) = \sup(fg)$ or $\lambda_{\min}(f,g) = \inf(fg)$. One can also observe that the norm of $T(g)^{1/2}T(f)T(g)^{1/2}$ is not always equal to $\|fg\|_\infty$ or $\|f\|_\infty\|g\|_\infty$. The situation is totally different from the case of a single Toeplitz operator $T(f)$ with bounded continuous real function as $\lambda_{\max}(f,1) = \sup(f)$ and $\lambda_{\min}(f,1) = \inf(f)$.

3 Application in Probability

Let (X_n) be a centered stationary real Gaussian process with bounded piecewise continuous spectral density $g \geq 0$ which means that

$$\mathbb{E}[X_j X_k] = \frac{1}{2\pi}\int_{\mathbb{T}} \exp(i(j-k)x)g(x)\,dx.$$

We assume in all the sequel that g is not the zero function. For any bounded piecewise continuous real function f on the torus $\mathbb{T}$, we are interested in the asymptotic behavior of

$$\mathscr{W}_n(f) = \frac{1}{2\pi n}\int_{\mathbb{T}} f(x)\left|\sum_{j=0}^{n} X_j \exp(ijx)\right|^2 dx. \tag{7}$$

The purpose of this section is to provide the last step in the analysis of the large deviation properties of $(\mathscr{W}_n(f))$ by establishing an LDP for $(\mathscr{W}_n(f))$ in the spirit of the original work of [1] or of Bryc and Dembo [4]. We refer the reader to the book of Dembo and Zeitouni [6] for the general theory on large deviations. The covariance matrix associated with the vector $X^{(n)} = (X_0,\ldots,X_n)^t$ is $T_n(g)$. Consequently, it immediately follows from (7) that

$$\mathscr{W}_n(f) = \frac{1}{n}X^{(n)t}T_n(f)X^{(n)} = \frac{1}{n}Y^{(n)t}T_n(g)^{1/2}T_n(f)T_n(g)^{1/2}Y^{(n)} \tag{8}$$

where the vector $Y^{(n)}$ has a Gaussian $\mathscr{N}(0, I_n)$ distribution. In order to investigate the large deviation properties of $(\mathscr{W}_n(f))$, it is necessary to calculate the normalized cumulant generating function given, for all $t \in \mathbb{R}$, by

$$L_n(t) = \frac{1}{n}\log \mathbb{E}\big[\exp(nt\mathscr{W}_n(f))\big].$$

For convenience and in all the sequel, we use of the notation that $\log t = -\infty$ if $t \leq 0$. We deduce from (8) and standard Gaussian calculation that for all $t \in \mathbb{R}$

$$\begin{aligned} L_n(t) &= -\frac{1}{2n}\log\det\big(I_n - 2tT_n(g)^{1/2}T_n(f)T_n(g)^{1/2}\big) \\ &= -\frac{1}{2n}\sum_{k=0}^{n}\log(1-2t\lambda_k^n), \end{aligned}$$

where $\lambda_0^n,\ldots,\lambda_n^n$ are the eigenvalues of $T_n(g)^{1/2}T_n(f)T_n(g)^{1/2}$. For all $t \in \mathbb{R}$, let

$$L_{fg}(t) = -\frac{1}{4\pi}\int_{\mathbb{T}} \log(1 - 2tf(x)g(x))\,dx,$$

and denote by I_{fg} its Fenchel-Legendre transform

$$I_{fg}(x) = \sup_{t\in\mathbb{R}}\{xt - L_{fg}(t)\}.$$

Furthermore, for all $x \in \mathbb{R}$, let

$$J_{fg}(x) = \begin{cases} I_{fg}(a) + \dfrac{1}{2\lambda_{\min}(f,g)}(x-a) & \text{if } x \in]-\infty, a] \\ I_{fg}(x) & \text{if } x \in]a, b[\\ I_{fg}(b) + \dfrac{1}{2\lambda_{\max}(f,g)}(x-b) & \text{if } x \in [b, +\infty[\end{cases} \tag{9}$$

where a and b are the extended real numbers given by

$$a = L'_{fg}\left(\frac{1}{2\lambda_{\min}(f,g)}\right)$$

if $\lambda_{\min}(f,g) < 0$ and $\lambda_{\min}(f,g) < \inf(fg)$, $a = -\infty$ otherwise, while

$$b = L'_{fg}\left(\frac{1}{2\lambda_{\max}(f,g)}\right)$$

if $\lambda_{\max}(f,g) > 0$ and $\lambda_{\max}(f,g) > \sup(fg)$, $b = +\infty$ otherwise. We immediately deduce from Theorem 1 of [1] together with Corollary 1, that an LDP holds for $(\mathscr{W}_n(f))$.

Theorem 2. *The sequence $(\mathscr{W}_n(f))$ satisfies an LDP with good rate function J_{fg}. More precisely, for any closed set $F \subset \mathbb{R}$*

$$\limsup_{n\to\infty} \frac{1}{n} \log \mathbb{P}(\mathscr{W}_n(f) \in F) \le -\inf_{x\in F} J_{fg}(x),$$

while for any open set $G \subset \mathbb{R}$

$$\liminf_{n\to\infty} \frac{1}{n} \log \mathbb{P}(\mathscr{W}_n(f) \in G) \ge -\inf_{x\in G} J_{fg}(x).$$

Remark 1. Denote by μ the derivative of L_{fg} at point zero

$$\mu = \frac{1}{2\pi}\int_{\mathbb{T}} f(x)g(x)dx.$$

Then, we have $J_{fg}(\mu) = 0$ and it follows from Theorem 2 that for all $x > \mu$

$$\lim_{n\to\infty} \frac{1}{n} \log \mathbb{P}(\mathscr{W}_n(f) \geq x) = -J_{fg}(x),$$

whereas for all $x < \mu$

$$\lim_{n\to\infty} \frac{1}{n} \log \mathbb{P}(\mathscr{W}_n(f) \leq x) = -J_{fg}(x).$$

4 An Illustrative Example

Let a and θ be two real numbers with $|\theta| < 1$ and consider the two bounded continuous real functions f and g given by

$$f(x) = a + \cos(x) \qquad \text{and} \qquad g(x) = \frac{1}{1 + \theta^2 - 2\theta \cos(x)}.$$

The goal of this section is to study the limiting set of eigenvalues of the sequence $(T_n(f)T_n(g))$. We clearly have $\|f\|_\infty = |a| + 1$ and $\|g\|_\infty = (1 - |\theta|)^{-2}$. The function g is simply the spectral density of a Gaussian autoregressive process [1]. If $\theta = 0$, $g = 1$ and the product $T_n(f)T_n(g)$ reduces to $T_n(f)$. Consequently, $\lambda_{\max}(f, 1) = a + 1$ and $\lambda_{\min}(f, 1) = a - 1$. If $\theta \neq 0$, denote

$$a_\theta = -\frac{(1+\theta)}{2\theta} \qquad \text{and} \qquad b_\theta = -\frac{(1-\theta)}{2\theta}.$$

It is more convenient to work with the inverse of $T_n(g)$. As a matter of fact, $T_n(g)^{-1}$ is a tridiagonal matrix quite similar to $T_n(g^{-1})$ except that, at the two diagonal corners of $T_n(g^{-1})$, the coefficient $1 + \theta^2$ is replaced by 1

$$T_n(g)^{-1} = \begin{pmatrix} 1 & -\theta & 0 & \dots \\ -\theta & 1+\theta^2 & -\theta & \dots \\ \dots & \dots & \dots & \dots \\ \dots & -\theta & 1+\theta^2 & -\theta \\ \dots & 0 & -\theta & 1 \end{pmatrix}.$$

It is not hard to see that $\det(T_n(g)^{-1}) = 1 - \theta^2$. In order to find the eigenvalues λ of the product $T_n(f)T_n(g)$, it is equivalent to calculate the zeros of its characteristic polynomial which correspond also to the zeros of $\det(M_n(t))$ where

$$M_n(t) = tT_n(f) - T_n(g)^{-1}$$

with $t = 1/\lambda$. As $T_n(f)$ and $T_n(g)^{-1}$ are both tridiagonal matrices, we can easily compute $\det(M_n(t))$. Via the same lines than in Lemma 11 of [1], we find that for n large enough, $M_n(t)$ is negative definite only on the domain $\mathscr{D} = \mathscr{D}_1 \cup \mathscr{D}_2$ with

$$\mathscr{D}_1 = \Big\{-2\theta^2 < p \leq -\theta^2 \text{ and } q^2 < -4\theta^2(p+\theta^2)\Big\},$$

$$\mathscr{D}_2 = \Big\{p < -2\theta^2 \text{ and } p < -|q|\Big\},$$

where $p = at-(1+\theta^2)$ and $q = t+2\theta$. In term of the variable λ, the inverses of the boundaries of $\mathscr{D}$ give the extrema of $\sigma(T(f)T(g))$ that is $\lambda_{\max}(f,g)$ and $\lambda_{\min}(f,g)$. After some tedious but straightforward calculations, we obtain three inverses of the boundaries

$$\frac{a-1}{(1+\theta)^2}, \qquad \frac{a+1}{(1-\theta)^2}, \qquad -\frac{1}{4\theta(1+a\theta)}.$$

Two of them coincide with $\inf(fg)$ and $\sup(fg)$. It only depends on the location of a with respect to $-(1+\theta^2)/(2\theta)$. The last one can be $\lambda_{\max}(f,g) > \sup(fg)$ or $\lambda_{\min}(f,g) < \inf(fg)$. It only depends on the sign of θ as well as on the location of a with respect to the interval $[a_\theta, b_\theta]$. More precisely, if $\theta > 0$ then $\lambda_{\max}(f,g) = \sup(fg)$ while

$$\lambda_{\min}(f,g) = \frac{1}{-4\theta(1+a\theta)} < \inf(fg) = \min\left(\frac{a-1}{(1+\theta)^2}, \frac{a+1}{(1-\theta)^2}\right)$$

if $a \in]a_\theta, b_\theta[$ and $\lambda_{\min}(f,g) = \inf(fg)$ otherwise. Moreover, if $\theta < 0$ then $\lambda_{\min}(f,g) = \inf(fg)$ while

$$\lambda_{\max}(f,g) = \frac{1}{-4\theta(1+a\theta)} > \sup(fg) = \max\left(\frac{a-1}{(1+\theta)^2}, \frac{a+1}{(1-\theta)^2}\right)$$

if $a \in]a_\theta, b_\theta[$ and $\lambda_{\max}(f,g) = \sup(fg)$ otherwise.

5 Toeplitz Operators and Functional Calculus

We will prove the following result which implies Corollary 1 for continuous functions.

Theorem 3. *Let f and g be two bounded continuous real functions with $g \geq 0$. For $\lambda \in \mathbb{R}$, the following properties are equivalent:*
(i) $\lambda \in \sigma(T(f)T(g))$,
(ii) λ is the limit of a sequence (λ_n) where $\lambda_n \in \sigma(T_n(f)T_n(g))$,
(iii) λ is the limit of a subsequence (λ_{n_k}) where $\lambda_{n_k} \in \sigma(T_{n_k}(f)T_{n_k}(g))$.

First, let us interpret the projection operators P_n and P as spectral projectors of the derivation operator and introduce the main ingredients of the proofs.

5.1 A Functional Point of View

We consider the Toeplitz operators $T(f)$ and $T_n(f)$ as the cut-off, in frequencies, of the operator of multiplication by f. To be more precise, let us introduce the Fourier transform, $\mathscr{F} : L^2(\mathbb{T}) \to \ell^2(\mathbb{Z})$, defined by

$$(\mathscr{F}u)_k = \widehat{u}_k = \frac{1}{2\pi}\int_{-\pi}^{\pi} u(x)e^{-ikx}dx.$$

The operator $\mathscr{F}$ is an isomorphism. We denote by $\mathscr{F}^{-1}$ its inverse, and we introduce the projections $\widehat{P}$ and $\widehat{P}_n$ as

$$\widehat{P} : \widehat{u} \in \ell^2(\mathbb{Z}) \longmapsto (\ldots, 0, 0, \widehat{u}_0, \widehat{u}_1, \ldots) \in \ell^2(\mathbb{Z})$$
$$\widehat{P}_n : \widehat{u} \in \ell^2(\mathbb{Z}) \longmapsto (\ldots, 0, 0, \widehat{u}_0, \widehat{u}_1, \ldots, \widehat{u}_n, 0, 0, \ldots) \in \ell^2(\mathbb{Z}).$$

On the other hand, if we identify $f \in L^\infty(\mathbb{T})$ with $L(f)$, the bounded operator defined on $L^2(\mathbb{T})$ by

$$u \in L^2(\mathbb{T}) \longmapsto fu \in L^2(\mathbb{T}),$$

we have

$$T(f) = PfP \quad \text{and} \quad T_n(f) = P_n f P_n,$$

with $P = \mathscr{F}^{-1}\widehat{P}\mathscr{F}$ and $P_n = \mathscr{F}^{-1}\widehat{P}_n\mathscr{F}$. In the following, we will systematically identify f with the operator $L(f)$. Since

$$\frac{1}{i}\frac{d}{dx}(e^{ikx}) = ke^{ikx},$$

the derivation operator D defined on

$$H^1(\mathbb{T}) = \left\{u \in L^2(\mathbb{T});\ \frac{d}{dx}u \in L^2(\mathbb{T})\right\} = \{u \in L^2(\mathbb{T});\ (k\hat{u}_k)_k \in \ell^2(\mathbb{Z})\}$$

by

$$D : u \in H^1(\mathbb{T}) \longmapsto \frac{1}{i}\frac{d}{dx}u \in L^2(\mathbb{T})$$

is self-adjoint on $L^2(\mathbb{T})$ and $\mathscr{F}D\mathscr{F}^{-1}$ is the diagonal operator $(k\delta_{k,j})_{k,j\in\mathbb{Z}}$. For any bounded Borel function φ, the bounded operator $\varphi(D)$ is defined with the help of the functional calculus for self-adjoint operators. It satisfies

$$\varphi(D) = \mathscr{F}^{-1}M(\varphi)\mathscr{F},$$

where $M(\varphi)$ is the operator

$$\widehat{u} \in \ell^2(\mathbb{Z}) \longmapsto (\ldots, \varphi(k)\widehat{u}_k, \ldots) \in \ell^2(\mathbb{Z}).$$

In particular, if $\mathbf{1}_I$ denotes the indicator function of the interval I, we have

$$\mathbf{1}_{[0,+\infty[}(D) = P \qquad \text{and} \qquad \mathbf{1}_{[0,n]}(D) = \mathbf{1}_{[0,1]}(n^{-1}\, D) = P_n.$$

Moreover, note that if $\operatorname{supp}(\varphi) \subset [a,b]$, we have the trivial properties

$$\mathbf{1}_{[a,b]}(D)\,\varphi(D) = \varphi(D) \qquad \text{and} \qquad \mathbf{1}_{[a,b]}(D)\, e^{ikx} = e^{ikx}\, \mathbf{1}_{[a-k,b-k]}(D).$$

In the rest of the paper, a function is a $o^c_{a\to b}(1)$ if, for each c fixed, the function goes to 0 as a tends to b. In the same way, a function is a $\mathcal{O}^c(1)$ if, for each c fixed, the function is a $\mathcal{O}(1)$.

5.2 A Commutator Estimate

In this subsection, we recall a standard result of the functional analysis. For $\rho \in \mathbb{R}$, we denote by $S^\rho(\mathbb{R})$ the class of functions φ in $C^\infty(\mathbb{R})$ such that

$$|\partial_s^k \varphi(s)| \le C_k \langle s\rangle^{\rho-k},$$

for $k \ge 0$. Here $\langle x\rangle = (1+|x|^2)^{1/2}$.

Lemma 1 (Lemma C.3.2 of [7]). *Let A, B be self-adjoint operators on a Hilbert space with B and $[A,B]$ bounded. If $\varphi \in S^\rho(\mathbb{R})$ with $\rho < 1$, then*

$$\|[\varphi(A), B]\| \le C_\varphi \|[A,B]\|.$$

Here, $[A,B] = AB - BA$ denotes the commutator. The constant C_φ only depends on φ.

Applying this lemma, we immediately obtain

Lemma 2. *Let $f \in C^0(\mathbb{T})$ and $\varphi \in S^\rho(\mathbb{R})$ with $\rho \le 0$. Then*

$$[\varphi(\varepsilon D), f] = o_{\varepsilon\to 0}(1).$$

Proof. By Weierstrass's theorem, there exist $f_k \in C^1(\mathbb{T})$ satisfying $f_k \to f$ in $L^\infty(\mathbb{T})$. Then, viewed as operators, we have $f_k \to f$. Remark that $[\varepsilon D, f_k] = -\varepsilon i f_k'$. From Lemma 1, we obtain

$$\|[\varphi(\varepsilon D), f_k]\| \le \varepsilon C_\varphi \|f_k'\|_\infty.$$

Then, using the assumption that φ is bounded,

$$\begin{aligned}[\varphi(\varepsilon D), f] &= [\varphi(\varepsilon D), f_k] + o_{k\to\infty}(1) = \mathscr{O}^k(\varepsilon) + o_{k\to\infty}(1)\\ &= o_{\varepsilon\to 0}(1),\end{aligned}$$

since $[\varphi(\varepsilon D), f]$ does not depend on k. □

5.3 Essential Spectrum of the Product of Toeplitz Operators

Here, we recall, in our setting, a consequence of a theorem of Coburn [5] concerning the essential spectrum of the product of Toeplitz operators. This result has been extended by Douglas to a more general framework (see [10, Theorem 4.5.10]). We shall give in Sect. 7 an alternative proof of the following theorem, more related to our approach.

Theorem 4 (Coburn). *Let f and g be two bounded continuous real functions with $g \geq 0$. The bounded self-adjoint operator $T(g)^{1/2}T(f)T(g)^{1/2}$ satisfies on* Im P

$$\sigma_{\text{ess}}\big(T(g)^{1/2}T(f)T(g)^{1/2}\big) = \big[\inf(fg), \sup(fg)\big].$$

Here, $\sigma_{\text{ess}}(A)$ denotes the essential spectrum of A.

In Theorem 4, the operator $T(g)^{1/2}T(f)T(g)^{1/2}$ is viewed as an operator on Im P. On $L^2(\mathbb{T})$, this operator is a block diagonal operator with respect to the orthogonal sum $L^2 = \operatorname{Im} P \oplus^{\perp} \operatorname{Im}(1-P)$ and is equal to 0 on $\operatorname{Im}(1-P)$. In particular, we have

Remark 2. If the operator $T(g)^{1/2}T(f)T(g)^{1/2}$ is viewed on $L^2(\mathbb{T})$, we have

$$\sigma_{\text{ess}}\big(T(g)^{1/2}T(f)T(g)^{1/2}\big) = \big[\inf(fg), \sup(fg)\big] \cup \{0\}.$$

6 Proof of Theorem 3

The goal of this section is to prove Theorem 3. First of all, one can observe that part (ii) clearly implies (iii). In the next subsection, we first show that (i) implies (ii).

6.1 The Implication (i) Gives (ii)

Lemma 3. *Let f and g be two bounded piecewise continuous real functions with $g \geq 0$. Then,*

$$T_n(g)^{1/2}T_n(f)T_n(g)^{1/2} \longrightarrow T(g)^{1/2}T(f)T(g)^{1/2}$$

strongly on $L^2(\mathbb{T})$. *If* λ *belongs to the spectrum of* $T(f)T(g)$ *on* Im P, *then there exists an eigenvalue* λ_n *of* $T_n(f)T_n(g)$ *on* Im P_n *such that* $\lambda_n \to \lambda$.

Proof. Since $P_n \longrightarrow P$, it follows from Lemma III.3.8 of [9] that for all $f \in L^\infty(\mathbb{T})$, $T_n(f) \longrightarrow T(f)$. In particular, from Problem VI.14 of [11] (see also Theorem VI.9 of [11]), $T_n(g)^{1/2} \longrightarrow T(g)^{1/2}$. Consequently, we deduce from Lemma III.3.8 of [9] that

$$T_n(g)^{1/2}T_n(f)T_n(g)^{1/2} \longrightarrow T(g)^{1/2}T(f)T(g)^{1/2}, \tag{10}$$

on $L^2(\mathbb{T})$. In particular, we obtain on Im P

$$T_n(g)^{1/2}T_n(f)T_n(g)^{1/2} + M(P - P_n) \longrightarrow T(g)^{1/2}T(f)T(g)^{1/2},$$

for all $M \in \mathbb{R}$. We choose $\mu = \|f\|_\infty \|g\|_\infty$ and $M = \mu + 1$. Therefore, it follows from Corollary VIII.1.6 together with Theorem VIII.1.14 of [9] that, for each λ belonging to the spectrum $\sigma(T(f)T(g)) = \sigma(T(g)^{1/2}T(f)T(g)^{1/2})$ on Im P, there exists an eigenvalue λ_n of the matrix

$$T_n(g)^{1/2}T_n(f)T_n(g)^{1/2} + M(P - P_n),$$

on Im P such that $\lambda_n \to \lambda$. As $\|T(g)^{1/2}T(f)T(g)^{1/2}\| \leq \mu$, we necessarily have $\lambda \in [-\mu, \mu]$ and then $M \geq |\lambda| + 1$. In particular, for n large enough, $M > |\lambda_n| + 1/2$. Therefore, λ_n is an eigenvalue of $T_n(g)^{1/2}T_n(f)T_n(g)^{1/2}$ on Im P_n because

$$T_n(g)^{1/2}T_n(f)T_n(g)^{1/2} + M(P - P_n) = T_n(g)^{1/2}T_n(f)T_n(g)^{1/2} \oplus^\perp M(P - P_n),$$

is a block diagonal operator with respect to the orthogonal sum Im $P =$ Im $P_n \oplus^\perp$ Im$(P - P_n)$. □

6.2 The Implication (iii) Gives (i)

Let λ_N be a sequence of eigenvalues of $T_N(f)T_N(g)$ such that $\lambda_N \to \lambda \in \mathbb{R}$. Here N is a subsequence of $\mathbb{N}$ and we have to show that λ is in the spectrum of $T(f)T(g)$. From Theorem 4, we know that $[\inf(fg), \sup(fg)]$ is always inside the spectrum of $T(f)T(g)$. Thus, we can assume that

$$\lambda \notin \big[\inf(fg), \sup(fg)\big]. \tag{11}$$

By Weierstrass's theorem, there exists a sequence of functions $(f_k) \in C^\infty(\mathbb{T})$ such that $f_k \to f$ in $L^\infty(\mathbb{T})$ and supp $\widehat{f_k} \subset [-k, k]$. We also consider (g_k) a sequence corresponding to g with the same properties mutatis mutandis. In particular, for all $n \in \mathbb{N}$,

$$T_n(f) = T_n(f_k) + o_{k\to\infty}(1) \qquad \text{and} \qquad T(f) = T(f_k) + o_{k\to\infty}(1). \tag{12}$$

Recall that, by definition, a $o_{k\to\infty}(1)$ is uniform with respect to n.

Finally, let $u_N \in \operatorname{Im} P_N$ be an eigenvector of $T_N(f)T_N(g)$ associated with λ_N and satisfying $\|u_N\| = 1$. From (12),

$$T_N(f)T_N(g)u_N = \lambda_N u_N = \lambda u_N + o_{N\to\infty}(1) \tag{13}$$

$$T_N(f_k)T_N(g_k)u_N = \lambda u_N + o_{N\to\infty}(1) + o_{k\to\infty}(1). \tag{14}$$

In the following, we denote $D_n = n^{-1}D$.

6.2.1 Localization of the Eigenvectors

Lemma 4. *Let $\varphi \in C_0^\infty(]0,1[)$. Then, in $L^2(\mathbb{T})$ norm,*

$$\varphi(D_N)u_N = o_{N\to\infty}(1).$$

Proof. From Lemma 2, we have

$$\begin{aligned}\varphi(D_N)T_N(f) &= \varphi(D_N)\mathbf{1}_{[0,1]}(D_N)f\mathbf{1}_{[0,1]}(D_N) = \varphi(D_N)f\mathbf{1}_{[0,1]}(D_N)\\ &= f\varphi(D_N)\mathbf{1}_{[0,1]}(D_N) + o_{N\to\infty}(1)\\ &= f\varphi(D_N) + o_{N\to\infty}(1).\end{aligned} \tag{15}$$

Applying two times this estimate, we obtain

$$\begin{aligned}\varphi(D_N)T_N(f)T_N(g)u_N &= f\varphi(D_N)T_N(g)u_N + o_{N\to\infty}(1)\\ &= fg\varphi(D_N)u_N + o_{N\to\infty}(1).\end{aligned}$$

Then, (13) gives

$$(fg-\lambda)\varphi(D_N)u_N = o_{N\to\infty}(1).$$

Since $\lambda \notin [\inf(fg), \sup(fg)]$, the function $(fg-\lambda)^{-1}$ belongs to $L^\infty(\mathbb{T})$ and the lemma follows from the last equation. □

Now, we take $\varphi \in C_0^\infty(]0,1[,[0,1])$ such that $\varphi = 1$ near $[\varepsilon, 1-\varepsilon]$ for $\varepsilon > 0$ small enough (we choose $\varepsilon = 1/8$). Let $\varphi^- \in C_0^\infty([-\varepsilon, 2\varepsilon],[0,1])$ and $\varphi^+ \in C_0^\infty([1-2\varepsilon, 1+\varepsilon],[0,1])$ be two functions such that

$$\varphi^- + \varphi + \varphi^+ = 1,$$

in the neighborhood of $[0,1]$. Set

$$u_N^{\pm} = \varphi^{\pm}(D_N)u_N = \varphi^{\pm}(D_N)\mathbf{1}_{[0,1]}(D_N)u_N. \tag{16}$$

As $\|u_N\| = 1$, it follows from Lemma 4 that

$$\|u_N^- + u_N^+\| = 1 + o_{N\to\infty}(1). \tag{17}$$

In particular, we can assume, up to the extraction of a subsequence, that

$$\forall N \quad \|u_N^-\| \geq 1/3 \qquad \text{or} \qquad \forall N \quad \|u_N^+\| \geq 1/3.$$

In the next section, we will suppose that

$$\|u_N^-\| \geq 1/3. \tag{18}$$

The case $\|u_N^+\| \geq 1/3$ follows essentially the same lines and is treated in Sect. 6.2.3. But before, we show that u_N^- and u_N^+ are both quasimodes of $T_N(f)T_N(g)$ (this means that they are eigenvectors modulo a small term).

Lemma 5. *We have*

$$T_N(f_k)T_N(g_k)u_N^{\pm} = \lambda u_N^{\pm} + o_{k\to\infty}(1) + o^k_{N\to\infty}(1).$$

Proof. As in (15), using Lemma 2, we get

$$\begin{aligned}
T_N(f_k)T_N(g_k)u_N^{\pm} &= \mathbf{1}_{[0,1]}(D_N)f_k\mathbf{1}_{[0,1]}(D_N)g_k\mathbf{1}_{[0,1]}(D_N)\varphi^{\pm}(D_N)u_N \\
&= \mathbf{1}_{[0,1]}(D_N)f_k\mathbf{1}_{[0,1]}(D_N)g_k\varphi^{\pm}(D_N)\mathbf{1}_{[0,1]}(D_N)u_N \\
&= \mathbf{1}_{[0,1]}(D_N)f_k\mathbf{1}_{[0,1]}(D_N)\varphi^{\pm}(D_N)g_k\mathbf{1}_{[0,1]}(D_N)u_N + o^k_{N\to\infty}(1) \\
&= \mathbf{1}_{[0,1]}(D_N)\varphi^{\pm}(D_N)f_k\mathbf{1}_{[0,1]}(D_N)g_k\mathbf{1}_{[0,1]}(D_N)u_N + o^k_{N\to\infty}(1) \\
&= \varphi^{\pm}(D_N)T_N(f_k)T_N(g_k)u_N + o^k_{N\to\infty}(1). \qquad (19)
\end{aligned}$$

The lemma follows from (14), (16) and the last identity. □

6.2.2 Concentration Near the Low Frequencies

Here, we assume (18) and we prove that u_N^-, viewed as an element of $\operatorname{Im} P$, is a quasimode of $T(f_k)T(g_k)$.

Lemma 6. *For* $4k \leq N$, *we have*

$$T(f_k)T(g_k)u_N^- = T_N(f_k)T_N(g_k)u_N^-.$$

Remark 3. In fact, for $4k \leq N \leq n$, we have

$$T_n(f_k)T_n(g_k)u_N^- = T_N(f_k)T_N(g_k)u_N^-.$$

Proof. Recall that, if u, v are two functions of $L^2(\mathbb{T})$ such that $\operatorname{supp}\widehat{u} \subset [a,b]$ and $\operatorname{supp}\widehat{v} \subset [c,d]$, then $\operatorname{supp}\widehat{uv} \subset [a+c, b+d]$. By definition,

$$T(f_k)T(g_k)u_N^- = Pf_k Pg_k P\varphi^-(D_N)u_N = Pf_k Pg_k P_N\varphi^-(D_N)u_N. \tag{20}$$

Since $\operatorname{supp}\widehat{g_k} \subset [-k,k]$ and $\operatorname{supp}\mathscr{F}(P_N\varphi^-(D_N)u_N) \subset [0, N/4]$, the Fourier transform of the function $g_k P_N\varphi^-(D_N)u_N$ is supported inside $[-k, N/4+k] \subset [-k, N]$. In particular,

$$Pg_k P_N\varphi^-(D_N)u_N = P_N g_k P_N\varphi^-(D_N)u_N, \tag{21}$$

and the Fourier transform of this function is supported inside $[0, N/4+k]$. As before, the Fourier transform of

$$f_k P_N g_k P_N\varphi^-(D_N)u_N$$

is supported inside $[-k, N/4+2k] \subset [-k, N]$. Then

$$Pf_k P_N g_k P_N\varphi^-(D_N)u_N = P_N f_k P_N g_k P_N\varphi^-(D_N)u_N. \tag{22}$$

The lemma follows from (20) to (22). □

From (12), Lemmas 5 and 6, we get

$$T(f)T(g)u_N^- = \lambda u_N^- + o_{k\to\infty}(1) + o^k_{N\to\infty}(1), \tag{23}$$

for $4k \le N$. If $\lambda \notin \sigma(T(f)T(g))$, the operator $T(f)T(g) - \lambda$ is invertible and then

$$u_N^- = o_{k\to\infty}(1) + o^k_{N\to\infty}(1).$$

From (18), we obtain $1/3 \le o_{k\to\infty}(1) + o^k_{N\to\infty}(1)$. Taking k large enough and then N large enough, it is clear that this is impossible. Thus,

$$\lambda \in \sigma(T(f)T(g)),$$

which implies Theorem 3 under Assumption (18).

6.2.3 Concentration Near the High Frequencies

We replace the Assumption (18) by $\|u_N^+\| \ge 1/3$. Let J be the isometry $f \mapsto \widetilde{f}$ in $L^2(\mathbb{T})$. One can observe that $J(uv) = J(u)J(v)$. Using the notation $P_{[a,b]} = \mathbf{1}_{[a,b]}(D)$, we have $P_{[a,b]}J = JP_{[-b,-a]}$ and $P_{[a,b]}e^{icx} = e^{icx}P_{[a-c,b-c]}$. Combining these identities with Lemma 5, we get

$$
\begin{aligned}
T_N(Jf_k)T_N(Jg_k)e^{iNx}(Ju_N^+) &= P_{[0,N]}(Jf_k)P_{[0,N]}(Jg_k)P_{[0,N]}e^{iNx}(Ju_N^+)\\
&= e^{iNx}P_{[-N,0]}(Jf_k)P_{[-N,0]}(Jg_k)P_{[-N,0]}(Ju_N^+)\\
&= e^{iNx}JP_{[0,N]}f_k P_{[0,N]}g_k P_{[0,N]}u_N^+\\
&= \lambda e^{iNx}(Ju_N^+) + o_{k\to\infty}(1) + o^k_{N\to\infty}(1). \qquad (24)
\end{aligned}
$$

In particular, $\widetilde{u_N^-} = e^{iNx}(Ju_N^+)$ satisfies $\|\widetilde{u_N^-}\| \geq 1/3$,

$$T_N(Jf_k)T_N(Jg_k)\widetilde{u_N^-} = \lambda\widetilde{u_N^-} + o_{k\to\infty}(1) + o^k_{N\to\infty}(1),$$

and the support of the Fourier transform of $\widetilde{u_N^-}$ is inside $[0, N/4]$. Hence, we can apply the method developed in the case $\|u_N^-\| \geq 1/3$. The unique difference is that f, g are replaced by $\widetilde{f}, \widetilde{g}$. Then, we obtain

$$\lambda \in \sigma\big(T(\widetilde{f})T(\widetilde{g})\big).$$

Theorem 3 follows from the following lemma and $\sigma(T(f)T(g)) = \sigma(T(g)T(f))$ (the spectrum of $T(f)T(g)$ is real and $(T(f)T(g) - z)^* = T(g)T(f) - \overline{z}$).

Lemma 7. *Let $f, g \in L^\infty(\mathbb{T})$. Then*

$$\sigma\big(T(\widetilde{f})T(\widetilde{g})\big) = \sigma\big(T(g)T(f)\big).$$

Proof. For A a bounded linear operator on L^2, we define A^t by

$$(A^t u, v) = (u, \overline{A\overline{v}}),$$

for all $u, v \in L^2$. Simple calculi give $f^t = f$, $P^t_{[a,b]} = P_{[-b,-a]}$, $(AB)^t = B^t A^t$ and then

$$T(f)^t = \big(P_{[0,+\infty[}fP_{[0,+\infty[}\big)^t = P_{]-\infty,0]}fP_{]-\infty,0]}.$$

By the same way, since $J = J^* = J^{-1}$,

$$JP_{]-\infty,0]}fP_{]-\infty,0]}J = P_{[0,+\infty[}\widetilde{f}P_{[0,+\infty[} = T(\widetilde{f}).$$

Combining these identities concerning ^{t} and J, we get

$$
\begin{aligned}
J(T(\widetilde{f})T(\widetilde{g}))^t J^{-1} &= J\big(P_{]-\infty,0]}\widetilde{g}P_{]-\infty,0]}\big)\big(P_{]-\infty,0]}\widetilde{f}P_{]-\infty,0]}\big)J\\
&= T(g)T(f). \qquad (25)
\end{aligned}
$$

Since $JA^tJ - z = J(A-z)^tJ$, A and JA^tJ have the same spectrum and the lemma follows. □

7 Proof of Theorem 4

We give here an alternative proof of Coburn's theorem. Let $\psi \in C^\infty(\mathbb{R})$ satisfying $\psi = 1$ near $[2, +\infty[$ and $\psi = 0$ near $]-\infty, 1]$. For $\varepsilon > 0$, we have on Im P

$$\begin{aligned} T(g)^{1/2}T(f)T(g)^{1/2} &= T(g)^{1/2}\psi(\varepsilon D)T(f)\psi(\varepsilon D)T(g)^{1/2} + \widetilde{R}_\varepsilon \\ &= T(g)^{1/2}\psi(\varepsilon D)f\psi(\varepsilon D)T(g)^{1/2} + \widetilde{R}_\varepsilon, \end{aligned} \tag{26}$$

where

$$\widetilde{R}_\varepsilon = T(g)^{1/2}(1-\psi(\varepsilon D))T(f)\psi(\varepsilon D)T(g)^{1/2}+T(g)^{1/2}T(f)(1-\psi(\varepsilon D))T(g)^{1/2},$$

is a self-adjoint operator of finite rank. Recall that if $A \geq 0$ is a bounded operator with $\|A\| \leq 1$, then

$$A^{1/2} = \sum_{j=0}^{+\infty} c_j(1-A)^j,$$

where $\|1 - A\| \leq 1$ and $\sum_{j\geq 0}|c_j| \leq 2 < +\infty$. On the other hand, Lemma 2 implies

$$\begin{aligned} T(g)\psi(\varepsilon D) &= PgP\psi(\varepsilon D) = Pg\psi(\varepsilon D) = P\psi(\varepsilon D)g + o_{\varepsilon\to 0}(1) \\ &= \psi(\varepsilon D)g + o_{\varepsilon\to 0}(1), \end{aligned} \tag{27}$$

Then, for a fixed $\delta > 0$ such that $\|T(g)\| \leq \|g\|_\infty < \delta^{-1}$, we have

$$\begin{aligned} T(g)^{1/2}\psi(\varepsilon D) &= \delta^{-1/2}T(\delta g)^{1/2}\psi(\varepsilon D) \\ &= \delta^{-1/2}\sum_{j=0}^{+\infty} c_j(1-T(\delta g))^j\psi(\varepsilon D) \\ &= \delta^{-1/2}\sum_{j=0}^{J} c_j(1-T(\delta g))^j\psi(\varepsilon D) + o_{J\to\infty}(1) \\ &= \delta^{-1/2}\psi(\varepsilon D)\sum_{j=0}^{J} c_j(1-\delta g)^j + o_{J\to\infty}(1) + o^J_{\varepsilon\to 0}(1) \\ &= \delta^{-1/2}\psi(\varepsilon D)(\delta g)^{1/2} + o_{J\to\infty}(1) + o^J_{\varepsilon\to 0}(1) \\ &= \psi(\varepsilon D)g^{1/2} + o_{\varepsilon\to 0}(1), \end{aligned} \tag{28}$$

since these quantities do not depend on J. Using this identity and its adjoint, (26) becomes

$$T(g)^{1/2}T(f)T(g)^{1/2} = \psi(\varepsilon D)fg\psi(\varepsilon D) + \widetilde{R}_\varepsilon + o_{\varepsilon\to 0}(1)$$
$$= T(fg) + R_\varepsilon + e_\varepsilon, \tag{29}$$

where $e_\varepsilon = o_{\varepsilon\to 0}(1)$ and

$$R_\varepsilon = \widetilde{R}_\varepsilon + (\psi(\varepsilon D) - 1)T(fg)\psi(\varepsilon D) + T(fg)(\psi(\varepsilon D) - 1),$$

is a self-adjoint operator of finite rank. In particular, e_ε is a self-adjoint operator. Since, on $\operatorname{Im} P$

$$\inf(fg) \le T(fg) \le \sup(fg),$$

we get $\sigma(T(fg)+e_\varepsilon) \subset [\inf(fg) - o_{\varepsilon\to 0}(1), \sup(fg) + o_{\varepsilon\to 0}(1)]$. As R_ε is of finite rank, we obtain, from Weyl's theorem [11, Theorem S.13],

$$\sigma_{\mathrm{ess}}\big(T(g)^{1/2}T(f)T(g)^{1/2}\big) = \sigma_{\mathrm{ess}}(T(fg) + e_\varepsilon)$$
$$\subset [\inf(fg) - o_{\varepsilon\to 0}(1), \sup(fg) + o_{\varepsilon\to 0}(1)].$$

As the essential spectrum of $T(g)^{1/2}T(f)T(g)^{1/2}$ does not depend on ε, we get

$$\sigma_{\mathrm{ess}}\big(T(g)^{1/2}T(f)T(g)^{1/2}\big) \subset [\inf(fg), \sup(fg)], \tag{30}$$

which is the first inclusion of Coburn's theorem.

Now, let $\varphi \in C^\infty([-1, 1], [0, 1])$ with $\|\varphi\|_{L^2} = 1$. For $x_0 \in \mathbb{T}$ and $\alpha, \beta \in \mathbb{N}$, we set

$$u = \alpha^{1/2}\varphi\big(\alpha(x - x_0)\big)e^{i\beta x} \quad \text{and} \quad v = Pu \in \operatorname{Im} P,$$

which satisfies $\|u\| = 1$. We have

$$(1 - P)u = \alpha^{1/2}\mathbf{1}_{]-\infty,0]}(D)e^{i\beta x}\varphi\big(\alpha(x - x_0)\big)$$
$$= \alpha^{1/2}e^{i\beta x}\mathbf{1}_{]-\infty,-\beta]}(D)\varphi\big(\alpha(x - x_0)\big)$$
$$= \alpha^{1/2}e^{i\beta x}\mathbf{1}_{]-\infty,-\beta]}(D)(D + i)^{-M}(D + i)^M\varphi\big(\alpha(x - x_0)\big)$$
$$= \mathcal{O}\big(\beta^{-M}\alpha^M\big), \tag{31}$$

in L^2 norm for any $M \in \mathbb{N}$. Moreover, for a continuous function ℓ, we have

$$\ell u = \ell(x_0)\alpha^{1/2}\varphi\big(\alpha(x - x_0)\big)e^{i\beta x} + o_{\alpha\to\infty}(1), \tag{32}$$

in L^2 norm. Using that $\|T(\ell)^{1/2}\| \le \|\ell\|_\infty^{1/2}$, for all function $\ell \in L^\infty$ with $\ell \ge 0$, we get

$$
\begin{aligned}
T(f)T(g)v = PfPgPu &= PfPgu + \mathcal{O}(\alpha\beta^{-1}) \\
&= g(x_0)PfPu + \mathcal{O}(\alpha\beta^{-1}) + o_{\alpha\to\infty}(1) \\
&= g(x_0)Pfu + \mathcal{O}(\alpha\beta^{-1}) + o_{\alpha\to\infty}(1) \\
&= (fg)(x_0)Pu + \mathcal{O}(\alpha\beta^{-1}) + o_{\alpha\to\infty}(1) \\
&= (fg)(x_0)v + \mathcal{O}(\alpha\beta^{-1}) + o_{\alpha\to\infty}(1). \qquad (33)
\end{aligned}
$$

Taking $\beta = \alpha^2 \to +\infty$, (31) implies $\|v\| = 1 + o_{\alpha\to\infty}(1)$. On the other hand, (33) leads to

$$T(f)T(g)v = (fg)(x_0)v + o_{\alpha\to\infty}(1).$$

Then, $(fg)(x_0) \in \sigma(T(f)T(g)) = \sigma(T(g)^{1/2}T(f)T(g)^{1/2})$. Therefore,

$$\big[\inf(fg), \sup(fg)\big] \subset \sigma\big(T(g)^{1/2}T(f)T(g)^{1/2}\big). \qquad (34)$$

Recall that the essential spectrum of a self-adjoint bounded operator on an infinite Hilbert space is never empty. Therefore, if $\inf(fg) = \sup(fg)$, (30) implies the theorem.

Assume now that $\inf(fg) < \sup(fg)$. Then $[\inf(fg), \sup(fg)]$ is an interval with non empty interior. From the definition of the essential spectrum, this interval is necessarily inside the essential spectrum of $T(g)^{1/2}T(f)T(g)^{1/2}$. This achieves the proof of the second inclusion of Coburn's theorem.

Acknowledgements The authors would like to thanks A. Böttcher for providing the reference of Roch and Silbermann. They also thank the anonymous referee for his careful reading of the paper.

References

1. B. Bercu, F. Gamboa, A. Rouault, Large deviations for quadratic forms of stationary Gaussian processes. Stoch. Process. Appl. **71**(1), 75–90 (1997)
2. A. Böttcher, B. Silbermann, *Introduction to Large Truncated Toeplitz Matrices* (Springer, New York, 1999)
3. A. Böttcher, B. Silbermann, *Analysis of Toeplitz Operators*, 2nd edn. Springer Monographs in Mathematics (Springer, Berlin, 2006); Prepared jointly with Alexei Karlovich
4. W. Bryc, A. Dembo, Large deviations for quadratic functionals of Gaussian processes. J. Theor. Probab. **10**(2), 307–332 (1997); Dedicated to Murray Rosenblatt
5. L.A. Coburn, The C^*-algebra generated by an isometry. Bull. Am. Math. Soc. **73**, 722–726 (1967)
6. A. Dembo, O. Zeitouni, *Large Deviations Techniques and Applications*, 2nd edn. Applications of Mathematics (New York), vol. 38 (Springer, New York, 1998)
7. J. Dereziński, C. Gérard, *Scattering Theory of Classical and Quantum N-Particle Systems*. Texts and Monographs in Physics (Springer, Berlin, 1997)
8. U. Grenander, G. Szegö, *Toeplitz Forms and Their Applications*. California Monographs in Mathematical Sciences (University of California Press, Berkeley, 1958)

9. T. Kato, *Perturbation Theory for Linear Operators*. Classics in Mathematics (Springer, Berlin, 1995); Reprint of the 1980 edition
10. N. Nikolski, *Operators, Functions, and Systems: An Easy Reading*, vol. 1, Hardy, Hankel, and Toeplitz (translated from the French by A. Hartmann), Mathematical Surveys and Monographs, vol. 92 (American Mathematical Society, RI, 2002)
11. M. Reed, B. Simon, *Methods of Modern Mathematical Physics. I*, 2nd edn. Functional Analysis (Academic, New York, 1980)
12. S. Roch, B. Silbermann, Limiting sets of eigenvalues and singular values of Toeplitz matrices. Asymptotic Anal. **8**, 293–309 (1994)
13. S. Serra-Capizzano, Distribution results on the algebra generated by Toeplitz sequences: A finite-dimensional approach. Linear Algebra Appl. **328**(1–3), 121–130 (2001)
14. H. Widom, Asymptotic behavior of block Toeplitz matrices and determinants. II. Adv. Math. **21**(1), 1–29 (1976)

Girsanov Theory Under a Finite Entropy Condition

Christian Léonard

Abstract This paper is about Girsanov's theory. It (almost) doesn't contain new results but it is based on a simplified new approach which takes advantage of the (weak) extra requirement that some relative entropy is finite. Under this assumption, we present and prove all the standard results pertaining to the absolute continuity of two continuous-time processes on $\mathbb{R}^d$ with or without jumps. We have tried to give as much as possible a self-contained presentation.
The main advantage of the finite entropy strategy is that it allows us to replace martingale representation results by the simpler Riesz representations of the dual of a Hilbert space (in the continuous case) or of an Orlicz function space (in the jump case).

Keywords Stochastic processes • Relative entropy • Girsanov's theory • Diffusion processes • Processes with jumps

AMS Classification: 60G07, 60J60, 60J75, 60G44

1 Introduction

This paper is about Girsanov's theory. It (almost) doesn't contain new results but it is based on a simplified new approach which takes advantage of the (weak) extra requirement that some relative entropy is finite. Under this assumption, we present and prove all the standard results pertaining to the absolute continuity of two continuous-time processes on $\mathbb{R}^d$ with or without jumps.

This article intends to look like lecture notes and we have tried to give as much as possible a self-contained presentation of Girsanov's theory. The author hopes that it

C. Léonard (✉)
Modal-X. Université Paris Ouest. Bât.G, 200 av. de la République, 92001 Nanterre, France
e-mail: christian.leonard@u-paris10.fr

C. Donati-Martin et al. (eds.), *Séminaire de Probabilités XLIV*, Lecture Notes in Mathematics 2046, DOI 10.1007/978-3-642-27461-9_20,

could be useful for students and also to readers already acquainted with stochastic calculus.

The main advantage of the finite entropy strategy is that it allows us to replace martingale representation results by the simpler Riesz representations of the dual of a Hilbert space (in the continuous case) or of an Orlicz function space (in the jump case). The gain is especially interesting in the jump case where martingale representation results are not easy, see [1]. Another feature of this simplified approach is that very few about exponential martingales is needed.

Girsanov's theory studies the relation between a reference process R and another process P which is assumed to be absolutely continuous with respect to R. In particular, it is known that if R is the law of an $\mathbb{R}^d$-valued semimartingale, then P is also the law of a semimartingale. In its wide meaning, this theory also provides us with a formula for the Radon-Nikodým density $\frac{dP}{dR}$.

In this article, we assume that the probability measure P has its relative entropy with respect to R:

$$H(P|R) := \begin{cases} E_P \log\left(\frac{dP}{dR}\right) \in [0,\infty] & \text{if } P \ll R \\ +\infty & \text{otherwise,} \end{cases}$$

which is finite, i.e.

$$H(P|R) = E_R\left[\frac{dP}{dR}\log\left(\frac{dP}{dR}\right)\right] < \infty. \tag{1}$$

In comparison, requiring $P \ll R$ only amounts to assume that

$$E_R\left(\frac{dP}{dR}\right) < \infty \tag{2}$$

since P has a finite mass. We are going to take advantage of the only difference between (1) and (2) which is the stronger integrability property carried by the extra term $\log \frac{dP}{dR}$.

A key argument of this approach is the variational representation of the relative entropy which is stated at Lemma 1.

A clear exposition of the general Girsanov theorems, with no explicit expression of $\frac{dP}{dR}$ in terms of the characteristics of the processes, is given in Protter's textbook [4]. The most complete results about Girsanov's theory for $\mathbb{R}^d$-valued processes, including explicit formulas for $\frac{dP}{dR}$, are available in Jacod's textbook [2]. An alternate presentation of this realm of results is also given in the later book by Jacod and Shiryaev [3]. A good standard reference in the continuous case is Revuz and Yor's textbook [6] about continuous martingales.

Next Sect. 2 is devoted to the statement of the main results. At Sect. 3, we state and prove the above mentioned variational representation of the relative entropy. At Sects. 4 and 5, we present the proofs of Theorems 1 and 2 which correspond to the continuous case. At Sect. 6, we give the proofs of Theorems 3 and 4 which correspond to the jump case.

2 Statement of the Results

We separate the cases when the sample paths are continuous and when they exhibit jumps.

2.1 *Continuous Processes in* $\mathbb{R}^d$

The paths to be considered are built on the time interval $[0, 1]$. An $\mathbb{R}^d$-valued continuous stochastic process is a random variable taking its values in the set

$$\Omega = C([0,1], \mathbb{R}^d)$$

of all continuous paths from $[0, 1]$ to $\mathbb{R}^d$. The canonical process $(X_t)_{t\in[0,1]}$ is defined by

$$X_t(\omega) = \omega_t, \quad t \in [0,1], \omega = (\omega_s)_{s\in[0,1]} \in \Omega.$$

In other words, $X = (X_t)_{t\in[0,1]}$ is the identity on Ω and $X_t : \Omega \to \mathbb{R}^d$ is the tth projection. The set Ω is endowed with the σ-field $\sigma(X_t; t \in [0,1])$ which is generated by the canonical projections. We also consider the canonical filtration $\Big(\sigma(X_{[0,t]}); t \in [0,1]\Big)$ where for each t, $X_{[0,t]} := (X_s)_{s\in[0,t]}$.

Let us give ourselves a reference probability measure R on Ω such that X admits the R-semimartingale decomposition

$$X = X_0 + B^R + M^R, \quad R\text{-a.s.} \tag{3}$$

This means that B^R is an adapted process with bounded variation sample paths R-a.s. and M^R is a local R-martingale, i.e. there exists an increasing sequence of stopping times $(\tau_k)_{k\geq 1}$ which converges to infinity R-a.s. and such that for each $k \geq 1$, the stopped process $t \mapsto M^R_{t\wedge\tau_k}$ is a uniformly integrable R-martingale.

As a typical example, one may think of the solution to the SDE (if it exists)

$$X_t = X_0 + \int_{[0,t]} b_s(X_{[0,s]})\, ds + \int_{[0,t]} \sigma_s(X_{[0,s]})\, dW_s, \quad 0 \leq t \leq 1 \tag{4}$$

where W is a Wiener process on $\mathbb{R}^d$ and $b : [0,1] \times \Omega \to \mathbb{R}^d$ and $\sigma : [0,1] \times \Omega \to \mathbf{M}_{d\times d}$ are locally bounded. In this situation, a natural localizing sequence $(\tau_k)_{k\geq 1}$ is the sequence of exit times from the Euclidean balls of radius k, $B^R_t = \int_0^t b_s(X_{[0,s]})\, ds$ has absolutely continuous sample paths R-a.s. and $M^R_t = \int_0^t \sigma_s(X_{[0,s]})\, dW_s$ has the quadratic variation

$$[M^R, M^R]_t = \int_0^t a_s\, ds \quad R\text{-a.s.} \tag{5}$$

where $a_t := \sigma_t\sigma_t^*(X_{[0,t]})$ takes its values in the set $\mathbf{S}_+$ of all positive semi-definite $d \times d$ matrices.

More generally, we assume that the quadratic variation of M^R is a process which is R-a.s. equal to a random element of the set $\mathscr{M}^{\mathrm{na}}_{\mathbf{S}_+}([0,1])$ of all bounded measures on $[0,1]$ with no atoms and taking their values in $\mathbf{S}_+$:

$$[M^R, M^R](dt) = A(dt) \in \mathscr{M}^{\mathrm{na}}_{\mathbf{S}_+}([0,1]), \quad R\text{-a.s.} \tag{6}$$

and also that

$$t \in [0,1] \mapsto [M^R, M^R]_t := A([0,t]) = A(t, X_{[0,t]}; [0,t]) \in \mathbf{S}_+, \quad R\text{-a.s.}$$

is an adapted process. The quadratic variation given at (6) might have an atomless singular part in addition to its absolutely continuous component $a_t\,dt$. This notation is concise: $A(dt)$ is random and for any $\mathbb{R}^d$-valued processes $\alpha, \beta, \alpha_t \cdot A(dt)\beta_t$ is the infinitesimal element of a measure on $[0,1]$. In particular, $t \mapsto \int_{[0,t]} A(ds)\beta_s \in \mathbb{R}^d$, $t \mapsto \int_{[0,t]} \alpha_s \cdot A(ds)\beta_s \in \mathbb{R}$ and the process $t \mapsto \int_{[0,t]} \beta_s \cdot A(ds)\beta_s \in \mathbb{R}$ is increasing. Summing up, R is a solution to the *martingale problem* $\mathrm{MP}(B^R, A)$. This means that the canonical process X is the sum (3) of a bounded variation adapted process B^R and a local R-martingale M^R whose quadratic variation is specified by A and (6). We write

$$R \in \mathrm{MP}(B^R, A)$$

for short.

Theorem 1 (Girsanov's theorem). *Let R and P be as above, satisfying in particular the finite entropy condition* (1)*. Then, P is the law of a semimartingale. More precisely, there exists an $\mathbb{R}^d$-valued adapted process β which is defined P-a.s. and such that*

$$E_P \int_{[0,1]} \beta_t \cdot A(dt)\beta_t < \infty \tag{7}$$

and defining

$$\widehat{B}_t := \int_{[0,t]} A(ds)\beta_s, \quad 0 \le t \le 1, \quad P\text{-a.s.} \tag{8}$$

one obtains

$$X = X_0 + B^R + \widehat{B} + M^P, \quad P\text{-a.s.} \tag{9}$$

where M^P is a local P-martingale such that $[M^P, M^P] = [M^R, M^R]$, P-a.s. In other words, $P \in \mathrm{MP}(B^R + \widehat{B}, A)$.

Remark 1.

(a) The process β only needs to be defined P-a.s. (and not R-a.s.) for the statement of Theorem 1 to be meaningful. In fact, its proof only gives the "construction" of a process β, P-almost everywhere.
(b) The process $\widehat{B}$ is well defined. Indeed, by Cauchy-Schwarz inequality, for any $\mathbb{R}^d$-valued process ξ,

$$\int_{[0,1]} |\xi_t \cdot A(dt)\beta_t| \leq \left(\int_{[0,1]} \beta_t \cdot A(dt)\beta_t\right)^{1/2} \left(\int_{[0,1]} \xi_t \cdot A(dt)\xi_t\right)^{1/2}$$
$$\in [0, \infty], \ P\text{-a.s.}$$

Looking at $A(\omega)$ with ω fixed as a matrix of measures, we see that $\sup\{\int_{[0,1]} \xi_t \cdot A(dt)\xi_t; \xi : |\xi_t| = 1, \forall t\}$ is bounded above by the sum of the total variations of the entries of A. Consequently, this supremum is finite P-a.s. On the other hand, as $E_P \int_{[0,1]} \beta_t \cdot A(dt)\beta_t < \infty$, we see that a fortiori $\int_{[0,1]} \beta_t \cdot A(dt)\beta_t < \infty$, P-a.s. It follows that $\int_{[0,1]} |A(dt)\beta_t| < \infty$, P-a.s. which means that $\widehat{B}$ is well defined.

(c) When the quadratic variation is given by (5), one retrieves the standard representation

$$\widehat{B}_t = \int_{[0,t]} a_s \beta_s \, ds.$$

It is then known that under the minimal assumption (2), Theorem 1 still holds true with

$$\int_{[0,1]} \beta_t \cdot a_t \beta_t \, dt < \infty, \qquad P\text{-a.s.}$$

instead of $E_P \int_{[0,1]} \beta_t \cdot a_t \beta_t \, dt < \infty$ under the assumption (1), see for instance [3, Chap. III].

We introduce a slight modification of the standard definitions of stochastic integral and local martingale.

Definition 1 (P-locality). Let $P \ll R \in \mathrm{P}(\Omega)$.

i) A process M is said to be a P-local R-martingale if there exists an increasing sequence of $[0, 1] \cup \{\infty\}$-valued stopping times $(\tau_k)_{k\geq 1}$ such that $\lim_{k\to\infty} \tau_k = \infty$, P-a.s. (and not R-a.s.) such that the stopped processes M^{τ_k} are R-martingales, for all $k \geq 1$.
ii) A process Y is said to be a P-local R-stochastic integral if there exists an increasing sequence of $[0, 1] \cup \{\infty\}$-valued stopping times $(\tau_k)_{k\geq 1}$ such that $\lim_{k\to\infty} \tau_k = \infty$, P-a.s. (and not R-a.s.) such that the stopped processes Y^{τ_k} are $L^2(R)$-stochastic integrals, for all $k \geq 1$.

For any probability Q on Ω, let us denote $Q_0 = (X_0)_\# Q$ the law of the initial position X_0 under Q.

Definition (Condition (U)). One says that $R \in \mathrm{MP}(B^R, A)$ satisfies the *uniqueness condition* (U) if for any probability measure R' on Ω such that the initial laws $R'_0 = R_0$ are equal, $R' \ll R$ and $R' \in \mathrm{MP}(B^R, A)$, we have $R = R'$.

It is known [1] that if the SDE (4) admits a unique solution, for instance if the coefficients b and σ are locally Lipschitz, then its law R satisfies (U).

Theorem 2 (The density dP/dR). *Let R and P be as above, satisfying in particular the finite entropy condition* (1). *Keeping the notation of Theorem 1, we have*

$$H(P_0|R_0)+\frac{1}{2}E_P\int_{[0,1]}\beta_t\cdot A(dt)\beta_t \le H(P|R).$$

If in addition it is assumed that R satisfies the uniqueness condition (U), *then*

$$H(P_0|R_0)+\frac{1}{2}E_P\int_{[0,1]}\beta_t\cdot A(dt)\beta_t = H(P|R),$$

P is the unique solution of the martingale problem $\mathrm{MP}(B^R+\widehat{B},A;P_0)$ *in the class of all probability measures which have a finite relative entropy with respect to R and*

$$\begin{aligned}\frac{dP}{dR}&=\mathbf{1}_{\{\frac{dP}{dR}>0\}}\frac{dP_0}{dR_0}(X_0)\exp\left(\int_{[0,1]}\beta_t\cdot dM_t^R-\frac{1}{2}\int_{[0,1]}\beta_t\cdot A(dt)\beta_t\right)\\&=\mathbf{1}_{\{\frac{dP}{dR}>0\}}\frac{dP_0}{dR_0}(X_0)\exp\left(\int_{[0,1]}\beta_t\cdot dM_t^P+\frac{1}{2}\int_{[0,1]}\beta_t\cdot A(dt)\beta_t\right)\end{aligned}$$

where M^R *and* M^P *are defined at* (3) *and* (9) *respectively and*

(i) $\int_{[0,1]}\beta_t\cdot dM_t^R=\int_{[0,1]}\beta_t\cdot(dX_t-dB_t^R)$ *is a P-local R-stochastic integral and a P-local R-martingale;*
(ii) $\int_{[0,1]}\beta_t\cdot dM_t^P=\int_{[0,1]}\beta_t\cdot(dX_t-dB_t^R-A(dt)\beta_t)$ *is a well-defined P-stochastic integral such that* $E_P\left(\int_{[0,1]}\beta_t\cdot dM_t^P\right)^2<\infty$.

Because of the prefactor $\mathbf{1}_{\{\frac{dP}{dR}>0\}}$, the formula for $\frac{dP}{dR}$ is meaningful when considering $\int_{[0,1]}\beta_t\cdot dM_t^R$ as a P-local R-stochastic integral and $\int_{[0,1]}\beta_t\cdot dM_t^P$ as a P-stochastic integral. Note that (7) implies that $\int_{[0,1]}\beta_t\cdot A(dt)\beta_t<\infty$, P-a.s.

Let us denote

$$Z_t:=\frac{dP_{[0,t]}}{dR_{[0,t]}},\quad t\in[0,1].$$

Since $R_{[0,t]}$ and $P_{[0,t]}$ are the push-forward of R and P by $X_{[0,t]}$, we have $Z_t=E_R(\frac{dP}{dR}\mid X_{[0,t]})$, which also shows that $(Z_t)_{t\in[0,1]}$ is a uniformly integrable R-martingale.

Corollary 1. *Suppose that the condition* (U) *is satisfied. Then, the R-martingale* $(Z_t)_{t\in[0,1]}$ *satisfies:* $dZ_t=Z_t\,\beta_t\cdot dM_t^R$, *R-a.s.*

Proof. By Theorem 2, $Z_t=\mathbf{1}_{\{Z_t>0\}}\exp\left(\int_{[0,t]}\beta_s\cdot dM_s^R-\frac{1}{2}\int_{[0,t]}\beta_s\cdot A(ds)\beta_s\right)$. It follows with Itô's formula applied to the exponential function that, when $Z_t>0$, $dZ_t=Z_t\beta_t\cdot dM_t^R$, P-a.s. On the other hand, for all $0\le s\le t\le 1$, we have $0=P_{[0,s]}(Z_s=0)=P_{[0,t]}(Z_s=0)=E_{R_{[0,t]}}(Z_t\mathbf{1}_{\{Z_s=0\}})$ which implies that $Z_t=0$ if $Z_s=0$ for some $0\le s\le t$, R-a.s. It follows that $dZ_t=0$ when $Z_t=0$, R-a.s. Hence, choosing any version of β on the P-negligible set where it is unspecified, $dZ_t = Z_t\ \beta_t \cdot dM_t^R$, R-a.s., which is meaningful R-a.s. because of the prefactor Z_t. □

2.2 Processes with Jumps in $\mathbb{R}^d$

The law of a process with jumps is a probability measure P on the canonical space

$$\Omega = D([0,1], \mathbb{R}^d)$$

of all left limited and right continuous (càdlàg) paths, endowed with its canonical filtration. We denote $X = (X_t)_{t\in[0,1]}$ the canonical process,

$$\Delta X_t = X_t - X_{t^-}$$

the jump at time t and $\mathbb{R}_*^d := \mathbb{R}^d \setminus \{0\}$ the set of all effective jumps.
A Lévy kernel is a random σ-finite positive measure

$$\overline{L}_\omega(dtdq) = \rho(dt)L_\omega(t,dq), \quad \omega \in \Omega$$

on $[0,1] \times \mathbb{R}_*^d$ where ρ is assumed to be a σ-finite positive *atomless* measure on $[0,1]$. As a definition, any Lévy kernel is assumed to be predictable, i.e. $L_\omega(t,dq) = L(X_{[0,t)}(\omega);t,dq)$ for all $t \in [0,1]$.
Let B be a bounded variation *continuous* adapted process.

Definition 2 (Lévy kernel and martingale problem).
We say that a probability measure P on Ω solves the martingale problem $\mathrm{MP}(B,\overline{L})$ if the integrability assumption

$$E_P \int_{[0,1]\times\mathbb{R}_*^d} (|q|^2 \wedge 1)\,\overline{L}(dtdq) < \infty \tag{10}$$

holds and for any function f in $\mathscr{C}_b^2(\mathbb{R}^d)$, the process

$$f(\tilde{X}_t) - f(\tilde{X}_0) - \int_{(0,t]\times\mathbb{R}_*^d} [f(\tilde{X}_{s^-}+q) - f(\tilde{X}_{s^-}) - \nabla f(\tilde{X}_{s^-})\cdot q]\,\mathbf{1}_{\{|q|\le 1\}}\overline{L}(dsdq)$$
$$- \int_{(0,t]\times\mathbb{R}_*^d} [f(\tilde{X}_{s^-}+q) - f(\tilde{X}_{s^-})]\,\mathbf{1}_{\{|q|>1\}}\overline{L}(dsdq)$$

is a local P-martingale, where $\tilde{X} := X - B$. We write this

$$P \in \mathrm{MP}(B,\overline{L})$$

for short. In this case, we also say that P admits the Lévy kernel $\overline{L}$ and we denote this property

$$P \in \mathrm{LK}(\overline{L})$$

for short.

If $P \in \mathrm{MP}(B, \overline{L})$, the canonical process is decomposed as

$$X_t = X_0 + B_t + (\mathbf{1}_{(0,t]}\mathbf{1}_{\{|q|>1\}}q) \odot \mu^X + (\mathbf{1}_{(0,t]}\mathbf{1}_{\{|q|\le 1\}}q) \odot \widetilde{\mu}^L, \quad P\text{-a.s.} \quad (11)$$

where

$$\mu^X := \sum_{t\in[0,1];\Delta X_t\neq 0} \delta_{(t,\Delta X_t)}$$

is the canonical jump measure, $\varphi(q) \odot \mu^X = \int_{[0,1]\times\mathbb{R}^d_*} \varphi \, d\mu^X = \sum_{t\in[0,1];\Delta X_t\neq 0} \varphi(\Delta X_t)$ and $\varphi(q) \odot \widetilde{\mu}^L$ is the P-stochastic integral with respect to the compensated sum of jumps

$$\widetilde{\mu}^L_\omega(dtdq) := \mu^X_\omega(dtdq) - \overline{L}_\omega(dtdq).$$

For short, we rewrite formula (11):

$$X = X_0 + B + (\mathbf{1}_{\{|q|>1\}}q) \odot \mu^X + (\mathbf{1}_{\{|q|\le 1\}}q) \odot \widetilde{\mu}^L, \quad P\text{-a.s.} \quad (12)$$

Definition 3 (Class $\mathscr{H}_{p,r}(P, \overline{L})$). Let P be a probability measure on Ω and $\overline{L}$ a Lévy kernel such that $P \in \mathrm{LK}(\overline{L})$. We say that a predictable integrand $h_\omega(t, q)$ is in the class $\mathscr{H}_{p,r}(P, \overline{L})$ if $E_P \int_{[0,1]\times\mathbb{R}^d_*} \mathbf{1}_{\{|q|\le 1\}} |h_t(q)|^p \, \overline{L}(dtdq) < \infty$ and $E_P \int_{[0,1]\times\mathbb{R}^d_*} \mathbf{1}_{\{|q|>1\}} |h_t(q)|^r \, \overline{L}(dtdq) < \infty$.
We denote $\mathscr{H}_{p,p}(P, \overline{L}) = \mathscr{H}_p(P, \overline{L})$.

We take our reference law R such that

$$R \in \mathrm{MP}(B^R, \overline{L})$$

for some adapted continuous bounded variation process B^R. The integrability assumption (10) means that the integrand $|q|$ is in $\mathscr{H}_{2,0}(R, \overline{L})$. This will be always assumed in the future. We introduce the function

$$\theta(a) = \log \mathbb{E} e^{a(N-1)} = e^a - a - 1, \quad a \in \mathbb{R}.$$

where N is a Poisson(1) random variable. Its convex conjugate is

$$\theta^*(b) = \begin{cases} (b+1)\log(b+1) - b & \text{if } b > -1 \\ 1 & \text{if } b = -1 \\ \infty & \text{otherwise} \end{cases}, \quad b \in \mathbb{R}.$$

Note that θ and θ^* are respectively equivalent to $a^2/2$ and $b^2/2$ near zero.

Theorem 3 (Girsanov's theorem. The jump case). *Let R and P be as above: $R \in \mathrm{MP}(B^R, \overline{L})$ and $H(P|R) < \infty$. Then, there exists a predictable nonnegative process $\ell : \Omega \times [0,1] \times \mathbb{R}^d_* \to [0,\infty)$ which is defined P-a.s., satisfying*

$$E_P \int_{[0,1]\times\mathbb{R}^d_*} \theta^*(|\ell - 1|)\, d\overline{L} < \infty, \tag{13}$$

such that $P \in \mathrm{MP}(B^R + \widehat{B}^\ell, \ell\overline{L})$ where

$$\widehat{B}^\ell_t = \int_{[0,t]\times\mathbb{R}^d_*} \mathbf{1}_{\{|q|\le 1\}}(\ell_s(q) - 1)q\,\overline{L}(dsdq), \quad t \in [0,1]$$

is well-defined P-a.s.

It will appear that, in several respects, $\log \ell$ is analogous to β in Theorem 1. Again, ℓ only needs to be defined P-a.s. and not R-a.s. for the statement of Theorem 3 to be meaningful. And indeed, its proof only provides us with a P-a.s.-construction of ℓ.

Corollary 2. *Suppose that in addition to the assumptions of Theorem 3, there exist some $a_o, b_o, c_o > 0$ such that*

$$E_R \exp\left(a_o \int_{[0,1]\times\mathbb{R}^d_*} \mathbf{1}_{\{|q|>c_o\}} e^{b_o|q|}\,\overline{L}(dtdq)\right) < \infty. \tag{14}$$

It follows immediately that $\mathbf{1}_{\{|q|>c_o\}}|q|$ is $R \otimes \overline{L}$-integrable so that the stochastic integral $q \odot \widetilde{\mu}^L$ is well-defined R-a.s. and we are allowed to rewrite (12) as

$$X = X_0 + B + q \odot \widetilde{\mu}^L, \quad R\text{-a.s.},$$

for some adapted continuous bounded variation process B.
Then, there exists a predictable nonnegative process $\ell : \Omega \times [0,1] \times \mathbb{R}^d_* \to [0,\infty)$ satisfying (13) such that

$$X = X_0 + B + \overline{B}^\ell + q \odot \widetilde{\mu}^{\ell L}, \quad P\text{-a.s.},$$

where

$$\overline{B}^\ell_t = \int_{[0,t]\times\mathbb{R}^d_*} (\ell_s(q) - 1)q\,\overline{L}(dsdq), \quad t \in [0,1]$$

is well-defined P-a.s. and the P-stochastic integral $q \odot \widetilde{\mu}^{\ell L}$ with respect to the Lévy kernel $\ell\overline{L}$ is a local P-martingale.

Remark 2.

(a) The energy estimate (13) is equivalent to: $\mathbf{1}_{\{0\le\ell\le 2\}}(\ell - 1)^2$ and $\mathbf{1}_{\{\ell\ge 2\}}\ell \log \ell$ are integrable with respect to $P \otimes \overline{L}$.

(b) Together with (13), (14) implies that the integral for $\overline{B}^{\ell}$ is well-defined since

$$E_P \int_{[0,1]\times\mathbb{R}^d_*} (\ell_t(q) - 1)|q|\,\overline{L}(dtdq) < \infty. \tag{15}$$

In the present context of processes with jumps, the uniqueness condition (U) becomes:

Definition (Condition (U)). One says that $R \in \mathrm{MP}(B^R, \overline{L})$ satisfies the *uniqueness condition* (U) if for any probability measure R' on Ω such that the initial laws $R'_0 = R_0$ are equal, $R' \ll R$ and $R' \in \mathrm{MP}(B^R, \overline{L})$, we have $R = R'$.

Theorem 4 (The density dP/dR**).** *Suppose that R and P verify $R \in \mathrm{MP}(B, \overline{L})$ and $H(P|R) < \infty$. With ℓ given at Theorem 3, we have*

$$H(P_0|R_0) + E_P \int_{[0,1]\times\mathbb{R}^d_*} (\ell \log \ell - \ell + 1)\, d\overline{L} \le H(P|R)$$

with the convention $0 \log 0 - 0 + 1 = 1$.
If in addition it is assumed that R satisfies the uniqueness condition (U)*, then*

$$H(P_0|R_0) + E_P \int_{[0,1]\times\mathbb{R}^d_*} (\ell \log \ell - \ell + 1)\, d\overline{L} = H(P|R),$$

P is the unique solution of the martingale problem $\mathrm{MP}(B^R + \widehat{B}^{\ell}, \ell\overline{L}; P_0)$ in the class of all probability measures which have a finite relative entropy with respect to R and

$$\begin{aligned} \frac{dP}{dR} &= \mathbf{1}_{\{\frac{dP}{dR}>0\}} \frac{dP_0}{dR_0}(X_0)\, \widetilde{\exp}\left(\log \ell \odot \widetilde{\mu}^L - \int_{[0,1]\times\mathbb{R}^d_*} \theta(\log \ell)\, d\overline{L} \right) \\ &= \mathbf{1}_{\{\frac{dP}{dR}>0\}} \frac{dP_0}{dR_0}(X_0)\, \widetilde{\exp}\left(\log \ell \odot \widetilde{\mu}^{\ell L} + \int_{[0,1]\times\mathbb{R}^d_*} (\ell \log \ell - \ell + 1)\, d\overline{L} \right). \end{aligned} \tag{16}$$

In formula (16)*, $\widetilde{\exp}$ indicates a shorthand for the rigorous following expression*

$$\begin{cases} \dfrac{dP}{dR} = \dfrac{dP_0}{dR_0}(X_0) Z^+ Z^- \quad \textit{with} \\ Z^+ = \mathbf{1}_{\{\frac{dP}{dR}>0\}} \exp\left([\mathbf{1}_{\{\ell\ge 1/2\}} \log \ell] \odot \widetilde{\mu}^L - \displaystyle\int_{\{\ell\ge 1/2\}} (\ell - \log \ell - 1) d\overline{L} \right) \\ \quad = \mathbf{1}_{\{\frac{dP}{dR}>0\}} \exp\left([\mathbf{1}_{\{\ell\ge 1/2\}} \log \ell] \odot \widetilde{\mu}^{\ell L} + \displaystyle\int_{\{\ell\ge 1/2\}} (\ell \log \ell - \ell + 1) d\overline{L} \right) \\ Z^- = \mathbf{1}_{\{\frac{dP}{dR}>0, \tau^- = \infty\}} \exp\left(-\displaystyle\int_{\{0\le \ell < 1/2\}} [\ell - 1] d\overline{L} \right) \displaystyle\prod_{0\le t\le 1; 0<\ell(t,\Delta X_t)<1/2} \ell(t, \Delta X_t) \end{cases} \tag{17}$$

where :

- *The inner term* $[\mathbf{1}_{\{\ell \geq 1/2\}} \log \ell] \odot \widetilde{\mu}^L$ *in* Z^+ *is a* P*-local* R*-martingale and a* P*-local* R*-stochastic integral;*
- *The inner term* $[\mathbf{1}_{\{\ell \geq 1/2\}} \log \ell] \odot \widetilde{\mu}^{\ell L}$ *in* Z^+ *is a well-defined* P*-stochastic integral such that* $E_P \left([\mathbf{1}_{\{1/2 \leq \ell \leq 2\}} \log \ell] \odot \widetilde{\mu}^{\ell L} \right)^2 < \infty$ *and* $E_P \left| [\mathbf{1}_{\ell \geq 2\}} \log \ell] \odot \widetilde{\mu}^{\ell L} \right| < \infty$;
- *The product in* Z^- *contains finitely many terms,* P*-a.s., (and is equal to 1 when it doesn't contain any term);*
- *We defined*

$$\tau^- := \sup_{n \geq 1} \inf \{t \in [0,1]; \ell(t, \Delta X_t) \leq 1/n\} \in [0,1] \cup \{\infty\},$$

with the convention that $\inf \emptyset = \infty$.

Note that although ℓ is only defined P-a.s., Z^+, Z^- and τ^- are well-defined since they contain the prefactor $\mathbf{1}_{\{\frac{dP}{dR} > 0\}}$.

Remark 3.

(a) Because of (13), the integral $\int_{\{\ell \geq 1/2\}} (\ell - \log \ell - 1) d\overline{L}$ inside Z^+ is finite P-a.s.
(b) Similarly, the product $\prod_{0 \leq t \leq 1; 0 < \ell(t, \Delta X_t) < 1/2} \ell(t, \Delta X_t)$ doesn't vanish P-a.s. because it is proved at Lemma 8 that $P(\tau^- = \infty) = 1$.
(c) Note that this product is well-defined in $[0, 1]$ since it contains P-a.s. at most countably many terms in $(0, 1/2]$. But, if it contains infinitely many such terms, it vanishes. Therefore, it contains P-a.s. finitely many terms.
(d) Since $\inf \{t \in [0,1]; \ell(t, \Delta X_t) = 0\} \geq \tau^-$, if $\ell(t, \Delta X_t) = 0$ for some $t \in [0, 1]$, then $\frac{dP}{dR} = 0$. Therefore, $\ell > 0$, P-a.s.
(e) If $\mathbf{1}_{\{\ell \geq 1/2\}} \log \ell$ is $R \otimes \overline{L}$-integrable, an alternate expression of $\frac{dP}{dR}$ is

$$\frac{dP}{dR} = \mathbf{1}_{\{\frac{dP}{dR} > 0, \tau^- = \infty\}} \frac{dP_0}{dR_0}(X_0) \exp\left(- \int_{[0,1] \times \mathbb{R}^d_*} (\ell - 1) \, d\overline{L} \right) \prod_{0 \leq t \leq 1} \ell(t, \Delta X_t).$$

(f) If $\mathbf{1}_{\{0 \leq \ell < 1/2\}} \log \ell$ is not $R \otimes \overline{L}$-integrable, then $\log \ell \odot \widetilde{\mu}^L$ is undefined and (16) with exp instead of $\widetilde{\exp}$ is meaningless and must be replaced by (17).
(g) On the other hand, if $\ell > 0$, R-a.s. and $E_R \int_{[0,1] \times \mathbb{R}^d_*} \theta(\log \ell) \, d\overline{L} < \infty$, then (16) gives the rigorous expression for $\frac{dP}{dR}$ with exp instead of $\widetilde{\exp}$.

For more details about the relationship between (16) and (17), see the discussion below Proposition 1 at the Appendix.

Corollary 3. *Suppose that the condition* (U) *is satisfied. Then, the* R*-martingale* $Z_t = \frac{dP_{[0,t]}}{dR_{[0,t]}}$ *satisfies:* $dZ_t = Z_{t-} [(\ell - 1) \odot d\widetilde{\mu}^L_t]$, R*-a.s.*

Its proof is similar to Corollary 1's one. For more details about the martingale Z of Corollary 3, see Proposition 1 at the Appendix.

3 Variational Representations of the Relative Entropy

Theorems 1 and 3's proofs rely on some variational representation of the relative entropy which is stated and proved below.

Lemma 1 (Variational representations of the relative entropy). *Let R be a probability measure on some space Ω. For any probability measure $P \in \mathrm{P}(\Omega)$ such that $H(P|R) < \infty$ we have*

$$\begin{aligned} H(P|R) \\ &= \sup\left\{\int u\,dP - \log\int e^u\,dR; u:\Omega\to[-\infty,\infty), \int e^u\,dR<\infty\right\} \quad (18)\\ &= \sup\left\{\int u\,dP - \log\int e^u\,dR; u \textit{ bounded}\right\}\end{aligned}$$

where we use the convention $e^{-\infty}=0$ and the functions u are assumed to be measurable.

About the first identity, $H(P|R) < \infty$ and $\int e^u\,dR < \infty$ imply that $\int u\,dP \in [-\infty,\infty)$.

Proof. The proof is based on Fenchel's inequality

$$ab \le (a\log a - a + 1) + (e^b - 1), \quad \forall a \ge 0, b\in[-\infty,\infty)$$

and its case of equality

$$ab = (a\log a - a + 1) + (e^b - 1) \Leftrightarrow e^b = a$$

where as a convention $0\log 0 = 0$, $e^{-\infty} = 0$ and $0\times(-\infty) = 0$.
Choosing $a = \frac{dP}{dR}(\omega)$, $b = u(\omega)$ and integrating with respect to R, we obtain $\int u\,dP - \int(e^u - 1)\,dR \le H(P|R) < \infty$ which proves the last statement of the lemma. With the case of equality, we also see that

$$H(P|R) = \sup\left\{\int u\,dP - \int(e^u-1)\,dR; u:\Omega\to[-\infty,\infty), \int e^u\,dR<\infty\right\}. \quad (19)$$

Now, let us take advantage of the unit mass of R and P :

$$\int(u+\beta)\,dP - \int(e^{(u+\beta)}-1)\,dR = \int u\,dP - e^\beta\int e^u\,dR + \beta + 1, \quad \forall\beta\in\mathbb{R}.$$

Thanks to the elementary identity $\log\alpha = \inf_{\beta\in\mathbb{R}}\{\alpha e^\beta - \beta - 1\}$, we obtain

$$\sup_{\beta\in\mathbb{R}}\left\{\int(u+\beta)\,dP - \int(e^{(u+\beta)}-1)\,dR\right\} = \int u\,dP - \log\int e^u\,dR.$$

Hence,

$$\sup\left\{\int u\,dP - \int (e^u - 1)\,dR; u : \Omega \to [-\infty, \infty), \int e^u\,dR < \infty\right\}$$

$$= \sup\left\{\int (u+\beta)\,dP - \int (e^{(u+\beta)} - 1)\,dR; \beta \in \mathbb{R}, u : \Omega \to [-\infty, \infty),\right.$$

$$\left.\int e^u\,dR < \infty\right\}$$

$$= \sup\left\{\int u\,dP - \log \int e^u\,dR; u : \Omega \to [-\infty, \infty), \int e^u\,dR < \infty\right\}.$$

With (19), this proves the first identity of the lemma.

The second one follows by monotone convergence, considering the approximation $(u_n)_{n\geq 1}$ of $\log \frac{dP}{dR}$ which is given by: $u_n := (-n) \wedge \log \frac{dP}{dR} \vee n$. □

4 Proof of Theorem 1

For the proof of Theorem 1 we need to exhibit a large enough family of exponential supermartingales.

Lemma 2 (Exponential supermartingales). *Let M be a local martingale, then*

$$Z_t^M = \exp\left(M_t - \frac{1}{2}[M,M]_t\right), \quad 0 \leq t \leq 1,$$

is also a local martingale and a supermartingale. In particular, $0 \leq E_R Z_1^M \leq 1$.

Proof. Recall Itô's formula

$$df(Y_t) = f'(Y_t)\,dY_t + \frac{1}{2} f''(Y_t)\,d[Y,Y]_t$$

which is valid for any $\mathscr{C}^2$ function f and any continuous semimartingale Y. Applying it to $Y_t = M_t - \frac{1}{2}[M,M]_t$ and $f(y) = e^y$, we obtain

$$dZ_t^M = Z_t^M\left(dM_t - \frac{1}{2}d[M,M]_t + \frac{1}{2}d[M,M]_t\right) = Z_t^M\,dM_t$$

which proves that Z^M is a local martingale. Since $Z^M \geq 0$, Fatou's lemma applied to the localized sequence $Z^M_{t\wedge\tau_k}$ as k tends to infinity tells us that Z^M is a R-supermartingale, with $(\tau_k)_{k\geq 1}$ an increasing sequence of stopping times which tends almost surely to infinity and localizes the local martingale M. In particular, $E(Z_1^M) \leq E(Z_0^M) = 1$. □

The standard notation for the supermartingale of Lemma 2 is

$$\mathscr{E}(M) := \exp\left(M - \frac{1}{2}[M, M]\right).$$

We are now ready for the proof of Theorem 1.

Proof (of Theorem 1). We start with some useful notation. Let Q be a probability measure on Ω; later we shall take $Q = R$ or $Q = P$. For any measurable function g on $[0,1] \times \Omega$, let us denote

$$(g, g)_A(\omega) := \int_{[0,1]} g_t(\omega) \cdot A_t(\omega; dt) g_t(\omega) \in [0, \infty]$$

and introduce the function space

$$\mathscr{G}(Q) := \{g : [0,1] \times \Omega \to \mathbb{R}^d; g \text{ measurable}, E_Q(g, g)_A < \infty\}$$

endowed with the seminorm $\|g\|_{\mathscr{G}(Q)} := (E_Q(g, g)_A)^{1/2}$. Identifying the functions with their equivalence classes when factorizing by the kernel of this seminorm, turns $\mathscr{G}(Q)$ into a Hilbert space. These equivalence classes are called $\mathscr{G}(Q)$-classes and with some abuse, we say that two elements of the same class are equal $\mathscr{G}(Q)$-almost everywhere. The relevant space of integrands for the stochastic integral is

$$\mathscr{H}^Q := \{h \in \mathscr{G}(Q); h \text{ adapted}\}.$$

Identity (3) says that $M^R = X - X_0 - B^R$ is a local R-martingale. For all $h \in \mathscr{H}^R$, let us denote the stochastic integral

$$h \cdot M_t^R := \int_0^t h_s \, dM_s^R, \quad t \in [0, 1].$$

By Lemma 2, $0 < E_R Z_1^{h \cdot M^R} \le 1$ for all $h \in \mathscr{H}^R$ and because of (18), for any probability measure P such that $H(P|R) < \infty$, we have

$$E_P\left(h \cdot M_1^R - \frac{1}{2}[h \cdot M^R, h \cdot M^R]_1\right) \le H(P|R), \ \forall h \in \mathscr{H}^R. \tag{20}$$

Note that, as $P \ll R$, $h \cdot M_1^R$ and $[h \cdot M^R, h \cdot M^R]_1$ which are defined R-a.s., are a fortiori defined P-a.s. With (6) and (20), we see that

$$E_P(h \cdot M^R) \le H(P|R) + \frac{1}{2} E_P(h, h)_A, \quad \forall h \in \mathscr{G}(P) \cap \mathscr{H}^R. \tag{21}$$

The notation $\mathscr{G}(P) \cap \mathscr{H}^R$ is a little bit improper. Indeed, $\mathscr{G}(P)$ is a set of equivalence classes with respect to the equality $\mathscr{G}(P)$-a.e., while $\mathscr{H}^R$ is a set of $\mathscr{G}(R)$-classes. But since $P \ll R$, keeping in mind that any $\mathscr{G}(P)$-class is the union of some $\mathscr{G}(R)$-classes, one can interpret $\mathscr{G}(P) \cap \mathscr{G}(R)$ as a set of $\mathscr{G}(P)$-classes. It is also clear that $\mathscr{G}(P) \cap \mathscr{H}^R = \mathscr{H}^P \cap \mathscr{H}^R$ which is a set of $\mathscr{G}(P)$-classes. Considering $-h$ in (21), we obtain for all $\lambda > 0$,

$$\left| E_P\left(\frac{h}{\lambda} \cdot M^R\right)\right| \leq H(P|R) + \frac{1}{2\lambda^2} E_P(h,h)_A, \quad \forall h \in \mathscr{H}^P \cap \mathscr{H}^R.$$

Let

$$S := \left\{ h : [0,1] \times \Omega \to \mathbb{R}^d ; h = \sum_{i=1}^k h_i \mathbf{1}_{]\!]S_i,T_i]\!]} \right\}$$

denote the set of all *simple* adapted processes h where k is finite and for all i, $h_i \in \mathbb{R}^d$ and $S_i \leq T_i$ are stopping times. As $S \subset \mathscr{H}^P \cap \mathscr{H}^R$, taking $\lambda = \|h\|_{\mathscr{G}(P)}$ in previous inequality, we obtain the keystone of the proof:

$$|E_P(h \cdot M^R)| \leq [H(P|R) + 1/2]\, \|h\|_{\mathscr{G}(P)}, \quad \forall h \in S.$$

This estimate still holds when $\|h\|_{\mathscr{H}(P)} = 0$. Indeed, for all real α, by (21) we see that $\alpha E_P(h \cdot M^R) \leq H(P|R) + \alpha^2/2\ E_P(h,h)_A = H(P|R)$. Letting $|\alpha|$ tend to infinity, it follows that $E_P(h \cdot M^R) = 0$.

Under the assumption that $H(P|R)$ is finite, this means that $h \mapsto h \cdot M^R$ is continuous on S with respect to the Hilbert topology of $\mathscr{H}^P$. As S is dense in $\mathscr{H}^P$, this linear form extends uniquely as a continuous linear form on $\mathscr{H}^P$. It also appears that this extension is again a stochastic integral *with respect to* P. We still denote this extension by $h \cdot M^R$.

As $h \mapsto h \cdot M^R$ is a continuous linear form on $\mathscr{H}^P$, we know by the Riesz representation theorem that there exists a unique $\beta \in \mathscr{H}^P$ such that

$$E_P(h \cdot M^R) = E_P(\beta, h)_A, \quad \forall h \in S.$$

In other words,

$$E_P \int_{[0,1]} h_t\, dM_t^P = 0, \quad \forall h \in S$$

where

$$M_t^P := M_t^R - \int_{[0,t]} A(ds)\beta_s = X_t - X_0 - B_t^R - \widehat{B}_t,$$

which means that M^P is a local P-martingale. □

5 Proof of Theorem 2

It relies on a transfer result which is stated below at Lemma 3. But we first need to introduce its framework and some additional notation. Let P be a probability measure on Ω such that $[X,X]=A$, P-a.s. and

$$X = X_0 + B + M^P, \quad P\text{-a.s.},$$

where B is a bounded variation process and M^P is a local P-martingale. Let γ be an adapted process such that $\int_{[0,1]} \gamma_t \cdot A(dt)\gamma_t < \infty$, P-a.s. We define

$$Z_t = \exp\left(\int_{[0,t]} \gamma_s \, dM_s^P - \frac{1}{2}\int_{[0,t]} \gamma_s \cdot A(ds)\gamma_s \right), \quad 0 \le t \le 1$$

and for all $k \ge 1$,

$$\sigma^k := \inf\left\{ t \in [0,1]; \int_{[0,t]} \gamma_s \cdot A(ds)\gamma_s \ge k \right\} \in [0,1] \cup \{\infty\},$$

with the convention $\inf \emptyset = \infty$.
We use the standard notation $Y_t^\tau = Y_{\tau \wedge t}$ for the process Y stopped at a stopping time τ. For all k, $P^k := X^{\sigma_k}{}_{\#}P$ is the push-forward of P with respect to the stopping procedure X^{σ_k}. Note that P^k and P match on the σ-field which is generated by $X_{[0,\sigma_k]}$.

Lemma 3. *Let P and γ as above. Then, for all $k \ge 1$, Z^{σ_k} is a genuine P-martingale and the measure*

$$Q^k := Z_1^{\sigma_k} P^k$$

is a probability measure on Ω which satisfies $Q^k \in \mathrm{MP}(\widehat{B}^{\sigma_k}, A^{\sigma_k})$ where $\widehat{B}_t^{\sigma_k} = \int_{[0,t\wedge\sigma_k]} A(ds)\gamma_s$ and M^k is a local Q^k-martingale.

Proof. Let us first show that Z^{σ_k} is a P^k-martingale.[1] The local martingale Z^{σ_k} is of the form $Z^{\sigma_k} = \mathcal{E}(N) := \exp(N - \frac{1}{2}[N,N])$ with N a local P^k-martingale such that $[N,N] \le k$, P^k-a.s. For all $p \ge 0$, since $\mathcal{E}(N)^p = \exp(pN - \frac{p}{2}[N,N])$ and $\mathcal{E}(pN) = \exp(pN - \frac{p^2}{2}[N,N]) \ge e^{pN} e^{-kp^2/2}$, we obtain

$$\mathcal{E}(N)^p \le e^{pN} \le e^{kp^2/2}\mathcal{E}(pN).$$

As a nonnegative local martingale, $\mathcal{E}(pN)$ is a supermartingale. We deduce from this that $E_{P^k}\mathcal{E}(pN) \le 1$ and

[1]It is a direct consequence of Novikov's criterion, but we prefer presenting an elementary proof which will be a guideline for a similar result with jump processes.

$$E_{P^k}\mathcal{E}(N)^p \le e^{kp^2/2}E_{P^k}\mathcal{E}(pN) \le e^{kp^2/2} < \infty.$$

Choosing $p > 1$, it follows that $\mathcal{E}(N)$ is uniformly integrable. In particular, this implies that

$$E_{P^k}\mathcal{E}(N)_1 = E_{P^k}\mathcal{E}(N)_0 = 1$$

and proves that Q^k is a probability measure.
Suppose now that the supermartingale $\mathcal{E}(N)$ is not a martingale. This implies that there exists $0 \le t < 1$ such that on a subset with positive measure, $E_{P^k}(\mathcal{E}(N)_1 \mid X_{[0,t]}) < \mathcal{E}(N)_t$. Integrating, we get $1 = E_{P^k}\mathcal{E}(N)_1 < E_{P^k}\mathcal{E}(N)_t$, which contradicts $E_{P^k}\mathcal{E}(N)_s \le E_{P^k}\mathcal{E}(N)_0 = 1, \forall s$: a consequence of the supermartingale property of $\mathcal{E}(N)$. Therefore, $\mathcal{E}(N)$ is a genuine P^k-martingale.

Let us fix $k \ge 1$ and show that Q^k is a solution to $\mathrm{MP}(\widehat{B}^{\sigma_k}, A^{\sigma_k})$. First of all, as it is assumed that $[X, X] = A$, P-a.s., we obtain $[X, X] = A^{\sigma_k}$, P^k-a.s. With $Q^k \ll P^k$, this implies that $[X, X] = A^{\sigma_k}$, Q^k-a.s.

Now, we check

$$X = X_0 + B^{\sigma_k} + \widehat{B}^{\sigma_k} + M^k \tag{22}$$

where M^k is a Q^k-martingale. Let τ be a stopping time and denote $F_t = \xi \cdot X_t^\tau$ with $\xi \in \mathbb{R}^d$. The martingale Z^{σ_k} is the stochastic exponential $\mathcal{E}(N)$ of $N_t = \int_{[0,t]} \mathbf{1}_{[0,\sigma_k]}(s)\gamma_s \cdot dM_s^P$. Hence, denoting $Z = Z^{\sigma_k}$, we have $dZ_t = Z_t \mathbf{1}_{[0,\sigma_k]}(t)\gamma_t \cdot dM_t^P$, $dF_t = \mathbf{1}_{[0,\tau]}(t)\xi \cdot (dB_t + dM_t^P)$ and $d[Z, F]_t = Z_t \mathbf{1}_{[0,\tau\wedge\sigma_k]}(t)\xi \cdot A(dt)\gamma_t$, P^k-a.s. Consequently,

$$\begin{aligned}
E_{Q^k}[\xi \cdot (X_\tau - X_0)] &\overset{(a)}{=} E_{P^k}[Z_\tau F_\tau - Z_0 F_0] \\
&\overset{(b)}{=} E_{P^k}\left[\int_{[0,\tau]} (F_t\, dZ_t + Z_t dF_t + d[Z, F]_t)\right] \\
&= E_{P^k}\left[\int_{[0,\tau]} F_t\, dZ_t + \int_{[0,\tau]} Z_t \xi \cdot (dB_t + dM_t^P)\right. \\
&\quad \left. + \int_{[0,\tau]} Z_t \xi \cdot A(dt)\gamma_t\right] \\
&\overset{(c)}{=} E_{P^k}\left[\int_{[0,\tau]} Z_t \xi \cdot dB_t + \int_{[0,\tau]} Z_t \xi \cdot A(dt)\gamma_t\right] \\
&\overset{(d)}{=} E_{Q^k}\left[\xi \cdot \int_{[0,\tau]} (dB_t + A(dt)\gamma_t)\right].
\end{aligned}$$

In order that all the above terms are meaningful, we choose τ such that it localizes F, B, M^P and $\xi \cdot A\gamma$. This is possible, taking for any $n \ge 1$, $\tau \le \tau_n = \min(\tau_n^F, \tau_n^B, \tau_n^M, \tau_n^\gamma)$ where $\tau_n^F = \inf\{t \in [0,1]; |X_t| \ge n\}$, $\tau_n^B = \inf\{t \in [0,1]; \int_{[0,t]} |dB_s| \ge n\}$, $\tau_n^\gamma = \inf\{t \in [0,1]; \int_{[0,t]} \gamma_s \cdot A(ds)\gamma_s \ge n\}$, and τ_n^M is a localizing sequence of the local martingale M^P. We have

$$\lim_{n\to\infty} \tau_n = \infty, \quad P^k\text{-a.s.} \tag{23}$$

We used the definition of Q^k and the martingale property of Z at (a) and (d), (b) is Itô's formula and (c) relies on the martingale property of Z and $(M^P)^\tau$. Finally, taking $\tau = \varsigma \wedge \tau_n$, we see that for any stopping time ς, any $n \geq 1$ and any $\xi \in \mathbb{R}^d$

$$E_{Q^k}[\xi \cdot (X_\varsigma^{\tau_n} - X_0^{\tau_n})] = E_{Q^k}\left[\xi \cdot \int_{[0,\varsigma\wedge\tau_n]} (dB_t + A(dt)\gamma_t)\right].$$

Taking (23) into account, this means that $X - X_0 - B - \widehat{B}$ is a local Q^k-martingale. We conclude remarking that for any process Y, we have $Y = Y^{\sigma^k}$, Q^k-a.s. This leads us to (22). □

Let us denote $P^\tau = X^\tau{}_\# P$ the law under P of the process X^τ which is stopped at the stopping time τ.

Lemma 4. *If R fulfills the condition (U), then for any stopping time τ, R^τ also fulfills it.*

Proof. Let us fix the stopping time τ. Our assumption on R implies that

$$X = X_0 + B + M, \quad R^\tau\text{-a.s.}$$

where $M = M^R$ is a local R-martingale and we denote $B = B^R$. Let $Q \ll R^\tau$ be given such that $Q_0 = R_0$ and

$$X = X_0 + B + M^Q, \quad Q\text{-a.s.}$$

where M^Q is a local Q-martingale. We wish to show that $Q = R^\tau$.
The disintegration

$$R = R_{[0,\tau]} \otimes R(\cdot \mid X_{[0,\tau]})$$

means that for any bounded measurable function F on Ω, denoting $F = F(X) = F(X_{[0,\tau]}, X_{(\tau,1]})$,

$$E_R(F) = \int_\Omega E_R[F(\eta, X_{(\tau,1]}) \mid X_{[0,\tau]} = \eta]\, R_{[0,\tau]}(d\eta).$$

Similarly, we introduce the probability measure

$$R' := Q_{[0,\tau]} \otimes R(\cdot \mid X_{[0,\tau]}).$$

To complete the proof, it is enough to show that R' satisfies

$$X = X_0 + B + M', \quad R'\text{-a.s.} \tag{24}$$

with M' a local R'-martingale. Indeed, the condition (U) tells us that $R' = R$, which implies that $R'^{\tau} = R^{\tau}$. But $R'^{\tau} = Q$, hence $Q = R^{\tau}$.

Let us show (24). Let $\xi \in \mathbb{R}^d$ and a stopping time σ be given. We denote $(\tau_n)_{n\geq 1}$ a localizing sequence of $M = M^R$ and $B = B^R$. Then,

$$
\begin{aligned}
&E_{R'}[\xi \cdot (X_\sigma^{\tau_n} - X_0^{\tau_n})] \\
&= E_{R'}[\mathbf{1}_{\{\tau \leq \sigma\}} \xi \cdot (X_\sigma^{\tau_n} - X_\tau^{\tau_n})] + E_Q[\xi \cdot (X_\sigma^{\tau_n} - X_0^{\tau_n})] \\
&= \int_\Omega E_R[\mathbf{1}_{\{\tau \leq \sigma\}} \xi \cdot (X_\sigma^{\tau_n} - X_\tau^{\tau_n}) \mid X_{[0,\tau]} = \eta]\, Q(d\eta) + E_Q[\xi \cdot (X_\sigma^{\tau_n} - X_0^{\tau_n})] \\
&= \int_\Omega E_R[\mathbf{1}_{\{\tau \leq \sigma\}} \xi \cdot (B_\sigma^{\tau_n} - B_\tau^{\tau_n}) \mid X_{[0,\tau]} = \eta]\, Q(d\eta) + E_Q[\xi \cdot (B_\sigma^{\tau_n} - B_0^{\tau_n})] \\
&= E_{R'}[\xi \cdot (B_\sigma^{\tau_n} - B_0^{\tau_n})]
\end{aligned}
$$

This means that (24) is satisfied (with the localizing sequence $(\tau_n)_{n\geq 1}$) and completes the proof of the lemma. □

For all $k \geq 1$, we consider the stopping time

$$
\tau_k = \inf\left\{t \in [0,1];\, \int_{[0,t]} \beta_s \cdot A(ds)\beta_s \geq k\right\} \in [0,1] \cup \{\infty\}
$$

where β is the process which is associated with P in Theorem 1 and as a convention $\inf \emptyset = \infty$. We are going to use this stopping time R-a.s. Since β is only defined P-a.s., we assume for the moment that P and R are equivalent measures: $P \sim R$.

Lemma 5. *Assume that $P \sim R$ and suppose that R satisfies the condition* (U). *Then, for all $k \geq 1$, on the stochastic interval $[\![0, \tau_k \wedge 1]\!]$ we have, R-almost everywhere*

$$
\mathbf{1}_{[\![0,\tau_k\wedge 1]\!]} \frac{dP}{dR} = \mathbf{1}_{[\![0,\tau_k\wedge 1]\!]} \frac{dP_0}{dR_0}(X_0) \exp\left(\int_{[0,\tau_k\wedge 1]} \beta_t \cdot dM_t^R - \frac{1}{2}\int_{[0,\tau_k\wedge 1]} \beta_t \cdot A(dt)\beta_t\right). \tag{25}
$$

Proof. By conditioning with respect to X_0, we see that we can assume without loss of generality, that $R_0 := (X_0)_\# R = (X_0)_\# P =: P_0$, i.e. $\frac{dP_0}{dR_0}(X_0) = 1$. Let $k \geq 1$. Denote $R^k = R^{\tau_k}$, $P^k = P^{\tau_k}$. Applying Lemma 3 with $\gamma = -\beta$ and remarking that $\widehat{B}_{-\beta} = -\widehat{B}_\beta$, we see that

$$
\begin{aligned}
Q^k := \mathscr{E}(-\beta \cdot M^P)_{\tau_k\wedge 1} P^k &\in \mathrm{MP}(\mathbf{1}_{[\![0,\tau_k]\!]}[(B^R + \widehat{B}_\beta) + \widehat{B}_{-\beta}], \mathbf{1}_{[\![0,\tau_k]\!]} A) \\
&= \mathrm{MP}(\mathbf{1}_{[\![0,\tau_k]\!]} B^R, \mathbf{1}_{[\![0,\tau_k]\!]} A).
\end{aligned}
$$

But, it is known with Lemma 4 that R^k satisfies the condition (U). Therefore,

$$
Q^k = R^k. \tag{26}
$$

Applying twice Lemma 3, we observe on the one hand that

$$\widetilde{P}^k := \mathscr{E}(\beta \cdot M^R)_{\tau_k \wedge 1} R^k \in \mathrm{MP}(\mathbf{1}_{[\![0,\tau_k]\!]}(B^R + \widehat{B}_\beta), \mathbf{1}_{[\![0,\tau_k]\!]} A), \tag{27}$$

and on the other hand that

$$\begin{aligned} \widetilde{Q}^k &:= \mathscr{E}(-\beta \cdot M^P)_{\tau_k \wedge 1} \widetilde{P}^k \\ &\in \mathrm{MP}(\mathbf{1}_{[\![0,\tau_k]\!]}[(B^R + \widehat{B}_\beta) - \widehat{B}_\beta], \mathbf{1}_{[\![0,\tau_k]\!]} A) = \mathrm{MP}(\mathbf{1}_{[\![0,\tau_k]\!]} B^R, \mathbf{1}_{[\![0,\tau_k]\!]} A). \end{aligned}$$

As for the proof of (26), the condition (U) which is satisfied by R^k leads us to $\widetilde{Q}^k = R^k$. Therefore, we see with (26) that $Q^k = \widetilde{Q}^k$, i.e. $\mathscr{E}(-\beta \cdot M^P)_{\tau_k \wedge 1} P^k = \mathscr{E}(-\beta \cdot M^P)_{\tau_k \wedge 1} \widetilde{P}^k$. And since $\mathscr{E}(-\beta \cdot M^P)_{\tau_k \wedge 1} > 0$, we obtain $P^k = \widetilde{P}^k$ which is (25). □

Remark 4. As a by-product of this proof, we obtain that for all $k \geq 1$, P^{τ_k} is the *unique* solution of the martingale problem $\mathrm{MP}(\mathbf{1}_{[\![0,\tau_k]\!]}(B^R + \widehat{B}_\beta), \mathbf{1}_{[\![0,\tau_k]\!]} A; P_0)$ in the class of all probability measures which are absolutely continuous with respect to R^{τ_k}.

We are ready to complete the proof of Theorem 2.

Proof (of Theorem 2. Derivation of $\frac{dP}{dR}$). Provided that R satisfies the condition (U), when $P \sim R$ we obtain the announced formula

$$\frac{dP}{dR} = \frac{dP_0}{dR_0}(X_0) \exp\left(\int_{[0,1]} \beta_t \cdot dM_t^R - \frac{1}{2} \int_{[0,1]} \beta_t \cdot A(dt) \beta_t \right), \tag{28}$$

letting k tend to infinity in (25), remarking that $\tau := \lim_{k \to \infty} \tau_k = \inf\{t \in [0,1];\ \int_{[0,t]} \beta_s \cdot A(ds) \beta_s = \infty\}$ and that (7) implies

$$\tau = \infty, \quad P\text{-a.s.} \tag{29}$$

and, since $P \sim R$, we also have $\tau = \infty$, R-a.s. Indeed, since $\tau(\omega) = \infty$, there is some $k_o \geq 1$ such that $\tau_{k_o}(\omega) = \infty$ and applying Lemma 5 with $k = k_o$: $\frac{dP}{dR}(\omega) = \frac{dP_0}{dR_0}(\omega_0) \exp\left(\int_{[0,1]} \beta_t \cdot dM_t^R - \frac{1}{2} \int_{[0,1]} \beta_t \cdot A(dt) \beta_t \right)(\omega) > 0$.

It also follows with Remark 4 that P is the unique solution of $\mathrm{MP}(B^R + \widehat{B}, A; P_0)$ in the class of all probability measures which are absolutely continuous with respect to R.

Now, we consider the general case when P might not be equivalent to R. The main idea is to approximate P by a sequence $(P_n)_{n \geq 1}$ such that $P_n \sim R$ for all $n \geq 1$, and to rely on our previous intermediate results. We consider

$$P_n := \left(1 - \frac{1}{n}\right) P + \frac{1}{n} R, \quad n \geq 1.$$

Clearly, $P_n \sim R$ and by convexity $H(P_n|R) \le (1-\frac{1}{n})H(P|R) + \frac{1}{n}H(R|R) \le H(P|R) < \infty$. More precisely, the function $x \in [0,1] \mapsto H(xP + (1-x)R|R) \in [0,\infty]$ is a finitely valued convex continuous and increasing. It follows that $\lim_{n\to\infty} H(P_n|R) = H(P|R)$.

As regards the uniqueness statement, suppose that P and Q are two solutions of $\mathrm{MP}(B^R + \widehat{B}, A; P_0)$ such that $H(P|R), H(Q|R) < \infty$. With $n = 2$, by linearity $P_2 = (P+R)/2$ and $Q_2 = (Q+R)/2$ are again solutions to the same martingale problem. But we already saw a few lines above that this implies that $P_2 = Q_2$. Therefore, $P = Q$.

It is clear that $\lim_{n\to\infty} P_n = P$ in total variation norm. Let us prove that the stronger convergence

$$\lim_{n\to\infty} H(P|P_n) = 0 \tag{30}$$

also holds. It is easy to check that $\mathbf{1}_{\{\frac{dP}{dR}\ge 1\}} dP/dP_n$ and $\mathbf{1}_{\{\frac{dP}{dR}\le 1\}} dP/dP_n$ are respectively decreasing and increasing sequences of functions. It follows by monotone convergence that

$$\begin{aligned}\lim_{n\to\infty} H(P|P_n) &= \lim_{n\to\infty} \int \log(dP/dP_n)\, dP \\ &= \lim_{n\to\infty} \int_{\{\frac{dP}{dR}\ge 1\}} \log(dP/dP_n)\, dP \\ &\quad + \lim_{n\to\infty} \int_{\{\frac{dP}{dR}< 1\}} \log(dP/dP_n)\, dP = 0.\end{aligned}$$

By Theorem 1, there exist two vector fields β^n and β which are respectively defined R-a.s. and P-a.s. such that $E_{P_n} \int_{[0,1]} \beta_t^n \cdot A(dt)\beta_t^n < \infty$, $E_P \int_{[0,1]} \beta_t \cdot A(dt)\beta_t < \infty$ and

$$dX_t = dB_t^R + A(dt)\beta_t^n + dM_t^{P_n},\ R\text{-a.s.};\quad dX_t = dB_t^R + A(dt)\beta_t + dM_t^P,\ P\text{-a.s.}$$

where M^{P_n} and M^P are respectively a local P_n-martingale and a local P-martingale. Therefore,

$$dM_t^{P_n} = dM_t^P + A(dt)(\beta_t - \beta_t^n), \quad P\text{-a.s.} \tag{31}$$

Extending β arbitrarily by $\beta = 0$ on the P-null set where it is unspecified, we know that

$$\exp\left(\int_{[0,t]} (\beta_s - \beta_s^n)\cdot dM_s^{P_n} - \frac{1}{2}\int_{[0,t]} (\beta_s - \beta_s^n)\cdot A(ds)(\beta_s - \beta_s^n)\right)$$

is a P^n-supermartingale. It follows with Lemma 1, (31) and a standard monotone convergence argument that

$$H(P|P_n) \geq E_P\left(\int_{[0,1]} (\beta_s - \beta_s^n)\cdot dM_s^{P_n} - \frac{1}{2}\int_{[0,1]} (\beta_s - \beta_s^n)\cdot A(ds)(\beta_s - \beta_s^n)\right)$$
$$= \frac{1}{2}E_P \int_{[0,1]} (\beta_s - \beta_s^n)\cdot A(ds)(\beta_s - \beta_s^n).$$

With (30), this shows the key estimate

$$\lim_{n\to\infty} E_P \int_{[0,1]} (\beta_s - \beta_s^n)\cdot A(ds)(\beta_s - \beta_s^n) = 0. \tag{32}$$

Since $H(P_n|R) < \infty$ and $P_n \sim R$, under the condition (U) we can invoke (28) to write

$$\frac{dP_n}{dR} = \frac{dP_{n,0}}{dR_0}(X_0)\exp\left(\int_{[0,1]} \beta_t^n \cdot dM_t^R - \frac{1}{2}\int_{[0,1]} \beta_t^n \cdot A(dt)\beta_t^n\right).$$

As $\lim_{n\to\infty} P_n = P$ in total variation norm, up to the extraction of a R-a.s.-convergent subsequence we have $\lim_{n\to\infty} dP_n/dR = dP/dR$ and $\lim_{n\to\infty} dP_{n,0}/dR = dP_0/dR_0$. On the other hand, (32) implies that P-a.s., $\lim_{n\to\infty} \frac{1}{2}\int_{[0,1]} \beta_t^n \cdot A(dt)\beta_t^n = \frac{1}{2}\int_{[0,1]} \beta_t \cdot A(dt)\beta_t$. It follows that

$$\frac{dP}{dR} = \mathbf{1}_{\{\frac{dP}{dR}>0\}}\frac{dP_0}{dR_0}(X_0)\exp\left(\int_{[0,1]} \beta_t \cdot dM_t^R - \frac{1}{2}\int_{[0,1]} \beta_t \cdot A(dt)\beta_t\right).$$

where (32) also implies that the limit of the stochastic integrals

$$\lim_{n\to\infty}\int_{[0,1]} \beta_t^n \cdot dM_t^R = \int_{[0,1]} \beta_t \cdot dM_t^R,\ P\text{-a.s.}$$

exists P-a.s. □

It remains to compute $H(P|R)$.

Proof (End of the proof of Theorem 2. Computation of $H(P|R)$). Let us first compute $H(P|R)$ when R satisfies (U). Remark that in the proof of Lemma 5, for all $k \geq 1$ the local $\widetilde{P}^k$-martingale $N^k = M^R - \widehat{B}$ which is behind (27) is a genuine martingale. It is a consequence of the first statement of Lemma 3. As $\widetilde{P}^k = P^k$, N^k is a genuine P^k-martingale. This still holds when $P \sim R$ fails. Indeed, this hypothesis has only been invoked to insure that τ_k is well-defined R-a.s. But in the present situation, τ_k only needs to be defined P-a.s. With (25), we have

$$H(P^k|R^k) = E_{P^k}\log\frac{dP^k}{dR^k}$$
$$\overset{(25)}{=} E_P\left(\log\frac{dP_0}{dR_0}(X_0)\right) + E_{P^k}\left(\int_{[0,1]} \beta_t \cdot dM_t^R - \frac{1}{2}\int_{[0,1]} \beta_t \cdot A(dt)\beta_t\right)$$

$$\stackrel{(27)}{=} H(P_0|R_0) + E_{P^k}\left(\int_{[0,1]} \beta_t \cdot (dN_t^k + d\widehat{B}_t) - \frac{1}{2}\int_{[0,1]} \beta_t \cdot A(dt)\beta_t\right)$$

$$\stackrel{(8)}{=} H(P_0|R_0) + \frac{1}{2}E_{P^k}\left(\int_{[0,1]} \beta_t \cdot A(dt)\beta_t\right) + E_{P^k}\left(\int_{[0,1]} \beta_t \cdot dN_t^k\right)$$

$$= H(P_0|R_0) + \frac{1}{2}E_P\left(\int_{[0,\tau^k\wedge 1]} \beta_t \cdot A(dt)\beta_t\right)$$

where the last equality comes from the P^k-martingale property of N^k. It remains to let k tend to infinity to see that

$$H(P|R) = H(P_0|R_0) + \frac{1}{2}E_P\left(\int_{[0,1]} \beta_t \cdot A(dt)\beta_t\right).$$

Indeed, because of (29) and since the sequence $(\tau_k)_{k\geq 1}$ is increasing, we obtain by monotone convergence that

$$\lim_{k\to\infty} E_P\left(\int_{[0,\tau^k\wedge 1]} \beta_t \cdot A(dt)\beta_t\right) = \frac{1}{2}E_P\left(\int_{[0,1]} \beta_t \cdot A(dt)\beta_t\right).$$

As regards the left hand side of the equality, with Lemma 1 and (29), we see that

$$\begin{aligned} H(P|R) &= \sup\{E_P u(X) - \log E_R e^{u(X)}; u \in L^\infty(P)\} \\ &= \sup_k \sup\{E_P u(X^{\tau^k}) - \log E_R e^{u(X^{\tau_k})}; u \in L^\infty(P)\} \\ &= \lim_{k\to\infty} H(P^k|R^k). \end{aligned}$$

It remains to check that, without the condition (U), we have

$$H(P|R) \geq H(P_0|R_0) + \frac{1}{2}E_P\left(\int_{[0,1]} \beta_t \cdot A(dt)\beta_t\right). \tag{33}$$

Let us extend β by $\beta = 0$ on the P-null set where it is unspecified and define

$$\tilde{u}(X) := \log \frac{dP_0}{dR_0}(X_0) + \int_{[0,\tau^k\wedge 1]} \beta_t \cdot dM_t^R - \frac{1}{2}\int_{[0,\tau^k\wedge 1]} \beta_t \cdot A(dt)\beta_t.$$

Choosing $\tilde{u}(X)$ at inequality $\stackrel{(i)}{\geq}$ below, thanks to an already used supermartingale argument, we obtain the inequality $\stackrel{(ii)}{\geq}$ below and

$$
\begin{aligned}
H(P^k|R^k) &\overset{(18)}{=} \sup\left\{\int u\,dP^k - \log\int e^u\,dR^k; u : \int e^u\,dR^k < \infty\right\}\\
&\overset{(i)}{\geq} \int \tilde{u}\,dP^k - \log\int e^{\tilde{u}}\,dR^k\\
&\overset{(ii)}{\geq} \int \tilde{u}\,dP^k\\
&\overset{(iii)}{=} H(P_0|R_0) + E_{P^k}\left(\int_{[0,\tau^k\wedge 1]} \beta_t\cdot d\widehat{B}_t - \frac{1}{2}\int_{[0,\tau^k\wedge 1]}\beta_t\cdot A(dt)\beta_t\right)\\
&\overset{(8)}{=} H(P_0|R_0) + \frac{1}{2}E_{P^k}\int_{[0,\tau^k\wedge 1]}\beta_t\cdot A(dt)\beta_t.
\end{aligned}
$$

Equality (iii) is a consequence of

$$
\tilde{u}(X) = \log\frac{dP_0}{dR_0}(X_0) + \int_{[0,\tau^k\wedge 1]}\beta_t\cdot(dM_t^P + d\widehat{B}_t) - \frac{1}{2}\int_{[0,\tau^k\wedge 1]}\beta_t\cdot A(dt)\beta_t,\quad P^k\text{-a.s.}
$$

which comes from Theorem 1. It remains to let k tend to infinity, to obtain as above with (29) that (33) holds true. This completes the proof of the theorem. □

6 Proofs of Theorems 3 and 4

We begin recalling Itô's formula. Let P be the law of a semimartingale

$$
dX_t = b_t\rho(dt) + dM_t^P
$$

with M^P a local P-martingale such that $M^P = q\odot\widetilde{\mu}^K$, P-a.s. That is $P\in \mathrm{LK}(\overline{K})$ for some Lévy kernel $\overline{K}$. For any f in $\mathscr{C}^2(\mathbb{R}^d)$ which satisfies:

(∗) *When localizing with an increasing sequence* $(\tau_k)_{k\geq 1}$ *of stopping times tending* P*-almost surely to infinity, for each* $k\geq 1$ *the truncated process* $\mathbf{1}_{\{|q|>1\}}\mathbf{1}_{\{t\leq\tau_k\}}[f(X_{t^-}+q)-f(X_{t^-})]$ *is a* $\mathscr{H}_1(P,\overline{K})$ *integrand,*

Itô's formula is

$$
\begin{aligned}
df(X_t) = &\left[\int_{\mathbb{R}^d_*}[f(X_{t^-}+q) - f(X_{t^-}) - \nabla f(X_{t^-})\cdot q]\,K_t(dq)\right]\rho(dt)\\
&+\nabla f(X_{t^-})\cdot b_t\,\rho(dt) + dM_t, \quad P\text{-a.s.}
\end{aligned}
\tag{34}
$$

where M is a local P-martingale. This identity would fail if ρ was not assumed to be atomless.

6.1 Proof of Theorem 3

Based on Itô's formula, we start computing a large family of exponential local martingales. Recall that we denote

$$a \mapsto \theta(a) := e^a - a - 1 = \sum_{n \geq 2} a^n/n!, \quad a \in \mathbb{R}.$$

Lemma 6 (Exponential martingale). *Let $h : \Omega \times [0,1] \times \mathbb{R}^d_* \to \mathbb{R}$ be a real valued predictable process which satisfies*

$$E_R \int_{[0,1]\times\mathbb{R}_*} \theta[h_t(q)]\,\overline{L}(dtdq) < \infty. \tag{35}$$

Then, h and $e^h - 1$ belong to $\mathscr{H}_{1,2}(R,\overline{L})$. In particular, $h \odot \widetilde{\mu}^L$ is a R-martingale. Moreover,

$$Z^h_t := \exp\Big(h \odot \widetilde{\mu}^L_t - \int_{(0,t]\times\mathbb{R}^d_*} \theta[h_s(q)]\,\overline{L}(dsdq)\Big), \quad t \in [0,1]$$

is a local R-martingale and a positive R-supermartingale which satisfies

$$dZ^h_t = Z^h_{t-}\,[(e^{h(q)} - 1) \odot d\widetilde{\mu}^L_t].$$

Proof. The function θ is nonnegative, quadratic near zero, linear near $-\infty$ and it grows exponentially fast near $+\infty$. Therefore, (35) implies that h and $e^h - 1$ belong to $\mathscr{H}_{1,2}(R,\overline{L})$. In particular, $M^h := h \odot \widetilde{\mu}^L$ is a R-martingale.
Let us denote $Y_t = M^h_t - \int_{(0,t]} \beta_s\,\rho(ds)$ where $\beta_t = \int_{\mathbb{R}^d_*} \theta[h_t(q)]\,L_t(dq)$. Remark that (35) implies that these integrals are almost everywhere well-defined. Applying (34) with $f(y) = e^y$ and $dY_t = -\beta_t\,\rho(dt) + dM^h_t$, we obtain

$$de^{Y_t} = e^{Y_{t-}}\Big[-\beta_t + \int_{\mathbb{R}^d_*} \theta[h_t(q)]\,L_t(dq)\Big]\rho(dt) + dM_t = dM_t$$

where M is a local martingale. We are allowed to do this because $(*)$ is satisfied. Indeed, with $f(y) = e^y$, $f(Y_{t-} + h_t(q)) - f(Y_{t-}) - f'(Y_{t-})h_t(q) = e^{Y_{t-}}\theta[h_t(q)]$ and if $Y^\sigma_t := Y_{t\wedge\sigma}$ is stopped at $\sigma := \inf\{t \in [0,1]; Y_t \notin C\} \in [0,1]\cup\{\infty\}$ for some compact subset C with the convention $\inf\emptyset = \infty$, we see with (35) and the fact that any path in Ω is bounded, that $\exp(Y^\sigma_{t-})\theta[h_t(q)]$ is in $\mathscr{H}_1(R,\overline{L})$. Now, choosing the compact set C to be the ball of radius k and letting k tend to infinity, we obtain an increasing sequence of stopping times $(\sigma_k)_{k\geq 1}$ which tends almost surely to infinity. This proves that $Z^h := e^Y$ is a local martingale.
We see that $dM_t = e^{Y_{t-}}\,d[(e^{h(q)-1}) \odot \widetilde{\mu}^L_t]$, keeping track of the martingale terms in the above differential formula:

$$
\begin{aligned}
de^{Y_t} &= e^{Y_{t-}}\big[\theta(\Delta Y_t) + dY_t\big] \\
&= e^{Y_{t-}}\Big[\theta[h_t(q)] \odot d\widetilde{\mu}_t^L + \Big(\int_{\mathbb{R}^d_*} \theta[h_t(q)]\, L_t(dq)\Big)\rho(dt) \\
&\quad -\beta_t\, \rho(dt) + h(q) \odot d\widetilde{\mu}_t^L\Big] \\
&= e^{Y_{t-}}\big[\theta[h_t(q)] \odot d\widetilde{\mu}_t^L + h_t(q) \odot d\widetilde{\mu}_t^L\big] \\
&= e^{Y_{t-}}\big[(e^{h_t(q)} - 1) \odot d\widetilde{\mu}_t^L\big].
\end{aligned}
$$

By Fatou's lemma, any nonnegative local martingale is also a supermartingale.

Proof (of Theorem 3). It follows the same line as the proof of Theorem 1. By Lemma 6, $0 < E_R Z_1^h \leq 1$ for all h satisfying the assumption (35). By (18), for any probability measure P such that $H(P|R) < \infty$, we have

$$
E_P\left(h \odot \widetilde{\mu}_1^L - \int_{[0,1]\times\mathbb{R}^d_*} \theta(h)\, d\overline{L}\right) \leq H(P|R).
$$

As in the proof of Theorem 1, see that

$$
|E_P(h \odot \widetilde{\mu}_1^L)| \leq (H(P|R) + 1)\|h\|_\theta, \quad \forall h
$$

where

$$
\|h\|_\theta := \inf\left\{a > 0; E_P \int_{[0,1]\times\mathbb{R}^d_*} \theta(h/a)\, d\overline{L} \leq 1\right\} \in [0,\infty] \tag{36}
$$

is the Luxemburg norm of the Orlicz space

$$
\begin{aligned}
L_\theta :=& \Big\{h : [0,1]\times\mathbb{R}^d_*\times\Omega \to \mathbb{R}; \text{measurable s.t. } E_P \int_{[0,1]\times\mathbb{R}^d_*} \theta(b_o|h|)\, d\overline{L} < \infty, \\
&\text{for some } b_o > 0\Big\}.
\end{aligned}
$$

It differs from the corresponding *small* Orlicz space

$$
S_\theta := \Big\{h : [0,1]\times\mathbb{R}^d_*\times\Omega \to \mathbb{R}; \text{measurable s.t.} E_P \int_{[0,1]\times\mathbb{R}^d_*} \theta(b|h|) d\overline{L} < \infty, \forall b \geq 0\Big\}
$$

because the function $\theta(|a|)$ grows exponentially fast.

We introduce the space $\mathscr{B}$ of all the bounded processes such that $E_P \int_{[0,1]\times\mathbb{R}^d_*} |h|\, d\overline{L} < \infty$, and its subspace $\mathscr{H} \subset \mathscr{B}$ which consists of the processes in $\mathscr{B}$ which are predictable. We have $\mathscr{B} \subset S_\theta$ and any h in $\mathscr{H}$ satisfies (35), which is the hypothesis of Lemma 6. Hence, (36) holds for all $h \in \mathscr{H}$ and, as $H(P|R) < \infty$, it tells us that the linear mapping $h \mapsto E_P(h \odot \widetilde{\mu}_1^L)$ is continuous on $\mathscr{H}$ equipped with the norm $\|\cdot\|_\theta$. Since the convex conjugate of the Young function $\theta(|a|)$ is $\theta^*(|b|)$, the dual space of $(S_\theta, \|\cdot\|_\theta)$[2] (see [5]), is isomorphic to

$$L_{\theta^*} := \left\{k : [0,1] \times \mathbb{R}^d_* \times \Omega \to \mathbb{R}; \text{measurable s.t. } E_P \int_{[0,1]\times\mathbb{R}^d_*} \theta^*(|k|)\, d\overline{L} < \infty\right\}.$$

Therefore, there exists some $k \in L_{\theta^*}$ such that

$$E_P h \odot \widetilde{\mu}^L_1 = E_P \int_{[0,1]\times\mathbb{R}^d_*} kh\, d\overline{L}, \quad \forall h \in \mathscr{H}. \tag{37}$$

Let us introduce the predictable projection k^{pr} of k which is defined by $k^{\mathrm{pr}}_t := E_P(k \mid X_{[0,t)})$, $t \in [0,1]$. As the space $\mathscr{B}$ is dense in S_θ,[3] $\mathscr{H}$ is dense in the subspace of all predictable processes in S_θ and it follows that any g and k in L_{θ^*} which both satisfy (37), share the same predictable projection: $g^{\mathrm{pr}} = k^{\mathrm{pr}}$. Consequently, there is a *unique* predictable process k in the space

$$\mathscr{K}(P) := \left\{k : [0,1]\times\mathbb{R}^d_*\times\Omega \to \mathbb{R}; \text{predictable s.t. } E_P \int_{[0,1]\times\mathbb{R}^d_*} \theta^*(|k|)\, d\overline{L} < \infty\right\}$$

which verifies (37).
As $\mathscr{H}$ is included in $\mathscr{H}_1(P, \overline{L})$, we have for all $h \in \mathscr{H}$, $h \odot \widetilde{\mu}^L - h \odot k\overline{L} = h \odot (\mu^X - \overline{L} - h \odot k\overline{L} = h \odot (\mu^X - \ell\overline{L})$ with $\ell := k + 1$. Consequently, (37) is equivalent to

$$E_P\big[h \odot (\mu^X - \ell\overline{L})\big] = 0, \quad \forall h \in \mathscr{H}, \tag{38}$$

which is the content of the theorem. It remains however to note that, being an expectation of the positive measure μ^X, $\ell\overline{L}$ is also a positive measure. Therefore, ℓ is nonnegative. This completes the proof of the theorem. □

6.2 Proof of Corollary 2

It is mainly a remark based on Hölder's inequality in Orlicz spaces.

Proof (of Corollary 2). We are under the exponential integrability assumption (14) and we denote $Z = \frac{dP}{dR}$. The finite entropy assumption (1) is equivalent to: Z belongs to the Orlicz space $L_{\theta^*}(R)$, i.e. $\|Z\|_{\theta^*,R} < \infty$. Hölder's inequality in Orlicz spaces[4] expressed with the Luxemburg norms (see (36)) gives us for any nonnegative random variable U: $E_P(U) = E_R(ZU) \le 2\|Z\|_{\theta^*,R}\|U\|_{\theta,R}$. This quantity is finite if $\|U\|_{\theta,R} < \infty$, and this is equivalent to $E_R(e^{a_o U}) < \infty$ for some

[2]This doesn't hold with L_θ instead of S_θ.

[3]In general, it is not dense in L_θ.

[4]It is an easy consequence of Fenchel's inequality: $|ab| \le \theta(|a|) + \theta^*(|b|)$, for all $a, b \in \mathbb{R}$.

$a_o > 0$. As a consequence, (14) implies that $E_P \int_{[0,1]\times\mathbb{R}^d_*} \mathbf{1}_{\{|q|\geq 1\}} e^{b_o|q|}\,\overline{L}(dtdq) < \infty$ for some b_o. But this is equivalent to: $\mathbf{1}_{\{|q|\geq 1\}}|q|$ belongs to the Orlicz space $L_\theta(P\otimes\overline{L})$. With (13) we see that $(\ell - 1)$ is in $L_{\theta^*}(P\otimes\overline{L})$ and by Hölder's inequality again, we obtain

$$E_P \int_{[0,1]\times\mathbb{R}^d_*} \mathbf{1}_{\{|q|\geq 1\}}|q||\ell(t,q)-1|\,\overline{L}(dtdq) < \infty.$$

The small jump part: $E_P \int_{[0,1]\times\mathbb{R}^d_*} \mathbf{1}_{\{|q|<1\}}|q||\ell(t,q)-1|\,\overline{L}(dtdq) < \infty$, is a direct consequence of Hölder's inequality in L_2. This proves (15).
We write symbolically

$$\widetilde{\mu}^L = \mu - \overline{L} = \mu - \ell\overline{L} + (\ell-1)\overline{L} = \widehat{\mu} + (\ell-1)\overline{L}.$$

Hence, $q \odot \widetilde{\mu}^L = q \odot \widehat{\mu} + \int (\ell-1)q\,d\overline{L}$ provided that all these terms are well defined. But, we have assumed that $q\odot\widetilde{\mu}^L$ is well-defined and we have just proved that $\int(\ell-1)q\,d\overline{L}$ is well-defined. Therefore, the remaining term is also well-defined and the proof is complete. □

6.3 Proof of Theorem 4

It is similar to the proof of Theorem 2. We begin with a transfer result in the spirit of Lemma 3. Let P be a probability measure on Ω such that

$$P \in \mathrm{MP}(B, \overline{K})$$

where B is a continuous bounded variation adapted process and $\overline{K}$ is some Lévy kernel

$$\overline{K}(dtdq) := \rho(dt)K(t;dq).$$

Let λ be a $[-\infty,\infty)$-valued predictable process on $[0,1]\times\mathbb{R}^d_*$ such that $\int_{\{\lambda\geq -1\}}\theta(\lambda)\,d\overline{K} < \infty$ and $\overline{K}(-\infty\leq\lambda<-1) < \infty$, P-a.s. We define for all $t\in[0,1]$,

$$Z_t = \widetilde{\exp}\left(\lambda\odot\widetilde{\mu}^K_t - \int_{[0,t]\times\mathbb{R}^d_*}\theta(\lambda)\,d\overline{K}\right) := Z^+_t Z^-_t$$

with

$$\begin{cases} Z^+_t = \exp\left(\lambda^+\odot\widetilde{\mu}^K_t - \int_{(0,t]\times\mathbb{R}^d_*}\theta(\lambda^+)d\overline{K}\right) \\ Z^-_t = \mathbf{1}_{\{t<\tau^\lambda\}}\exp\left(\sum_{0\leq s\leq t}\lambda^-(s,\Delta X_s) - \int_{(0,t]\times\mathbb{R}^d_*}(e^{\lambda^-}-1)\,d\overline{K}\right)\end{cases}$$

where

$$\lambda^+ = \mathbf{1}_{\{\lambda \geq -\alpha\}}\lambda, \qquad \lambda^- = \mathbf{1}_{\{-\infty \leq \lambda < -\alpha\}}\lambda$$

with $\alpha > 0$, $e^{-\infty} = 0$ and $\tau^\lambda = \inf\{t \in [0,1], \lambda(t, \Delta X_t) = -\infty\}$. Remark that, although Z^+ and Z^- both depend on the choice of α, their product $Z = Z^+ Z^-$ doesn't depend on $\alpha > 0$. For all $j, k \geq 1$, we define

$$\sigma_j^k := \inf\left\{ t \in [0,1]; \int_{[0,t]\times\mathbb{R}^d_*} \theta(\lambda^+)\, d\overline{K} \geq k \text{or} \lambda(t, \Delta X_t) \notin [-j, k] \right\} \in [0,1] \cup \{\infty\}$$

and $P_j^k := X^{\sigma_j^k}{}_{\#} P$.

Lemma 7. *Let P and λ be as above. Then, for all $j, k \geq 1$, $Z^{\sigma_j^k}$ is a genuine P-martingale and the measure*

$$Q_j^k := Z_1^{\sigma_j^k} P^k$$

is a probability measure on Ω which satisfies

$$Q_j^k \in \mathrm{MP}\left(B^{\sigma_j^k} + \widehat{B}^{\sigma_j^k}, \mathbf{1}_{[\![0,\sigma_j^k]\!]} e^\lambda \overline{K} \right)$$

where

$$\widehat{B}_t = \int_{[0,t]\times\mathbb{R}^d_*} \mathbf{1}_{\{|q|\leq 1\}}(e^\lambda - 1) q\, d\overline{K}, \quad t \in [0,1]. \tag{39}$$

Note that $\widehat{B}_t$ might not be well defined in the general case. Only the stopped processes $\widehat{B}^{\sigma_j^k}$ are asserted to be meaningful.

Proof. Let us fix $j, k \geq 1$. We have $Z^{\sigma_j^k} = \widetilde{\exp}(\lambda_j^k \odot \widetilde{\mu}^K - \int_{[0,1]\times\mathbb{R}^d_*} \theta(\lambda_j^k)\, d\overline{K})$ with $\lambda_j^k = \mathbf{1}_{[\![0,\sigma_j^k]\!]}\lambda$ which is predictable since λ is predictable and $\mathbf{1}_{[\![0,\sigma_j^k]\!]}$ is left continuous. We drop the subscripts and superscripts j, k and write $\lambda = \lambda_j^k$, $\lambda^+ = (\lambda_j^k)^+$, $\lambda^- = (\lambda_j^k)^-$, $Z^{\sigma_j^k} = Z$ for the remainder of the proof. By the definition of σ_j^k, we obtain with this simplified notation

$$\int_{[0,1]\times\mathbb{R}^d_*} \theta(\lambda^+)\, d\overline{K} \leq k, \qquad -j \leq \lambda \leq k, \quad P_j^k\text{-a.s.} \tag{40}$$

Let us first prove that Z is a P_j^k-martingale. Since it is a local martingale, it is enough to show that

$$E_{P_j^k} Z_1^p < \infty, \qquad \text{for some } p > 1.$$

Choosing $\alpha = j$ in the definition of $(Z^{\sigma_j^k})^+$ and $(Z^{\sigma_j^k})^-$, we see that $Z^{\sigma_j^k} = (Z^{\sigma_j^k})^+ = Z^+ = \mathscr{E}((e^{\lambda^+} - 1) \odot \widetilde{\mu}^K)$. For all $p \geq 0$,

$$(Z^+)^p = \exp\left(p\lambda^+ \odot \widetilde{\mu}^K - p \int_{[0,1]\times\mathbb{R}_*^d} \theta(\lambda^+)\, d\overline{K}\right) \leq \exp(p\lambda^+ \odot \widetilde{\mu}^K)$$

and

$$\begin{aligned}\mathscr{E}((e^{p\lambda^+} - 1) \odot \widetilde{\mu}^K) &= \exp\left(p\lambda^+ \odot \widetilde{\mu}^K - \int_{[0,1]\times\mathbb{R}_*^d} \theta(p\lambda^+)\, d\overline{K}\right)\\ &\geq e^{p\lambda^+ \odot \widetilde{\mu}^K} / C(k,p)\end{aligned}$$

for some finite deterministic constant $C(k,p) > 0$. To derive $C(k,p)$, we must take account of (40) and rely upon the inequality $\theta(pa) \leq c(k,p)\theta(a)$ which holds for all $a \in (-\infty, k]$ and some $0 < c(k,p) < \infty$. With this in hand, we obtain

$$(Z^+)^p \leq e^{p\lambda^+ \odot \widetilde{\mu}^K} \leq C(k,p)\mathscr{E}((e^{p\lambda^+} - 1) \odot \widetilde{\mu}^K).$$

We know with Lemma 6 that $\mathscr{E}((e^{p\lambda^+} - 1) \odot \widetilde{\mu}^K)$ is a nonnegative local martingale. Therefore, it is a supermartingale. We deduce from this that $E_{P_j^k}\mathscr{E}((e^{p\lambda^+} - 1) \odot \widetilde{\mu}^K) \leq 1$ and

$$E_{P_j^k}(Z^+)^p \leq C(k,p) E_{P_j^k}\mathscr{E}((e^{p\lambda^+} - 1) \odot \widetilde{\mu}^K) \leq C(k,p) < \infty.$$

Choosing $p > 1$, it follows that $\mathscr{E}((e^{\lambda^+} - 1) \odot \widetilde{\mu}^K)$ is uniformly integrable. We conclude as in Lemma 3's proof that $\mathscr{E}((e^{\lambda} - 1) \odot \widetilde{\mu}^K)$ is a genuine P_j^k-martingale.

Now, let us show that

$$Q_j^k \in \mathrm{LK}\left(\mathbf{1}_{[\![0,\sigma_j^k]\!]} e^{\lambda}\overline{K}\right).$$

Let τ be a finitely valued stopping time and f a measurable function on $[0,1] \times \mathbb{R}_*^d$ which will be specified later. We denote $F_t = \sum_{0\leq s\leq t\wedge\tau} f(s, \Delta X_s)$ with the convention that $f(t,0) = 0$ for all $t \in [0,1]$. By Lemma 6, the martingale Z satisfies $dZ_t = \mathbf{1}_{[\![0,\sigma_j^k]\!]}(t) Z_{t-}[(e^{\lambda} - 1) \odot \widetilde{\mu}^K]$. We have also $dF_t = \mathbf{1}_{[\![0,\tau]\!]}(t) f(t, \Delta X_t)$ and $d[Z,F]_t = \mathbf{1}_{[\![0,\sigma_j^k\wedge\tau]\!]}(t) Z_{t-}(e^{\lambda(\Delta X_t)} - 1) f(t, \Delta X_t)$, P_j^k-a.s. Consequently,

$$\begin{aligned}&E_{Q_j^k} \sum_{0\leq t\leq\tau} f(t, \Delta X_t)\\ &= E_{P_j^k}(Z_\tau F_\tau - Z_0 F_0)\\ &= E_{P_j^k} \int_{[0,\tau]} (F_t\, dZ_t + Z_t dF_t + d[Z,F]_t)\end{aligned}$$

$$
\begin{aligned}
&= E_{P_j^k}\left[\int_{[0,\tau]} F_t\, dZ_t + \sum_{0\le t\le\tau} Z_{t^-} f(t,\Delta X_t) + \sum_{0\le t\le\tau} Z_{t^-}(e^{\lambda(t,\Delta X_t)}-1) f(t,\Delta X_t)\right]\\
&= E_{P_j^k} \sum_{0\le t\le\tau} Z_{t^-} e^{\lambda(t,\Delta X_t)} f(t,\Delta X_t)\\
&= E_{P_j^k} \int_{[0,\tau]\times\mathbb{R}^d_*} Z_{t^-} f(t,q) e^{\lambda(t,q)}\, \overline{K}(dt dq)\\
&= E_{Q_j^k} \int_{[0,\tau]\times\mathbb{R}^d_*} f(t,q) e^{\lambda(t,q)}\, \overline{K}(dt dq).
\end{aligned}
$$

We are going to choose τ such that the above terms are meaningful. For each $n \ge 1$, consider $\tau_n := \inf\{t \in [0,1]; \sum_{0\le s\le t\wedge\tau} |f(s,\Delta X_s)| \ge n\}$ and take f in $L_1(P_j^k \otimes \overline{K})$ to obtain $\lim_{n\to\infty} \tau_n = \infty$, P_j^k-a.s. and a fortiori Q_j^k-a.s. It remains to take $\tau = \sigma \wedge \tau_n$ with any stopping time σ to see that the Lévy kernel of Q_j^k is $e^{\lambda}\overline{K} = e^{\lambda_j^k}\overline{K}$.

It remains to compute the drift term. Let us denote $X_t^* := \sum_{0\le s\le t} \mathbf{1}_{\{|\Delta X_s|>1\}} \Delta X_s$ the cumulated sum of large jumps of X, and $X^\Delta := X - X^*$ its complement. Let τ be a finitely valued stopping time and take $G_t = \xi \cdot X^\Delta_{t\wedge\tau}$ with $\xi \in \mathbb{R}^d$. We have $dG_t = \mathbf{1}_{[\![0,\tau]\!]}(t)\xi \cdot (dB_t + (\mathbf{1}_{\{|q|\le 1\}} q) \odot d\widetilde{\mu}_t^K)$ and $d[Z,G]_t = \mathbf{1}_{[\![0,\sigma_j^k\wedge\tau]\!]}(t) Z_{t^-}(e^{\lambda(\Delta X_t)} - 1)\mathbf{1}_{\{|\Delta X_t|\le 1\}} \xi \cdot \Delta X_t$, P_j^k-a.s. Therefore,

$$
\begin{aligned}
&E_{Q_j^k}[\xi \cdot (X_\tau^\Delta - X_0^\Delta)]\\
&= E_{P_j^k}[Z_\tau G_\tau - Z_0 G_0]\\
&= E_{P_j^k}\left[\int_{[0,\tau]} (G_t\, dZ_t + Z_t dG_t + d[Z,G]_t)\right]\\
&= E_{P_j^k}\Big[\int_{[0,\tau]} G_t\, dZ_t + \int_{[0,\tau]} Z_{t^-} \xi \cdot (dB_t + (\mathbf{1}_{\{|q|\le 1\}} q) \odot d\widetilde{\mu}_t^K)\\
&\quad + \sum_{0\le t\le\tau} Z_{t^-} \mathbf{1}_{\{|\Delta X_t|\le 1\}} (e^{\lambda(t,\Delta X_t)} - 1)\xi \cdot \Delta X_t\Big]\\
&= E_{P_j^k}\left[\int_{[0,\tau]} Z_{t^-} \xi \cdot dB_t + \sum_{0\le t\le\tau} Z_{t^-} \mathbf{1}_{\{|\Delta X_t|\le 1\}} (e^{\lambda(t,\Delta X_t)} - 1)\xi \cdot \Delta X_t\right]\\
&= E_{P_j^k}\left[\int_{[0,\tau]} Z_{t^-} \xi \cdot dB_t + \int_{[0,\tau]} Z_{t^-} \left\{\int_{\mathbb{R}^d_*} \mathbf{1}_{\{|q|\le 1\}} (e^{\lambda(t,q)} - 1)\xi \cdot q K_t(dq)\right\}\rho(dt)\right]\\
&= E_{Q_j^k} \int_{[0,\tau]} \xi \cdot \left(dB_t + \left\{\int_{\mathbb{R}^d_*} \mathbf{1}_{\{|q|\le 1\}} (e^{\lambda(t,q)} - 1) q\, K_t(dq)\right\} \rho(dt)\right)
\end{aligned}
$$

where we take $\tau = \tau_n := \inf\{t \in [0,1]; |X_t| \geq n\}$ which tends to ∞ as n tends to infinity. This shows that the drift term of X under Q_j^k is $(B+\widehat{B})^{\sigma_k}$ where $\widehat{B}$ is given at (39) and the stopped process $\widehat{B}^{\sigma_k}$ is well-defined.

As a first step, it is assumed that $P \sim R$ for the stopping times τ_j^k, τ_j and τ^- to be defined (below) R-a.s. and not only P-a.s.

Following the proofs of Lemmas 4 and 5, except for minor changes (but we skip the details), we arrive at analogous results:

(i) If R fulfills the uniqueness condition (U), then for any stopping time τ, R^τ also fulfills (U).
(ii) If $P \sim R$, then for any $j,k \geq 1$, we have

$$\mathbf{1}_{[\![0,\tau_j^k\wedge 1]\!]}\frac{dP}{dR} = \mathbf{1}_{[\![0,\tau_j^k\wedge 1]\!]}\frac{dP_0}{dR_0}(X_0)\exp\left((\mathbf{1}_{(0,\tau_j^k\wedge 1]}\log\ell)\odot\widetilde{\mu}^L\right.$$
$$\left.-\int_{(0,\tau_j^k\wedge 1]\times\mathbb{R}^d_*}\theta(\log\ell)\,d\overline{L}\right)$$

where

$$\tau_j^k := \inf\left\{t\in[0,1];\int_{[0,t]\times\mathbb{R}^d_*}\mathbf{1}_{\{\ell>1/2\}}\theta(\log\ell)\,d\overline{L}\geq k \text{ or } \log\ell(t,\Delta X_t)\right.$$
$$\left.\notin[-j,k]\right\}\in[0,1]\cup\{\infty\}.$$

For the proof of (ii), we use Lemma 6 where $\lambda = \log\ell$ plays the same role as β in Lemma 5, and we go backward with $-\lambda$ which corresponds to ℓ^{-1}.

We fix j, and let k tend to infinity to obtain with (13) that

$$\lim_{k\to\infty}\tau_j^k = \tau_j := \inf\{t\in[0,1];\ell(t,\Delta X_t) < e^{-j}\}\in[0,1]\cup\{\infty\},\quad P\text{-a.s.}$$

and therefore R-a.s. also. More precisely, this increasing sequence is stationary after some time: there exists $K(\omega)<\infty$ such that $\tau_j^k(\omega)=\tau_j(\omega)$, for all $k\geq K(\omega)$. It follows that for all $j\geq 1$,

$$\mathbf{1}_{[\![0,\tau_j\wedge 1]\!]}\frac{dP}{dR} = \mathbf{1}_{[\![0,\tau_j\wedge 1]\!]}\frac{dP_0}{dR_0}(X_0)\exp\left((\mathbf{1}_{(0,\tau_j\wedge 1]}\log\ell)\odot\widetilde{\mu}^L\right.$$
$$\left.-\int_{(0,\tau_j\wedge 1]\times\mathbb{R}^d_*}\theta(\log\ell)\,d\overline{L}\right). \tag{41}$$

Lemma 8. *We do not assume that $P\sim R$ and we extend ℓ by $\ell=1$ on the P-negligible subset where it is unspecified. Defining $\tau^- := \sup_{j\geq 1}\tau_j$, we have $P(\tau^-=\infty)=1$.*

Proof. For all $j \geq 1$, we have $\tau^- \leq 1 \Rightarrow \sum_{t\leq 1} \mathbf{1}_{\{\ell(t,\Delta X_t)\leq e^{-j}\}} \geq 1$. Therefore,

$$P(\tau^- \leq 1) \leq P\left(\sum_{t\leq 1} \mathbf{1}_{\{\ell(t,\Delta X_t)\leq e^{-j}\}} \geq 1\right) \leq E_P \sum_{t\leq 1} \mathbf{1}_{\{\ell(t,\Delta X_t)\leq e^{-j}\}}$$

$$\stackrel{\checkmark}{=} E_P \int_{[0,1]\times\mathbb{R}^d_*} \mathbf{1}_{\{\ell\leq e^{-j}\}}\, \ell d\overline{L} \leq e^{-j} E_P \overline{L}(\ell \leq e^{-j}) \leq e^{-j} E_P \overline{L}(\ell \leq 1/2)$$

where we used (38) at the marked equality. The result will follow letting j tend to infinity, provided that we show that $E_P \overline{L}(\ell \leq 1/2) < \infty$.
But, we know with (13) that $E_P \int_{[0,1]\times\mathbb{R}^d_*} \theta^*(|\ell - 1|)\, d\overline{L} < \infty$. Hence, $E_P \overline{L}(\ell \leq 1/2) \leq E_P \int_{[0,1]\times\mathbb{R}^d_*} \theta^*(|\ell - 1|)\, d\overline{L}/\theta^*(1/2) < \infty$ and the proof is complete. □

Lemma 9. *Assume $P \sim R$. Let R_j and P_j be the laws of the stopped process $X^{\tau_j\wedge 1}$ under R and P respectively. Then, under the condition* (U) *we have for all $j \geq 1$*

$$H(P_j|R_j) = H(P_0|R_0) + E_P \int_{(0,\tau_j\wedge 1]\times\mathbb{R}^d_*} (\ell \log \ell - \ell - 1)\, d\overline{L}.$$

Proof. We denote R^k_j and P^k_j the laws of the stopped process $X^{\tau^k_j\wedge 1}$ under R and P respectively. With the expression of $\frac{dP}{dR}$ on $[\![0, \tau^k_j \wedge 1]\!]$ we see that

$$\begin{aligned}
&H(P^k_j|R^k_j)\\
&= H(P_0|R_0) + E_{P^k_j}\left((\mathbf{1}_{(0,\tau^k_j\wedge 1]} \log \ell) \odot \widetilde{\mu}^L - \int_{(0,\tau^k_j\wedge 1]\times\mathbb{R}^d_*} \theta(\log \ell)\, d\overline{L}\right)\\
&= H(P_0|R_0) + E_{P^k_j}\left((\mathbf{1}_{(0,\tau^k_j\wedge 1]} \log \ell) \odot \widetilde{\mu}^{\ell L} + \int_{(0,\tau^k_j\wedge 1]\times\mathbb{R}^d_*} [(\ell - 1) - \theta(\log \ell)] d\overline{L}\right)\\
&= H(P_0|R_0) + E_{P_j} \int_{(0,\tau^k_j\wedge 1]\times\mathbb{R}^d_*} (\ell \log \ell - \ell - 1)\, d\overline{L}
\end{aligned}$$

where we invoke Lemma 7 at the last equality. We complete the proof letting k tend to infinity. □

Proof (Conclusion of the proof of Theorem 4).
When $P \sim R$, by Lemma 8, P-almost surely there exists j_o large enough such that for all $j \geq j_o$, $\tau_j = \infty$ and (41) tells us that

$$\frac{dP}{dR} = \frac{dP_0}{dR_0}(X_0) \exp\left((\log \ell) \odot \widetilde{\mu}^L - \int_{[0,1]\times\mathbb{R}^d_*} \theta(\log \ell)\, d\overline{L}\right)$$

and also that the product appearing in Z^- contains P-almost surely a finite number of terms which are all positive. Note that we do not use any limit result for stochastic or standard integrals; it is an immediate ω-by-ω result with a stationary sequence. This is the desired expression for $\frac{dP}{dR}$ when $P \sim R$.

Let us extend this result to the case when P might not be equivalent to R. We proceed exactly as in Theorem 2's proof and start from (30): $\lim_{n\to\infty} H(P|P_n) = 0$ where $P_n := (1-1/n)P + R/n$, $n \geq 1$. Let us write $\lambda = \log \ell$ and $\lambda^n = \log \ell^n$ which are well-defined P-a.s. Thanks to Theorem 3, we see that

$$\begin{aligned} H(P|P_n) &\geq E_P\left((\lambda - \lambda^n) \odot \widetilde{\mu}^{\ell^n L} - \int_{[0,1]\times\mathbb{R}^d_*} \theta(\lambda - \lambda^n)\, \ell^n d\overline{L}\right) \\ &= E_P\left((\lambda^n - \lambda) \odot \widetilde{\mu}^{\ell L} + \int_{[0,1]\times\mathbb{R}^d_*} [\ell/\ell^n \log(\ell/\ell^n) - \ell/\ell^n + 1]\, \ell^n d\overline{L}\right) \\ &= E_P \int_{[0,1]\times\mathbb{R}^d_*} [\ell^n/\ell - \log(\ell^n/\ell) - 1]\, d\ell\overline{L} \\ &= E_P \int_{[0,1]\times\mathbb{R}^d_*} \theta(\lambda^n - \lambda)\, d\ell\overline{L} \end{aligned}$$

which leads to the entropic estimate analogous to (32):

$$\lim_{n\to\infty} E_P \int_{[0,1]\times\mathbb{R}^d_*} \theta(\lambda^n - \lambda)\, d\ell\overline{L} = 0. \tag{42}$$

Taking the difference between $\log(dP_n/dR) = \lambda^n \odot \widetilde{\mu}^L - \int_{[0,1]\times\mathbb{R}^d_*} \theta(\lambda^n)\, d\overline{L}$ and the logarithm of the announced formula (16) for dP/dR on the set $\{\frac{dP}{dR} > 0\}$, we obtain

$$(\lambda^n - \lambda) \odot \widetilde{\mu}^{\ell L} - \int_{[0,1]\times\mathbb{R}^d_*} \theta(\lambda^n - \lambda)\, d\ell\overline{L}, \quad P\text{-a.s.}$$

and the desired convergence follows from (42). Note that $\theta(a) = a^2/2 + o_{a\to 0}(a^2)$. This completes the proof of (16).

As in the proof of Theorem 2, we obtain the announced formula for $H(P|R)$ under the condition (U) with Lemmas 8 and 9, and the corresponding general inequality follows from choosing

$$\tilde{u}(X) := \log \frac{dP_0}{dR_0}(X_0) + (\mathbf{1}_{(0,\tau_j^k \wedge 1]} \log \ell) \odot \widetilde{\mu}^L - \int_{(0,\tau_j^k\wedge 1]\times\mathbb{R}^d_*} \theta^*(\log \ell)\, d\overline{L}$$

in the variational representation formula (18), and then letting k and j tend to infinity. □

Appendix. An Exponential Martingale with Jumps

Next proposition is about exponential martingale with jumps. We didn't use it during the proofs of this paper. But we give it here for having a more complete picture of the Girsanov theory.

In this result, integrands h are considered which may attain the value $-\infty$. This is because with $h = \log \ell$, $h = -\infty$ corresponds to $\ell = 0$.

Proposition 1 (Exponential martingale). *Let $h : \Omega \times [0,1] \times \mathbb{R}^d_* \to [-\infty, \infty)$ be an extended real valued predictable process which may take the value $-\infty$ and satisfies*

$$E_R \int_{[0,1]\times\mathbb{R}_*} \mathbf{1}_{\{h_t(q)\geq -1\}} \theta[h_t(q)]\, \overline{L}(dtdq) < \infty, \tag{43}$$

$$E_R \int_{[0,1]\times\mathbb{R}_*} \mathbf{1}_{\{h_t(q)< -1\}}\, \overline{L}(dtdq) < \infty. \tag{44}$$

Let us introduce the stopping time

$$\tau^h := \inf\{t \in [0,1]; h(\Delta X_t) = -\infty\} \in [0,1] \cup \{\infty\}$$

and the convention $e^{-\infty} = 0$.
Then, $e^h - 1$ is in $\mathcal{H}_{1,2}(R, \overline{L})$ and

$$Z_t^h := \mathbf{1}_{\{t<\tau^h\}} \widetilde{\exp}\Big(h \odot \widetilde{\mu}_t^L - \int_{(0,t]\times\mathbb{R}^d_*} \theta[h_s(q)]\, \overline{L}(dsdq)\Big), \quad t \in [0,1] \tag{45}$$

is a local R-martingale and a nonnegative R-supermartingale which satisfies

$$dZ_t^h = \mathbf{1}_{\{t\leq\tau^h\}} Z_{t^-}^h\, [(e^{h(q)} - 1) \odot d\widetilde{\mu}_t^L]. \tag{46}$$

The standard notation is $Z^h := \mathscr{E}([e^h - 1] \odot \widetilde{\mu}^L)$, the stochastic exponential of $[e^h - 1] \odot \widetilde{\mu}^L$. Some details are necessary to make precise the sense of the inner stochastic integral $h \odot \widetilde{\mu}_t^L$ in the expression of Z_t^h. We denote

$$h^+ := \mathbf{1}_{\{h\geq -1\}} h \in \mathbb{R}$$
$$h^- := \mathbf{1}_{\{h< -1\}} h \in [-\infty, 0].$$

Under the assumption (43), $h^+ \odot \widetilde{\mu}^L$ is well defined as a stochastic integral. On the other hand, (44) implies that $h^-(t, \Delta X_t)$ has R-a.s. finitely many jumps. It follows that $\sum_{0\leq s\leq t} h^-(s, \Delta X_s)$ is meaningful for all $t < \tau^h$. But the integral $\int_{(0,t]\times\mathbb{R}^d_*} h_s^-(q)\, \overline{L}(dsdq)$ might not be defined under (44) and $h^- \odot \widetilde{\mu}_t^L = \sum_{0\leq s\leq t} h^-(s, \Delta X_s) - \int_{(0,t]\times\mathbb{R}^d_*} h_s^-(q)\, \overline{L}(dsdq)$ is meaningless in this case. Nevertheless, the full expression in the exponential $\zeta(h) := h \odot \widetilde{\mu}^L - \int \theta(h)\, d\overline{L}$

is defined as follows. We put $\zeta(h^-) := \sum_{0\le s\le t} h^-(s,\Delta X_s) - \int_{(0,t]\times\mathbb{R}^d_*} [e^{h^-_s(q)}-1]\,\overline{L}(dsdq)$ which is well defined under (44) and is obtained by cancelling the terms $\int_{(0,t]\times\mathbb{R}^d_*} h^-_s(q)\,\overline{L}(dsdq)$. As $\theta(0) = 0$, we have $\zeta(h) = \zeta(h^+ + h^-) = \zeta(h^+) + \zeta(h^-)$ and for all $t \in [0,1]$,

$$\begin{cases} Z^h_t = Z^{h^+}_t Z^{h^-}_t \quad \text{with} \\ Z^{h^+}_t := \exp\left(h^+ \odot \widetilde{\mu}^L_t - \int_{(0,t]\times\mathbb{R}^d_*} \theta[h^+_s(q)]\,\overline{L}(dsdq)\right), \\ Z^{h^-}_t := \mathbf{1}_{\{t<\tau^h\}} \exp\left(\sum_{0\le s\le t} h^-(s,\Delta X_s) - \int_{(0,t]\times\mathbb{R}^d_*} [e^{h^-_s(q)}-1]\,\overline{L}(dsdq)\right). \end{cases} \tag{47}$$

This is what is meant by the concise expression (45).

Proof. Now, we consider the general case where h may attain the value $-\infty$ and (35) is weakened by (43) and (44). We use the decomposition (47) and write $Z^+ = Z^{h^+}$ and $Z^- = Z^{h^-}$ for short. Clearly, Z^+ and Z^- do not jump at the same times and $d[Z^+, Z^-] = \Delta Z^+ \Delta Z^- = 0$. Hence,

$$dZ_t = Z^+_{t-} dZ^-_t + Z^-_{t-} dZ^+_t. \tag{48}$$

The h^+-part enters the framework of Lemma 6 and we have

$$dZ^+_t = Z^+_{t-}\left([e^{h^+} - 1] \odot \widetilde{\mu}^L\right). \tag{49}$$

Let us look at the h^--part. We need to compute dZ^-_t. For all $t < \tau^h$, put

$$Y^-_t = \sum_{0\le s\le t} h^-(s,\Delta X_s) - \int_{(0,t]\times\mathbb{R}^d_*} [e^{h^-_s(q)} - 1]\,\overline{L}(dsdq).$$

Then, with the convention that $h^-(t,0) = 0$, $dY^-_t = h^-(t,\Delta X_t) - \gamma_t\,\rho(dt)$ with $\gamma_t = \int_{\mathbb{R}^d_*} [e^{h^-_t(q)} - 1]\,L_t(dq)$, $\Delta Y^-_t = h^-(t,\Delta X_t)$ and with Itô's formula, we arrive at

$$\begin{aligned} de^{Y^-_t} &= e^{Y^-_{t-}}\left([e^{\Delta Y^-_t} - 1] + dY^-_t - \Delta Y^-_t\right) = e^{Y^-_{t-}}\left([e^{h^-(t,\Delta X_t)} - 1] - \gamma_t\,\rho(dt)\right) \\ &= e^{Y^-_{t-}}\left([e^{h^-} - 1] \odot d\widetilde{\mu}^L_t\right). \end{aligned}$$

It follows that

$$dZ^-_t = Z^-_{t-}\left([e^{h^-} - 1] \odot d\widetilde{\mu}^L_t\right), \quad t < \tau^h. \tag{50}$$

At $t = \tau^h$, by the Definition (47) of Z^-, we have

$$dZ^-_{|t=\tau^h} = -Z^-_{(\tau^h)-} = Z^-_{(\tau^h)-} \times [e^{-\infty} - 1]$$

which is (50) at $t = \tau^h$ with the convention $e^{-\infty} = 0$. This provides us with

$$dZ_t^- = \mathbf{1}_{\{t \le \tau^h\}} Z_{t^-}^- \left([e^{h^-} - 1] \odot d\widetilde{\mu}_t^L \right).$$

Together with (48) and (49), this proves (46) which implies that Z^h is a local R-martingale.

By Fatou's lemma, any nonnegative local martingale is also a supermartingale.

References

1. J. Jacod, Multivariate point processes: Predictable representation, Radon-Nikodým derivatives, representation of martingales. Z. Wahrsch. verw. Geb. **31**, 235–253 (1975)
2. J. Jacod, *Calcul stochastique et problèmes de martingales*. Lecture Notes in Mathematics, vol. 714 (Springer, Berlin, 1979)
3. J. Jacod, A.N. Shiryaev, *Limit Theorems for Stochastic Processes*. Grundlehren der mathematischen Wissenshaften, vol. 288 (Springer, Berlin, 1987)
4. P.E. Protter, *Stochastic Integration and Differential Equations*. Applications of Mathematics. Stochastic Modelling and Applied Probability, vol. 21, 2nd edn. (Springer, Berlin, 2004)
5. M.M. Rao, Z.D. Ren, *Theory of Orlicz Spaces*. Pure and Applied Mathematics, vol. 146 (Marcel Dekker, New York, 1991)
6. D. Revuz, M. Yor, *Continuous Martingales and Brownian Motion*. Grundlehren der Mathematischen Wissenschaften, vol. 293, 3rd edn. (Springer, Berlin, 1999)

Erratum to Séminaire XXVII

Michel Émery and Marc Yor

As pointed out by Vilmos Prokaj, the contribution *On the Lévy transformation of Brownian motions and continuous martingales* in Séminaire XXVII, by the two of us and our late friend Lester E. Dubins, contains a serious error: the proposition page 128 is false. The computation (not given in the article) which leads to this proposition is not sound, and does not seem to be mendable. Fortunately, Remark c) pp. 127–128, which contains this false statement, is completely independent from the rest of the article, so chaos does not propagate.

We thank Vilmos Prokaj for this observation, and apologize to all other readers—if any.

M. Émery
IRMA, CNRS et Université Unique de Strasbourg, 7 rue René Descartes, 67084 Strasbourg Cedex, France
e-mail: emery@math.unistra.fr

M. Yor
LPMA, Université Pierre et Marie Curie, boîte courrier 188, 4 place Jussieu, 75252 Paris Cedex 05, France

C. Donati-Martin et al. (eds.), *Séminaire de Probabilités XLIV*, Lecture Notes in Mathematics 2046, DOI 10.1007/978-3-642-27461-9,

Erratum to Séminaire XXXV

Michel Émery and Walter Schachermayer

In Séminaire XXXV, our contribution *On Vershik's standardness criterion and Tsirelson's notion of cosiness* contains a material error: in Lemma 17 on top of page 284, we forgot to write that the random variable R is assumed to be *simple.* Simplicity is used in the proof, but not mentioned in the statement.

This error bears no consequence on the rest of the article, for two reasons. Firstly, Lemma 17 is used only once, in the proof of Lemma 18, and the r.v. R there is simple; so the whole argument is sound. And secondly, Lemma 17 remains true if R is not simple; but this is a consequence of Lemma 18, whose proof uses the simple case (Lemma 17) as a preliminary step.

We apologize to any reader whose admiration for Vershik's truly marvellous theory might have been hindered.

M. Émery
IRMA, CNRS et Université Unique de Strasbourg, 7 rue René Descartes,
67084 Strasbourg Cedex, France
e-mail: emery@math.unistra.fr

W. Schachermayer
Fakultät für Mathematik, Universität Wien, Nordbergstraße 15, 1090 Wien, Österreich
e-mail: walter.schachermayer@univie.ac.at

LECTURE NOTES IN MATHEMATICS Springer

Edited by J.-M. Morel, B. Teissier; P.K. Maini

Editorial Policy (for Multi-Author Publications: Summer Schools / Intensive Courses)

1. Lecture Notes aim to report new developments in all areas of mathematics and their applications - quickly, informally and at a high level. Mathematical texts analysing new developments in modelling and numerical simulation are welcome. Manuscripts should be reasonably selfcontained and rounded off. Thus they may, and often will, present not only results of the author but also related work by other people. They should provide sufficient motivation, examples and applications. There should also be an introduction making the text comprehensible to a wider audience. This clearly distinguishes Lecture Notes from journal articles or technical reports which normally are very concise. Articles intended for a journal but too long to be accepted by most journals, usually do not have this "lecture notes" character.

2. In general SUMMER SCHOOLS and other similar INTENSIVE COURSES are held to present mathematical topics that are close to the frontiers of recent research to an audience at the beginning or intermediate graduate level, who may want to continue with this area of work, for a thesis or later. This makes demands on the didactic aspects of the presentation. Because the subjects of such schools are advanced, there often exists no textbook, and so ideally, the publication resulting from such a school could be a first approximation to such a textbook. Usually several authors are involved in the writing, so it is not always simple to obtain a unified approach to the presentation.
For prospective publication in LNM, the resulting manuscript should not be just a collection of course notes, each of which has been developed by an individual author with little or no coordination with the others, and with little or no common concept. The subject matter should dictate the structure of the book, and the authorship of each part or chapter should take secondary importance. Of course the choice of authors is crucial to the quality of the material at the school and in the book, and the intention here is not to belittle their impact, but simply to say that the book should be planned to be written by these authors jointly, and not just assembled as a result of what these authors happen to submit.
This represents considerable preparatory work (as it is imperative to ensure that the authors know these criteria before they invest work on a manuscript), and also considerable editing work afterwards, to get the book into final shape. Still it is the form that holds the most promise of a successful book that will be used by its intended audience, rather than yet another volume of proceedings for the library shelf.

3. Manuscripts should be submitted either online at www.editorialmanager.com/lnm/ to Springer's mathematics editorial, or to one of the series editors. Volume editors are expected to arrange for the refereeing, to the usual scientific standards, of the individual contributions. If the resulting reports can be forwarded to us (series editors or Springer) this is very helpful. If no reports are forwarded or if other questions remain unclear in respect of homogeneity etc, the series editors may wish to consult external referees for an overall evaluation of the volume. A final decision to publish can be made only on the basis of the complete manuscript; however a preliminary decision can be based on a pre-final or incomplete manuscript. The strict minimum amount of material that will be considered should include a detailed outline describing the planned contents of each chapter.
Volume editors and authors should be aware that incomplete or insufficiently close to final manuscripts almost always result in longer evaluation times. They should also be aware that parallel submission of their manuscript to another publisher while under consideration for LNM will in general lead to immediate rejection.

4. Manuscripts should in general be submitted in English. Final manuscripts should contain at least 100 pages of mathematical text and should always include
   - a general table of contents;
   - an informative introduction, with adequate motivation and perhaps some historical remarks: it should be accessible to a reader not intimately familiar with the topic treated;
   - a global subject index: as a rule this is genuinely helpful for the reader.

   Lecture Notes volumes are, as a rule, printed digitally from the authors' files. We strongly recommend that all contributions in a volume be written in the same LaTeX version, preferably LaTeX2e. To ensure best results, authors are asked to use the LaTeX2e style files available from Springer's web-server at
   ftp://ftp.springer.de/pub/tex/latex/svmonot1/ (for monographs) and
   ftp://ftp.springer.de/pub/tex/latex/svmultt1/ (for summer schools/tutorials).
   Additional technical instructions, if necessary, are available on request from:
   lnm@springer.com.

5. Careful preparation of the manuscripts will help keep production time short besides ensuring satisfactory appearance of the finished book in print and online. After acceptance of the manuscript authors will be asked to prepare the final LaTeX source files and also the corresponding dvi-, pdf- or zipped ps-file. The LaTeX source files are essential for producing the full-text online version of the book. For the existing online volumes of LNM see:
   http://www.springerlink.com/openurl.asp?genre=journal&issn=0075-8434.
   The actual production of a Lecture Notes volume takes approximately 12 weeks.

6. Volume editors receive a total of 50 free copies of their volume to be shared with the authors, but no royalties. They and the authors are entitled to a discount of 33.3 % on the price of Springer books purchased for their personal use, if ordering directly from Springer.

7. Commitment to publish is made by letter of intent rather than by signing a formal contract. Springer-Verlag secures the copyright for each volume. Authors are free to reuse material contained in their LNM volumes in later publications: a brief written (or e-mail) request for formal permission is sufficient.

Addresses:

Professor J.-M. Morel, CMLA,
École Normale Supérieure de Cachan,
61 Avenue du Président Wilson, 94235 Cachan Cedex, France
E-mail: morel@cmla.ens-cachan.fr

Professor B. Teissier, Institut Mathématique de Jussieu,
UMR 7586 du CNRS, Équipe "Géométrie et Dynamique",
175 rue du Chevaleret,
75013 Paris, France
E-mail: teissier@math.jussieu.fr

For the "Mathematical Biosciences Subseries" of LNM:

Professor P. K. Maini, Center for Mathematical Biology,
Mathematical Institute, 24-29 St Giles,
Oxford OX1 3LP, UK
E-mail : maini@maths.ox.ac.uk

Springer, Mathematics Editorial I,
Tiergartenstr. 17,
69121 Heidelberg, Germany,
Tel.: +49 (6221) 4876-8259
Fax: +49 (6221) 4876-8259
E-mail: lnm@springer.com